Die Kunst of Phonons
Lectures from the Winter
School of Theoretical Physics

Die Kunst of Phonons
Lectures from the Winter School of Theoretical Physics

Edited by
Tadeusz Paszkiewicz and
Krzysztof Rapcewicz

University of Wrocław
Wrocław, Poland

Plenum Press • New York and London

Library of Congress Cataloging-in-Publication Data

Winter School of Theoretical Physics (29th : 1993 : Kudowa Zdrój, Poland)
  Die Kunst of phonons : lectures from the Winter School of Theoretical Physics / edited by Tadeusz Paszkiewicz and Krzysztof Rapcewicz.
      p.   cm.
  Includes bibliographical references and index.
  ISBN 0-306-44677-4
  1. Phonons--Congresses.   I. Paszkiewicz, T. II. Rapcewicz, Krzysztof. III. Title.
  QC176.8.P5W56 1993
  530.4'16--dc20   94-8070
                                                                    CIP

Proceedings of the Twenty-Ninth Winter School of Theoretical Physics, held February 15–27, 1993, in Kudowa Zdrój, Poland

ISBN 0-306-44677-4

© 1994 Plenum Press, New York
A Division of Plenum Publishing Corporation
233 Spring Street, New York, N.Y. 10013

All rights reserved

No part of this book may be reproduced, stored in a retrieval system, or transmitted in any form or by any means, electronic, mechanical, photocopying, microfilming, recording, or otherwise, without written persmission from the Publisher

Printed in the United States of America

FOREWORD

Between 15 and 27 February 1993, the XXIX Winter School of Theoretical Physics was held in the Hotel Gwarek in Kudowa Zdrój, Poland. The present volume contains the proceedings of the Winter School.

In keeping with the name *Die Kunst of Phonons*, the School was dedicated to the physics of phonons. Historically the field of phonons has been a conglomeration of various directions related only by the fact that phonons were involved. The subject has matured greatly and it is now possible to speak of a field of phonon physics proper. Further, it is possible to speak of many of the phenomena seen in phonon physics as being, truly, beautiful. As the object of the book is to present tutorial reviews of the state of the art (Kunst) together with the fact that the state of the art often times reveals beautiful phenomena (more Kunst) we chose a title that reflected this and was also arresting, forcing the prospective reader to stop and reconsider his preconceptions about phonon physics. The use of the german word (with its echoes of Bach's "die Kunst der Fuge") enhances the effect on the prospective reader.

The School was the second Winter School organized by the Institute of Theoretical Physics of the University of Wrocław on this topic and, to certain extent, reviewed the progress that has been achieved in this field over the past six years. The previous XXIII School was held in 1987 at Karpacz, the dominant theme of which was the dynamics of phonons, the kinetics of gases of phonons and their interactions with electrons being treated in less detail during the XXIII School.

The first week of the XXIX School was dedicated chiefly to the interaction of 3D phonons with low dimensional gases of charge carriers (1D and 2D gases of electrons and holes). The study of these interactions provides a deeper understanding of the properties of condensed matter. Simultaneously, the results of this research are of importance for the development of modern technology. Two–dimensional electron gases are used in field–effect transistors (used in satelite communications) and, at room temperatures, the interaction of electrons with phonons is the most important factor limiting the speed of these transistors. The newest experimental and theoretical results relating to low dimensional gases of charge carriers were presented in a series of lectures by P.N. Butcher, L.J. Challis, W. Dietsche, M.V. Entin, A.I. Kopeliovich, A. Shik and V. Karpus.

In the interaction of low dimensional gases of charge carriers with phonons, phonon focussing plays a very important role. A. Every presented a very detailed review of aspects of phonon focussing. In the future, this phenomenon will most certainly be used as a technique for the nondestructive testing of technically valuable materials, *e.g.* composites and building materials. For surface waves, the phenomenon of phonon focussing has been predicted and studied theoretically. This has now been seen experimentally by Kolomenskii and Maznev. The latter author attended the School and presented these results.

During the first week, there were also three lectures on the physics of fullerenes. The lecturers were K. Prassides, K. Rapcewicz and J. Stankowski. Lectures devoted to the physics of molecular crystals and glasses were also delivered by V.G. Manzhelii and L.A. Turski.

The second week was dedicated to, among other things, phonon–mediated detectors of elementary particles. The lecturers were H. Kraus and R. Gaitskell. Construction and study of such detectors is the subject of very vigorous research in contemporary experimental particle physics. It is hoped that with the help of such detectors, the neutrino and the particles comprising dark matter will be observed.

Another, important topic covered in the second week was the kinetic of phonons. M. Meissner and P. Strehlow presented research on the thermalisation of phonon pulses at low temperatures and gave a theoretical discussion of it. In these extraordinarily subtle experiments, M. Meissner has proven the existence of a hierarchy of relaxation times which are related to local equilibrium states. S. Ivanov discussed the problem of the crossover from ballistic to diffusive motion of phonon beams while I.I. Tartakovskii and D. Kazakovtsev presented very interesting experimental results on the evolution of hot spots and theoretical attempts at understanding. Hot spots occur on the surface of a crystal as a result of the crystal being irradiated by intense beams of particles or radiation. It is expected that a hot spot should occur inside a crystal as a result of the thermalisation of the high energy elementary particles.

A.M. Kosevich presented results of calculations of the density of states of energy for layered structures. The method used by him does not require phonons and is not limited to harmonic and perturbation methods. Recent progress in using the effective potential method in the classical Monte Carlo calculations of the thermodynamic properties of crystals was reviewed by G.K. Horton while the subtleties of quasimomenta in condensed media were discussed by A. Thellung.

Many of the lecturers also gave introductory lectures to the participants of the Kindergarten of Theoretical Physics and we would like to express our appreciation on behalf of the participants of the Kindergarten.

For the first time in recent memory, two poster sessions were held during the course of the School. This allowed the participants of the School to present their researches to the lecturers of the School, thereby providing them with a chance to discuss with and receive advice on their work from leading researchers in the field.

The XXIX School was generously sponsored by the University of Wrocław, the State Commettee for Scientific Research, and the Polish Academy of Sciences. The Stefan Batory Fundation and the WE-Hereaus-Stiftung, which is administered by the Deutsche Physikalische Gesellschaft, sponsored all of the participants from the new republics of the former Soviet Union.

The papers in these proceedings underwent a refereeing process which was mainly undertaken during the School itself. We are grateful to the lecturers of the School for their efforts. We also wish to thank W. Gańcza, Cz. Jasiukiewicz, G. Jastrzębski, J. Lorenc, Z. Petru, P. Siemion, Z. Strycharski, M. Wilczyński, and R. Zossel for their untiring help during the School.

Finally we would like to express our gratitude to Ms. A. Jadczyk for her herculean efforts in assisting us with the technical aspects of the preparation of the manuscripts.

Tadeusz Paszkiewicz
Krzysztof Rapcewicz

30 May 1993
Wrocław

CONTENTS

PHONONS: GENERAL

Recent Progress in Using the Effective Potential Method 1
    G.K. Horton and E.R. Cowley

Momentum and Quasimomentum in the Physics of Condensed Matter ... 15
    A. Thellung

Partial Frequency Distribution Functions and the Problem of the Localization of Atomic Vibrations in Real and Multilayered Crystals 33
    A.M. Kosevich, E.S. Syrkin, and S.B. Feodosyev

Selfconsistent Approach to Phonon Localization in Harmonic Crystals with Defects 47
    I.Ya Polishchuk, A.L. Burin, and L.A. Maksimov

Bulk Phonon-Polaritons in Reststrahlen Region of A^3B^5 Compound Superlattice 51
    R. Brazis, R. Narkowicz, and L. Safonova

PHONON FOCUSING

Thermal Phonon Imaging 55
    A.G. Every

Phonon Imaging at Ultrasonic Frequencies: The Dynamic Response of Anisotropic Solids 73
    A.G. Every, K.Y. Kim, and W. Sachse

Phonon Patterns of Cubic Crystals Monte Carlo Simulation Program 87
    W.M. Gańcza and T. Paszkiewicz

Surface Phonon Focusing at Ultrasonic Frequencies 99
    Al.A Kolomenskii and A.A. Maznev

NONEQUILIBRIUM PHONONS

Time-Dependent Specific Heat of Crystals and Glasses at Low
    Temperatures 105
    N. Sampat M. Meissner

Optical Studies of Nonequilibrium Phonons in Semiconductors 113
    A.V. Akimov, A.A. Kaplyanskii, and E.S. Moskalenko

Monte-Carlo Calculated Nonequilibrium Phonon Pulses in GaAS 129
    D.V. Kazakovtsev, B.A. Danilchenko, and I.A. Obukhov

Influence of Sample Temperature and Pumping Intensity on the Processes
    of Hot Spot Formation and Degradation 135
    A.A. Maksimov, D.A. Pronin, and I.I. Tartakovskii

Ballistic Phonon Propagation in AT-Cut Quartz 139
    B. Sujak-Cyrul, J. Szczepański, and T. Tyc

Effect of Resonant Scattering by Paramagnetic Centers
    on the Propagation of Nonequilibrium Phonons 143
    K.L. Aminov

Singularities of the Heat Conductivity in Thin Dielectric Slabs 153
    J. Czerwonko and M.I. Kaganov

INTERACTION OF BULK PHONONS
WITH LOW DIMENSIONAL GASES OF CARRIERS

Acoustic Phonon Interaction with Two-Dimensional Electron and Hole
    Systems 159
    L.J. Challis and A.J. Kent

Phonon Emission and Absorption Experiments in the Quantum-Hall Regime 189
    F. Dietzel, U. Klass, W. Dietsche, and K. Ploog

Phonon Measurements of the Energy Gap in the Fractional Quantum Hall
    State 201
    R.H. Eyles, C.J. Mellor, A.J. Kent, L.J. Challis, S. Kravchenko,
    N. Zinov'ev, and M. Henini

Response of a Two-Dimensional Electron Gas to Pulses of a Phonon Field . 205
    R.N. Gurzhi, A.I. Kopeliovich, and T. Paszkiewicz

Phonon-Drag Effect in 1-Dimensional Electron Gases 211
    D. Lehmann

The Thermoelectric Behaviour of Two Dimensional Electron and Hole
    Gases and Quantum Point Contacts 219
    P.N. Butcher, T.M. Fromhold, R.J. Barraclough, P.J. Rogers, B.L. Gal-
    lagher, J.P. Oxely, and M. Henini

Energy Relaxation via Acoustic Phonons in 2D and 1D Electron Systems . 233
    A. Shik

A Theory of the Suppression of the Electron-Phonon Interaction 243
    M.V. Entin and O.V. Kibis

Carrier Capture by Quantum Wells via 3D→2D and 2D→2D Channels . . 251
    V. Karpus

PHONON MEDIATED DETECTORS OF ELEMENTARY PARTICLES

Application of Phonon Physics to Cryogenic Detectors 263
    H. Kraus

Cryogenic Particle Detectors: Phonon Physics in Niobium 297
    R. Gaitskell

MOLECULAR CRYSTALS

Phonon Scattering and Heat Transfer in Simple Molecular Crystals 321
    V.G. Manzhelii and V.A. Konstantinov

Fullerenes 333
    K. Prassides

Theoretical Investigations of the Orientational Ordering Transition
    in Solid C_{60}. 353
    K. Rapcewicz

Dynamics Close to the Glass Transition 365
    Ł. Turski

ELECTRON SYSTEMS

The Surface Impedance and the Slab Conductivity of Metals Beyond
    the Relaxation Time Approximation Transition 381
    J. Czerwonko, M.I. Kaganov, and G.Ya Lyubarski

Linear Thermoelastic Generation of Ultrasound in Metals 399
    M.I. Kaganov and A.N. Vasil'ev

Electron-Phonon Coupling in Strongly Correlated Electron Systems 407
    Z.K. Petru and N.M. Plakida

Magnetophonon Resonances on Three Phonon Modes
    in $Zn_xCd_yHg_{1-x-y}Te$ Epitaxial Layers 415
    J. Polit, E. Sheregii, A. Andruchiv, and P. Sidorchuk

A Note on the Measurement of Intersubband Relaxation Time
   by an Infrared Bleaching Technique 419
   M. Załużny

Participants 425

Index 427

RECENT PROGRESS IN USING THE EFFECTIVE POTENTIAL METHOD

George K. Horton and *E. Roger Cowley

Serin Physics Laboratory
Rutgers - the State University
Piscataway, NJ 08855-0849
*Department of Physics
Camden College of Arts and Sciences
Rutgers - the State University
Camden NJ, 08102-1205

1. INTRODUCTION

The effective potential method has been developed over the last eight years, starting from an original simple approximation of Feynman's [1]. A substantially better approximation was introduced by Riccardo Giachetti and Valerio Tognetti [2, 3] and by Feynman and Kleinert [4]. In both original and the improved methods, the partition function is obtained in a classical form but with an effective potential. The application to fairly realistic systems has been made by Tognetti and his co-workers, including Alessandro Cuccoli and Ruggero Vaia [7], and by our group at Rutgers, which has included Shudun Liu, Zizhong Zhu, Dominic Acocella, and Eugene Freidkin. We have shown that the use of this effective potential in a classical Monte Carlo calculation (EPMC) yields thermodynamic properties for the heavier inert-gas solids which agree closely at high temperatures with classical Monte Carlo results, at low temperatures with anharmonic perturbation theory, and succeed in interpolating smoothly between the two [8, 9, 10]. Here we shall first review the general idea of the method, and then we discuss two recent applications which test the applicability of the method further. Finally we shall speculate a little about the future prospects of the method.

2. FORMALISM

The effective potential method is a way of including quantum mechanical effects in a classical statistical mechanics formalism [1, 2, 3, 4, 7]. It is based on the path-integral form of the partition function

$$Z = e^{-\beta F} = \int \prod_{i=1}^{N} \mathcal{D}\vec{r}_i(\tau) e^{-S[\vec{r}(\tau)]/\hbar}, \qquad (1)$$

where $\vec{r}(\tau) \equiv \vec{r}_1(\tau), \vec{r}_2(\tau), \ldots, \vec{r}_N(\tau)$, $\beta = 1/k_B T$, and N is the number of atoms. The action S is

$$S[\vec{r}(\tau)] = \int_0^{\beta\hbar} [\sum_{i=1}^{N} \frac{1}{2} m \dot{\vec{r}}_i^2(\tau) + V(\vec{r}(\tau))] d\tau, \qquad (2)$$

where $V(\vec{r})$ is the potential. The path integral is over all paths $\vec{r}(\tau)$ with the same end point and beginning point, and then over all beginning points. You can think of it as an integral over the classical paths of the particles, together with an integral over the quantum mechanical fluctuations of the particles around their classical positions. The integration variable τ has the dimension of time. The average value of the particle positions is

$$\bar{\vec{r}}_i = \frac{1}{\beta\hbar} \int_0^{\beta\hbar} \vec{r}_i(\tau) d\tau. \qquad (3)$$

and this corresponds to the classical position.

The Quantum Monte Carlo method also starts from the path-integral form of the partition function. In QMC the integration over τ is approximated by a summation over a discrete set of M values of τ [5, 6]. The calculation of thermodynamic averages such as the energy and the pressure can then be set up in the form of a classical Monte Carlo calculation but for a system of $N \times M$ particles, representing the positions of the N actual particles at the M values of τ. Provided M can be made large enough, the results are exact within the usual statistical uncertainties of a Monte Carlo simulation. However, $N \times M$ can quickly become a very large number so that the calculation is very demanding of computer time.

The effective potential method depends on a variational principle based on the inequality [1]

$$F < F_0 + \frac{1}{\beta} \langle S - S_0 \rangle_0. \qquad (4)$$

The final subscript 0 indicates that the average is weighted by a trial action. If the trial action differs from the true action only in the potential energy, this reduces to the more familiar inequality

$$F < F_0 + \langle V - V_0 \rangle_0. \qquad (5)$$

The trial action containing adjustable parameters is introduced, and the corresponding trial free energy is calculated. The parameters are then varied to minimize the trial free energy. Feynman [1] originally introduced a very simple trial action consisting of the kinetic energy and the value of the potential at the average point on the path

$$S_0[\vec{r}(\tau)] = \int_0^{\beta\hbar} [\sum_{i=1}^{N} \frac{1}{2} m \dot{\vec{r}}_i^2(\tau) + W(\bar{\vec{r}})] d\tau. \qquad (6)$$

This firstly satisfies the requirement that the path integral can be evaluated analytically, and it has a second very important feature. When the integral over all paths with a given average has been performed, the value of $W(\bar{r})$ which minimises the trial free energy also has the effect of making $\langle S - S_0 \rangle_0$ disappear. The partition function Z_0 then has the appearance of a classical partition function with the potential replaced by an effective potential:

$$Z_0 = (\frac{m}{2\pi\beta\hbar^2})^{3N/2} \int e^{-V_{eff}(\bar{r})} \prod_{i=1}^{N} d^3\bar{r}_i. \tag{7}$$

While the original idea was a brilliant illustration of the relationship between the path-integral formalism and conventional quantum statistical mechanics, its quality was not good enough to make it a serious tool. A much better variational function was proposed, independently, by Riccardo Giachetti and Valerio Tognetti [2, 3] and by Feynman and Kleinert [4]. They noted that, if the trial action includes terms which are quadratic in the displacements of the particles around their average postions, the path integrals can still be evaluated analytically, and the variational procedure still leads to an effective potential. The trial action is now

$$S_0[\vec{r}(\tau)] = \int_0^{\beta\hbar} [\sum_{i=1}^{N} \frac{1}{2} m\dot{\vec{r}}_i^2(\tau) + W(\bar{r}) + \frac{1}{2}\sum_{i,j=1}^{N}\sum_{\alpha\beta} \phi_{\alpha\beta}(ij,\bar{r})u_\alpha(i,\tau)u_\beta(j,\tau)]d\tau. \tag{8}$$

It is convenient to add and subtract the true potential so that the effective potential can be written

$$V_{eff}(\bar{r}) = V(\bar{r}) + \Delta V(\bar{r}), \tag{9}$$

where

$$\Delta V(\bar{r}) = K(\bar{r}) - V(\bar{r}) - \frac{1}{2\beta}\sum_s (f_s \coth f_s - 1) + \frac{1}{\beta}\sum_s \ln(\frac{\sinh f_s}{f_s}). \tag{10}$$

The sums are over the $3N$ normal modes of the trial quadratic function, labelled by an index s. $K(\bar{r})$ is a smeared potential, quite similar to the smeared quantities which arise in self-consistent phonon theory. However, the smearing represents only the blurring of the particle paths due to quantum mechanical fluctuations and disappears in the classical limit. There is some similarity between the expression for ΔV and the free energy in first-order self-consistent theory. In fact, the terms in ΔV correspond to the differences between the quantum mechanical and classical values of terms in F_{SC1}, with the proviso that the nature of the smearing is different in the SC1 and effective potential cases. Finally, the force constants are found, from the variational principle, to be the smeared values of the second derivatives of the actual potential. The average of any function $A(\bar{r})$ of the positions of the atoms can be put in the form

$$\langle A \rangle = [(2\pi^3)\det \mathbf{D}]^{-\frac{1}{2}} \int A(\bar{r}+\vec{u})\exp(-\frac{1}{2}\sum_{i\alpha j\beta} u_\alpha^i [\mathbf{D}^{-1}]_{\alpha\beta}^{ij} u_\beta^j)d^{3N}u, \tag{11}$$

where

$$D_{\alpha\beta}^{ij} = \frac{\hbar}{mN}\sum_s e_{i\alpha}^s e_{j\beta}^s [\coth(f_s) - 1/f_s]/\omega_s. \tag{12}$$

The averages of the potential and of the force constants can both be calculated in this way; $f_s = \beta\hbar\omega_s$, $\beta = \frac{1}{kT}$.

In most applications so far, this formalism has been considered too complicated to be applied exactly. In particular, at a general value of $\bar{\vec{r}}$, the force constants do not have the periodicity of the lattice, and a $3N \times 3N$ matrix would need to be diagonalised to give the frequencies ω_s. An exception to this is the model ferroelectric crystal which we describe later. As a result of the very simple, and artificial, form of the model potential, each term in the effective potential depends on the position of only one atom, and can be tabulated once and for all. In more realistic cases, the effective potential must be calculated at each instantaneous configuration. Since the correction term disappears in the high temperature limit while at low temperatures the atoms remain close to their equilibrium positions, it is usual to make the approximation of evaluating the force constants and frequencies at the equilibrium positions and to carry out some sort of expansion around these positions. This is called the Low-Coupling-Approximation. At the equilibrium positions, the force constants have the full periodic symmetry of the crystal. In consequence, the normal modes are labelled by a wave vector $\vec{q}$ and a polarization index j, the frequencies can be calculated as the eigenvalues of N 3×3 matrices instead of one $3N \times 3N$ matrix and the elements of $\mathbf{D}$ take the form

$$D_{\alpha\beta} = \frac{\hbar}{mN} \sum_{\vec{q}j} e_\alpha(\vec{q}j) e_\beta(\vec{q}j)(1 - \cos\vec{q}.\vec{R}_{ij})[\coth(f_{\vec{q}j}) - 1/f_{\vec{q}j}]/\omega_{\vec{q}j}. \qquad (13)$$

In addition, in all applications of the method so far, the smeared quantities have been evaluated by series expansion in powers of a quantum renormalization parameter, essentially in powers of the elements of $\mathbf{D}$. The details have been given elsewhere [3, 7, 8, 9]. In a first order EPMC theory the expansion is truncated at the first term. In this case, the bare quasi-harmonic frequencies are used in the calculation of the effective potential. In a second or higher order theory, the frequencies must be calculated self-consistently, since the renormalization factor itself depends on the frequencies. We can carry out a completely self-consistent second-order theory. The equations clearly simplify drastically at zero degrees. At this temperature we can obtain an approximate third-order theory.

As the temperature goes to zero, the free energy in the effective potential method is given entirely by the value of the effective potential itself. Furthermore, the values of the functions $D_{\alpha\beta}$ are equal to the SC1 values, and the effective potential is equal to the SC1 free energy. Also, note that the effective potential V_{eff} is expressed in terms of the actual potential $V(\vec{r})$ plus a correction, and the trial action is involved only in the calculation of this correction. As the temperature increases the whole of the correction term in Eq. (5) goes to zero and we are left with an exact classical expression for the partition function.

3. AN APPLICATION TO NEON

One of our recent applications [11] was to apply the effective-potential Monte Carlo technique to a simple model of neon. For comparison purposes we also carried out calculations using the Improved Self Consistent Phonon theory [12], and using a full Quantum Monte Carlo technique. This work was done in collaboration with Alex Maradudin, Arthur McGurn and Dick Wallis.

In all of the calculations we used a nearest neighbour Lennard-Jones potential with parameters $m = 21.9914$, $\epsilon = 72.09 \times 10^{-16} ergs$, $\sigma = 2.7012 Å$. We believe that the use of this model potential with parameters determined by properties of the solid at $T = 0K$ is preferable to the use of a two-body potential with parameters based on gas properties [13], since it contains some compensation for the omitted many-body forces. In the figures the calculated thermodynamic properties are expressed in terms of the Lennard-Jones parameters e.g. the nearest neighbour distance is expressed in terms of σ.

For the effective potential and ISC formalisms we have found the zero-pressure atomic spacing as a function of temperature, and have calculated the internal energy and heat capacity at that spacing. In the effective potential calculations we did not use the most general form for the force constants but made the assumption

$$\phi_{\alpha\beta}(ij,\bar{\bar{r}}) = \Omega(ij,\bar{\bar{r}})\delta_{\alpha\beta}. \tag{14}$$

Cuccoli et al. have used a more general form [13] but the advantage of the simpler version used here is that we can solve the self-consistent equations to higher order more easily. The effective potential calculations are done with 108 atoms and 4.2 million configurations in the Monte Carlo simulations. The main reason for repeating the earlier ISC calculations was to obtain values for the internal energy. The results are shown in Figs. 1-2. Since the QMC calculation is much more time consuming, we did not try to zero the pressure in that case, but instead used the spacings calculated by the second-order effective potential method. We expected that the pressures would then be small, and that turned out to be the case, though except at the highest temperature they are statistically not zero. From the calculated pressures and the compressibility we could then calculate a better zero pressure spacing. We performed QMC runs for 32 and 108 atom samples, at three temperatures, 7K, 13K, and 22K. In the reduced units ϵ/k_B these are 0.134, 0.249, and 0.421 respectively. We used a variety of Trotter numbers (that is, values of M), ranging from 10 to 40, and in each case used a large enough value that the results were converged to within the statistical uncertainties. The number of single particle moves attempted in a run varied from 13 to 40 million.

The best QMC results, i.e. those for 108 atoms and with the highest Trotter numbers, for the three temperatures, are plotted in the figures with the effective potential and ISC values. The three numerical techniques we have used all yield remarkably similar results. This overall agreement gives us confidence that the statistical uncertainties which we estimate for QMC by the usual type of method are in fact a good indication of the reliability of those results. We can then gauge the other techniques by their agreement with QMC theory. At the lowest temperature, the ISC results lie within the statistical uncertainties of QMC. The energy remains accurate at all temperatures, but at the highest temperature the ISC pressure becomes inaccurate. This is all very plausible. ISC improves on the first-order self-consistent theory (SC1) in an ad hoc fashion. It is gratifying that it agrees as well as it does, and not disturbing that deviations appear at higher temperatures.

The EPMC method is designed to pass smoothly into an exact classical calculation at high temperatures, and the agreement with QMC is best at 22K, and worst at 7K. Note that at 22K neon is still a long way from being classical. We believe that at zero degrees the physical content of the effective potential method is identical with SC1, any differences in numerical results arising from the different treatment of the smearing i.e. truncated series expansion as opposed to numerical integration. The rise in the lattice spacing at low temperatures is certainly an artifact of the expansion procedure.

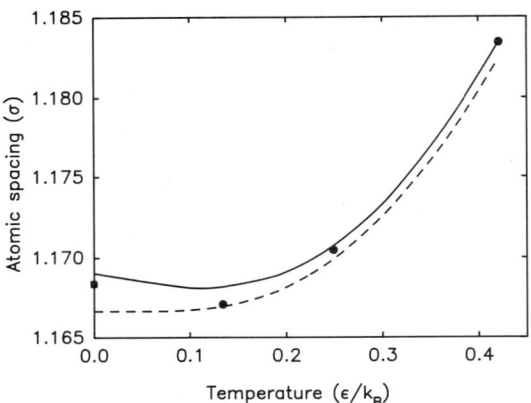

Fig. 1. Nearest neighbour distance in Ne^{22}, expressed in units of the hard sphere radius σ. Circles are QMC results, dashed line is ISC and solid line is second-order EPMC. The square at 0K is the third-order effective potential result

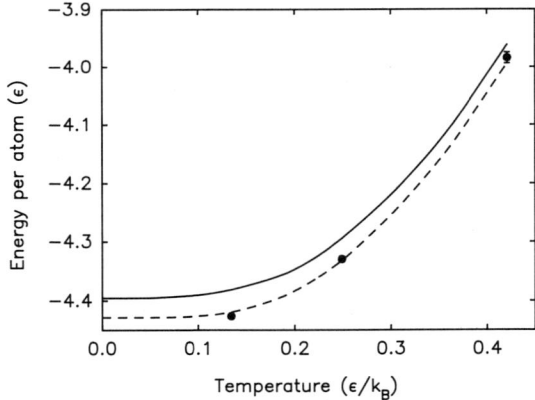

Fig. 2. Internal energy per atom in Ne^{22}, in units of the well-depth ϵ. Circles are QMC results, dashed line is ISC, and solid line is second-order EPMC.

As we have previously noted [9], the rise is less in the second-order theory than in the first-order theory, and the third-order point we can calculate at $T = 0K$ shows almost no rise. We have plotted this single third-order point in Fig. 1. At finite temperatures, the use of the effective potential in a Monte Carlo simulation transcends SC1, and, as we have said, it becomes exact both at the classical limit, and in the first correction term in the Wigner expansion [14]. Already at 7K the effective potential value of the heat capacity is much better than the SC1 value.

The relative cost of the calculations is pertinent information when the choice of method is being made. The QMC calculations described here required approximately forty hours of time on a Cray-YMP computer, the EPMC calculations at the same three temperatures used about thirty minutes, and the ISC calculations were done on a Hewlett-Packard mini-computer but would have occupied the Cray for a few seconds. The EPMC and ISC calculations yield values for the heat capacity, which we were not able to get from QMC calculations, and ISC theory also gave the entropy and compressibility.

To summarize, it is clear that for a quantum solid like Ne, thermal and elastic properties can be obtained cheaply and accurately using ISC theory, with some small inaccuracies near melting. The EPMC theory is at the limit of its usefulness here. For even more pronounced quantum solids, such as He, a different evaluation of the effective potential, not using a series expansion, is required and this seems feasible. The QMC approach is very accurate but correspondingly expensive in computer time.

4. A MODEL FERROELECTRIC

The effective potential method should be valuable for any lattice dynamics problem which involves large anharmonicity at temperatures too low for a classical approximation to be valid. Many ferroelectric materials display structural phase transitions, which certainly involve anharmonic atomic displacements, at temperatures such that kT is comparable with, or smaller than, typical phonon energies. We are confident that the effective potential method will be useful in this field. However, before the method is applied to a realistic, three-dimensional, and therefore complicated, system, it is appropriate to test it out on a simpler model where the new problems associated with this type of application can be discovered. We have therefore applied the effective potential method to an extremely simple one-dimensional model ferroelectric for which a full quantum mechanical calculation has been performed [15]. The model consists of a set of one-dimensional anharmonic oscillators, coupled by a long-ranged interaction. In the limit that the interaction is of infinite range the Hamiltonian takes the molecular field form, and an exact solution is in principle possible. It has been shown that the mean-field approximation, which is equivalent to the self-consistent phonon approximation for this model, can give seriously misleading results for the phase transition. We find that the effective potential method gives identical results with the mean-field approximation at zero degrees but, in contrast to the mean-field approximation, gives good values for the transition temperatures, the nature of the transitions, and the temperature dependence of the order parameter, over a wide range of model values [16].

Following Gillis and Koehler [15], we consider a system of particles, labelled by an index l, moving in one dimension with a potential energy containing quadratic and quartic terms and interacting with each other with force constants $\chi(ll')$. The complete Hamiltonian is

$$\mathcal{H}^{\pm} = \sum_l \left(-\frac{1}{2}\lambda^2 \frac{d^2}{du_l^2} \pm 4u_l^2 + 4u_l^4\right) - \frac{1}{2}\sum_{ll'}\chi(ll')u_l u_{l'}. \tag{15}$$

The interaction is assumed to be infinitely long-ranged so that each particle moves in a molecular field Hamiltonian

$$\mathcal{H}_l^{\pm} = -\frac{1}{2}\lambda^2 \frac{d^2}{du_l^2} \pm 4u_l^2 + 4u_l^4 - \chi \bar{u} u_l, \tag{16}$$

with

$$\chi = \sum_{l'} \chi(ll'). \tag{17}$$

If the ± sign in the quadratic term is positive the model describes a typical displacive ferroelectric, while if it is negative the potential has a double minimum. In that case, if the quantum parameter λ is small the model displays an order-disorder transition, which gradually becomes more displacive in character as λ increases. Gillis and Koehler expanded the wave function of the particle in terms of a basis set of 50 optimized harmonic oscillator functions, and used the lowest 50 eigenvalues in a calculation of the thermal expectation of the displacement $\langle u \rangle$. We have repeated their calculations using 200 eigenvalues and wave functions in the partition function and it is these results that are shown on our figures. They differ significantly from Gillis and Koehler's results. The average was found for a range of values of the parameter $\bar{u}$, and at each temperature the self-consistent solution with $\langle u \rangle$ equal to $\bar{u}$ was selected or interpolated. The calculation was carried out for a range of values of the coupling parameter χ, and for two values of λ, 0.2 and 1.0, which, in the dimensionless units used, correspond to small and large zero point motion respectively.

A classical calculation of the average displacement is very simple, requiring only a one-dimensional numerical quadrature:

$$\langle u \rangle = \frac{\int u e^{-\beta V} du}{\int e^{-\beta V} du}. \tag{18}$$

A value is assumed for the parameter $\bar{u}$ in the molecular field Hamiltoniam, and $\langle u \rangle$ is calculated. The value of $\bar{u}$ is then adjusted in a systematic way until the calculated average agrees with the initial value. To use the effective potential approach we simply replace the potential energy V in the above equation by the effective potential. To justify this, we include in the Hamiltonian an additional term $-Fu_l$. The average value of u_l is then the derivative of the Free Energy with respect to F, evaluated at $F=0$. Since a linear term such as this does not change the form of ΔV, the above result is obtained.

We can also calculate the susceptibility of the crystal using the same approach. The susceptibility is defined as the derivative of $\langle u \rangle$ with respect to F, evaluated at F equal to zero. The result is

$$\frac{\partial \bar{u}}{\partial F} = \frac{\beta\{\langle u^2 \rangle_{cl} - \langle u \rangle^2\}}{1 - \beta\chi\{\langle u^2 \rangle_{cl} - \langle u \rangle^2\}}, \tag{19}$$

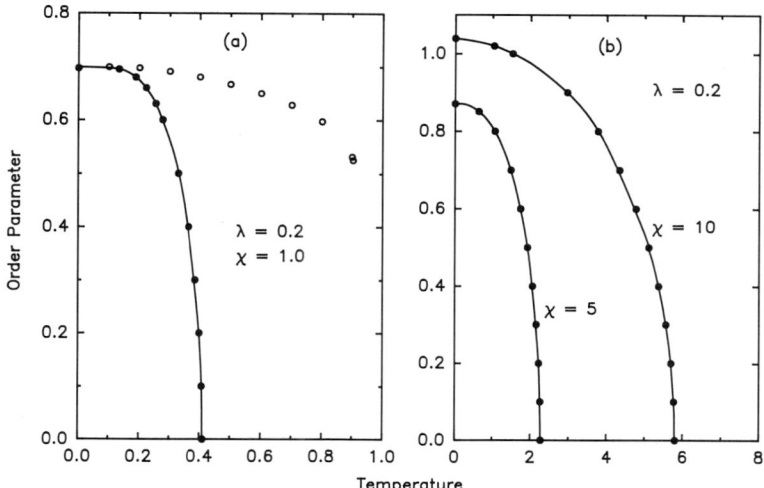

Fig. 3. The order parameter $\langle u \rangle$ as a function of temperature for the order-disorder case, with $\lambda = 0.2$, (a) for $\chi = 1$, (b) for $\chi = 5$ and $\chi = 10$. The lines are the present calculation and filled circles are exact results from reference [15]. In (a) the mean-field approximation results, also from reference [15], are shown as hollow circles.

where

$$\langle u^2 \rangle_{Cl} = \frac{\int u^2 e^{-\beta V_{eff}} du}{\int e^{-\beta V_{eff}} du}. \tag{20}$$

This quantity is of the form of a classical average of u^2, except that it contains V_{eff}. It is however not equal to the mean-square-displacement $\langle u^2 \rangle$, but is only a part of it. To find the complete expression we can include yet another extra term in the Hamiltion, of the form Cu_i^2. The mean-square displacement is then obtained as the derivative of the Free Energy with respect to the parameter C, evaluated at $C = 0$. The result is

$$\langle u^2 \rangle = \langle u^2 \rangle_{Cl} + \langle \frac{\hbar}{2m\omega}\{\coth(\hbar\omega/2kT) - 2kt/\hbar\omega\}\rangle, \tag{21}$$

where ω is the single frequency arising in the effective potential for this simple model. The extra term arises because the correction term in the effective potential is itself a function of C. Many other time-independent thermal averages can be calculated by similar methods.

The calculated values of the order parameter $\langle u \rangle$ for a range of values of χ and λ are shown in Figs. 3 and 4 for the order-disorder case (negative quadratic term in potential energy) and in Figs. 5 and 6 for the displacive case (positive quadratic term). In each case the solid line is the present result, and the filled points are the exact (numerical) results taken from Gillis and Koehler, as modified by us [15]. In Figs. 4 and 6 we have also plotted the classical result as the dashed line. These were included to show the magnitude of the quantum corrections. In Figs. 3(a) and 6(b) the hollow circles show the results of the mean-field-approximation (also given by Gillis and Koehler), to give some idea of the variable nature of that approximation. In all cases the present

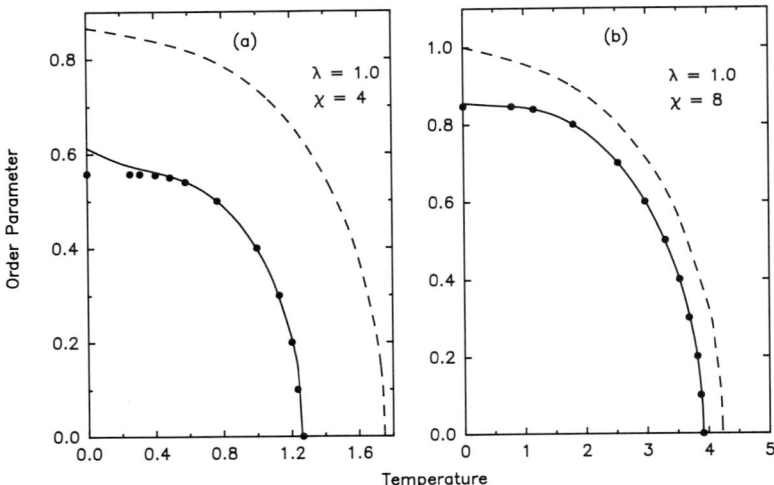

Fig. 4. As Fig. 3, for the case $\lambda = 1.0$, (a) for $\chi = 4$ and (b) for $\chi = 8$. The dashed lines show the results of a classical calculation (i.e. $\lambda = 0$).

calculation coincides with the mean-field result at zero degrees but is an improvement, sometimes a very large improvement, at finite temperatures.

Fig. 7 shows the susceptibilities for two cases, one each of the order-disorder and displacive transitions. Since we are calculating only time-independent quantities, we do not have values for vibration frequencies *per se*, but the inverse susceptibilites, shown as dashed lines, can be interpreted as illustrating soft mode behaviour.

In most of the cases shown in Figs. 3-6 the effective potential values are almost indistinguishable from the exact quantum mechanical values. The largest discrepancies are close to T_c for small λ and large χ. This is the region where the quantum mechanical sum over states is most difficult to converge. We have, in fact, repeated some of Gillis and Koehler's calculations but with extended ranges for the sums over states, and have found excellent agreement with the effective potential results in this region. There is also some disagreement for the order-disorder model (double potential well) for small χ and large λ, at low temperatures. Here the ground state wavefunction is poorly represented by the single shifted Gaussian peak of the mean-field approximation.

It will certainly be possible to elaborate the simple model used here to make it more realistic. The next step will be to include a short range interaction between the atoms. Koehler and Gillis [17] have made this extension, but could do so only in a classical approximation. The effective potential approach includes quantum mechanical effects accurately while retaining most of the simplicity of a classical calculation. The possible applications of the technique seem to be unlimited.

5. FUTURE DEVELOPMENTS

In all calculations on more or less realistic three dimensional solids, some version

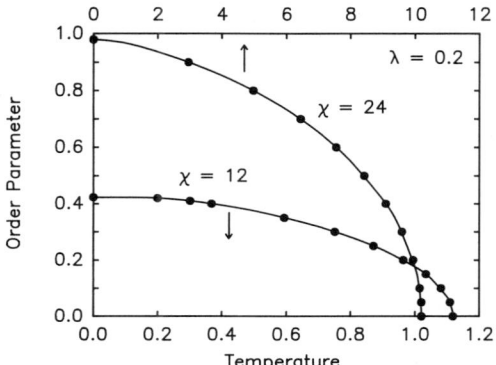

Fig. 5. The order parameter for the displacive model, with $\lambda = 0.2$, and $\chi = 12$ and 24. The lines are the present calculation, and filled circles are exact results from reference [15], and modified by us.

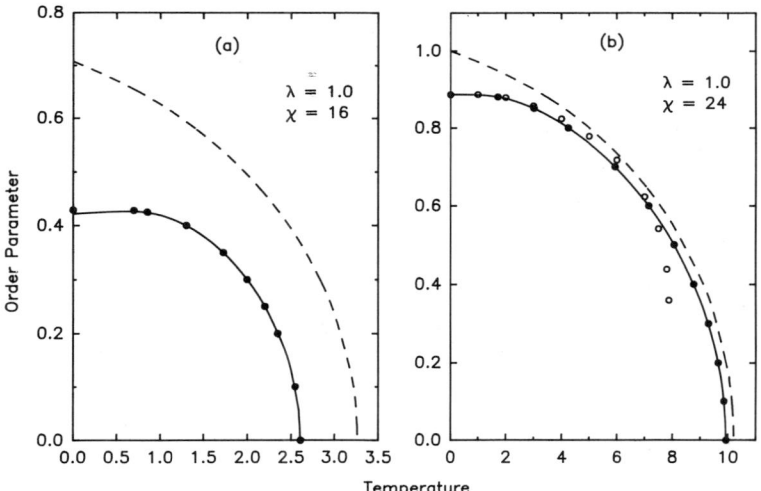

Fig. 6. As Fig. 5, for the case $\lambda = 1.0$, (a) for $\chi = 16$ and (b) for $\chi = 24$. The dashed lines show the results of a classical calculation. In (b) the mean-field approximation results are shown as hollow circles.

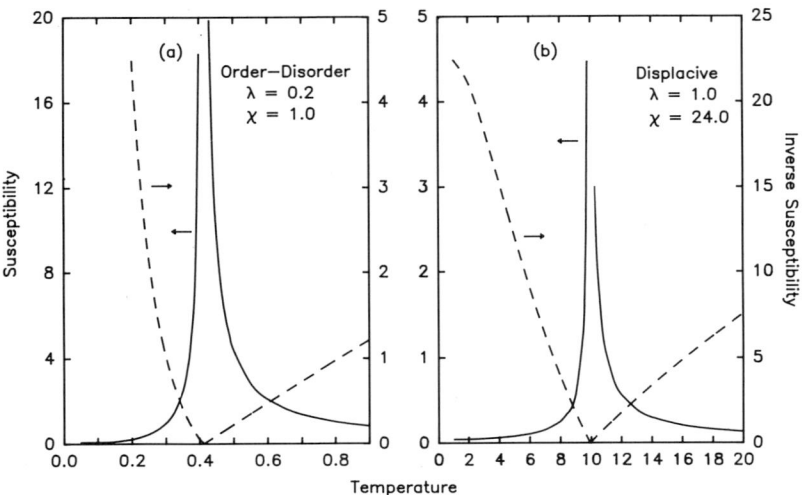

Fig. 7. The susceptibilty (solid line) and its inverse (dashed line) (a) for an order-disorder model with $\lambda = 0.2$ and $\chi = 1.0$, (b) for a displacive model with $\lambda = 1.0$ and $\chi = 24$.

of the Low Coupling Approximation has been used. The reason for this was that a full calculation of the effective potential requires the diagonalization of a $3N \times 3N$ matrix to calculate the normal mode frequencies of the effective potential, at every position of the atoms. If a single atom is moved, in a Monte Carlo step, the whole calculation has to be repeated. In the Low Coupling Approximation, not only is the $3N \times 3N$ diagonalization replaced by N 3×3 diagonalizations, but also as the atoms make small displacements the change in the effective potential is calculated as a series expansion. This introduces an additional uncertainty into the results. To what extent are the discrepancies results of this approximation? The only completely satisfactory way to answer this question is to avoid making the approximation. However, the calculation is then considerably more difficult, so that it has not been clear that it was even possible. A student of ours, Dominic Acocella, has been programming the full calculation for the last year and now has a working program. We expect that he will shortly have some results which will give a good answer to the question of the reliability of the low coupling approximation. It does seem clear that his calculations take more time - but how much more? Therefore, if he finds that the low coupling approximation is *not* adequate, the time needed for the full effective potential method as compared to a Quantum Monte Carlo calculation will become crucial. We are waiting eagerly for his results.

There is no doubt that, as the speed of readily available computers increases, full Quantum Monte Carlo calculations will become more common. The accuracy of such calculations is limited by the attainable values of the Trotter number M and by the fact that Quantum Monte Carlo calculations do not readily conform to the requirements of massively parallel computers such as the CM machines. There are a number of schemes for accelerating the convergence of such a calculation, sometimes called Fourier acceleration techniques. The name arises from the fact that a particular value of M corresponds to including the lowest M components in a Fourier expansion of the trajectory in the path integral. Doll, Coalson, and Freeman [18] have developed an approximate correction for the omitted components, based on the free-particle form of the action. They show that their correction reduces to the original Feynman form of the effective potential if M is set to zero, i.e. if the correction is used for the whole expression. They point out that a similar correction can be developed using the harmonic oscillator form for the density matrix. This work actually slightly predated the development of the effective potential formalism. It seems that the more elaborate correction will give a formalism which will smoothly link the Quantum Monte Carlo method with a small value of M into the effective potential method used to calculate a remainder. This may, in fact, turn out in the long term to be the most valuable use of the effective potential procedure.

A post script is also relevant. It is possible to use the effective potential formalism to obtain moments of the correlation function. Using a continued fraction technique it is then possible to obtain phonon frequencies and widths as long as the phonon involved is a reasonably narrow and smooth Lorentzian. This approach has been used by S. Aubry [19] and is currently being studied further by [20] for a nearest neighbour Lennard-Jones linear chain and by our group for the ferroelectric model discussed in Sect. 4.

ACKNOWLEDGEMENTS

This research was supported in part by the U.S. National Science Foundation under grants No. DMR-88-08756 and DMR-92-02907. We acknowledge grants of computer time from the Pittsburgh Supercomputer Center under grant No. DMR890022P and from the San Diego Supercomputer Center. We also thank the Rutgers Research Council for a grant and Dr. R. Vaia for sending us a reprint.

REFERENCES

[1] R.P. Feynman, *Statistical Mechanics* (W.A. Benjamin, 1972; Addison-Wesley, 1988).

[2] R. Giachetti and V. Tognetti, Phys. Rev. Lett. **55**, 912 (1985).

[3] R. Giachetti and V. Tognetti, Phys. Rev. B **33**, 7647 (1986).

[4] R.P. Feynman and H. Kleinert, Phys. Rev. A **34**, 5080 (1986).

[5] A.R. McGurn, P. Ryan, A.A. Maradudin, and R.F. Wallis, Phys. Rev. B **40**, 2407 (1989).

[6] M. Takahashi and M. Imada, Proc. Phys. Soc. Jpn. **53**, 912 (1985).

[7] A. Cuccoli, V. Tognetti, and R. Vaia, Phys. Rev. B **41**, 9588 (1990).

[8] S. Liu, G.K. Horton, and E.R. Cowley, Phys. Lett. A **152**, 79 (1991).

[9] S. Liu, G.K. Horton, and E.R. Cowley, Phys. Rev. B **44**, 11714 (1991).

[10] Z. Zhu, S. Liu, G.K. Horton, and E.R. Cowley, Phys. Rev. B **45**, 7122 (1992).

[11] S. Liu, G.K. Horton, E.R. Cowley, A.R. McGurn, A.A. Maradudin, and R.F. Wallis, Phys. Rev. B **45**, 9716 (1992).

[12] V.V. Goldman, G.K. Horton, and M.L. Klein, J. Low Temp. Phys. **1**, 391 (1969).

[13] A. Cuccoli, A. Macchi, M. Neumann, V. Tognetti, and R. Vaia, preprint.

[14] E.P. Wigner, Phys. Rev. **40**, 749 (1932).

[15] N.S. Gillis and T.R. Koehler, Phys. Rev. B **9**, 3806 (1974).

[16] E.R. Cowley and G.K. Horton, Ferroelectrics **136**, 157 (1992).

[17] T.R. Koehler and N.S. Gillis, Phys. Rev. B **13**, 4183 (1976).

[18] J.D. Doll, R.D. Coalson, and D.L. Freeman, Phys. Rev. Letters **55**, 1 (1985).

[19] S. Aubry, Journal of Chemical Physics, **62**, 3217 (1975). Ibid, **64**, 3392 (1976).

[20] A. Cuccoli, V. Tognetti, A.A. Maradudin, A.R. McGurn and R. Vaia, Phys. Rev. B **46**, 8839, (1992).

MOMENTUM AND QUASIMOMENTUM IN THE PHYSICS OF CONDENSED MATTER

A. Thellung

Institut für Theoretische Physik der Universität Zürich
Winterthurerstrasse 190, CH–8057 Zürich, Switzerland

Abstract

In the first part of these lectures the formulations of hydrodynamics by Euler and by Lagrange are laid out. Eulerian phonons carry momenta, in contrast to Lagrangian phonons, whose momenta are zero, but the momentum density of the fluid is the same. The proof is much shorter than in earlier papers.

The second, main, part starts from the familiar concept of the quasimomentum of phonons and electrons in solids and proceeds to quasimomentum (in contrast to ordinary momentum) in the classical theory of elasticity and electrodynamics of continuous media. Conservation of quasimomentum is proven for arbitrarily nonlinear and anisotropic homogeneous media; it even holds for dispersive media, whose energy depends on higher than the first derivatives of the fields. If the difference between the local (Eulerian) coordinates (used in electrodynamics) and the material (Lagrangian) coordinates (used in the theory of elasticity) is properly taken into account a new term in the expression for the quasimomentum of photoelastic media appears. Various applications of quasimomentum conservation are mentioned and a new kind of radioelectric effect is put forward.

1. INTRODUCTION

In the first part of these lectures I want to point out a few things about the momentum of phonons in condensed matter physics, particularly in hydrodynamics. In the second, main, part I want to show that, in addition to energy and momentum, there exists in condensed matter another, independent, integral of the motion, called quasimomentum, provided the medium is macroscopically homogeneous, but otherwise it may be arbitrarily nonlinear and anisotropic. In a solid it is a more interesting quantity than ordinary momentum, because the latter, $\sum_i m_i \mathbf{v}_i$ for mass points or $\int d^3x \rho \mathbf{v}$ in a continuous medium, is connected with mass transport and is usually zero in a solid. On the other hand, quasimomentum conservation is a most useful tool in analyzing a number of physical effects, e.g. when the interaction of sound or light with thermal phonons or conduction electrons plays a role.

In sections 2–5 problems concerning the momentum of phonons in hydrodynamics are discussed. In section 2 we briefly lay out the two descriptions of hydrodynamic motion due to Euler and to Lagrange. Both descriptions allow a canonical formalism. Quantization is straightforward and the eigenvalues of the harmonic parts of the Hamiltonians lead to the concept of phonons. Eulerian phonons have momentum $\hbar\mathbf{k}$ ($\mathbf{k}$ = wave vector) (section 3), Lagrangian phonons have zero momentum (section 4). In spite of this difference the momentum density of the fluid, as observed in a laboratory frame, which is a fundamental quantity in two–fluid hydrodynamics, turns out to be the same (section 5). This should of course be so since the formulations by Euler and Lagrange are equivalent, but it is fun to see how this result comes about. The proof presented here is much shorter than in earlier papers[1]. The distinction between Eulerian and Lagrangian coordinates laid out in this first part of the lectures will also be important for the second part.

The second part is mainly devoted to quasimomentum, also called pseudomomentum or crystal momentum. In section 6 we remind the reader of the common definition of quasimomentum for phonons and electrons in the quantum theory of solids. In section 7.1 we derive a conservation law for a vectorial quantity directly from the equations of motion of a homogeneous elastic medium. That conservation law is valid under very general circumstances and - as shown in section 7.3 - even if the elastic energy depends on higher than first order derivatives of the displacement field, giving rise to dispersion. The connection, via Noether's theorem, with the invariance properties of the Lagrangian is established in section 7.2. The physical meaning of the conserved vector quantity is found from coupling acoustic waves to thermal phonons, and the term "quasimomentum" is justified (section 7.4). In section 8 we consider electrodynamics in macroscopic bodies and find again a conserved vector quantity in arbitrarily nonlinear and anisotropic, but homogeneous dielectric media. Interaction with Bloch electrons in a metal shows that the conserved quantity is the quasimomentum. The difference between momentum and quasimomentum in electrodynamics is made clear in section 9 by considering the interaction between electromagnetic and elastic waves. Since in the theory of elasticity one works in material (Lagrangian) coordinates, whereas electrodynamics is formulated in local (Eulerian) coordinates, there are some subtleties. If one takes this distinction exactly into account a new term appears in the quasimomentum density. Finally, in section 10, practical applications are mentioned and (a new version of) the radioelectric effect is discussed.

2. EULER'S AND LAGRANGE'S FORMULATIONS OF HYDRODYNAMICS

A <u>Lagrangian</u> (material) coordinate $\mathbf{y}$ labels a given particle of the fluid (independent of its present position), it equals e.g. its equilibrium position when the fluid is at rest. An <u>Eulerian</u> (local) coordinate $\mathbf{x}$ specifies a fixed point in space (independent of which fluid particle passes through that point). The connection between $\mathbf{x}$ and $\mathbf{y}$ is made by saying that $\mathbf{x}$ is the position at time t of the fluid particle whose equilibrium position is $\mathbf{y}$. Therefore the relation between $\mathbf{x}$ and $\mathbf{y}$ is

$$\mathbf{x} = \mathbf{y} + \mathbf{u}(\mathbf{y}, t) , \qquad (2.1)$$

where $\mathbf{u}$ is the displacement vector of the fluid particle labeled by $\mathbf{y}$. In Lagrange's formulation, $\mathbf{y}$ and t are taken as the independent variables; in Euler's formulation, the independent variables are **x** and t. The velocity field in Lagrange's formulation is

$$\mathbf{v}_L(\mathbf{y},t) = \frac{\partial \mathbf{x}(\mathbf{y},t)}{\partial t} = \frac{\partial \mathbf{u}(\mathbf{y},t)}{\partial t}, \qquad (2.2)$$

whereas Euler's velocity field $\mathbf{v}_E$ at **x**, t is defined as

$$\mathbf{v}_E(\mathbf{x},t) \equiv \mathbf{v}_L(\mathbf{y}(\mathbf{x},t),t), \qquad (2.3)$$

assuming that the inverse function of (2.1), $\mathbf{y}(\mathbf{x},t)$, exists. The mass contained in an element of volume d^3y and in the corresponding d^3x is given by

$$\rho_0 \, d^3y = \rho \, d^3x. \qquad (2.4)$$

ρ is the actual mass density observed in the laboratory system; ρ_0 is the equilibrium density and is constant in a homogeneous medium.

In <u>Euler's</u> formulation the dependent variables are the velocity field $\mathbf{v}(\mathbf{x},t)$ and the density $\rho(\mathbf{x},t)$; the pressure p is assumed to be a given function of ρ (adiabatic law). The basic equations[2] are Euler's equation of motion,

$$\partial \mathbf{v}/\partial t + (\mathbf{v} \cdot \nabla)\mathbf{v} = -\frac{1}{\rho}\nabla p, \qquad (2.5)$$

and the continuity equation,

$$\partial \rho/\partial t + \nabla \cdot (\rho \mathbf{v}) = 0. \qquad (2.6)$$

In <u>Lagrange's</u> formulation the dependent variables are $\mathbf{x}(\mathbf{y},t)$, the position of the fluid particle **y** at time t, and the density $\rho(\mathbf{y},t)$, which can be expressed in terms of derivatives of $\mathbf{x}(\mathbf{y},t)$ (see (2.8) below). The pressure p is again assumed to be a given function of ρ. The basic equations[2] are Lagrange's equations of motion

$$\frac{\partial^2 x_i}{\partial t^2}\frac{\partial x_i}{\partial y_k} = -\frac{1}{\rho}\frac{\partial p}{\partial y_k} \qquad (2.7)$$

and the equation for the density

$$\rho/\rho_0 = [\det(\partial x_i/\partial y_k)]^{-1}, \qquad (2.8)$$

which follows from (2.4), $\det(\partial x_i/\partial y_k)$ being the Jacobian.

In fluid mechanics one usually prefers Euler's form; in solid state physics one normally works with Lagrangian coordinates. Note that the <u>linearized</u> versions of the equations of motion (2.5) and (2.7) (with $(\partial \mathbf{x}/\partial t)_y = \mathbf{v}$) are formally the same.

3. EULERIAN PHONONS

Since we are only interested in (longitudinal) phonons it is sufficient to consider vortex-free motion; we write the velocity field in the form

$$\mathbf{v}_E = -\nabla \phi + \mathbf{U}_E \qquad (3.1)$$

(ϕ = velocity potential, $\mathbf{U}_E$ = uniform translation). Euler's equation of motion can be obtained from a canonical formalism.[3] The Hamiltonian is

$$\mathcal{H}_E = \int_V d^3x \left\{ \frac{1}{2} \nabla_E \rho \nabla_E + E_p(\rho) \right\}, \tag{3.2}$$

where

$$E_p(\rho) = -\rho \int_{1/\rho_0}^{1/\rho} (p - p_0) \, d\left(\frac{1}{\rho}\right) \tag{3.3}$$

is the density of the potential energy. ϕ and ρ are canonically conjugate fields. The canonical field equations

$$\dot{\phi} = \frac{\delta \mathcal{H}_E}{\delta \rho}, \quad \dot{\rho} = -\frac{\delta \mathcal{H}_E}{\delta \phi} \tag{3.4}$$

yield the Bernoulli equation (whose gradient gives (2.5)) and the continuity equation (2.6).

Quantization is straightforward,[3] as ρ and ϕ fulfill the canonical commutation relations

$$[\rho(\mathbf{x}), \phi(\mathbf{x}')] = \frac{\hbar}{i} \delta(\mathbf{x} - \mathbf{x}'). \tag{3.5}$$

The harmonic part of the Hamiltonian can be diagonalized and leads to the concept of phonons. It is formed by the terms up to second order when $\mathcal{H}_E$ is expanded in powers of $\nabla \phi$ and $\rho - \rho_0$. The higher order (anharmonic) terms are treated as a perturbation. (3.2) and (3.1) yield

$$\mathcal{H}_E = \int_V d^3x \left\{ \frac{1}{2} \rho \mathbf{U}_E^2 - \frac{1}{2} \mathbf{U}_E \left[\rho (\nabla \phi) + (\nabla \phi) \rho \right] + \right.$$

$$\left. + \frac{1}{2} \rho_0 (\nabla \phi)^2 + \frac{c^2}{2\rho_0} (\rho - \rho_0)^2 + \text{anharmonic terms} \right\}, \tag{3.6}$$

where $c^2 = (dp/d\rho)_{\rho=\rho_0}$ is the square of the sound velocity. Assuming periodic boundary conditions in a cube of volume V, we use a Fourier decomposition of ϕ and ρ,

$$\left.\begin{aligned}
\phi(\mathbf{x}) &= V^{-\frac{1}{2}} \sum_\mathbf{k} \sqrt{\frac{c\hbar}{2\rho_0 k}} \left(a_\mathbf{k} + a^+_{-\mathbf{k}} \right) e^{i\mathbf{k}\mathbf{x}}, \\
\rho(\mathbf{x}) - \rho_0 &= V^{-\frac{1}{2}} \sum_\mathbf{k} \sqrt{\frac{\rho_0 \hbar k}{2c}} \left(a^+_\mathbf{k} - a_{-\mathbf{k}} \right) e^{-i\mathbf{k}\mathbf{x}}.
\end{aligned}\right\} \tag{3.7}$$

As a consequence of (3.5), $a^+_\mathbf{k}$ and $a_\mathbf{k}$ fulfill the commutation relations characteristic of creation and annihilation operators,

$$\left[a_\mathbf{k}, a^+_{\mathbf{k}'} \right] = \delta_{\mathbf{k},\mathbf{k}'}. \tag{3.8}$$

Consequently $a^+_\mathbf{k} a_\mathbf{k}$ has the eigenvalues

$$n_\mathbf{k} = 0, 1, 2, \ldots. \tag{3.9}$$

Then the harmonic part of $\mathcal{H}_E$ is diagonal,

$$\mathcal{H}_E = \frac{1}{2}\rho_0 V \mathbf{U}_E^2 + \sum_{\mathbf{k}} n_{\mathbf{k}}(\epsilon_{\mathbf{k}} + \hbar \mathbf{k} \cdot \mathbf{U}_E) + \qquad (3.10)$$

$$+ \text{anharmonic terms}$$

with $\epsilon_{\mathbf{k}} = \hbar\omega_{\mathbf{k}} = \hbar c|\mathbf{k}|$. The zero point energy has been omitted. Obviously, $n_{\mathbf{k}}$ is the number of phonons of wave vector $\mathbf{k}$, energy $\epsilon'_{\mathbf{k}} = \hbar\omega'_{\mathbf{k}}$, and frequency $\omega'_{\mathbf{k}} = \omega_{\mathbf{k}} + \mathbf{k}\cdot\mathbf{U}_E$, accounting for the Doppler shift. The total momentum becomes

$$\mathbf{P}_E = \int_V d^3x \frac{1}{2}(\rho\mathbf{v}_E + \mathbf{v}_E\rho) = \rho_0 V \mathbf{U}_E + \sum_{\mathbf{k}} n_{\mathbf{k}} \hbar \mathbf{k} . \qquad (3.11)$$

An important quantity in view of two–fluid hydrodynamics is the spatially averaged velocity,

$$\mathbf{u}_{sE} \equiv \frac{1}{V}\int_V d^3x\, \mathbf{v}_E = \mathbf{U}_E . \qquad (3.12)$$

(3.11) and (3.12) are exact. They show that <u>Eulerian phonons carry momenta $\hbar\mathbf{k}$, but give no contribution to the average velocity</u>.

4. LAGRANGIAN PHONONS

Since the linearized equations of motion in Euler's and Lagrange's descriptions are formally the same, the harmonic part of the Hamiltonian can be chosen to be formally the same. So we write in analogy to (3.1)

$$\mathbf{v}_L = \mathbf{U}_L - \nabla_{\mathbf{y}}\phi , \qquad (4.1)$$

$$\mathcal{H}_L = \int_V d^3y \left\{ \frac{1}{2}\rho_0 \mathbf{v}_L^2 + \frac{c^2}{2\rho_0}(\rho - \rho_0)^2 + \right. \qquad (4.2)$$

$$\left. + \text{different anharmonic terms} \right\},$$

where again ϕ and ρ are canonically conjugate fields. Another possibility would be to take the Hamiltonian leading to (2.7) (see the first part of a paper by Tyabji[4]) in the harmonic approximation. The conclusions would be the same.

The Fourier decomposition is analogous to (3.7), but now in <u>Lagrangian</u> coordinates, $\propto e^{\pm i\mathbf{k}\cdot\mathbf{y}}$. Quantization yields the following expressions for energy, total momentum and average velocity:

$$\mathcal{H}_L = \frac{1}{2}\rho_0 V \mathbf{U}_L^2 + \sum_{\mathbf{k}} n_{\mathbf{k}} \epsilon_{\mathbf{k}} + \text{anharmonic terms} , \qquad (4.3)$$

$$\mathbf{P}_L = \int_V d^3y\, \rho_0 \mathbf{v}_L = \rho_0 V \mathbf{U}_L , \qquad (4.4)$$

$$\mathbf{u}_{sL} \equiv \frac{1}{V}\int_V d^3x\, \mathbf{v}_L(\mathbf{y}(\mathbf{x},t),t) = \frac{1}{V}\int_V d^3y\, \frac{\rho_0}{\rho} \mathbf{v}_L(\mathbf{y},t) =$$

$$\frac{1}{V}\int_V d^3y \left[1 - \frac{\rho - \rho_0}{\rho_0} + \ldots\right] \mathbf{v}_L = \mathbf{U}_L - \frac{1}{\rho_0 V}\sum_{\mathbf{k}} n_{\mathbf{k}}\, \hbar\mathbf{k} + \ldots, \quad (4.5)$$

where use has been made of (2.4). Note that $\mathbf{u}_{sL}$ is defined in the same way as in (3.12): it is the average of the velocity in the laboratory system, to be taken in <u>local</u> coordinates. Eqs. (4.4) and (4.5) show that <u>a Lagrangian phonon carries no momentum, but gives a contribution</u> $\left(-\frac{1}{\rho_0 V}\hbar\mathbf{k}\right)$ <u>to the average velocity.</u>

5. MOMENTUM DENSITY

According to (3.11) and (3.12) the momentum density in Euler's case is

$$\frac{1}{V}\mathbf{P}_E = \rho_0\, \mathbf{u}_{sE} + \frac{1}{V}\sum_{\mathbf{k}} n_{\mathbf{k}}\, \hbar\, \mathbf{k}. \quad (5.1)$$

According to (4.4) and (4.5) the momentum density in Lagrange's case is

$$\frac{1}{V}\mathbf{P}_L = \rho_0\, \mathbf{U}_L = \rho_0\, \mathbf{u}_{sL} + \frac{1}{V}\sum_{\mathbf{k}} n_{\mathbf{k}}\, \hbar\mathbf{k}. \quad (5.2)$$

Since $\mathbf{u}_{sE}$ and $\mathbf{u}_{sL}$ are defined in the same way (see (3.12) and (4.5)) Eqs. (5.1) and (5.2) show that the momentum density is the same in the two cases, as it should be. This is the crucial point in the proof[1] that Eulerian and Lagrangian phonons lead to the same two-fluid equations in He II.

If one tries to apply the methods of sections 3 - 5 to an elastic solid one obtains similar results for the longitudinal phonons. But the transverse phonons carry no momenta nor do they contribute to the average velocity. On the other hand, only the sum of the wave vectors of <u>all</u> phonons is a conserved quantity (if umklapp-processes are negligible). This is an indication that in solids, instead of momentum, one should study another quantity, which will be dealt with in the following sections: quasimomentum.

6. QUASIMOMENTUM IN THE QUANTUM THEORY OF SOLIDS

In the quantum theory of crystal lattices the phonons are commonly defined in Lagrangian coordinates. An important quantity is $\hbar\mathbf{k}$ ($\mathbf{k}$ = wave vector), the quasimomentum [5-7] of a phonon, also called pseudomomentum[8] or crystal momentum.[5,9] Like in hydrodynamics, the ordinary momentum of a Lagrangian phonon is zero. For the total quasimomentum

$$\mathbf{P}_q = \sum_{\mathbf{k},\alpha} n_{\mathbf{k},\alpha}\, \hbar\mathbf{k} \quad (6.1)$$

($n_{\mathbf{k},\alpha}$ = number of phonons of wave vector $\mathbf{k}$ and polarization index α) one has a "quasi conservation-law": In interaction processes, due to the anharmonic terms in the Hamiltonian, $\mathbf{P}_q$ is either conserved (normal processes) or changed by a term $\hbar \mathbf{K}$ ($\mathbf{K} \neq 0$, umklapp–processes), where $\mathbf{K}$ is a vector of the reciprocal lattice.

At low temperatures umklapp–processes are negligible and in a very pure crystal the quasimomentum is conserved. If the wavelength of the phonons λ is large compared to the interatomic distance a,

$$\lambda \gg a, \qquad (6.2)$$

the atomistic structure can be neglected and the crystal can be treated as a continuous medium. For acoustic waves the condition (6.2) is always fulfilled. In the following sections we shall investigate quasimomentum in continuous media.

Expression (6.1) can also be understood as the quasimomentum of the Bloch electrons in a conductor; α then stands for the spin and the band index. Again, normal and umklapp processes are possible, and $\hbar \mathbf{k}$ is not a true momentum, the latter being $m\, \partial \omega_{\mathbf{k}\alpha}/\partial \mathbf{k}$ (m = electron mass).

7. THEORY OF ELASTICITY OF CONTINUOUS MEDIA

In this section we investigate the quasimomentum in the general (linear or nonlinear) theory of a (macroscopically) homogeneous elastic medium.

The equation of motion[10] reads

$$\rho_0\, \ddot{u}_i = \frac{\partial}{\partial y_l} \frac{\partial E}{\partial u_{i,l}}. \qquad (7.1)$$

y_l are the Lagrangian (material) coordinates, $\mathbf{u}(\mathbf{y}, t)$ is the displacement vector, the dot means the time derivative (of course at constant $\mathbf{y}$), $u_{i,l} = \partial u_i / \partial y_l$. $E(\{u_{i,l}\})$ is the density of the (internal) elastic energy; it is an arbitrary function of the $u_{i,l}$.

7.1. QUASIMOMENTUM CONSERVATION

Proceeding as in Ref.,[11] we multiply (7.1) by $u_{i,n}$ and sum over the index i (without writing the summation sign). The left–hand side of the ensuing equation can be written

$$\rho_0\, \ddot{u}_i\, u_{i,n} = \frac{\partial}{\partial t}(\rho_0\, \dot{u}_i\, u_{i,n}) - \rho_0\, \dot{u}_i\, \dot{u}_{i,n} \qquad (7.2)$$

and the right–hand side

$$\left(\frac{\partial}{\partial y_l} \frac{\partial E}{\partial u_{i,l}}\right) u_{i,n} = \frac{\partial}{\partial y_l}\left(\frac{\partial E}{\partial u_{i,l}} u_{i,n}\right) - \frac{\partial E}{\partial u_{i,l}} \frac{\partial^2 u_i}{\partial y_n \partial y_l}. \qquad (7.3)$$

The last term in (7.2) equals $-\dfrac{\partial}{\partial y_n}\left(\dfrac{1}{2}\rho_0\, \dot{\mathbf{u}}^2\right)$ and the last term in (7.3) equals $-\partial E/\partial y_n$, provided the density ρ_0 and the elastic constants in E do not depend on $\mathbf{y}$, i.e. the medium is homogeneous. Then we end up with

$$-\frac{\partial}{\partial t}(\rho_0 \dot{u}_i u_{i,n}) + \frac{\partial}{\partial y_l}\left[\frac{\partial E}{\partial u_{i,l}} u_{i,n} + \delta_{n,l}\left(\frac{1}{2}\rho_0\, \dot{\mathbf{u}}^2 - E\right)\right] = 0. \qquad (7.4)$$

This has the form of a continuity equation; therefore (with suitable boundary conditions)

$$P_{qn} = -\int_V d^3y\, \rho_0\, \dot{u}_i\, u_{i,n} = const. \qquad (7.5)$$

is a conserved quantity. In Section 7.4 we shall show that it has to be interpreted as the quasimomentum of the elastic medium. This conservation law was already found in a one-dimensional linear system by Gilbert and Mollow,[12] in a non-local theory by Kobussen,[13] for the 3-dimensional linearized theory of elasticity by Kobussen and Paszkiewicz[14] and generally by Peierls.[8]

7.2. NOETHER'S THEOREM

An obvious question is: Which symmetry property of an elastic medium is, via Noether's theorem, responsible for the conservation law (7.4)? We shall see in this section that it is the following property: If all the fields (in our case the elastic deformation pattern) are spatially displaced by a vector ε, but <u>without</u> shifting the medium, then in a coordinate frame also shifted by ε the course of physical events will be the same if the medium is homogeneous.

<u>Noether's theorem</u> can be started as follows.[15] If a simultaneous infinitesimal transformation of the coordinates y_μ ($\mu = 0, 1, 2, 3$; $y_0 = t$) and the fields ψ_i ($i = 1, \ldots, n$),

$$y'_\mu = y_\mu + \delta y_\mu , \tag{7.6a}$$

$$\psi'_i(y') = \psi_i(y) + \delta \psi_i(y) \tag{7.6b}$$

leaves the action integral

$$W = \int_G d^4y \, L(y, \psi_i(y), \psi_{i,\mu}(y)) \tag{7.7}$$

invariant (integration over an arbitrary region G) then the following continuity equation holds:

$$\frac{\partial}{\partial y_\mu} \left[L \, \delta y_\mu + \frac{\partial L}{\partial \psi_{i,\mu}} \bar{\delta} \psi_i \right] = 0 , \tag{7.8}$$

where $\bar{\delta}\psi_i$ is defined as the field variation at a fixed point y,

$$\bar{\delta}\psi_i(y) = \psi'_i(y) - \psi_i(y) , \tag{7.9}$$

such that

$$\delta\psi_i(y) = \bar{\delta}\psi_i(y) + \psi_{i,\mu}\delta y_\mu . \tag{7.10}$$

We apply Noether's theorem to two special cases.

(a) <u>L only depends on the fields ψ_i and their derivatives</u>, not explicitly on y (spatially homogeneous medium, no explicit time dependence). A displacement in space-time of the fields by an arbitrary infinitesimal displacement vector ε_ν, together with a simultaneous shift of the integration variables (and the region of integration) by ε_ν,

$$\left. \begin{array}{c} y'_\nu - y_\nu = \delta y_\nu = \varepsilon_\nu \\ \psi'_i(y') - \psi_i(y) = \delta\psi_i(y) = 0 \end{array} \right\} \tag{7.11}$$

leaves L and d^4y and therefore W invariant. According to (7.10) and (7.11)

$$\bar{\delta}\psi_i(y) = -\psi_{i,\nu}\varepsilon_\nu , \tag{7.12}$$

and if we take just one of the ϵ_ν's different from zero, (7.8) takes the form

$$\frac{\partial}{\partial y_\mu}\left[L\,\delta_{\mu\nu} - \frac{\partial L}{\partial \psi_{i,\mu}}\,\psi_{i,\nu}\right] = 0\,. \tag{7.13}$$

In the case of an elastic medium the ψ_i's are the displacement fields u_i, and the Lagrangian density is

$$L = \frac{1}{2}\,\rho_0\,\dot{\mathbf{u}}^2 - E\left(\{u_{i,k}\}\right)\,. \tag{7.14}$$

For $\nu = n = 1,2,3$ (7.13) yields (7.4), the conservation law for quasimomentum. For $\nu = 0$, (7.13) yields the energy conservation law.[11]

(b) L does not depend on the fields ψ_i themselves, only on their derivatives and possibly explicitly on the coordinates (inhomogeneous system). Then L (as well as d^4y and W) is invariant under the transformation

$$\delta y_\mu = 0, \qquad \delta \psi_i = \bar{\delta}\psi_i = \epsilon_i = const.\,, \tag{7.15}$$

and (7.8) gives

$$\frac{\partial}{\partial y_\mu}\left(\frac{\partial L}{\partial \psi_{i,\mu}}\right) = 0 \qquad \forall\, i\,. \tag{7.16}$$

In the case of an elastic medium, (7.16) becomes $\delta\mathbf{u} = \bar{\delta}\mathbf{u} = \boldsymbol{\epsilon}$. This is a rigid displacement of all the material particles, i.e. a shift of the deformation pattern <u>and</u> the medium. With (7.14), Eq. (7.17) yields the equation of motion

$$\frac{\partial}{\partial t}\,(\rho_0\,\dot{u}_i) - \frac{\partial}{\partial y_l}\frac{\partial E}{\partial u_{i,l}} = 0\,, \tag{7.17}$$

which means conservation of ordinary momentum

$$\mathbf{P} = \int d^3y\,\rho_0\,\dot{\mathbf{u}} = const. \tag{7.18}$$

Note that it is also valid for inhomogeneous media.

7.3. GENERALIZED THEORY OF ELASTICITY

Usually one assumes the elastic energy E to depend only on the first derivatives of the displacement field $\mathbf{u}$. We are now going to show – what was stated without proof in Ref.[11] – that the foregoing methods can easily be generalized to the case where the elastic energy depends also on higher derivatives of $\mathbf{u}$. Such a medium shows dispersion.

We assume a Lagrangian density of the form

$$L = \frac{1}{2}\,\rho_0\,\dot{\mathbf{u}}^2 - E\left(u_{i,k},\ u_{i,k,l},\ u_{i,k,l,m}\right) \tag{7.19}$$

with

$$u_{i,k} = \frac{\partial u_i}{\partial y_k}\,,\qquad u_{i,k,l} = \frac{\partial^2 u_i}{\partial y_k \partial y_l}\,,\,\ldots\,.$$

The equation of motion, which at the same time gives the conservation law for ordinary momentum, reads

$$\frac{\partial}{\partial t}\,(\rho_0\,\dot{u}_i) + \frac{\partial \pi_{ik}}{\partial y_k} = 0\,, \tag{7.20}$$

$$\pi_{ik} = -\frac{\partial E}{\partial u_{i,k}} + \frac{\partial}{\partial y_l}\frac{\partial E}{\partial u_{i,k,l}} - \frac{\partial^2}{\partial y_l \partial y_m}\frac{\partial E}{\partial u_{i,k,l,m}}. \tag{7.21}$$

Quasimomentum conservation can be derived, as before, either by multiplying the equation of motion by $u_{i,n}$ and taking the sum over i, or from Noether's theorem. The result is

$$-\frac{\partial}{\partial t}(\rho_0\, \dot{u}_i\, u_{i,n}) + \frac{\partial}{\partial y_k} j_{nk} = 0. \tag{7.22}$$

Thus, the quasimomentum is the same as before, but the flux tensor is generalized to

$$j_{nk} = \delta_{nk}\, L + u_{i,n}\left[\frac{\partial E}{\partial u_{i,k}} - \frac{\partial}{\partial y_l}\frac{\partial E}{\partial u_{i,k,l}} + \frac{\partial^2}{\partial y_l \partial y_m}\frac{\partial E}{\partial u_{i,k,l,m}}\right] +$$
$$+ u_{i,l,n}\left[\frac{\partial E}{\partial u_{i,k,l}} - \frac{\partial}{\partial y_m}\frac{\partial E}{\partial u_{i,k,l,m}}\right] + u_{i,l,m,n}\frac{\partial E}{\partial u_{i,k,l,m}}. \tag{7.23}$$

7.4. PHYSICAL MEANING OF THE CONSERVED QUANTITY

In this section we are going to justify the name "quasimomentum" given to the conserved quantity $\mathcal{P}_q$ (7.5).

For an elastic medium in the harmonic approximation the significance of $\mathcal{P}_q$ can easily be found if the system is quantized. The Hamiltonian can be diagonalized by methods similar to the ones used in Sect. 3. $\mathcal{P}_q$ in equation (7.5) becomes simultaneously diagonal and takes indeed the form (6.1). This was already shown by Kobussen and Paszkiewicz.[16]

In order to get the general interpretation of $\mathcal{P}_q$ (7.5), independent of specific examples and independent of quantization, we consider the interaction of the elastic medium with thermal phonons.[11]

Even at low temperatures the wavelengths of the thermal phonons are small compared to the wavelengths of acoustic waves. The acoustic waves can be treated in a continuum theory as classical motion. As in the two–fluid theory of He II [17,18] the elastic medium forms the background or bearer medium in which the high–frequency (thermal) phonons move about. The latter can be understood as wavepackets (small in comparison to the acoustic wavelength), moving in a slowly (adiabatically) varying background. Their frequencies $\omega_{\mathbf{k}\alpha}$ become functions of $\mathbf{y}$ and t via their dependence on the strain tensor $u_{i,k}(\mathbf{y},t)$. The Boltzmann equation for the phonon distribution function $N = n_{\mathbf{k}\alpha}(\mathbf{y},t)$ takes the form [17,18,19]

$$\frac{\partial N}{\partial t} + \frac{\partial \omega}{\partial \mathbf{k}}\frac{\partial N}{\partial \mathbf{y}} - \frac{\partial \omega}{\partial \mathbf{y}}\frac{\partial N}{\partial \mathbf{k}} = \left(\frac{\partial N}{\partial t}\right)_{coll}. \tag{7.24}$$

We consider an ideal crystal at low temperatures so that umklapp–processes can be neglected and the sum of the phonon wave vectors is conserved in the collisions. Multiplication of (7.25) by $\hbar \mathbf{k}$ and summation over $\mathbf{k}$ and α gives zero on the right–hand side. One obtains a balance equation for $\int d^3k \sum_\alpha N\hbar\mathbf{k}$ with a source term.[11]

On the other hand the equation of motion of an elastic medium is also modified in the presence of phonons. It takes the form[11]

$$\rho_0\, \ddot{u}_i = \frac{\partial}{\partial y_l}\frac{\partial E}{\partial u_{i,l}} + \frac{\partial}{\partial y_l}\int d^3k \sum_\alpha N \frac{\partial(\hbar\omega)}{\partial u_{i,l}}. \tag{7.25}$$

Multiplication by $-u_{i,n}$ and summation over i yields a balance equation for $-\rho_0 \dot{u}_i u_{i,n}$ of the form (7.4) with a source term[11]. When the two balance equations are added the source terms cancel and one obtains the following conservation law

$$\frac{\partial}{\partial t} \mathcal{P}_{q\,n} + \frac{\partial}{\partial y_l} F_{nl} = 0 \tag{7.26}$$

with the flux tensor

$$F_{nl} = \frac{\partial E}{\partial u_{i,l}} u_{i,n} + \delta_{nl}\left(\frac{1}{2}\rho_0 \dot{u}^2 - E\right) +$$
$$+ \int d^3k \sum_\alpha \left[N \hbar k_n \frac{\partial \omega}{\partial k_l} + N \hbar \frac{\partial \omega}{\partial u_{i,l}} u_{i,n} \right]. \tag{7.27}$$

The density of the conserved quantity is now given by

$$\mathcal{P}_{q\,n} = -\rho_0 \dot{u}_i u_{i,n} + \int d^3k \sum_\alpha N \hbar k_n . \tag{7.28}$$

Since the second term is the quasimomentum density of the phonons one obviously has to call the first term the quasimomentum density of the elastic medium (in contrast to the density of the ordinary momentum, $\rho_0 \dot{u}_n$).

The same conclusion could have been drawn from considering the interaction of the strain field with conduction electrons.

8. ELECTRODYNAMICS IN MACROSCOPIC MEDIA

We start from Maxwell's equations in a macroscopic medium

$$\nabla \times \mathbf{H} - \frac{1}{c}\dot{\mathbf{D}} = \frac{4\pi}{c}\mathbf{j}_{c'}, \tag{8.1a}$$

$$\nabla \times \mathbf{E} + \frac{1}{c}\dot{\mathbf{B}} = 0, \tag{8.1b}$$

$$\nabla \cdot \mathbf{B} = 0, \tag{8.1c}$$

$$\nabla \cdot \mathbf{D} = 4\pi \rho_c . \tag{8.1d}$$

They are written in local (Eulerian) coordinates. In this section the dot therefore means the time derivative at constant $\mathbf{x}$. In a dielectric medium, which will be considered later, the conduction current and charge densities, $\mathbf{j}_c$ and ρ_c, are zero. We allow for arbitrary (nonlinear) constitutive relations between the fields $\mathbf{E}, \mathbf{D}, \mathbf{H}, \mathbf{B}$, but for simplicity we assume them not to involve spatial or temporal derivatives. This means that we neglect dispersion. Dispersion could be taken into account in a way similar to what we did in section 7.3. We furthermore assume that an (internal) energy density exists; its differential is given by[20,21]

$$dU = \frac{1}{4\pi}\mathbf{E} \cdot d\mathbf{D} + \frac{1}{4\pi}\mathbf{H} \cdot d\mathbf{B} . \tag{8.2}$$

Therefore, U has to be considered as a function of $\mathbf{D}$ and $\mathbf{B}$. It is in general not of the simple form $\frac{1}{8\pi}\left(\frac{1}{\epsilon}\mathbf{D}^2 + \frac{1}{\mu}\mathbf{B}^2\right)$. Then, from Maxwell's equations a balance equation for the energy follows,

$$\dot{U} + \nabla \cdot \mathbf{S} = -\mathbf{j}_c \cdot \mathbf{E} \tag{8.3}$$

with the Poynting vector
$$\mathbf{S} = \frac{c}{4\pi}\,\mathbf{E}\times\mathbf{H}\,. \tag{8.4}$$

It was shown by Gurevich and Thellung [22] that under the same general assumptions, but for a homogeneous medium (no explicit **x**-dependence in the constitutive relations), a balance equation for a vectorial quantity can be derived in the form

$$\frac{\partial}{\partial t}\left(\frac{1}{4\pi c}\mathbf{D}\times\mathbf{B}\right)_n + \frac{\partial}{\partial x_l}T_{nl} = -\rho_c\,E_n - \frac{1}{c}(\mathbf{j}_c\times\mathbf{B})_n \tag{8.5}$$

with

$$T_{nl} = \delta_{nl}\left[\frac{1}{4\pi}(\mathbf{E}\cdot\mathbf{D} + \mathbf{H}\cdot\mathbf{B}) - U\right] - \frac{1}{4\pi}(E_n D_l + H_n B_l)\,. \tag{8.6}$$

From (8.5) one can immediately see that $\frac{1}{4\pi c}\mathbf{D}\times\mathbf{B}$ has to be interpreted as the electromagnetic quasimomentum density. The argument is the following. The wave vector **k** of a Bloch electron is known to change with time according to the law[6,9,23]

$$\hbar\,\dot{\mathbf{k}} = e\,\mathbf{E} + \frac{e}{c}\,\mathbf{v}\times\mathbf{B} \tag{8.7}$$

(e = charge, **v** = velocity of the electron), at least as long as Landau quantization of the orbits can be neglected. Since $\hbar$ times the sum of the wave vectors of all the electrons in a unit volume is the electronic quasimomentum density, $\mathcal{P}_q^e$, it follows from (8.7) that its rate of change due to the electromagnetic fields is

$$\left(\frac{\partial \mathcal{P}_q^e}{\partial t}\right)_{em} = \rho_c\mathbf{E} + \frac{1}{c}\mathbf{j}_c\times\mathbf{B}\,, \tag{8.8}$$

where ρ_c is the charge density of the electrons, and the sum of $e\mathbf{v}$ over all the electrons per unit volume is the conduction current density $\mathbf{j}_c$. Comparison of (8.8) with (8.5) shows that $\frac{1}{4\pi c}\mathbf{D}\times\mathbf{B}$ has to be interpreted as quasimomentum density. This fact will be confirmed in section 9, when we couple the electromagnetic fields to an elastic medium.

In a dielectric medium, $\rho_c = \mathbf{j}_c = 0$, the source terms on the right-hand sides of (8.3) and (8.5) vanish and energy and quasimomentum are conserved. In the literature these conservation laws are usually proven only for linear relations between the fields **E**, **D** and **H**, **B**.[20,21]

The conservation laws for energy and quasimomentum can also be obtained via Noether's theorem. This was shown by Schoeller and Thellung.[24]

When field quantization is applied in the linear case $\mathbf{D} = \epsilon\mathbf{E}$, $\mathbf{B} = \mu\mathbf{H}$, the total quasimomentum $\int d^3x\,\frac{1}{4\pi c}\mathbf{D}\times\mathbf{B}$ equals $\hbar$ times the sum of all the photon wave-vectors.

9. PHOTOELASTICITY

In this section we investigate electromagnetic fields interacting with the displacement field of an elastic homogeneous dielectric medium. In that case the constitutive relations between $\mathbf{E}, \mathbf{D}, \mathbf{H}, \mathbf{B}$ will involve the strain tensor $u_{i,k}$ and the velocity field $\dot{\mathbf{u}}$, and the elastic "constants" will depend on the electromagnetic fields. Eq. (8.2) has to be generalized to

$$dU = \frac{1}{4\pi} \mathbf{E} \cdot d\mathbf{D} + \frac{1}{4\pi} \mathbf{H} \cdot d\mathbf{B} + \left(\frac{\partial U}{\partial u_{i,\mu}}\right)_{\mathbf{D},\mathbf{B}} du_{i,\mu}. \tag{9.1}$$

Here U is the total (internal) energy density, i.e. electromagnetic + elastic potential + interaction energy density, not including the kinetic energy of the elastic medium. As a consequence, our former conservation equations (7.4) and (8.5) (with $\rho_c = \mathbf{j}_c = 0$) will contain source terms. When they are added the source terms are expected to cancel. However, there is a difficulty in comparing equations of Sect. 7 with equations of Sect. 8: The elastic equations are expressed in Lagrangian coordinates $\mathbf{y}$, whereas the electrodynamic equations are formulated in Eulerian coordinates $\mathbf{x}$. In order to avoid long and complicated expressions, Gurevich and Thellung[22] only considered the case where the difference between Eulerian and Lagrangian coordinates, giving rise to higher order corrections, could be neglected ($\mathbf{u}$ in the equation $\mathbf{x} = \mathbf{y} + \mathbf{u}$ being very small). Then one indeed obtains a conservation law:

$$\dot{P}_{q\,n}^{total} + \frac{\partial}{\partial x_l} T_{nl}^{total} = 0 \tag{9.2}$$

$$P_{q\,n}^{total} = \frac{1}{4\pi c}(\mathbf{D} \times \mathbf{B})_n - \rho_0\, \dot{u}_i\, u_{i,n} \tag{9.3}$$

$$T_{nl}^{total} = \delta_{nl}\left[\frac{1}{4\pi}(\mathbf{E}\cdot\mathbf{D} + \mathbf{H}\cdot\mathbf{B})\right] - \frac{1}{2}\rho\,\dot{u}^2 - U\right] - $$
$$-\frac{1}{4\pi}(E_n D_l + H_n B_l) + \left(\frac{\partial U}{\partial u_{m,l}}\right)_{\mathbf{D},\mathbf{B}} u_{m,n}. \tag{9.4}$$

Since we know that $[-\rho_0\,\dot{u}_i\,u_{i,n}]$ is the quasimomentum density of the elastic medium we see from (9.3) again that $\frac{1}{4\pi c}\mathbf{D} \times \mathbf{B}$ is the quasimomentum density of the electromagnetic fields.

The difference between Eulerian and Lagrangian coordinates should be of no importance in most cases of practical interest, but it is, of course, interesting in principle. To take it into account one has to transcribe elastic equations from material to local coordinates or electrodynamic equations from local to material coordinates. For the elastic equations this was done by Kobussen[13] and by Peierls.[8] Maxwell equations in material coordinates were obtained by Lax and Nelson[34]. For the combined elastic and electrodynamic equations the transformation has been carried out by Schoeller and Thellung.[24] They found the conservation equations for energy, quasimomentum and momentum in local and material coordinates, still valid for nonlinear and anisotropic media. We quote the results in local coordinates.

Energy conservation:

$$\left(\frac{\partial}{\partial t}\right)_{\mathbf{x}}\left(\frac{1}{2}\rho\,\dot{u}^2 + U\right) + \frac{\partial}{\partial x_k}\left[-\frac{1}{\Delta}\frac{\partial x_k}{\partial y_m}\dot{u}_l\left(\frac{\partial \bar{U}}{\partial u_{l,m}}\right)_{\mathbf{D},\mathbf{B},\mathbf{p}_u} + \dot{u}_k\left(\frac{1}{2}\rho\,\dot{u}^2 + U\right) + \frac{c}{4\pi}(\mathbf{E}\times\mathbf{H})_k\right] = 0 \tag{9.5}$$

Here U is the total energy density without the kinetic energy $\frac{1}{2}\rho\dot{u}^2$, Δ is the Jacobian det $(\partial x_i/\partial y_k)$, $\bar{U} = \Delta U$, and $\mathbf{p_u}$ is the momentum conjugate to $\mathbf{u}$ in the canonical formalism.

Quasimomentum conservation:

$$\left(\frac{\partial}{\partial t}\right)_{\mathbf{x}} \left[-\rho\,\dot{u}_i u_{i,n} + \frac{1}{4\pi c}(\mathbf{D}\times\mathbf{B})_n - u_{i,n}\frac{1}{4\pi c}(\mathbf{E}\times\mathbf{H} - \mathbf{D}\times\mathbf{B})_i\right] + \frac{\partial}{\partial x_k} T'_{nk} = 0 \ . \quad (9.6)$$

Momentum conservation:

$$\left(\frac{\partial}{\partial t}\right)_{\mathbf{x}} \left[\rho\,\dot{u}_n + \frac{1}{4\pi c}(\mathbf{E}\times\mathbf{H})_n\right] + \frac{\partial}{\partial x_k} T''_{nk} = 0 \quad (9.7)$$

The flux tensors T'_{nk} and T''_{nk} are rather involved expressions; they are explicitly given in Ref. [24]. Eq. (9.6) shows that the quasimomentum density consists of three terms: the elastic term, $-\rho\,\dot{u}_i\,u_{i,n}$, the electromagnetic term, $\frac{1}{4\pi c}(\mathbf{D}\times\mathbf{B})_n$, and a mixed term, $-u_{i,n}\frac{1}{4\pi c}(\mathbf{E}\times\mathbf{H}-\mathbf{D}\times\mathbf{B})_i$. This last term is new. It was missing in the former literature because electrodynamics and elasticity were studied separately or because (in our first paper [22]) the difference between local and material coordinates was neglected.

Here a historical remark may be appropriate. In 1908 Minkowski[25] proved the conservation law for $\mathbf{D}\times\mathbf{B}$ in the linear electrodynamics of continuous media; he called it momentum. In 1909 Abraham[26] concluded from the required symmetry of the energy–momentum tensor that $\mathbf{E}\times\mathbf{H}$ must be the momentum. This started a big controversy in the literature. The solution was found 1971 by Blount,[27] 1973 by Gordon,[28] and 1976 by Peierls,[29] who showed that $\mathbf{E}\times\mathbf{H}$ is the momentum and $\mathbf{D}\times\mathbf{B}$ the pseudomomentum (=quasimomentum). Our results (9.6) and (9.7) are in accordance with this interpretation.

10. APPLICATIONS, RADIOELECTRIC EFFECTS

Quasimomentum conservation allows for a simple and very general analysis of various physical phenomena. Let us mention: Nonlinear elasticity and nonlinear optics; interaction between first and second sound in solids; the acoustoelectric[30] and acoustothermal[31] effects, where a sound wave is damped by energy and quasimomentum transfer to conduction electrons and thermal phonons; photoelasticity; the radioelectric effect[32,22] and photomagnetism[33] (quasimomentum transfer from electromagnetic waves to electrons).

As an example, we discuss the radioelectric effect in more detail. In Sect. 8 we have seen that the quasimomentum transfer per unit time and unit volume from the electromagnetic fields to the conduction electrons is given by (8.8). For simplicity let us now assume the linear relations

$$\mathbf{D} = \epsilon\mathbf{E}, \quad \mathbf{B} = \mu\mathbf{H} \ . \quad (10.1)$$

Since the force exerted on a charge e is $e\mathbf{E} + \frac{e}{c}\mathbf{v} \times \mathbf{B}$, one should replace $\mathbf{E}$ in the usual form of Ohm's law, $\mathbf{j}_c = \sigma \mathbf{E}$, by $\mathbf{E} \times \frac{1}{c} <\mathbf{v}> \times \mathbf{B}$, where $<\mathbf{v}>$ is the average velocity of the electrons, $<\mathbf{v}> = \mathbf{j}_c/\rho_c$. Then, Ohm's law takes the form (in the simplest case)

$$\mathbf{j}_c = \sigma \left(\mathbf{E} + \frac{1}{c\rho_c} \mathbf{j}_c \times \mathbf{B} \right) \tag{10.2}$$

and the quasimomentum transfer, (8.8), can be written

$$\left(\frac{\partial \mathbf{P}_q^e}{\partial t} \right)_{em} = \rho_c \left(\mathbf{E} + \frac{1}{c\rho_c} \mathbf{j}_c \times \mathbf{B} \right) = \frac{\rho_c}{\sigma} \mathbf{j}_c = \rho_c \left(\mathbf{E} + \frac{\sigma}{c\rho_c} \mathbf{E} \times \mathbf{B} \right), \tag{10.3}$$

where in the last step terms of third order in the fields have been omitted.

Assume now a plane monochromatic electromagnetic wave propagating in a conductor. It will be spatially damped (Jackson,[20] Sect. 7.7). Such a wave can be generated by a plane monochromatic light wave in vacuum, normally incident on the surface of a conductor. The time average of $\mathbf{E}$ over a period will vanish, but the time average of $\mathbf{E} \times \mathbf{B}$ (which is proportional to the intensity of the wave) is different from zero, and so is the time average of $(\partial \mathbf{P}_q^e/\partial t)_{em}$ and of $\mathbf{j}_c$. This means that quasimomentum is continually being transferred to the electrons, dragging them in the direction in which the wave propagates. Now, for a conducting slab of finite thickness there are two possibilities:

(a) It is insulated. Then surface and volume charges will build up, giving rise to an electrostatic field $\mathbf{E}_{stat}$, which drives the electrons in the opposite direction until a stationary state is reached. This is the usual radioelectric effect (see Ref. [22,32] for further references).

(b) The back side of the slab is connected to the front side by a superconductor going around the slab. Then there is no voltage between the two surfaces, but a stationary current will be established.

Both cases can be treated by starting from (10.3) (with $\mathbf{B} = \mu \mathbf{H}$) and taking the time average (indicated by a bar),

$$\bar{\mathbf{E}} + \frac{\sigma \mu}{c\rho_c} \overline{\mathbf{E} \times \mathbf{H}} = \frac{1}{\sigma} \bar{\mathbf{j}}_c . \tag{10.4}$$

For the time average of $\mathbf{E}$ only the static part contributes, for the time average of $\mathbf{E} \times \mathbf{H}$ only the oscillating part; $\overline{\mathbf{E} \times \mathbf{H}}$ can be expressed by the Poynting vector (8.4). Hence

$$\mathbf{E}_{stat} + \frac{4\pi\sigma\mu}{c^2\rho_c} \bar{\mathbf{S}} = \frac{1}{\sigma} \bar{\mathbf{j}}_c \tag{10.5}$$

In the time-averaged continuity equation,

$$\dot{\bar{\rho}}_c + \nabla \cdot \bar{\mathbf{j}}_c = 0 , \tag{10.6}$$

$\dot{\bar{\rho}}_c$ vanishes in the stationary case. For the simple geometry of a wave propagating in the z-direction in a slab situated between $z = 0$ and $z = L$, we have

$$\frac{\partial}{\partial z} \bar{j}_z^c = 0 , \quad \bar{j}_z^c = const. \tag{10.7}$$

In case (a)
$$\bar{j}_z^c = 0, \qquad E_z^{stat} = -\frac{4\pi\sigma\mu}{c^2\rho_c}\bar{S}_z. \tag{10.8}$$

The z-dependence of $\bar{S}_z$ is $\propto e^{-2\kappa z}$ with $\kappa = \dfrac{2\pi\sigma}{c}\sqrt{\dfrac{\mu}{\epsilon}}$ for a poor conductor (Jackson,[20] Sect. 7.7). If $\kappa L \gg 1$ the wave is practically completely absorbed and the voltage across the slab is

$$V = -\int_0^L dz\, E_z^{stat} = \frac{4\pi\sigma\mu}{c^2\rho_c}\frac{1}{2\kappa}\bar{S}_z(0). \tag{10.9}$$

$\bar{S}_z(0)$ is the intensity of the wave at the inner side of the surface of incidence, which is the difference of the intensities of the incident and the reflected waves outside. Eqs. (10,8) and (10.9) have been derived in two different ways in Ref. [22].

In case (b), with a superconducting short–circuit around the slab, the voltage across the slab vanishes and integration of (10.5) over z form 0 to L gives (for complete absorption)

$$\int_0^L dz\, E_z^{stat} = 0, \qquad \bar{j}_z^c = \frac{1}{L}\frac{4\pi\sigma^2\mu}{c^2\rho_c}\frac{1}{2\kappa}\bar{S}_z(0). \tag{10.10}$$

We might call this a radioelectric effect of the second kind. For a good conductor $\kappa = \sqrt{2\pi\omega\mu\sigma}/c$ (Jackson,[20] Sect. 7.7) The result will be somewhat modified by the anomalous skin effect, but it seems feasible that in a metal this current could be measured.

Acknowledgments

It is a great pleasure to thank Tadeusz Paszkiewicz and the organizers of the Kudowa Winter School for their invitation and their warm hospitality, and for interesting conversations, particularly with M.I. Kaganov and V. Vasilev.

Many of the results presented here were found in collaboration with Vadim Gurevich and Herbert Schoeller. I am most grateful to them for many helpful discussions, in particular a fruitful discussion with Vadim Gurevich in Turku, Finland, about the new radioelectric effect. I would also like to express my gratitude to Reino Laiho and the University of Turku for their kind hospitality.

REFERENCES

[1] A. Thellung, Ann. Phys. **127**, 289(1980); "Physics of Phonons", edited by T. Paszkiewicz, Springer Lecture Notes in Physics, Vol. **285**, 208 (1987)

[2] H. Lamb, "Hydrodynamics", Cambridge Univ. Press, London /New York, 1957

[3] R. Kronig and A. Thellung, Physica **18**, 749 (1952)

[4] S.F. Tyabji, Proc. Cambr. Phil. Soc.**50**, 449 (1954)

[5] C. Kittel, "Quantum Theory of Solids", Wiley, New York (1963)

[6] A.A. Abrikosov, "Fundamentals of the Theory of Metals", North-Holland, Amsterdam (1988)

[7] L.D. Landau and E.M. Lifshitz, "Statistical Physics", part one, Pergamon, Oxford (1980)

[8] R. Peierls, in "Highlights of Condensed-Matter Theory", ed. by F. Bassani, F. Fumi and M.P. Tosi, p. 237, North-Holland, Amsterdam (1985)

[9] N.W. Ashcroft and N.D. Mermin, "Solid State Physics", Holt, Rinehart and Winston, New York (1976)

[10] L. D. Landau and E.M. Lifshitz, "Theory of Elasticity", Pergamon, Oxford (1970)

[11] V.L. Gurevich and A. Thellung, Phys. Rev. **B 42**, 7345 (1990)

[12] I.M. Gilbert and B.R. Mollow, Am. J. Phys. **36**, 822 (1968)

[13] J.A. Kobussen, Helv. Phys. Acta **49**, 599 (1976)

[14] J.A. Kobussen and T. Paszkiewicz, Helv. Phys. Acta **54**, 383 (1981)

[15] N.N. Bogoliubov and D.V. Shirkov, "Introduction to the Theory of Quantized Fields", Wiley, New York (1980)

[16] J.A. Kobussen and T. Paszkiewicz, Helv. Phys. Acta **54**, 395 (1981)

[17] J. de Boer, in "Liquid Helium", edited by G. Careri, Academic, New York (1963)

[18] I.M. Khalatnikov, "Introduction to the Theory of Superfluidity", Benjamin, New York (1965)

[19] V.L. Gurevich, "Transport in Phonon Systems", North-Holland, Amsterdam (1986)

[20] J.D. Jackson, "Classical Electrodynamics", Wiley, New York (1962)

[21] L.D. Landau and E.M. Lifshitz, "Electrodynamics of Continuous Media", Pergamon, Oxford (1960)

[22] V.L. Gurevich and A. Thellung, Physica **A 188**, 654 (1992)

[23] R.E. Peierls, "Quantum Theory of Solids", Clarendon, Oxford (1955)

[24] H. Schoeller and A. Thellung, Ann. Phys. **220**, 18 (1992)

[25] H. Minkowski, Nachr. Ges. Wiss Göttingen, 53 (1908), Math. Ann. **68**, 473 (1910)

[26] M. Abraham, Rend. Circolo Mat. Palermo **28**, 1 (1909); **30**, 33 (1910)

[27] E.I. Blount, Bell Teleph. Lab. Tech. Memo. 38139-9 (1971)

[28] J.P. Gordon, Phys. Rev. **A 8**, 14 (1973)

[29] R. Peierls, Proc. Roy. Soc. **A 347**, 475 (1976)

[30] G. Weinreich, Phys. Rev. **107**, 317 (1957)

[31] Y.V. Gulyaev and A. G. Kozorezov, Sov. Phys. Solid State **18**, 82 (1976)

[32] A.D. Wieck, H. Sigg and K. Ploog, Phys. Rev. Lett. **64**, 463 (1990)

[33] V. L. Gurevich, R. Laiho and A.V. Lashkul, Phys. Rev. Lett. **69**, 180(1992)

[34] M. Lax and D.F. Nelson, Phys. Rev. **B 13**, 1777 (1976)

PARTIAL FREQUENCY DISTRIBUTION FUNCTIONS AND THE PROBLEM OF THE LOCALIZATION OF ATOMIC VIBRATIONS IN REAL LAYERED AND MULTILAYERED CRYSTALS

A.M. Kosevich, E.S. Syrkin, and S.B. Feodosyev

Institute for Low Temperature Physics and
Engineering of the Ukraine Academy of Sciences
310164 Kharkov, Ukraine

INTRODUCTION

Recently much attention has been directed towards the investigation of the various physical characteristics of polyatomic crystal lattices containing a large number of different atoms in their unit cells. HTS–materials, numerous ferroelectrics, ferroelastics, polymers, biopolymers, intercalated compounds as well as a series of other materials interesting for both their purely scientific and technical aspects are examples of compounds of this type. In many cases these compounds possess multilayer crystal structures. A large lattice parameter along the direction normal to the layers, which is a characteristic of these crystals, greatly weakens the influence of the crystalline regularity of the atoms spatial distribution perpendicular to the layers. Such a large crystal lattice period affects various physical properties of these structures so dramatically that they acquire features more typical of disordered systems. In particular this is the origin of the special features of the phonon spectra and related physical properties. The special character of these features is manifested first of all by the fact that the atomic vibrations in each layer, along different crystallographic directions, contribute differently to the phonon spectrum of a multilayered crystal. Such contributions to the total density of crystal vibrations are characterized by partial distribution functions.

PARTIAL FUNCTION OF FREQUENCY DISTRIBUTION

Here we discuss the dynamics of a multiatomic crystal lattice with a unit cell containing q atoms. Different atoms in the unit cell will be enumerated by a subscript s $(s = 1, 2, \ldots, q)$.

The dynamic equation for the field of atomic harmonic displacements in an ideal crystal may be written as

$$d^2\mathbf{u}/dt^2 + \hat{L}\mathbf{u} = 0 ,$$

where the displacement vector **u** is a function of the discrete arguments **r** and s: $\mathbf{u} = \mathbf{u}^s(\mathbf{r},t)$; $\mathbf{r} = \sum_{k=1}^{3} n_k \mathbf{a}_k$, n_k are integers, $\mathbf{a}_k$ are the basic translational lattice parameters ($k=1,2,3$); the symbol $\hat{L}\mathbf{u}$ in the coordinate representation stands for the expression

$$\left(\hat{L}\mathbf{u}\right)_i^s = \sum_{\mathbf{r}'s'k} L_{ik}^{ss'}(\mathbf{r},\mathbf{r}')\, u_k^{s'}(\mathbf{r}') \, . \tag{1}$$

For an ideal unbounded homogeneous lattice, the matrix elements of $\hat{L}$ have the form $L_{ik}^{ss'}(\mathbf{r}-\mathbf{r}')$. The elements of the operator $\hat{L}$ are related to the matrix of force constants by the simple rule:

$$L_{ik}^{ss'}(\mathbf{r},\mathbf{r}') = \phi_{ik}^{ss'}(\mathbf{r},\mathbf{r}') \Big/ (m_s m_{s'})^{1/2} \, , \tag{2}$$

where $\phi_{ik}^{ss'}(\mathbf{r},\mathbf{r}')$ are the elements of the matrix of force constants[1,2]; m_s is the mass of the sublattice atom s.

The normal vibration of an atom $(\mathbf{r},s)$ may be presented as:

$$\mathbf{u}^s(\mathbf{n}) = m_s^{-1/2}\mathbf{e}^s(\mathbf{k},p)\, exp\,[i(\mathbf{k},\mathbf{r})],$$

where p is the number of the vibrational branch ($p = 1,2,\ldots,3q$); $\mathbf{k}$ is a quasiwave vector; $\mathbf{e}^s(\mathbf{k},p)$ is the unit vector of polarization for this branch. The vector $\mathbf{k}$ determines the vibration frequency ω as $\omega = \omega_p(\mathbf{k})$, and this dependence is called the dispersion law of the p-th vibrational branch.

Let us introduce the partial distribution function $g_{ip}^s(\omega)$ which takes into account the vibrations of the sublattice atoms s along the crystallographic direction i of the p-th branch of vibrations in the following way:

$$g_{ip}^s(\lambda) = \frac{V_0}{(2\pi)^3} \oint_{\lambda_p(\mathbf{k})=\lambda} \frac{dS_p}{|\nabla\lambda_p|} \, |e_i^s(p,\mathbf{k})|^2 \, , \tag{3}$$

where $\lambda = \omega^2$; V is the volume of the crystal; the integration is carried out over the isofrequency surface $\lambda_p(\mathbf{k}) = \lambda = constant$; dS_p is its surface element.

Then, the partial function for the squared frequency distribution $g_i^s(\lambda)$, characterizing the contribution to the phonon density of all the vibrations of the sublattice atom s along the crystallographic direction i is

$$g_i^s(\lambda) = \sum_p g_{ip}^s(\lambda) \, . \tag{4}$$

The function $g_i^s(\lambda)$ is an important characteristic of the vibrations which determines the thermodynamical properties of the crystal. For example, it can be used in calculations of the mean squared displacement of the s-th atom along the crystallographic direction i

$$\left\langle [x_i^s(T)]^2 \right\rangle \equiv \left\langle u_s^i u_s^i \right\rangle = \frac{\hbar}{2m_s} \int \frac{d\lambda}{\sqrt{\lambda}} \, cth\left(\frac{\hbar\sqrt{\lambda}}{2kT}\right)\, g_i^s(\lambda) \, ,$$

where the integration is carried out over the total frequency spectrum of the crystal.

The distribution function of the squared frequencies commonly used in the dynamics of a crystal lattice is related to the partial distribution function in the following way: the distribution function of the squared frequencies vibration mode p is

$$g_p(\lambda) \equiv \frac{V_0}{(2\pi)^3} \oint_{\lambda_p(k)=\lambda} \frac{dS_p}{|\nabla\lambda_p|} = \sum_{s=1}^{q}\sum_{i=1}^{3} g_{ip}^s(\lambda) \, , \tag{5}$$

and the complete distribution function of squared frequencies is

$$g(\lambda) \equiv \sum_{p=1}^{3q} g_p(\lambda) = \sum_{p=1}^{3q} \sum_{s=1}^{q} \sum_{i=1}^{3} g_{ip}^s(\lambda) . \tag{6}$$

Note that the function $g(\lambda)$ is normalized to unity.

Equations (3)–(6) completely determine the partial distribution function $g_i^s(\lambda)$ and reveal its physical meaning.

An analytical calculation of the partial distribution function (4) requires the exact knowledge of the dispersion laws as well as the **k**–dependence of the polarization vector and as a result this is not a constructive procedure. However a method exists which allows one to find partial distribution functions and which involves neither dispersion laws nor the vibration polarization vectors. This method was developed by V. Peresada [3] in 1967. A summary of it may be found in the Refs. [4,5].

This method uses the fact that the operator $\hat{L}$, written in the form (2), may be presented as the sum of operators with simple spectra. Consider a unit displacement of the R–th atom in the system from the equilibrium position along a vector **h**. Denote this displacement as $\mathbf{h}_0$ and using the operator $\hat{L}$, one constructs the so–called cyclic subspace $H_\mathbf{h}$ built on the vector sequence

$$\mathbf{h}_0, \hat{L}\mathbf{h}_0, \hat{L}^2\mathbf{h}_0, \hat{L}^3\mathbf{h}_0, \ldots . \tag{7}$$

It is easy to show that the subspace $H_\mathbf{h}$ built in this way is orthogonal to all the possible oscillatory motions of the system in which the atom considered does not oscillate along the **h** direction. In the subspace $H_\mathbf{h}$ the operator $\hat{L}$ generates a new operator $\hat{L}_\mathbf{h}$ with simple spectra which takes the form of the Jacobi matrix in the basis built by the successive orthogonalization procedure of the vectors (7):

$$\hat{L}_\mathbf{h} = \begin{bmatrix} a_0 & b_0 & 0 & 0 & 0 & \cdots \\ b_0 & a_1 & b_1 & 0 & & \cdots \\ 0 & b_1 & a_2 & b_2 & & \cdots \\ & & \cdots & & & \\ 0 & 0 & \cdots & b_{n-1} & a_n & b_n & 0 & 0 & \cdots \\ 0 & 0 & \cdots & 0 & b_n & a_{n+1} & b_{n+1} & 0 & \cdots \\ & & \cdots & & & & & & \end{bmatrix} . \tag{8}$$

It is known in functional analysis that the matrix elements $\left(\mathbf{h}, f(\hat{L})\mathbf{h}\right)$ of the operator $\hat{L}$ may be written in the form [6]:

$$\left(\mathbf{h}, f(\hat{L})\mathbf{h}\right) = \int f(\lambda) \, \rho_\mathbf{h}(\lambda) d\lambda , \tag{9}$$

where

$$\rho_\mathbf{h}(\lambda) = \frac{d}{d\lambda} \left(\mathbf{h}, \hat{E}_\lambda \mathbf{h}\right) . \tag{10}$$

Here $\hat{E}$ is the orthogonal spectral function of the $\hat{L}$ operator (the λ's are the eigenvalues of the operator $\hat{L}$, simultaneously they are the squares of the oscillation eigenfrequencies). We shall call the function $\rho_{\mathbf{h}}(\lambda)$ the spectral density of the operator $\hat{L}$ generated by the vector $\mathbf{h}$.

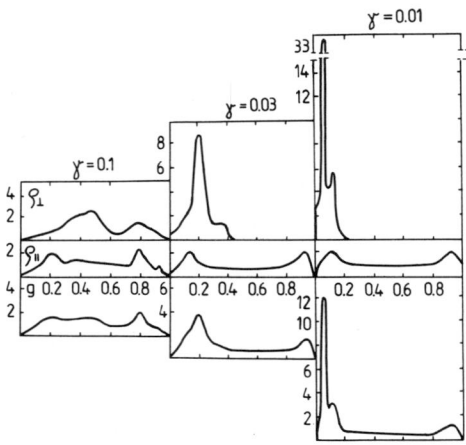

Fig. 1

There is the following very simple connection between Peresada's spectral densities and the partial distribution functions (4)

$$3qg_i^s(\lambda) = \rho_i^s(\lambda), \qquad (11)$$

where $\rho_i^s(\lambda)$ is the spectral density, generated by a single displacement of an atom s along the crystallographic direction i. In particular, the distribution function of the squared frequencies $g(\lambda)$ is equal to the arithmetic mean of the spectral densities, generated by the linearly independent displacement vectors of all the atoms in the unit cell.

It is impossible to find an analytic form of the function $\rho_{\mathbf{h}}(\lambda)$ in the general case. However, an effective procedure may be suggested to calculate $\rho_{\mathbf{h}}(\lambda)$ and integrals of the type (9).

The Jacobi matrix (8) possesses the following important property: its elements a_n and b_n tend to certain limiting values as $n \to \infty$. For example, if the spectral density $\rho(\lambda)$ of the operator $\hat{L}$ is confined to the interval $[0, \lambda_m]$, we can write

$$\lim_{n \to \infty} a_n = \lambda_m/2, \qquad \lim_{n \to \infty} b_n = \lambda_m/4. \qquad (12)$$

This fact allows one to avoid the explicit calculation of the matrix elements a_n and b_n for large n. The spectral density corresponding to the Jacobi matrices with the first

$2n + 1$ elements different from their limiting values can be presented in the form [7]:

$$\rho(\lambda) = -\pi^{-1} |P_{n+1}(\lambda) - b_n \, P_n(\lambda) \, K(\lambda)|^2 \, Im \, K(\lambda) \, . \tag{13}$$

Here $P_n(\lambda)$ are n-th degree polynomials in λ which satisfy the recurrence relations:

$$b_n \, P_{n+1}(\lambda) = (\lambda - a_n) \, P_n(\lambda) - b_{n-1} \, P_{n-1}(\lambda) \, , \tag{14}$$

under the initial conditions $P_{-1}(\lambda) = 0$, $P_0(\lambda) = 1$, $K(\lambda)$ is a continued fraction

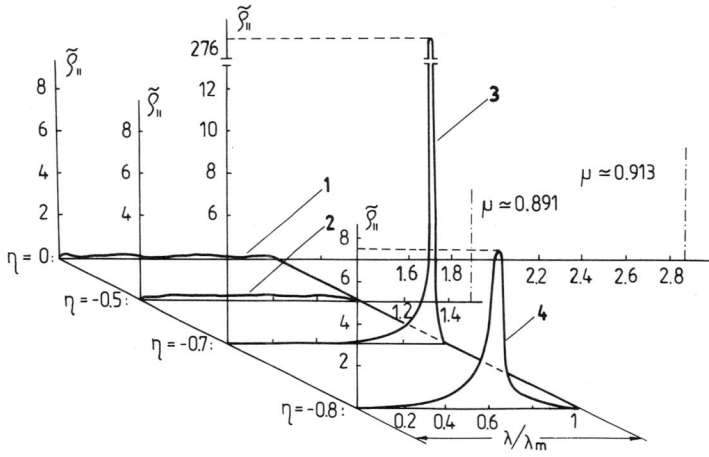

Fig. 2

corresponding to the Jacobi matrix with all elements equal to their limiting values. When the conditions (12) are satisfied, $K(\lambda)$ has the form:

$$K(\lambda) = 4\lambda_m^{-2} \left\{ 2\lambda - \lambda_m - 2i \left[\lambda \left(\lambda_m - \lambda \right) \right]^{1/2} \right\} \, . \tag{15}$$

If the spectrum of the system contains localized states lying outside the interval $(0, \lambda)$, their energies correspond to singularities of the functions $\rho(\lambda)$ and can be determined from the equation

$$|P_{n+1}(\lambda) - b_n P_n(\lambda) K(\lambda)|^2 = 0 \, . \tag{16}$$

Methods similar to Peresada were developed later by other authors (see, e.g., Refs. [8,9]). The Peresada method is especially convenient for numerical calculations of the dynamical and thermodynamical characteristics of real crystals.

PECULIAR FEATURES OF THE PHONON SPECTRA OF STRONGLY ANISOTROPIC CRYSTALS AND THE LOCALIZATION OF LIGHT IMPURITY VIBRATIONS

In order to demonstrate the possibilities of obtaining some important information about the phonon spectra by use of the partial distribution functions we consider as an example a monoatomic crystal lattice having tetragonal symmetry and consisting of weakly interacting atomic layers. Each layer is displaced by half a lattice constant with respect to the previous one along both coordinate axis in the basal plane. Such a structure permits us to take into account only the central interaction between the nearest interlayer neighbours avoiding a crystal instability. Describing the intralayer interaction we take into consideration only the central interaction between the first and second neighbours.

The matrix of force constants in the accepted model takes the following form

$$\phi_{ik}(\pm a, 0, 0) = -\alpha\, \delta_{ix}\, \delta_{kx}\,, \tag{17}$$

$$\phi_{ik}(\pm a, \pm a, 0) = -\alpha' \begin{pmatrix} 1 & 1 & 0 \\ 1 & 1 & 0 \\ 0 & 0 & 0 \end{pmatrix}, \tag{18}$$

$$\phi_{ik}(\pm a/2, \pm a/2, \pm a\epsilon/2) = -\gamma\alpha \begin{pmatrix} 1 & 1 & \epsilon \\ 1 & 1 & \epsilon \\ \epsilon & \epsilon & \epsilon^2 \end{pmatrix}, \tag{19}$$

where a is the interatomic distance in the basal plane: the interatomic distance along the C-axis is $a\epsilon$; the force constants α and α' describe the central interaction of the first and the second neighbours in the basal plane, respectively; the force constant $\gamma\alpha$ characterizes the interlayer interaction (for a strongly anisotropic crystal $\gamma \ll 1$).

The matrix $\phi_{ik}(0,0,0)$ is obtained from the condition of translation invariance. The other matrices of the force constants can be obtained from (17)–(19) with the symmetry operations of the spatial group D_{4h}.

It is easy to show that for one of the vibration modes ($p = 3$) the components of the polarization vector $e_x(\mathbf{k}, 3)$ and $e_y(\mathbf{k}, 3)$ are proportional to γ. Therefore it follows from definitions (3) and (5) that

$$g_3(\lambda) = g_{3,z}(\lambda) + \mathcal{O}\left(\gamma^2\right). \tag{20}$$

On the other hand, the values of the components of the polarization vectors in the two other branches $e_z(\mathbf{k}, 1)$ and $e_z(\mathbf{k}, 2)$ are also proportional to γ. Therefore it follows from (3) and (4) that

$$\sum_{p=1}^{3} g_{zp}(\lambda) = g_{3,z}(\lambda) + \mathcal{O}\left(\gamma^2\right). \tag{21}$$

Denoting the spectral density generated by a single displacement of a lattice atom along the direction normal to the layers as $\rho_\perp(\lambda) = 3g_z(\lambda)$, we can obtain from (11), (20) and (21):

$$\left| g_3(\lambda) - \rho_\perp(\lambda)/3 \right| = \mathcal{O}\left(\gamma^2\right). \tag{22}$$

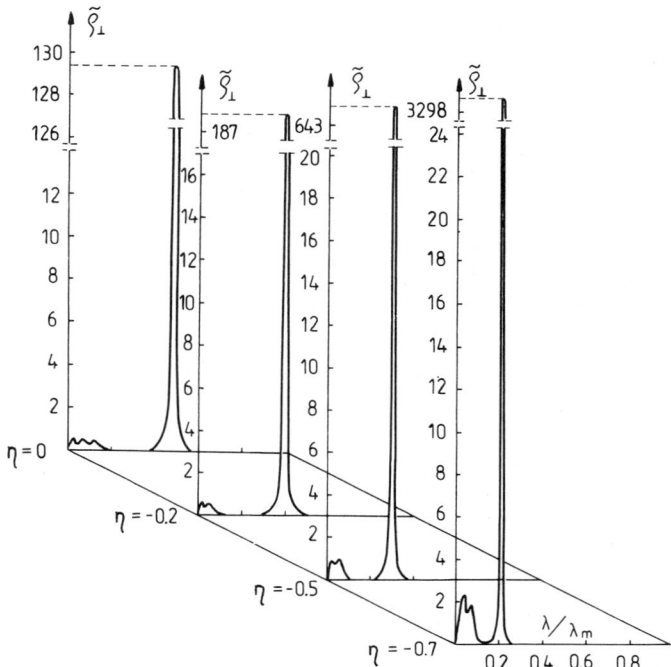

Fig. 3

Analogously we can obtain the estimate

$$\left|g_{1,2}(\lambda) - \rho_\|(\lambda)/3\right| = \mathcal{O}\left(\gamma^2\right), \tag{23}$$

where $\rho_\|(\lambda)$ is the spectral density generated by a single displacement along the basal plane: $\rho_\|(\lambda) = 3g_x(\lambda) = 3g_y(\lambda)$.

Consequently the spectral densities in the invariant subspaces of a strongly anisotropic crystal ($\gamma \ll 1$) may be considered with high accuracy to be proportional to the square frequency distribution functions of the related branches of the phonon spectrum. It is possible to compare the results obtained by the Jacobi matrices technique with those obtained by traditional methods.

Note that the vibration spectrum of the model under consideration can be analyzed in some limiting cases analytically. It may be shown under the condition $\gamma \ll 1$ that the vibrations with the displacement vector **u** in the XOY plane are independent of vibrations with a displacement vector **u**, parallel to the axis OZ, i.e. perpendicular to the layers. The vibration of the vector **u** in the layer plane has two branches with strongly anisotropic dispersion laws. The dispersion laws of such branches for $ak_1 \ll 1$ and $ak_2 \ll 1$ can be written in a simple form:

$$\lambda_1(\mathbf{k}) = s_1^2 k_1^2 + s_2^2 k_2^2 + \omega_*^2 \sin^2 q, \tag{24}$$

$$\lambda_2(\mathbf{k}) = s_2^2 k_1^2 + s_1^2 k_2^2 + \omega_*^2 \sin^2 q , \qquad (25)$$

where $s_1^2 = a^2(\alpha + 2\alpha')/m$; $s_2^2 = 2\alpha' a^2/m$; $\omega_*^2 = 16\gamma\alpha/m$.

It is easy to see that the isofrequency surfaces $\lambda(\mathbf{k}) = \omega^2 = constant$ are closed if $\omega^2 < \omega_*^2$ but they are open if $\omega^2 > \omega_*^2$. Consequently the topology of the isofrequency surfaces changes at the point $\omega^2 = \omega_*^2$ and the value $\lambda = \omega_*^2$ determines the squared frequency at which there is a $3D$ van Hove singularity.

In the interval of the squared frequencies $\omega_*^2 \ll \lambda \ll \lambda_m$ we can simplify the dispersion laws $\lambda_{1,2}(\mathbf{k})$ by omitting small terms proportional to γ. Then the dispersion laws for the first branch has the following two-dimensional form:

$$\lambda_1(\mathbf{k}) = \omega_1^2 \sin^2(ak_1/2) + \omega_t^2 [1 - \cos(ak_1)\cos(ak_2)] , \qquad (26)$$

where $\omega_1^2 = 4\alpha/m$ and $\omega_t^2 = 8\alpha'/m$. The second dispersion law for displacements in the basal plane has a similar form. Thus we have a situation in which crystal vibrations behave as two–dimensional vibrations and posses $2D$ dispersion laws. It is known that the phonon density of a $2D$ dispersion law has logarithmic van Hove singularities [1,2] at the squared frequencies ω_t^2 and $\omega_1^2 \simeq \lambda_m - \omega_t^2$.

Let us investigate the change of the spectral densities $\rho_\perp(\lambda)$, $\rho_\parallel(\lambda)$ and the total squared frequency distribution function $g(\lambda)$ of the layered crystal caused by increasing the anisotropy of the interatomic interaction (Fig. 1). For weak anisotropy ($\gamma = 0.1$) the function $g(\lambda)$ has a form that is typical for crystals with an isotropic interatomic interaction.

The centre of gravity of the spectral density $\rho_\perp(\lambda)$ shifts to the low frequency region of the phonon spectrum if the interaction anisotropy is increased. The density of states $\rho_\perp(\lambda)$ is localized in a narrow region at low frequencies. The width of this region is equal practically to $\omega_*^2 \epsilon^2$. For the same reason, the total density of states $g(\lambda)$ has a sharp maximum in the low frequency region. The spectral density $\rho_\parallel(\lambda)$ has a form typical for a $2D$–phonon density if $\gamma \ll 1$. We see two maxima on this diagram for $\gamma = 0.01$. They reflect the $2D$ van Hove singularities discussed above. For $\lambda \ll \omega_*^2 \epsilon^2$ the total density of states $g(\lambda)$ has two dimensional character as well.

The spectral density peculiarities of an ideal strongly anisotropic crystal associated with the quasi two–dimensional form of the function $\rho_\parallel(\lambda)$ and the localization of the function $\rho_\perp(\lambda)$ in the narrow low frequency region must influence the vibrational characteristics of the same crystals with defects. It is known that the conditions for the occurrence of local vibrations in $2D$–systems differ sufficiently from those in $3D$–systems. In particular the threshold for local vibration creation in $2D$–systems is absent.

Consider a change of the spectral densities caused by a light substitutional impurity. Denote the spectral density generated by a single displacement of the impurity atom in the direction i as $\tilde{\rho}_i(\lambda)$. If the local vibration with frequency $\tilde{\omega}_i$ ($\tilde{\omega}_i^2 > \lambda_m$) arises in the phonon spectrum of the defect crystal then the function $\tilde{\rho}_i(\lambda)$ may be written in the following form:

$$\tilde{\rho}_i(\lambda) = \rho_i^*(\lambda) + \mu_i \delta\left(\lambda - \tilde{\omega}_i^2\right) , \qquad (27)$$

where the function $\rho_i^*(\lambda)$ is the continuous part of the impurity's spectral density; μ_i is the so called intensity of the local vibration. As the spectral densities are normalized to unity we can write:

$$\mu_i = 1 - \int_0^{\lambda_m} \rho_i^*(\lambda) \, d\lambda . \qquad (28)$$

The functions $\tilde{\rho}_\parallel(\lambda)$ and $\tilde{\rho}_\perp(\lambda)$ are the spectral densities generated by the displacements of the light impurity atom in the directions along the layers and normal to them, respectively. Let us analyse the change of those functions with a decrease of the impurity force constants. Denote by η the relative change of the force constants, which characterises the interaction between the impurity atom and its neighbours.

The change of the function $\tilde{\rho}_\parallel(\lambda)$ with decreasing η is shown in Fig. 2. The vibrations of the impurity atom for $\eta = 0$ (the isotopic impurity) are concentrated mainly at a local frequency. The frequency and intensity of the local vibrations are lowered with a decrease in the magnitude of the force constants. The local frequency approaches the upper edge of the continuous spectrum. If this frequency merges with the continuous spectrum a sharp peak emerges. This state is shown as curve 3 ($\eta = -0.7$) in that region of spectrum where the density of states of the layered crystal has a 3D form and is small. The curve 4 ($\eta = -0.8$) corresponds to the case of a localized state in the interval of frequencies where the ideal density of states is similar to that in a 2D crystal (see Fig. 1.).

The spectral density $\rho_\perp(\lambda)$ reveals the same evolution tendencies (Fig. 3.). As was mentioned above this spectral density is generated by the displacement of the

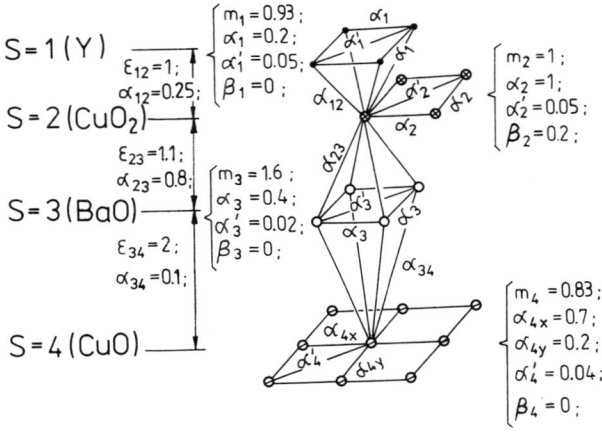

Fig. 4

atom in the direction of the weak bond of the layered crystal. The spectral density of such an ideal crystal is concentrated in a narrow region of low frequencies and that is why the creation of a true local vibration is very difficult. The light impurity forms a sharp maximum inside the continuous spectrum. The corresponding vibration is not delocalized because vibrations with the same frequency in the direction under consideration are practically absent in an ideal crystal (see Fig.1.).

Thus a two–parametric substitution impurity may form localized states in the "intermediate" region of the continuous spectrum. Note that the frequencies of the localized vibrations which have been previously investigated are situated either outside of the continuous spectrum (local vibrations) or in the low frequency region of the continuous spectrum (quasi–local vibrations).

THE PARTIAL DISTRIBUTION FUNCTIONS OF A MULTILAYER CRYSTAL MODEL ANALOGOUS TO HTS 1–2–3 COMPOUNDS AND THEIR ANALYSIS

Numerous phonon spectrum calculations for a crystal of the Y–Ba–Cu–O–type exist (see e.g., Ref. [10]), but according to authors knowledge a qualitative analysis of the role of the multiple layers in the formation of the peculiarities of the vibrational spectra of such structures is absent. To realise such an analysis we propose first of all to discuss a simplified model of the crystal, which takes into account its multilayeredness in a proper way. As is known, crystals of this type consist of alternating layers of atoms Y, "molecules" CuO_2, "molecules" BaO and "molecules" CuO. We assume that each of these layers is "monoatomic" with the mass of the "atoms" equal to

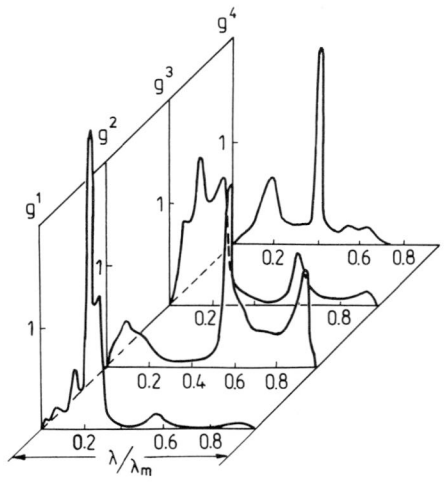

Fig. 5

the sum of the atom masses of this layer inside the unit cell. This model presents a polyatomic orthorombic lattice, consisting of alternating layers of the atoms of different types, each layer displaced half a lattice constant with respect to the previous one along both coordinate axes in the basal plane (see Fig. 4.). Such a structure permits us to describe the interlayer interactions accounting for only the central interaction between the nearest neighbours avoiding thereby a crystal instability. Describing the intralayer interaction we take into account the interaction between the first and second neighbours, both central and noncentral.

The matrices of the force constants in the accepted model take the following form:

$$\phi_{ik}^{ss'}(\pm a, 0, 0) = -\delta_{ss'}[\alpha_{sx}\,\delta_{ix}\delta_{kx} + \beta_s\,(\delta_{iy}\delta_{ky} + \delta_{iz}\delta_{kz})] \,, \tag{29}$$

$$\phi_{ik}^{ss'}(0, \pm a, 0) = -\delta_{ss'}[\alpha_{sy}\,\delta_{iy}\delta_{ky} + \beta_s\,(\delta_{ix}\delta_{kx} + \delta_{iz}\delta_{kz})] \,, \tag{30}$$

$$\phi_{ik}^{ss'}(\pm a, \pm a, 0) = -\delta_{ss'} \begin{pmatrix} \alpha'_s & \xi'_s & 0 \\ \xi'_s & \alpha'_s & 0 \\ 0 & 0 & -\beta/2 \end{pmatrix} , \tag{31}$$

$$\phi_{ik}^{ss'}(\pm a/2, \pm a/2, \pm a\epsilon/2) = -\alpha_{s,s+1}\,\delta_{s,s'-1} \begin{pmatrix} 1 & 1 & \epsilon_{s,s'} \\ 1 & 1 & \epsilon_{s,s'} \\ \epsilon_{s,s'} & \epsilon_{s,s'} & \epsilon_{s,s'}^2 \end{pmatrix}, \qquad (32)$$

where a is the lattice constant in the basal plane (001); parameter $\epsilon_{s,s'}$ describes the "tension" of the lattice along the C-axis between the neighouring layers s and s', the force constants α_{sx} and α_{sy} describe the central nearest–neighbour interaction in the basal plane filled with atoms from sublattice s along the crystallographic directions a and b respectively. In the model under consideration the interaction along these axes is accepted to be different only for the CuO–layer ($s=4$), for the other layers we take $\alpha_{sx} = \alpha_{sy} \equiv \alpha_s$. The parameter β_s describes the noncentral interaction of the first and second neighours in the basal plane (it will be taken nonzero for the layer CuO_2, $s=2$).

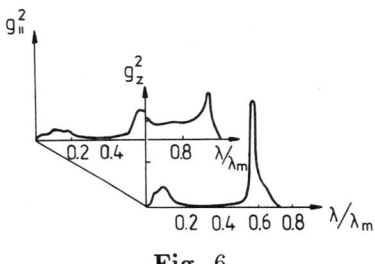

Fig. 6

The central interaction between next–nearest–neighbours is described by $(\alpha'_s + \xi'_s)/2$. The force constants $\alpha_{s,s+1}$ characterise the central interlayer interaction. The relation between the elements of the matrices (29) and (31), describing the noncentral intralayer interaction results from the symmetry requirements of the elastic moduli tensor (for comments see, e.g., Ref. [11]).

This lattice is stable for all positive α_{si}, $\alpha'_{s'}$, $\alpha_{s,s'}$, and $0 < \beta_s < \alpha_s$. For the parameters of interatomic interactions, we have the following relation, $\xi'_s = \alpha'_s + \beta_s/2$.

The calculations were carried out using the values of the parameters which characterise the structure and force constants for a HTS of the 1-2-3–type [12]. The values of these parameters are presented in Fig. 4. In the figure the force constants characterising the central interaction along the appropriate directions are shown by the segments connecting the atoms. Besides the central interaction in the CuO_2–layer ($s=2$) a noncentral one between the nearest and next nearest neighbours are taken into account. Fig. 5. presents the functions $g^s(\lambda) = \sum_{i=1}^{3} g_i^s(\lambda)$, i.e. the contributions to $g(\lambda)$ from each of the layers producing this structure: Y, CuO_2, BaO and CuO, we see that the phonon density contribution of the yttrium sublattice to the phonon density of the Y–Ba–Cu–O crystal practically does not differ from the contribution of the weakly coupled dopant monolayer. The functions $g^s(\lambda)$, of CuO_2–layer ($s=2$) and the CuO) layer ($s=4$) reveal distinct quasilocal maxima in the "intermediate" range of the phonon spectra.

It is well known that electron and Cooper pairs scattering by localized lattice vibrations results in significant changes in the electron - phonon interaction. If a real crystal with symmetry close to cubic symmetry and a weak anisotropy in the interatomic interaction, has a few atoms in the unit cell, the frequencies of the localized states lie either out of the continuous spectral band (localized vibrations), or inside its low–frequency range (quasilocal vibrations). Such localized states are induced by crystal imperfections, which are absent in ideal crystals. In multilayer crystals, quasilocal vibrations can appear without any defects as well and are associated with weakly dispersed optical modes. These vibrations weakly interact with other phonon modes, so the related sharp peak of the frequency density does not vanish and the vibrations are not delocalized. The frequencies of such vibrations are defined in a controllable way and depend on the mass of the atoms in the adjacent layers. As a result they can appear in any part of the phonon spectrum.

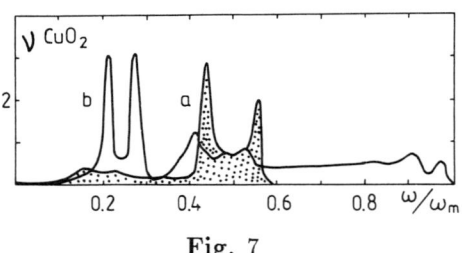

Fig. 7

Thus an essential peculiarity of the phonon spectrum of the multilayer structures is the appearance of these peaks located in an arbitrary part of the continuous frequency band. It turns out that these structures behave like intermediate materials between crystals and disordered systems and as a result they manifest the properties characteristic of both of these families.

Fig. 6. presents the partial functions of the squared frequency distribution $g_i^s(\lambda)$ for the layer CuO_2. The functions $g_x^2(\lambda)$ and $g_y^2(\lambda)$ in CuO_2 layer practically do not differ from each other because of the tetragonal symmetry of this layer and the adjacent ones (Fig. 6a. shows their sum $g_\parallel^2(\lambda)$). The partial distribution function $g_z^2(\lambda)$ has a sharp maximum at $\lambda \approx 0.58\lambda_m$ (Fig. 6b.).

According to Refs [3,4], the vibrations of the different invariant subspaces are orthogonal and consequently do not interact. As a result, after the summation of all the contributions into the density of states in this "intermediate" part of the phonon spectrum, the sharp maximum on the nonzero background does not vanish and the discussed vibrations do not delocalize. This is proved by the distinct "quasilocal" maxima of the functions $g_s(\lambda)$ (Fig. 6b) at frequencies not typical for normal quasilocal vibrations.

PARTIAL FREQUENCY DISTRIBUTION FUNCTIONS FOR A REALISTIC MODEL; THE HTS–CRYSTALS

In the preceding section we considered a simple model of a multilayer crystal. But that model could not take into account the complicated structure of the single layers consisting of several different atoms. The method used allows us to investigate the partial frequency distribution functions generated by the atoms of each type displaced in an arbitrary direction. We consider a polyatomic model adequate to the real structure of a multilayer crystal of a 1–2–3 HTS type. Taking into consideration only the central interaction between atoms, we use the magnitudes of the force constants calculated on the basis of the data published in the Ref. [12].

The obtained results are shown in Figs. 7–10. Fig. 7. presents the functions $\nu(\omega) = 2\omega g(\omega^2)$ for the layer CuO_2 in two different cases:

i) curve (a) corresponds to the simplified model, considered in preceding section;

ii) curve (b) corresponds to the model taking into account all the atomic interactions

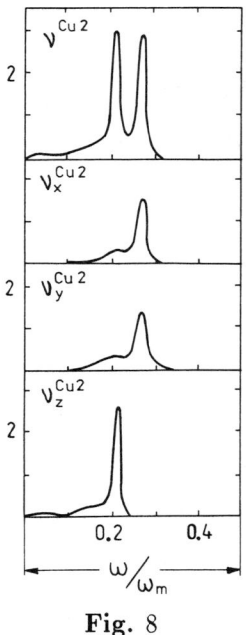

Fig. 8

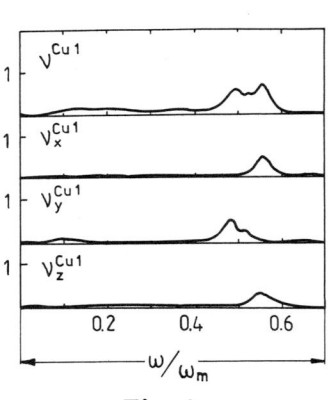

Fig. 9

in the layer. We see that the vibration localization remains, but the distinct peaks of the localized vibrations shift to the low frequency region. The density of states of the oxygen atoms are concentrated in the same frequency region where the density of states of CuO_2 "molecules" are concentrated according to the calculations on the basis of the simplified model. The vibration frequencies of the more heavy atoms Cu shift to the lower end of the spectrum. Fig. 8. presents the functions $\nu_i^{Cu2}(\omega)$ ($i = x, y, z$) describing the contribution of the Cu2 atoms located in the CuO_2 layer to density of states $\left[(\nu^{Cu2}(\omega) = \nu_x^{Cu2}(\omega) + \nu_y^{Cu2}(\omega) + \nu_z^{Cu2}(\omega)\right]$. We see vibration localization in each direction. Fig. 9. presents the analogous contribution of the Cu1 atoms, e.g.

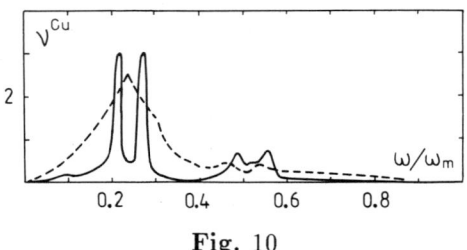

Fig. 10

the atoms in the CuO layers. The total contribution of the vibrations of the Cu atoms is presented in Fig. 10. The latter characteristic can be measured by neutron experiments [13]. The data of the neutron experiments are shown as a dash line.

Acknowledgments

We would like to thank Professor I. Natkaniec for giving us the results of his neutron experiments [13] prior publication.

REFERENCES

[1] A. A. Maradudin, E.W. Montroll, G.H. Weiss "Theory of lattice dynamics in the harmonic approximation", Academic Press, New York (1963).

[2] A.M. Kosevich, "The theory of crystal lattice", Vishtcha Shkola, Kharkov (1988) (in Russian).

[3] V.I. Peresada, Zh. Exper. Teor. Fiz., **53**, 605(1967).

[4] V.I. Peresada, E.S. Syrkin, Surf. Sci. **54**, 293(1976).

[5] A.A. Maradudin, Modern Problem Surf. Phys., I–st Int. School on Condensed Matter Phys., Publishing House of the Bulgarian Academy of Sciences, Sofia, 1981.

[6] N.I. Akhieser, I.M. Glasman, "The Theory of Linear Operators", CIITL, Moscow, 1950 (in Russian).

[7] V.I. Peresada, V.N. Afanasyev, V.S. Borovikov, Fiz. Nizkih Temperatur, **1**, 461(1975).

[8] R. Haydock, in Sol. St. Phys. **v. 35** (ed. H. Ehrenreich et al.) Academic Press, (1980).

[9] C.M.M. Nex, J. Phys. **A 11**, 653(1978).

[10] R. Feile, Physica, **C 159**, 1(1989).

[11] S.B. Foedosyev, E.S. Syrkin, Fiz. Nizkih Temperatur, **9**, 528(1983).

[12] S. Mass. T. Yasuda, Y. Horie, M. Kusaba, T. Fukami, J. Phys. Soc. Japan, **57**, 1024(1988).

[13] I. Natkaniec, private communication.

SELFCONSISTENT APPROACH TO PHONON LOCALIZATION IN HARMONIC CRYSTALS WITH DEFECTS

I.Ya. Polishchuk, A.L. Burin, L.A. Maksimov

Russian Scientific Center Kurchatov Institute
123182, Moscow, Russia

The anomalous behaviour of the low-temperature properties of amorphous insulators stems from the particular nature of low-frequency elementary excitations, in particular from the presence of localized modes of a two-level system[1] or fractons[2]. The excitation of a two-level system results from tunnelling, while fractons arise in percolation systems with ruptured bonds. It was hypothesized in Ref. 3 that localized excitations might arise under far less stringent conditions, especially, in harmonic lattices with heavy defects. Here we examine this idea in detail.

We begin with a scalar model of the vibrations of a cubic harmonic crystal with isotopic impurities, described by the Hamiltonian

$$H = H_O + H_{int}, \quad H_O = \frac{1}{2}\sum_i p_i^2 + \frac{1}{2}\sum_{i,j} A_{ij}\, u_i\, u_j,$$

$$H_{int} = \frac{1}{2}\left(m^{-1} - 1\right)\sum_i c_i\, p_i^2.$$

Here the index i enumerates the sites of a monoatomic cubic lattice, and u_i and p_i are the atom displacement and momentum operators, respectively, A_{ij} are the force constants, m is the mass of defect, $c_i = 1$ if the corresponding site is occupied by a defect and $c_i = 0$ in the opposite case, while the remaining parameters of the ideal crystal (the mass of the atom, the lattice constant, and the sound velocity) are all assumed to be equal to unity.

We consider the square of the retarded Green's function,

$$P_{ij} = \left[G_{ij}^+(t)\right]^2 = G_{ij}^+(t) G_{ji}^-(-t),$$

where

$$G_{ij}^\pm(t) = \mp i\theta(\pm t)\langle [u_i(t), u_j]\rangle.$$

For the Fourier transform of P_{ij} averaged over the positions of the defects we have the following expression:

$$P(k,\Omega) = \int \frac{d\omega}{2\pi} \sum_{pp'} \Phi_{pp'}, \quad \Phi_{pp'} = \overline{C_{p_+',p_+'}^+(\omega_+) G_{p_-',p_-}^-(\omega_-)}, \tag{1}$$

Die Kunst of Phonons, Edited by T. Paszkiewicz and
K. Rapcewicz, Plenum Press, New York, 1994

where the vectors $p_\pm = p \pm k/2$, $\omega_\pm = \omega \pm \Omega/2$, and the bar denotes an average over positions of the impurities. Using a version of the cross technique proposed in Ref. 3, we find the following expression for $\Phi_{pp'}$:

$$\Phi_{pp'} = G^+_{p_+}(\omega_+) G^-_{p_-}(\omega_-) \left(\delta_{pp'} + \sum_{p''} W_{pp''} \Phi_{p''p'} \right) , \qquad (2)$$

where $G^\pm_{p_\pm}(\omega_\pm) = (\omega_\pm^2 - p_\pm^2 - \Sigma_{p_\pm}(\omega \pm i0))^{-1}$, $\Sigma_p(\omega \pm i0) = \Delta_p(\omega) \mp i\Gamma_p(\omega)$ is the self–energy and $W_{pp'} = W_{pp'}(k, \Omega, \omega)$ is an irreducible vertex. We rewrite Eq. (2) in the equivalent form

$$\left[2\omega\Omega + 2p\,k + \Sigma_p(\omega_+) - \Sigma_p(\omega_-) \right] \Phi_{pp'} = \Delta G_p \left(\delta_{pp'} + \sum_{pp''} W_{pp''} \Phi_{p''p'} \right) , \qquad (3)$$

where $\Delta G_p = G^+_{p_+}(\omega_+) - G^-_{p_-}(\omega_-)$.
In the dipole approximation (cf. Ref. 4) we can write

$$\sum_{p'} \Phi_{pp'} = 2\pi i\, \Delta G_p / \tilde\omega \left(\Phi_0 + 3p\, \Phi_1 \hat k / \tilde\omega \right), \qquad \Phi_\sigma = \Phi_\sigma(k, \omega, \Omega), \quad (\sigma = 0, 1), \qquad (4)$$

where $\tilde\omega = [\omega^2 - \Delta_0(\omega)]^{1/2}$, $\hat k = k/|k|$. and $\Sigma_0(\omega + i0) = \Delta_0(\omega) - i\Gamma_0(\omega)$ is the momentum independent self–energy calculated by ignoring interference effects[3] (see also expression (12) below. From (3) we then find the algebraic system of equations

$$-\omega\,\Omega\,\Phi_0 + k\,\Phi_1 = -i\tilde\omega/4\pi , \qquad (5')$$

$$-\omega\,\Omega\,\Phi_1 + k\,\omega^2\,\Phi_0/3 - i\,M\,\Phi_1 = 0 , \qquad (5'')$$

where

$$M = \Gamma(\omega) + 3\pi\tilde\omega^{-3} \sum_{pp'} \Delta G_p \left(p\,\hat k \right) W_{pp'} \Delta G_{p'} \left(p'\,\hat k \right) \qquad (6)$$

is the memory function. In deriving Eq. (5), we used the Ward identity

$$\Sigma^+_{p_+}(\omega_+) - \Sigma^-_{p_-}(\omega_-) = \sum_{p'} W_{pp'} \Delta G_{p'} ,$$

which is valid for our model of the crystal. From equations (4) and (5) we find for Φ_0 the expression which is usually derived in the interacting-mode theory[5]:

$$\sum_{pp'} \Phi_{pp'} \simeq \Phi_0 = \frac{\tilde\omega/4\pi\omega}{-i\Omega + k^2/3 \dfrac{(\tilde\omega/\omega)^2}{-i\Omega + M/\omega}} \simeq 1/4\pi \frac{\tilde\omega/\omega}{-i\Omega + Dk^2} , \qquad (7)$$

To calculate the memory function M, we use the approximation of fan diagrams[6], making use of the results of Ref. 3:

$$M \simeq \Gamma_0(\omega) + \frac{8\pi^2 \Gamma_0(\omega)}{\tilde\omega^2\,\omega} \sum_q \frac{\theta\left(\Gamma_0(\omega) - \tilde\omega\,q\right)}{-i\Omega + D_0(\omega)\,q^2} , \qquad q = |p + p'| . \qquad (8)$$

where
$$D_0(\omega) = \tilde{\omega}^2/(3\omega \Gamma_0(\omega)).$$

Using (7) and (8), we find for the phonon diffusion coefficient an expression which is of the same form as the corresponding expression in the theory of weak localization of electrons:[6]

$$D = D_0(\omega)\left(1 - \frac{12\pi \Gamma_0(\omega)}{\tilde{\omega}^3} \int \frac{\theta\left(\Gamma_0(\omega) - \tilde{\omega} q\right)}{q^2 - \frac{i\Omega}{D_0(\omega)}} q^2 dq\right). \quad (9)$$

In the limit $\Omega \to 0$, we get

$$D = D_0(\omega)\left[1 - 12\pi \left(\Gamma_0(\omega)/\tilde{\omega}^2\right)^2\right]. \quad (10)$$

This expression is meaningful as long as the attenuation $\Gamma_0(\omega)$ is sufficiently small; the localization threshold is determined by the finite value $\Gamma_0(\omega) = (12\pi)^{-1/2}\tilde{\omega}^2$ in this case. In the localization region, in which expression (10) is no longer meaningful, we can make use of the self-consistent localization theory proposed by Vollhard and Wölfle[4] to replace $D_0(\omega)$ by D in the integral in (9). Solving the resulting self-consistent equation, we find

$$D \simeq -i\Omega\, R_c^2, \qquad R_c = \frac{6\pi^2 \Gamma_0(\omega)\tilde{\omega}^2}{12\pi\left[\Gamma_0(\omega)/\tilde{\omega}^2\right]^2 - 1}. \quad (11)$$

To calculate the quantity $\Gamma_0(\omega)$ in (10) and (11), we use the self-consistent expression for the self energy which was derived in Ref. 3:

$$\Sigma_0(\omega + i0) = \frac{n\,\omega^2\left(m^{-1} - 1\right)}{m^{-1} + \left(m^{-1} - 1\right)\omega^2\left[\left(3/4\pi^4\right)^{1/3} + i\left(\omega^2 - \Sigma_0(\omega + i0)/4\pi\right)^{1/2}\right]}. \quad (12)$$

Here n is the concentration of the impurities. The density of states in this system is given by[3]

$$g(\omega) = \omega/2\pi\, \mathrm{Re}\left[\omega^2 - \Sigma(\omega + i0)\right]^{1/2}. \quad (13)$$

Ignoring the imaginary part of the denominator in (12), and substituting the result into (13), we find that the expression in square brackets is negative in the frequency interval (ω_0, ω_*) (here $\omega_0 = (4\pi^4/3)^{1/6}(m-1)^{-1/2}$ is the quasilocal frequency corresponding to the pole of expression (12), at $\omega_* = \omega_0\left[1 + n(m-1)\right]^{1/2}$). Consequently, there is a gap in this interval.[7] In the interval $(0, \omega_0)$ the vibration spectrum is acoustic, with $\omega = (1 + nm)^{-1/2}k$ and

$$\Delta_0(\omega) \simeq -\frac{n\omega^2(m-1)}{1 - (\omega/\omega_0)^2}, \qquad \Gamma_0(\omega) = \frac{n\omega^4(m-1)^2\left[\omega^2 - \Delta_0(\omega)\right]^{1/2}}{\left[1 - (\omega/\omega_0^2)\right]^2}. \quad (14')$$

In the interval $\omega_* < \omega \ll \omega_D$, the vibration spectrum takes the form of an optical branch, with $\omega^2 = \omega_*^2 + k^2$ and

$$\Delta_0(\omega) \simeq \omega_*^2, \qquad \Gamma_0(\omega) = \omega_*^4\, \tilde{\omega}/n . \qquad (14'')$$

Using (10)–(14), we find the following expression for the diffusion coefficient

$$D \simeq D_0 \cdot \begin{cases} 1 - (\omega/\omega')^6, & \omega \ll \omega_0; \quad \omega' = \left[(1-m)^2 n\right]^{1/2}, \\ 1 - (\omega/\omega'), & \omega \lesssim \omega_C'; \quad \omega_C' = \omega_0 \left(1 - n^{1/3}\right), \\ 1 - (\tilde{\omega}_C''/\tilde{\omega}), & \omega \geq \omega_C'; \quad \omega_C'' = \omega_* \left\{1 - \pi^3 \left[(m-1)^1 + n\right]^3 / n\right\} . \end{cases}$$

$$(15)$$

The acoustic modes are thus localized in the interval (ω, ω_0), the localization radius is $R_C' \sim (\omega - \omega_C')^{-1}$. The optical modes are localized in the interval (ω_*, ω_C''), with a localization length $R'' \sim (\omega_C'' - \omega)^{-1}$. In contrast to the analysis of the present paper, the analysis of Ref. 8 ignored the singularities of the vibration spectrum near the quasilocal frequency, and the sole localization threshold found for acoustic modes[8] was erroneously identified with the quantity ω' (see (15)), which always lies above the boundary of the acoustic branch, ω_0 (i.e., in the gap). In Ref. 8, however, there was no discussion of the existence of a gap, the existence of an optical branch, or the properties of modes.

REFERENCES

[1] Anderson P.W. et al., Phil. Mag., **1**, 25 (1972).

[2] Alexander S. et al. Phys. Rev., **B28**, 4615 (1983).

[3] Polishchuk I.Ya. et al., Zh. Eksp. Teor. Fiz. **91**, 259 (1988).

[4] (Sov. Phys. JETP **67**, 362 (1988)).

[5] Vollhardt D Wolfle P., Phys. Rev., **B22**, 4666 (1980).

[6] Götze W., Solid State Comm., **27**, 1393 (1978).

[7] Gor'kov L.P. et al Pisma Zh. Eksp. Teor. Fiz. **30**, 248, (1979), (Sov. Phys. JETP Lett. **30**, 228 (1979)).

[8] Kosevich A.M., Physical mechanics of real crystals, Kiev, Naukova Dumka, 1981 (in Russian).

[9] Akkermans E., Meynard R., Phys. Rev., **B32**, 7850 (1986).

BULK PHONON–POLARITONS IN RESTSTRAHLEN REGION OF A^3B^5 COMPOUND SUPERLATTICE

R. Brazis[#][*], R. Narkowicz[*] and L. Safonova[#]

[#]Semiconductor Physics Institute, Gostauto 11, 2600 Vilnius, Lithuania

[*]Polish University at the Association of Polish Scientists of Lithuania, p.o. box 823, 2066 Vilnius

Superlattices composed of layers whose dielectric functions $\varepsilon(\nu)$ are of the opposite sign are known to support the propagation of specific electromagnetic modes which are related to a bulk–type polarization [1–3]. The term, bulk modes, is chosen in order to distinguish between the relevant excitations and the interface modes, encountered under similar conditions $\varepsilon(\nu)_1 \varepsilon(\nu)_2 < 0$. The existence of bulk modes has been confirmed experimentally in InSb/Al$_2$O$_3$ structures in the millimeter wave region [1], where the values of $\varepsilon\nu$ for InSb are negative due to the free–carrier plasma (or magnetoplasma) contribution.

The present study is aimed at determining the bulk mode dispersion in superlattices exhibiting negative $\varepsilon(\nu)$ values due to ionic polarization. The GaAs/AlAs superlattice is taken as an example, neglecting damping and free–carrier effects.

An effective dielectric function of the form

$$\varepsilon_{1,2}(\nu) = \varepsilon_\infty \left(\nu_{L1,2}^2 - \nu^2\right)\left(\nu_{T1,2}^2 - \varepsilon^2\right)^{-1} \qquad (1)$$

is used for the material characterization where ε, ε_L and ε_T–are the wavenumbers of photons LO – and TO – phonons, and subscripts 1 and 2 refer to AlAs and GaAs, respectively. The parameters are listed in the Table 1 together with the values of the filling factor $\alpha_{1,2} = d_{1,2}/(d_1 + d_2)$, where $d_{1,2}$ is the layer thickness.

Let us consider the case when the wavevector **k** is directed along the layers, i.e. $k = k_\parallel$, $k_\perp = 9$ (Fig. 1, insert). Supposing $d_{1,2} < \gamma_{1,2}^{-1}$ where

$$\gamma_{1,2} = \left(k^2 - 4\pi^2\nu^2\varepsilon_{1,2}\right)^{1/2} \qquad (2)$$

Table 1. Superlattice layer material parameters

	ε_∞	ν_T	ν_L	α
AlAs	8.16	362	400	0.9, 0.33, 0.1
GaAs	11.1	270	292	0.1, 0.67, 0.9

is the amplitude decay constant perpendicular to the interface, one can integrate the microscopic fields over the superlattice spatial period in order to get an anisotropic continuous medium (ACM) approximation in which the bulk mode dispersion equation turns out to be quite simple:

$$k^2 = 4\pi^2 \nu^2 \varepsilon_1 \varepsilon_2 (\alpha_1 \varepsilon_2 + \alpha_2 \varepsilon_1)^{-1} \qquad (3)$$

These modes are represented in Fig. 1. Strictly speaking the bulk modes refer to the saturation values ν_{b+} and ν_{b-} only, falling in the Reststrahlen bands of AlAs and GaAs (Fig. 1,. shaded).

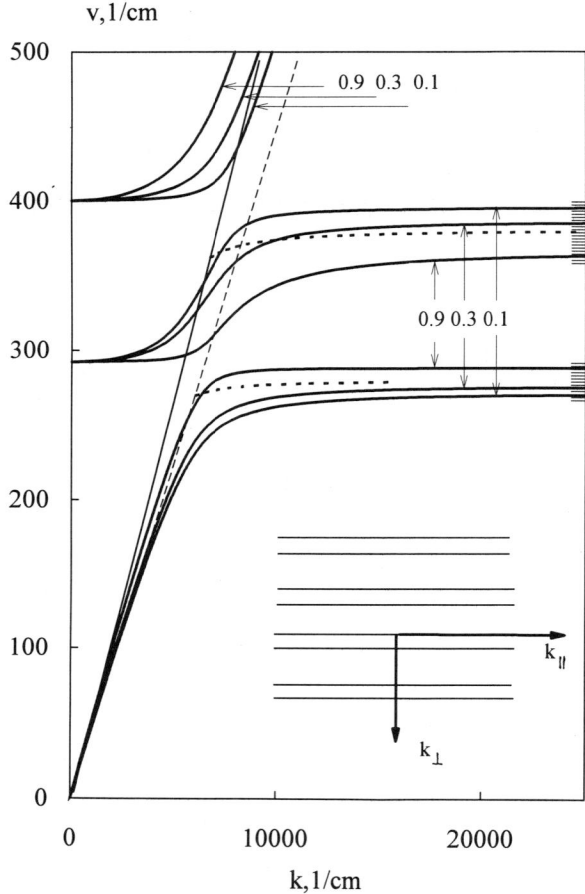

Fig. 1. Dispersion branches of phonon–polaritons propagating along the layers ($k = k_\|$, $k_\perp = 0$) in an AlAs/GaAs superlattice (curves 1 to 3), and the interface mode dispersion branches at a single AlAs/GaAs interface (dotted lines), and the "optic" and "acoustic" asymptotes (straight lines). The AlAs filling factor values are used to denote the corresponding curves.

The values of ν_{b+} and ν_{b-} are determined from Eq. (3) as

$$\nu_{b+} = \left(A + (A^2 - C)^{1/2}\right)^{1/2},$$
$$\nu_{b-} = \left(A - (A^2 - C)^{1/2}\right)^{1/2},$$
(4)

where

$$A = \left(\alpha_1 \left(\nu_{1T}^2 + \nu_{2L}^2\right)\varepsilon_2 + \alpha_2 \left(\nu_{2T}^2 + \nu_{2L}^2\right)\varepsilon_1\right) / \left(2\left(\alpha_1\epsilon_2 + \alpha_2\varepsilon_1\right)\right),$$

$$C = \left(\alpha_1 \nu_{1T}^2 \nu_{2L}^2 \varepsilon_2 + \alpha_2 \nu_{1L}^2 \nu_{2T}^2 \varepsilon_1\right) / \left(\alpha_1\epsilon_2 + \alpha_2\varepsilon_1\right)$$

If the GaAs layers are vanishingly thin ($d_1 \to 0$), the wavenumber ν_{b+} tends to ν_{2T}, and ν_{b-} tends to ν_{1L}. In the opposite limit of thin AlAs layers ν_{b+} tends to ν_{2L}, and ν_{b-} approaches ν_{1T}.

If the layers are thick in the sense that $d_{12} \gg \gamma_{1,2}^{-1}$ then the fields in adjacent layers do not interact, and the superlattice supports the propagation of surface modes along the AlAs/GaAs interfaces. Surface mode dispersion characteristics

$$k^2 = 4\pi^2 \nu^2 \varepsilon_1 \varepsilon_2 \left(\varepsilon_1 + \varepsilon_2\right)^{-1} \quad (5)$$

are represented in Fig. 1 by dotted lines which terminate at the asymptotes of the "optic" and "acoustic" branches of the bulk modes. The lower interface mode appears in a limited area of the $(\nu - k)$-plane, and it is not preserved in a small–period superlattice. On the contrary, the upper interface mode is expected to be the only mode of a superlattice which exists in the Reststrahlen band the extremely short wave limit.

With an increase of k the bulk mode frequency should gradually approach that of the interface mode $(\nu_{b+} \to \nu_s)$. This transition anticipated in the region of $k \geq (d_1 + d_2)^{-1}$, will exhibit negative or positive dispersion depending on whether the AlAs layer is this or thick as compared to the layer of GaAs.

Besides these three principal branches, one can get an unlimited number of Bragg–type modes, if k is non–zero, and the slab–guided modes. All these modes, as well as the transition region between the bulk, are beyond the scope of the ACM approximation.

The bulk and the interface modes are expected to be distinguished in Raman scattering experiments.

REFERENCES

[1] R. Brazis and L. Safonova, Electromagnetic waves in layered semiconductor–dielectric periodic structures in dc magnetic fields, Int. J. Infrared and MMW Waves, bf 8, 449 (1987)

[2] R. Brazis and L. Safonova, Resonance and absorption band in the classical magnetoactive semiconductor–insulator superlattice, Proc. Soc. Photo–Opt. Instrum. Eng., **1029**, 74 (1988)

[3] R. Brazis and L. Safonova, Electromagnetic wave dispersion in semiconductor–dielectric periodic structures below the plasma frequency, Litovskii fizicheskii sbornik, **3**, 285 (1991)

THERMAL PHONON IMAGING

A.G. Every

Physics Department
University of the Witwatersrand
P.O. Wits 2050, South Africa

I. INTRODUCTION

The development of phonon imaging by Northrop and Wolfe (1979,1980) arose from efforts to understand the dynamics of electron-hole droplets in semiconductors (Hensel and Dynes, 1979) and intensity measurements in ballistic heat pulse experiments (Taborek and Goodstein, 1979; Marx et al., 1978). Since its inception, phonon imaging has had a major impact on the field of phonon transport. It has led to the consolidation of earlier ideas on phonon focusing (Maris, 1971; McCurdy et al., 1970; Philip and Viswanathan, 1978; Rösch and Weis, 1976) and has opened up a broad new area of research. Over time it has emerged as a powerful tool that has found a range of applications in solid state physics, and significant progress has been made in its adaptation to particle detection.

In essence, phonon imaging is the study of the highly anisotropic ballistic or near ballistic phonon flux patterns in crystals and the information they convey about the processes in which these phonons participate - the generation and detection mechanisms, anharmonic decay, point defect and dislocation scattering, transmission and reflection at boundaries, transmission through superlattices, interactions with 2D and 3D electron states, large wavevector dispersion and so on. These spatial intensity patterns are created first and foremost by phonon focusing, which is a consequence of the bulk anisotropy of the medium and to a lesser extent by surface directivity, which is dependent on the surface conditions and the mechanisms by which the phonons are generated and detected. The distinctive features in these flux patterns become progressively washed out as the phonons experience increasing probability of scattering in their passage through the crystal, and much can be inferred about the nature of the scattering processes from the changes that occur.

In phonon imaging the dominant phonon frequencies lie somewhere in the range 10^{11} - 10^{12} Hz, depending on the methods of generation and detection. Bolometric detection tends to favour frequencies near the lower end of this range, in which case the wavelengths are much larger than the lattice spacing. and the phonons conform well to continuum elasticity theory. For the main part this talk will be confined to the subject of long wavelength acoustic phonons and the images they give rise to. Tunnel junction techniques, on the other hand, select frequencies towards the higher end of this range, where the wavelengths are comparable with the lattice spacing and dispersive effects are pronounced. The progress in large wavevector phonon imaging has been reviewed in a number of recent articles (Tamura, 1986; Wolfe, 1989) and I will only briefly touch on this topic towards the end of this lecture.

A common assumption in the theoretical modelling of phonon images is that the phonon can be regarded as a wavepacket of negligible dimensions moving along a trajectory or ray. Much of this talk will be concerned with Monte Carlo simulations based on large numbers of phonon rays, and the relation of their statistical properties to the geometry of the acoustic slowness and ray surfaces. The ray approximation is valid in the asymptotic far field. Maris (1983), on the basis of calculations going beyond the ray approximation, has predicted the existence of diffraction effects in thermal phonon imaging, but the spatial resolution required to observe these effects lies at the limits of what is presently attainable, and to date there have been no reported observations.

In the last few years the attention of a number of investigators has turned to ultrasonic experiments which are loosely analogous to phonon imaging. Here, because of the much lower frequencies and longer wavelengths, the long sought after diffraction effects are clearly evident. Their interpretation involves the consideration of the precise elastodynamical Green's function, and not simply its asymptotic form. My second lecture will describe the present status of these ultrasonic investigations and suggest some future directions that they might take. The knowledge base derived from thermal phonon imaging has been a valuable source of guidance in these recent investigations, and in time one may expect to see a counterflow of ideas and techniques.

In the course of this lecture there are two points I would like to make. The first is that in phonon imaging one has for the first time a tool with which the full grandeur of the theory of crystal acoustics developed by Christoffel, Rayleigh, Musgrave, Fedorov, Alshits and others can be examined experimentally. With this as the motivation, I will devote part of the lecture to a brief review of this theory before proceeding to describe the role that it plays in phonon imaging. The second point concerns the complex patterns of caustics found in phonon images, and their dependence on the elastic constants of the solid and other parameters. This behaviour provides an ideal testing ground and source of challenge for elementary catastrophe theory, and I will describe some of the results that have emerged in this regard.

II. PRINCIPLE OF PHONON IMAGING

The principle of phonon imaging is shown in Fig. 1. A focused laser beam is used to heat a small spot on the surface of a single crystal specimen maintained at an ambient temperature of $\approx 2K$ (Eichele et al., 1982, use the focused electron beam of a scanning electron microscope). At low excitation powers the thermal phonons emanating from this heated region undergo negligible bulk scattering and travel ballistically through the crystal to the opposite face where they are detected with a superconducting bolometer or tunnel junction detector. If the heat source is of short duration the detector signal consists of a number of separate pulses, not always all well resolved in practice, corresponding to the arrival of phonons of different polarization having group velocity vectors in the source-detector direction. The magnitudes of these pulses vary considerably with direction, particularly those of the transverse phonon branches, and it is the spatial dependence of the transmitted phonon intensity that is the main concern of phonon imaging. The usual method of constructing a phonon image is to raster scan the source over the front surface and record the boxcar integrated detector signal as a 256×256 pixel gray-scale image.

Variants on the phonon imaging technique have been developed that employ a fixed source and image the arriving flux. Eisenmenger (1980) has produced a direct visual display of the flux incident on a Si surface using the fountain effect in superfluid ^{4}He.

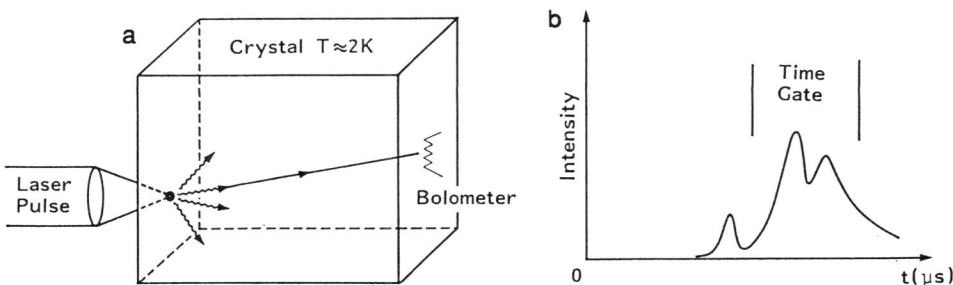

Figure 1. (a) Principle of phonon imaging. (b) Typical ballistic heat pulse profile.

Schreyer et al. (1984) have developed a spatially resolving tunnel junction detector and Kent et al. (1990) have developed a spatially resolving CdS detector which is able to operate at high magnetic fields. A new element has been added to phonon imaging through the study of 2DEG's (Karl et al., 1988; Kent, 1992), which has raised the issue of momentum focusing (Jasiukiewicz et al., 1991). The applications of phonon imaging to particle detection are reviewed elsewhere in this volume.

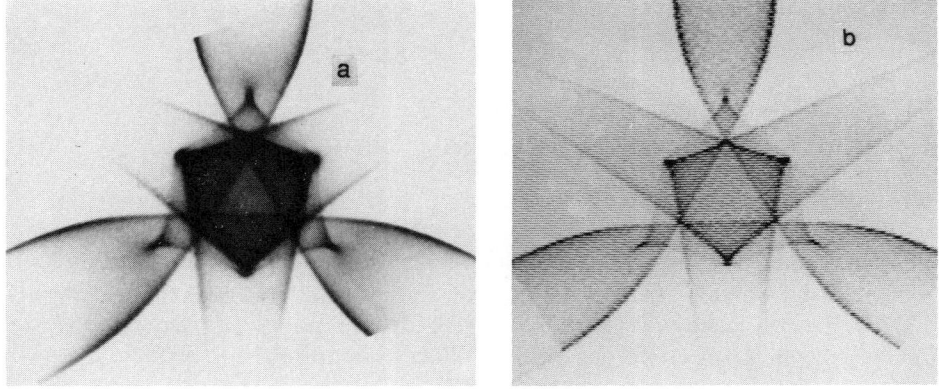

Figure 2. Phonon image for a [111]-cut CaF_2 crystal. Darkness is a measure of the phonon intensity. (a) Measured. (b) Monte Carlo simulation.

By way of example, Fig. 2(a) shows a phonon image of CaF_2 obtained by Hurley and Wolfe (1985) using laser heating and bolometric detection. The pronounced non-uniformity in the phonon intensity in this image is largely due to the bulk focusing of long wavelength acoustic phonons. The sharp structures are caustics and are well accounted for on the basis of geometrical acoustics, as we will show. The broadening of these singularities is attributable mainly to the finite size of the source (which generally exceeds the size of the irradiated area because of hot spot effects), detector and raster interval, which dominate over diffraction broadening at this resolution. For more details on the phonon imaging technique see Northrop and Wolfe (1985) and Wolfe (1993).

III. WAVES IN ANISOTROPIC SOLIDS

The starting point in our interpretation of phonon images is the wave equation for an elastically anisotropic solid (Fedorov, 1968: Musgrave; 1970; Auld, 1973)

$$\rho \frac{\partial^2 u_r}{\partial t^2} = C_{r\ell sm} \frac{\partial^2 u_s}{\partial x_\ell \partial x_m}. \tag{1}$$

Here $\mathbf{u}(\mathbf{x},t)$ is the particle displacement field, ρ is the density and $C_{r\ell sm}$ is the elastic modulus tensor of the medium. This equation admits plane wave solutions of the form

$$u_r = U_r \exp\{i(\mathbf{k}.\mathbf{x} - \omega t)\}, \tag{2}$$

where $\mathbf{U}$, the polarization vector, $\mathbf{k}$, the wavevector and ω, the angular frequency are related by the linear equations

$$(C_{r\ell sm} k_\ell k_m - \rho \omega^2 \delta_{rs}) U_s = 0. \tag{3}$$

The secular equation

$$\Omega(\mathbf{k},\omega) = \mid C_{r\ell sm} k_\ell k_m - \rho\omega^2 \delta_{rs} \mid = 0, \tag{4}$$

represents the dispersion relation of the medium. The constant frequency surfaces defined by Eq. (4) are all identical in shape, and scale in size with ω. This fact is exploited by dividing Eq. (3) through by k^6 to arrive at the Christoffel equations

$$(\Gamma_{rs} - \rho v^2 \delta_{rs}) U_s = 0, \tag{5}$$

where $\Gamma_{rs} = C_{r\ell sm} n_\ell n_m$ is the Christoffel tensor, $\mathbf{n} = \mathbf{k}/k$ is the wave normal and $v = \omega/k$ is the phase velocity. The corresponding secular equation

$$\mid \Gamma_{rs} - \rho v^2 \delta_{rs} \mid = 0, \tag{6}$$

is cubic in v^2 and the three roots may be obtained using standard numerical procedures or expressed in Cardan's trigonometric closed form (Every, 1980). The associated eigenvectors $\mathbf{U}^{(n)}$ are mutually orthogonal and, with rare exceptions, the one corresponding to the largest phase velocity is quasi-longitudinal, while the other two modes are quasi-transverse.

A. The Slowness Surface

The slowness surface plays a central role in the analysis of many crystal acoustics phenomena, including phonon focusing, reflection and transmission at surfaces, surface waves and plate modes. It is a centrosymmetric surface of three sheets representing the directional dependence of the acoustic slowness $s = 1/v$ and as such is the locus of slowness vectors $\mathbf{s} = \mathbf{k}/\omega = \mathbf{n}/v$ with respect to $\mathbf{n}$. The slowness surface is identical in shape to the the constant frequency surfaces. The equation for the slowness surface is obtained from Eq. (4) by dividing by ω^6

$$\Lambda(\mathbf{s}) = \mid C_{r\ell sm} s_\ell s_m - \rho \delta_{rs} \mid = 0, \tag{7}$$

and is of degree 6 in $\mathbf{s}$. A computationally efficient method of constructing the slowness surface is to vary $\mathbf{n}$, solve the cubic Eq. (6) for v^2 and thereby obtain s as a function of direction. However, when a physical surface features in a problem, phased matched partial waves are usually sought, and this entails setting the in-plane component $s_\parallel$ of the slowness to some value, and solving the sextic Eq. (7) for the normal component $s_\perp$.

An indication of the information that can be conveyed by slowness surfaces is provided by Fig. 3. Fig. 3(a) shows the ST sheet of the slowness surface of the tetragonal crystal $Ag_2SO_4 \cdot 4NH_3$ (Every, 1988), and Fig. 3(b) shows the corresponding sheet for the trigonal crystal sapphire (Every et al., 1984). The pattern of short lines in Fig. 3(a) represents the polarization field for this transverse branch. A striking feature of the polarization field are the singularities it contains. These singularities are located at acoustic axes, i.e. points of degeneracy where the FT and ST sheets of the slowness surface meet. The general nature of these polarization field singularities has been the subject of a number of papers (see Alshits and Lothe, 1979; Alshits and Shuvalov, 1984; Alshits et al., 1985). In this and my second lecture I will be describing some of the observable consequences of the characteristic field patterns surrounding these singularities.

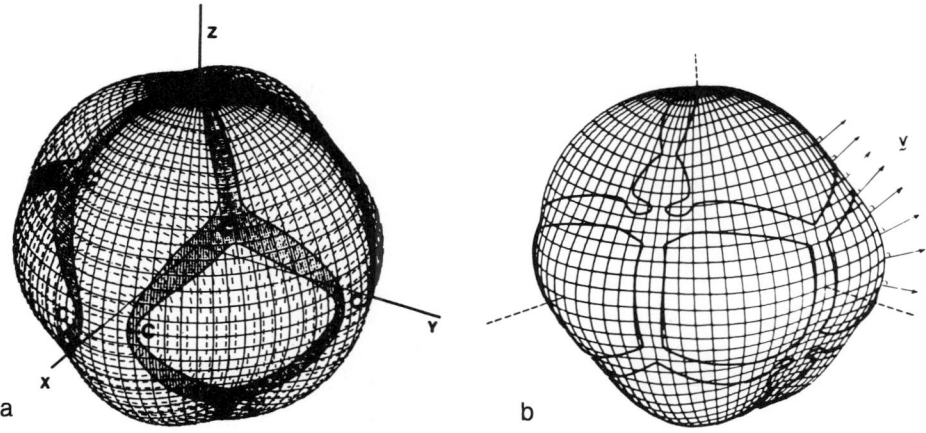

Figure 3. (a) ST sheet of the slowness surface of $Ag_2SO_4 \cdot 4NH_3$. (b) Corresponding surface for sapphire.

Two common forms of degeneracy are conical points such as those labelled C and C' in Fig. 3(a), and tangential degeneracies, of which there is one visible on the z-axis in Fig. 3(a). Conical points are structuraly stable to perturbations of the elastic constants. The polarization field in the vicinity of a conical point undergoes a rotation of either π (as is the case with points C) or $-\pi$ (as is the case with points C') in tracing out a circuit around that point, and is undefined precisely at the conical point. The ST and FT polarization fields both undergo the same $\pm\pi$ rotation, but the two fields are rotated by $\pi/2$ with respect to one another as required by orthogonality. On 3-fold axes of rotational symmetry there are always conical points of the $-\pi$ type.

Tangential degeneracies generically only occur at 4-fold and 6-fold symmetry axes. It is evident in Fig. 3(a) that the ST polarization field forms a circumferential pattern in the region of the tangential degeneracy on the z-axis, undergoing a rotation of 2π in a circuit around that point. The FT polarization field, being orthogonal to that of the ST branch, forms a radial pattern around the z-axis. This characteristic behaviour is fairly common in tetragonal and trigonal crystal and is a universial feature of hexagonal crystals. A -2π rotation of the polarization field at a 4-fold axis is also possible, but it is less common. A tangential degeneracy is structurally unstable to perturbations of the elastic constants that lower the symmetry. Depending on the change in symmetry, it may transform into 2 or 4 conical points preserving the net rotational character of the polarization field in that region. In fact the pairs of conical points C next to the X- and Y-axes would coalesce into a tangential degeneracy if the elastic constants were changed slightly so as to conform to cubic symmetry.

IV. ENERGY PROPAGATION AND FOCUSING

The velocity characterising acoustic energy transport or signal propagation is the group velocity (Auld, 1973; Fedorov,1968)

$$\mathbf{V} = \nabla_{\mathbf{k}}\, \omega(\mathbf{k}), \tag{8}$$

which, while differing from the phase velocity, is related to it by

$$\mathbf{V} \cdot \mathbf{n} = v. \tag{9}$$

There are various methods for calculating $\mathbf{V}$, all of which involve expressing it parametrically in terms of quantities that are derivable from Christoffel's equation. The method by which many of the results in this talk have been obtained, is to express v as an homogeneous function of $\mathbf{n}$ in Eq.(6) and obtain

$$\mathbf{V} = \frac{\partial v}{\partial \mathbf{n}}, \tag{10}$$

by implicit differentiation (Fedorov, 1968).

On elementary mathematical grounds it follows from Eq. (8) that $\mathbf{V}$ is normal to the constant frequency surface and therefore also to the slowness surface, from which it follows that

$$\mathbf{V} = \frac{\nabla_{\mathbf{s}}\Lambda(\mathbf{s})}{\mathbf{s}\cdot\nabla_{\mathbf{s}}\Lambda(\mathbf{s})}. \tag{11}$$

Fig. 3(b) shows a number of group velocity vectors in relation to the ST slowness sheet of sapphire. These might, for the sake of argument, represent the group velocity vectors of the phonons comprising a heat pulse. It is evident that where the curvature of the slowness surface is least, the group velocity vectors are most strongly concentrated in direction, and consequently the energy flux in that direction is greatest. This effect is known as phonon focusing, and is the most important source of non-uniformity of the intensity in phonon images. For a review of phonon focusing, see Maris (1986). A useful measure of focusing is given by the phonon enhancement factor (Maris, 1971)

$$A = \left| \frac{\delta\Omega_s}{\delta\Omega_V} \right|, \quad (12)$$

where $\delta\Omega_s$ is the solid angle subtended by an infinitesimal cone of slowness vectors and $\delta\Omega_V$ is the solid angle subtended by their associated group velocity vectors. From differential geometry it is readily established that (Northrop and Wolfe, 1980; Lax and Narayanamurti, 1980; Every, 1981)

$$A^{-1} = s^3 V |K|, \quad (13)$$

where $K = L_1 L_2$ is the Gaussian curvature of the slowness surface and L_1 and L_2 are the two principal curvatures, which can be positive or negative. K can take on any real value and so A, which is inversely proportional to $|K|$, ranges from zero to infinity.

Determining the net phonon intensity in a particular direction in physical space, even if it is a high symmetry direction, is not in general a trivial matter. The reason for this is that the mapping from the slowness surface to directions in real space is not one-to-one, and all points on the slowness surface which contribute to the flux in the particular direction have to be found numerically and A determined for each of them and summed. Since in phonon imaging one is interested in the phonon intensity not merely in a single direction but for a broad range of directions, the method described in the next section represents a more convenient approach.

A. Monte Carlo Calculated Phonon Intensity Diagrams

Monte Carlo methods provide the simplest and most efficient means of generating theoretical phonon intensity patterns of crystals. A common point of departure in these computer simulations, which is modified if there is need to take into account surface directivity or dispersion effects, is to assume that the phonons emanating from the heat source have a uniform distribution of wave normals. A random uniform distribution of say 10^6 wave normals are generated, the group velocities for each of them is computed, and the points where the rays meet the viewing surface are sorted into a 2D array of say 256×256 bins and the result represented as a gray scale image. Fig. 2(b) shows the theoretical focusing pattern of CaF_2 generated in this way. As can be seen, it is in very good agreement with the measured image of CaF_2 in Fig. 2(a).

Over the last decade or so phonon images have been reported for a large number of different crystals, including Ge (Northrop and Wolfe, 1979,1980; Metzger and Huebener, 1988), Si (Shields et al., 1989; Metzger et al., 1985), GaAs (Northrop et al., 1985; Held et al., 1989), LiF (Northrop et al., 1983), InSb (Hebboul and Wolfe, 1986), α−quartz (Eichele et al., 1982; Koos and Wolfe, 1984b), sapphire (Eichele et al., 1982; Every et al., 1984) and $LiNbO_3$ (Koos and Wolfe, 1984a,1984b), and by and large there is good agreement between the Monte Carlo simulations and experiment. Additional factors such as surface directivity, piezoelectric coupling and dispersion, are included in the calculations in some cases. On the basis of extensive numerical simulations, classification schemes have been devised for the focusing patterns of cubic crystals (Every,

1981; Every and Stoddart, 1985; Hurley and Wolfe, 1985) and tetragonal crystals (Every, 1988). With the aid of these, one can in principle infer the elastic constant ratios of a crystal from its phonon focusing pattern.

B. Caustics and Anticaustics

The most striking aspect of the phonon images in Fig. 2 is the complex pattern of caustics they contain. These are associated with lines of zero Gaussian curvature, or parabolic lines, on the slowness surface (Northrop and Wolfe, 1980; Hensel and Dynes, 1979). The bold lines in Figs. 3(a) and 3(b) are parabolic lines separating regions of positive and negative curvature on the two slowness surfaces. The vanishing of K results in A being infinite. The occurrence of parabolic lines and their associated caustics is a common but not universal feature of crystals. A sufficient, but not necessary, condition for their existence is the presence of one or more conical points. The reason for this is that in the region of a conical point at least one of the two principal curvatures of the outer of the two slowness sheets in contact must be negative, as is evident in Fig. 3(a). Since conical points occur at all 3-fold axes, all cubic and trigonal crystals display caustics. Hexagonal crystals do not possess conical points, but nevertheless display caustics if their anisotropy is large enough. Because the equation of the slowness surface is of degree 6, the inner longitudinal sheet must be entirely convex and thus cannot give rise to caustics. This constraint falls away in piezoelectric solids.

Elementary catastrophe theory provides a framework within which the topology of caustic patterns can be interpreted (Taborek and Goodstein, 1980). One is dealing here with a gradient mapping (Eq. (11)) from a two dimensional manifold of state variables (the slowness surface) to a two dimensional manifold of control parameters (the viewing surface in a phonon image or the unit sphere of directions). Caustics are singular points of the mapping, i.e. points where the Jacobian of the mapping is zero. Generically the only two catastrophes that can occur (Berry, 1976) are the fold catastrophe (the line caustics) and the cusp catastrophe (a number of cusps are evident in Fig. 2). However, one can take a broader point of view and regard the elastic constants as additional control parameters - there are after all many crystals to choose from, and their elastic constants can be varied to some extent by changing the temperature and pressure etc., and so an extensive domain in elastic constant space can in principle be explored. Also, on venturing into the dispersive regime, the frequency becomes yet another control parameter (Armbruster and Dangelmayr, 1983). In this expanded control parameter space there are many higher dimensional catastrophes that occur. From this point of view, a given phonon image can be regarded as occupying a two dimensional section of a higher dimensional control parameter space. Thus one can understand why it is that in phonon images slightly unfolded sections of catastrophes such as the hyperbolic umbilic and butterfly are often recognisable (Every, 1981; Chernosatonskii and Novikov, 1986).

The symmetry of crystals is another contributing factor to the higher dimensionality of phonon focusing caustics. A notable case is that of hexagonal crystals which exhibit transverse isotropy in their acoustic properties. Fig. 4 shows meriadian sections of the slowness surfaces of representative hexagonal media (Every, 1986b). Point 2 lies on a parabolic line which forms a circle around the z-axis. All rays associated with this circular parabolic line point along the z-axis and thus the entire caustic consists of a single point. This effect is associated with the phenomenon of external conical refraction. Lowering the symmetry by means of a change in the elastic constants, however small, causes the unfolding of this point caustic into a pattern of connected line and cusp caustics.

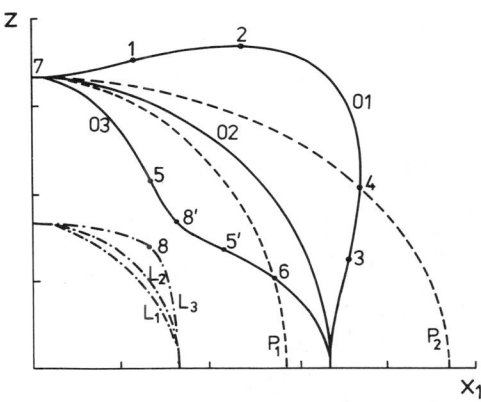

Figure 4. Meridian sections of the slowness surfaces of three representative transversely isotropic solids. L: longitudinal, P: pure transverse, Q: quasi-transverse.

The curvature of the slowness surface on approach to a conical point in general diverges and thus A vanishes. This goes hand in hand with the fact that the rays associated with an infinitesimal region around the conical point are spread out into a circular or elliptical cone, which is known as internal conical refraction. The ellipse or circle of zero phonon intensity on the viewing surface can be regarded as an anticaustic. For certain classes of fairly highly anisotropic crystals a parabolic line will weave back and forth several times through a conical point between the ST and FT sheets of the slowness surface. In physical space this yields a set of caustics which meet up tangentially with the circular anticaustic, fading in intensity to zero as they do so (Every, 1981). This topological feature is present in the images in Fig. 2. Even more complex phenomena occur if one considers also the variation of the elastic constants (Every, 1986b).

When a 4-fold axis of symmetry is removed by a perturbation to the elastic constants, the tangential degeneracy existing there transforms into a pair of neighbouring conical points and this is accompanied by the formation of caustics. Lowering a 6-fold axis to 3-fold on the other hand spawns a set of four conical points and accompanying caustics.

In hexagonal crystals, for certain ranges of elastic constants, the pure-T and quasi-T sheets of the slowness surface intersect along circles around the c-axis (see Fig. 4, points 4 and 6). When the symmetry is lowered these lines of wedge-shaped degeneracy are removed, leaving a discrete sets of conical points This is accompanied by the appeerence of a set of connecting, approximately circular caustics. This global formation of caustics is not readily accommodated within the framework of elementary catastrophe theory, which is concerned with the local topology of caustics rather than their global topology. Modified versions of these globally evolved caustics are recognisable in the focusing patterns of many crystals (Every, 1986b).

C. Phonon Focusing in Piezoelectric Solids

In piezoelectric media the effect of the coupling between stress and electric fields on acoustic wave propagation is taken into account by replacing the elastic constants by a set of piezoelectrically stiffened elastic moduli, which are themselves functions of the wave normals (Auld, 1973)

$$C_{r\ell sm} = C_{r\ell sm}^{\mathbf{E}} + \frac{e_{r\ell i}n_i e_{smj}n_j}{\epsilon_{pq}^s n_p n_q}, \qquad (14)$$

where $C_{r\ell sm}^{\mathbf{E}}$ is the elastic modulus tensor at constant electric field, $e_{r\ell i}$ is the piezoelectric stress tensor and ϵ_{pq}^s is the permittivity tensor at constant strain.

In strongly piezoelectric materials the electromechanical coupling can significantly alter the wavespeeds (by up to 28% in the case of LiNbO$_3$, for example) and have a dramatic effect on the phonon focusing pattern. This has been demonstrated experimentally by Koos and Wolfe (1984a, 1984b) for LiNbO$_3$ and Every and McCurdy (1987) have computed the effects of piezoelectric stiffening on the focusing patterns of a wide range of different crystals. One of the consequences of piezoelectric stiffening is that the equation of the slowness surface is raised in degree from 6 to 8 (see e.g. Every and Neiman, 1992), thereby permitting negative curvature in the innermost sheet of the slowness surface, and hence longitudinal mode caustics. Two strongly piezoelectric crystals which possess this property are Rochelle salt and Ba$_2$NaNb$_5$O$_{15}$ (Every and McCurdy, 1987). Piezoelectric stiffening also has the effect, in certain crystal classes of lowering the acoustic symmetry (Every, 1987a).

V. THE RAY SURFACE

Temporal information in phonon imaging, such as the number of distinct pulses travelling in each direction and their velocities of propagation, is contained in the acoustic ray surface, which is defined as the locus of the extremities of all group velocity vectors. It has the most direct physical significance of the acoustic surfaces in that it depicts the location of highest energy concentration at unit time after release of this energy, e.g. as a heat pulse or acoustic emission event, at the origin. Equivalently, the ray surface can be defined as the envelope of all plane wave fronts propagated outwards from the origin for unit time at their phase velocities. For an impulsive or discontinuous force applied at the origin, the ray surface represents the location of singularities of various orders in the wavefield. For a periodic disturbance applied at the origin it is a surface of constant phase in the wave field.

The ray surface is an algebraically much more complicated surface than the slowness surface, being of degree as high as 150 (Fedorov, 1968). Consequently, the only feasible method for constructing this surface is as the locus of group velocity vectors derived parametrically from the slowness surface or otherwise. This algebraic complexity is reflected in the topology of the ray surface, which commonly displays a complex pattern of folds. Monte Carlo methods provide a simple way of depicting the ray surface, through the construction of plane sections and projected ray plots (phonon intensity diagrams). To construct a section, a large number of rays are generated and those lying within some predetermined small angle of the sectioning plane are represented graphically by dots. A (001)-section of the ray surface of TeO$_2$ generated in this way is shown in Fig. 5(a), and Fig. 5(b) shows a number of sections containing the [001]-axis and at various angles ϕ to the [100]-axis, while Fig. 5(c) shows a polar plot of the distribution of the rays (Every, 1992b). The cusps in the ray surface sections all clearly match up with the caustics in the ray distributions.

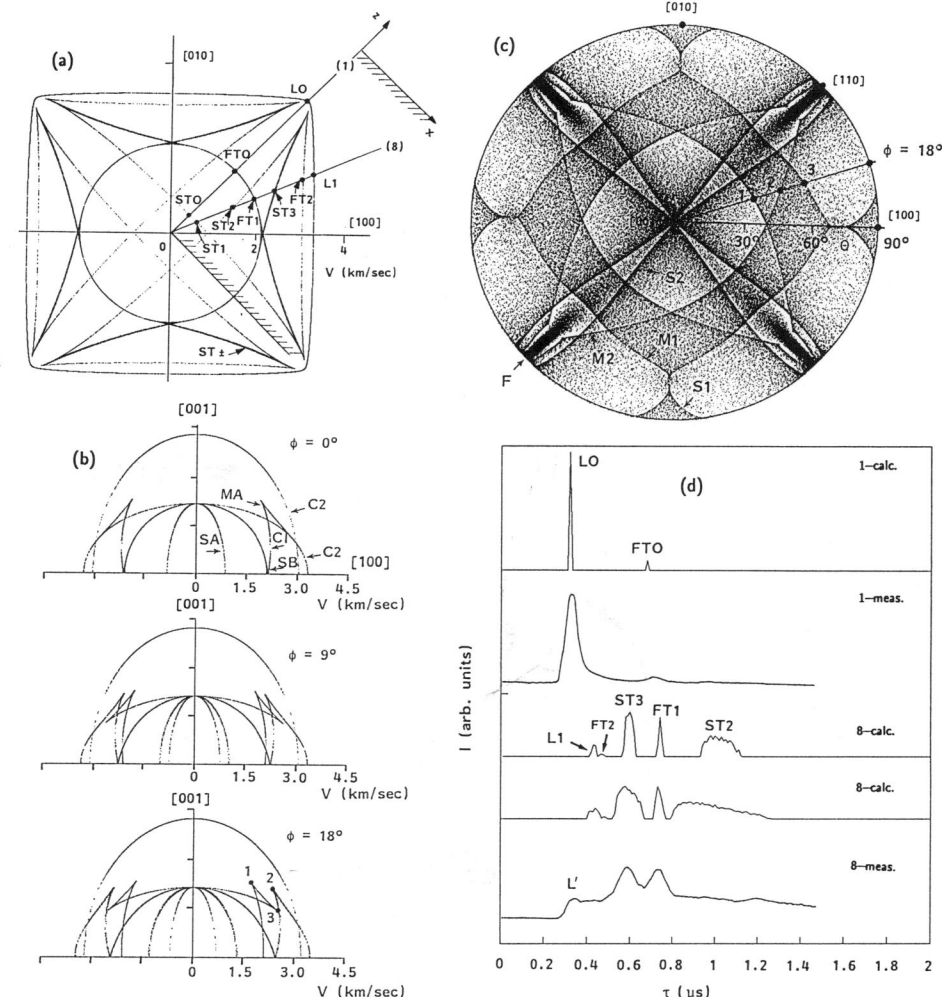

Figure 5 The ray surface of TeO$_2$. (a) (001)-section. (b) Various sections containing the [001]-axis. (c) Polar phonon intensity diagram. (d) Computed and measured heat pulse profiles in directions 1 and 8.

An artifact of the Monte Carlo method of generating ray surface sections, which turns out to be an advantage, is that the traces are not thin solid lines but slightly smeared out clusters of points, the spacing of which is a measure of the focusing, and the lateral spread of which is an indication of the velocity spread for rays located in a small cone around each direction. This velocity spread is an important factor determining the temporal widths of ballistic heat pulses, since in actual experiments the source and detector subtend finite solid angles at each other. Fig. 5(d) compares heat pulse profiles in TeO$_2$ measured by Hurley et al. (1986) with Monte Carlo simulations (Every, 1992b). There is good agreement as regards the time of arrival of the various pulses, their magnitudes and their widths, and these quantities are all readily understood on the basis of the wave section in Fig. 5(a). In a similar way, Danilchenko and Slutskii (1988) have accounted for the arrival times of ballistic heat pulses near the [001]-axis of GaAs on the basis of the complex folding of the ray surface there.

VI. SURFACES IN PHONON IMAGING

Physical surfaces and interfaces play an important role in phonon imaging, because this is where the phonons are usually generated and detected Also, reflection and/or transmission can be involved in some way in the formation of a phonon image. The emission and absorption processes to a large extent govern the distribution of momenta of the phonons that reach the detector, and much effort has been devoted to deciphering the form of this momentum selectivity from phonon images. The role that phonon imaging has played in the study of phonon emission and absorption by 2DEG's is taken up by other contributors to these lecture notes and will not be discussed here. Very often in phonon imaging the phonons are generated in a thin metal film overlayer on the surface of the sample. The $\mathbf{k}$−dependent transmittance across the interface can have a significant influence on the detected phonon intensity, and this is augmented by a similar effect at the interface between the crystal and detector. The nature of the

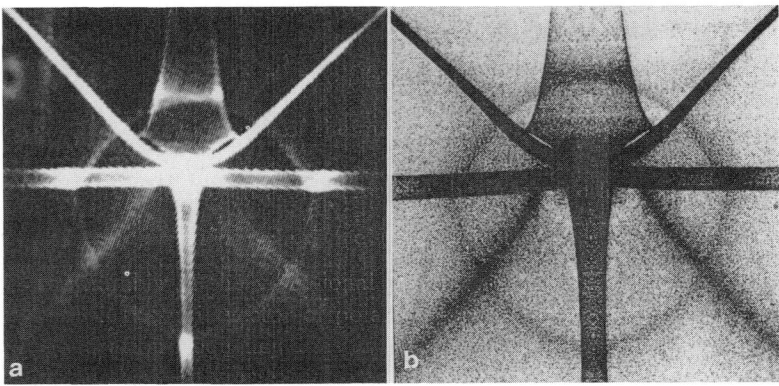

Figure 6. (a) Measured phonon image for $[1\bar{1}02]$-cut sapphire. Brightness is a measure of phonon intensity. (b) Monte Carlo simulation. Darkness is a measure of the phonon intensity.

transmittance depends on the quality of the bonding between the crystal and overlayer, and is altered if an intermediate layer or multilayer is introduced at this interface. A brief discussion of some of these effects follows.

A. Surface Wave Resonance Enhanced Transmission

Figure 6(a) shows a phonon image obtained by Every et al. (1984) with a highly polished $[1\bar{1}02]$−cut sapphire crystal. The approximately circular halo-like feature is a surface directivity effect and is located very close to the critical cone for mode conversion from ST to L waves at the sapphire surface. Its presence is an indication of the fact that there is rather poor adhesion of the metal heater film and/or the bolometer to the crystal surface. The quasi-free crystal surface supports a resonance L surface mode, akin to a pseudo-surface wave, which has a relatively small bulk ST wave component concentrated in the region of the critical cone. This surface mode provides a channel for the resonance excitation of these critical cone ST phonons.

There is no visible presence of an analogous feature at the FT to L critical cone. The reason for this is that the polarization vectors of the ST phonons featuring in Fig. 6(a) are directed approximately radially outwards from the centre of that image, while those of the FT modes are approximately circumferential, as if there were a 4-fold axis here. This is not fortuitous, but due to the fact that it requires only a small change in the elastic constants of sapphire to render the $[1\bar{1}02]-$ direction an axis of 4-fold acoustic symmetry. The ST modes, being sagittally polarized can undergo mode conversion to L at the surface and so participate in critical cone resonance transmisson, whereas the FT modes, being shear horizontally (SH) polarized are unable to undergo mode conversion and so do not participate in this resonance. Figure 6(b) shows a Monte Carlo image for sapphire calculated on the assumption that the metal overlayer acts as a weak damping mechanism on the otherwise free sapphire surface (Every, 1986a). The measured and computed phonon images in Fig. 6 are in good agreement.

Resonance transmission of ST phonons has also been observed in diamond (Hurley et al., 1984), where it is much narrower and more intense. This is because the effective Poisson's ratio of diamond is exceptionally small and as a consequence the surface skimming bulk L wave nearly satisfies the free surface boundary conditions on its own. The same consideration with regard to the polarization field applies as with sapphire, except that in this case there is a genuine 4-fold axis here, and there is no visible FT resonance. Pseudo-surface waves and L mode resonances are a common feature of many crystal cuts and some striking consequences have been predicted for their phonon intensity patterns (Every, 1986a).

B. Phonon Focusing in Transmission and Reflection

Phonon imaging has proved to be an effective method for studying phonon reflection and transmission at surfaces, distinguishing whether this takes place specularly or diffusely, and identifying mode conversion. Reflection phonon images have been measured in sapphire by Northrop and Wolfe (1984) and Wichard and Dietsche (1992) have observed analogous effects in silicon. Höss et al. (1990) have laid the foundation for transmission phonon imaging in experiments that reveal the the specular transmission of thermal phonons from a Ge crystal into an overlayer of MgO and accompanying critical cone cutoff effects.

In the case where transmission and reflection are specular and confined to a set of parallel surfaces in a thick multilayered sample, each ray path through the sample preserves the value of $s_\parallel$, whatever the mode conversion sequence. The theory of phonon imaging can be formulated in a convenient way to take advantage of this fact (Every, 1992a). Consider transmission through a single slab-shaped sample of unit thickness and with surfaces normal to the z-axis, from a small heated region located at the origin on the lower surface to a point $\mathbf{R}$ on the upper surface. The transmitted phonon intensity can be shown to be inversely proportional to the Jacobian

$$J = \frac{\partial(X, Y)}{\partial(s_x, s_y)} = \frac{\partial^2 s_z}{\partial s_x^2} \cdot \frac{\partial^2 s_z}{\partial s_y^2} - \left(\frac{\partial^2 s_z}{\partial s_x \partial s_y}\right)^2, \tag{15}$$

of the mapping from $\mathbf{s}_\parallel = (s_x, s_y)$ to $\mathbf{R}_\parallel = (X, Y)$, the in-plane component of $\mathbf{R}$, where $s_z = s_z(s_x, s_y)$ represents the explicit equation for the slowness surface.

Equation (15) is readily generalized to apply to transmission, via any mode sequence, through a multilayered sample by replacing s_z by

$$s_z^T(\mathbf{s}_\parallel) = \sum_i \alpha_i[\pm s_z^i(\mathbf{s}_\parallel)], \qquad (16)$$

where α_i are weighting factors proportional to the individual layer thicknesses and + and − apply to transmission in the forward and reverse directions respectively. A reflection phonon image is a special case of this. This result assumes that the individual layers are thick enough that there is no coherent interference between different overlapping rays and that evanescent waves play no role. The effective slowness surface, described by Eq. 16, has the α_i as additional control parameters, and through these there is greatly expanded scope for the higher order catastrophes. A limiting factor is that all the individual segments in a ray path must be real. There is a closed contour in the $\mathbf{s}_\parallel$ plane outside of which at least one of the s_z^i is complex and there is no transmission. This cutoff in the transmission occurring at the critical contour has been observed by Höss et al. (1990), who also describe the frustration of this effect when the layer thickness is comparable to the dominant phonon wavelength.

C. Transmission through Superlattices

When the layer thicknesses in a periodic multilayered sample are comparable to the dominant phonon wavelength, the partial waves in each layer superpose coherently, and the sample takes on the nature of a superlattice having a folded vibrational dispersion relation within a reduced Brillouin zone. Band gaps appear in the transmission spectrum both within and at the boundary of the Brillouin zone. Phonon imaging has proved to be a sensitive method of measuring the dependence of these stopbands on direction. Hurley et al. (1987) and Tamura et al. (1988) have investigated phonon transmission through a 40 period $In_{0.15}Ga_{0.85}As/AlAs$ superlattice of 20Å lattice spacing mounted on a $400\mu m$ thick GaAs wafer, using a PbBi tunnel junction detector tunable in the range 650-850 GHz. Their images show striking features due to these stopbands, and they are able to account well for their results with computer simulations. The subject of phonon transmission through periodic, quasiperiodic and random superlattices has been reviewed by Tamura (1990).

VII. OTHER APPLICATIONS OF PHONON IMAGING

In this lecture the emphasis has been on how the fundamentals of acoustic wave propagation in anisotropic solids impact on the calculation of phonon images. Attention has been confined to bulk focusing and surface effects. There are important aspects of phonon imaging that have barely been touched on, such as its use as a tool for studying phonon scattering and the propagation of large wavevector dispersive phonons, and for the imaging of defect structures. This section is an attempt, albeit inadequate, to redress the balance and draw attention to these major developments

A. Large Wavevector Dispersive Phonon Imaging

As a rule, when tunnel junction techniques are used in phonon imaging, phonons with wavevectors extending well into the Brillouin zone are detected, and dispersive effects are pronounced. With increasing $\mathbf{k}$ the deviation from linear behaviour in the $\omega(\mathbf{k})$ relation is to start with not quadratic but cubic in $\mathbf{k}$, and this delayed onset of dispersion accounts for the considerable success of continuum elasticity theory in accounting for bolometer-based phonon imaging (discrepancies due to dispersion, where they have been found, have been fairly small, see Held et al., 1989; Metzger and Huebener, 1988). An exception to this rule occurs at acoustic axes in certain crystals lacking a center of inversion, where there is a quadratic term in the dispersion relation, and this can cause significant localised distortion of the focusing pattern, even at relatively low frequencies (Every, 1987b).

In the dispersive regime successive constant frequency surfaces are no longer identical in shape and there is a different focusing pattern associated with each frequency. These different focusing patterns can be observed using a suitable set of tunnel junction detectors tuned to different frequencies. As was recognised from the first reports on dispersive phonon imaging (Dietsche et al., 1981; Northrop, 1982), this is a good testing ground for lattice dynamics models, since a phonon image provides information on the dispersion relation for a broad range of directions, which is otherwise difficult to obtain. In particular, phonon imaging is able to discriminate between lattice dynamics models which give comparable account of inelastic neutron scattering data (see e.g. Northrop et al., 1985; Hebboul and Wolfe, 1986, 1989). The considerable progress that has taken place in this field has been reviewed by Wolfe (1989) and Tamura (1986).

B. Bulk Scattering

The distinctive features in phonon images survive a moderate amount of bulk phonon scattering, and the changes they undergo in the process are a valuable source of information on the nature of the scattering. The first conclusive experimental evidence of thermal phonon scattering by the dislocation flutter mechanism was obtained with phonon imaging (Northrop et al., 1983). Phonon-phonon scattering shows up most clearly in phonon images at high excitation levels, which leads to the formation of a "hot spot", somewhat larger than the irradiated area, in which the high frequency phonons are down converted before escaping. This leads to a broadening of the caustics (Shields and Wolfe, 1989). Isotope scattering in Si has been studied by Shields et al. (1989, 1991) in experiments with a notched specimen. They observed intense channelling structures persisting beyond the notch, which they were able to account for on the basis of Monte Carlo simulations which incorporate the effects of isotope scattering. Similar effects have been observed in GaAs (Ramsbey et al., 1988), but in that case there is some uncertainty as to the role of defect scattering (Tamura, 1992; Held et al., 1989).

C. Imaging of Defect Structures

Huebener and coworkers in a number of papers (see e.g. Huebener, 1992, for a review) have adapted phonon imaging as a 3D tomographic technique for imaging defect structures in crystals. They accomplish this by using either a single movable bolometer or a pair of fixed bolometers attached to one side of the specimen to detect phonons which are generated by an electron beam which is scanned over the opposite surface (Doderer et al., 1988; Metzger et al., 1985). A comparison of the images obtained by detection at different positions yields information on structural inhomogeneities in the sample. They have used these techniques to image, for example, oxide precipitates in Si (Metzger et al., 1985) and laser drilled holes in sapphire (Doderer et al., 1988) with a spatial resolution of better than $10\mu m$.

VIII. REFERENCES

Alshits, V.I. and J. Lothe, 1979, Sov. Phys.-Crystallogr. **24**, 387; 393.
Alshits, V.I. and A.L. Shuvalov, 1984, Sov. Phys.-Crystallogr. **29**, 373.
Alshits, V.I., A.V. Sarychev and A.L. Shuvalov, 1985, Sov. Phys. JETP **62**, 531.
Armbruster, D. and G. Dangelmayr, 1983, Z. Phys. **B52**, 87.
Auld, B.A., 1973, Acoustic Fields and Waves in Solids (Wiley, New York).
Berry, M.V., 1976, Adv. Phys. **25**, 1.
Chernosatonskii, L.A. and V.V. Novikov, 1986, Phys. Lett. **117**, 349.
Danilchenko, B.A. and M.I. Slutskii, 1988, Sov. Phys.- Solid State **30**, 21.
Dietsche, W., G.A. Northrop and J.P. Wolfe, 1981, Phys. Rev. Lett. **47**, 660.
Doderer, T., E. Held, W. Klein and R.P. Huebener, 1988, Z. Phys. **B72**, 41.
Eichele, R., R.P. Huebener and H. Seifert, 1982, Z. Phys. **B48**, 89.
Eisenmenger, W., 1980, Phonon Detection by the Fountain Pressure in Superfluid Helium Films, in: Phonon Scattering in Condensed Matter, ed. H.J. Maris (Plenum, New York).
Every, A.G., 1980, Phys. Rev. **B22**, 1746.
Every, A.G., 1981, Phys. Rev. **B24**, 3456.
Every, A.G., G.L. Koos and J.P. Wolfe, 1984, Phys. Rev. **B29**, 2190.
Every, A.G. and A.J. Stoddart, 1985, Phys. Rev. **B32**, 1319.
Every, A.G., 1986a, Phys. Rev. **B33**,2719.
Every, A.G., 1986b, Phys. Rev. **B34**, 2852.
Every, A.G., 1987a, J.Phys.**C20**, 2973.
Every, A.G., 1987b, Phys. Rev. **B36**, 1448.
Every, A.G. and A.K. McCurdy, 1987, Phys. Rev. **B36**, 1432.
Every, A.G., 1988, Phys. Rev. **B37**, 9964.
Every, A.G. and V.I. Neiman, 1992, J. Appl. Phys. **71**, 6018.
Every, A.G., 1992a, Phys. Rev. **B45**, 5270.
Every, A.G., 1992b, Z. Phys. **B89**, 139.
Fedorov, F.I., 1968, Theory of Elastic Waves in Crystals (Plenum, New York).
Hebboul, S.E. and J.P. Wolfe, 1986, Phys. Rev. **B34**, 3948; Z. Phys. **B73**, 437.
Hebboul, S.E. and J.P. Wolfe, 1989, Z. Phys. **B74** 35.
Held, E., W. Klein and R.P. Huebener, 1989, Z. Phys. **B75**, 17.
Hensel, J.C. and R.C. Dynes, 1979, Phys. Rev. Lett. **43**, 1033.
Höss, C., J.P. Wolfe and H. Kinder, 1990, Phys. Rev. Lett. **64**, 1134.
Huebener, R.P, 1992, Phonon Imaging by Electron Beam Scanning, in: 7th Int. Conf. on Phonon Scattering in Condensed Matter.
Hurley, D.C., A.G. Every and J.P. Wolfe, 1984, J. Phys. **C17**, 3157.
Hurley, D.C. and J.P. Wolfe, 1985, Phys. Rev. **B32**, 2568.
Hurley, D.C., J.P. Wolfe and K.A. McCarthy, 1986, Phys. Rev. **B33**, 4189.
Hurley, D.C., S. Tamura, J.P. Wolfe and H. Morkoc, 1987, Phys. Rev. Lett. **58**, 2446.
Jasiukiewicz, C., D. Lehmann and T. Paskiewicz, 1991, Z. Phys. **B84**, 73; 1992, **B86**, 225.
Karl, H., W. Dietsche, A. Fisher and K. Ploog, 1988, Phys. Rev. Lett. **61**, 2360.
Kent, A.J., G.A. Hardy, P. Hawker and D.C. Hurley, 1990, A Spatially Resolving Heat Pulse Phonon Detector, in: Phonons89, ed. S. Hunklinger, W. Ludwig and G. Weiss (World Scientific, Singapore).
Kent, A.J., 1992, Phonon Imaging in Two-Dimensional Electron Systems, 7th Int. Conf. on Phonon Scattering in Condensed Matter.

Koos, G.L. and J.P. Wolfe, 1984a, Phys. Rev. **B29**, 6015.
Koos, G.L. and J.P. Wolfe, 1984b, Phys. Rev. **B30**, 3470.
Lax, M. and V. Narayanamurti, 1980, Phys. Rev. **B22**, 4876.
Maris, H.J., 1971, J. Acoust. Soc. Am. **50**, 812.
Maris, H.J., 1983, Phys. Rev. **B28**, 7033.
Maris, H.J., 1986, Phonon Focusing, in: Nonequilibrium Phonons in Nonmetallic Crystals, ed. W. Eisenmenger and A.A. Kaplyanskii (North Holland, Amsterdam).
Marx, D., J. Buck, K. Lassmann and W. Eisenmenger, 1978, J. de Phys. **C6**, Supp. to No. **8**, 1015.
McCurdy, A.K., H.J. Maris and C. Elbaum, 1970, Phys. Rev. **B2**, 4077.
Metzger, W., R.P. Huebener, R.J. Haug and H.U. Habermeier, 1985, App. Phys. Lett. **47**, 1051.
Metzger, W. and R.P. Huebener, 1988, Z. Phys. **B73**, 33.
Musgrave, M.J.P., 1970, Crystal Acoustics (Holden-Day, San Francisco).
Northrop, G.A. and J.P. Wolfe, 1979, Phys. Rev. Lett. **43**, 1424.
Northrop, G.A. and J.P. Wolfe, 1980, Phys. rev. **B22**, 6196.
Northrop, G.A., 1982, Phys. Rev. **B26**, 903.
Northrop, G.A., E.J. Cotts, A.C. Anderson and J.P. Wolfe, 1983, Phys. Rev. **B27**, 6395.
Northrop, G.A. and J.P. Wolfe, 1984, Phys. Rev. Lett. **52**, 2156.
Northrop, G.A. and J.P. Wolfe, 1985, Phonon Imaging: Theory and Applications, in: Nonequilibrium Phonon Dynamics, ed. W.E. Bron (Plenum, New York).
Northrop, G.A., S.E. Hebboul and J.P. Wolfe, 1985, Phys. Rev. Lett. **55**, 95.
Philip, J and K. Viswanathan, 1978, Phys. Rev. **B17**, 4969.
Ramsbey, M.J., J.P. Wolfe and S. Tamura, Z. Phys. **B73**, 167.
Rösch, F. and O. Weis, 1976, Z. Phys. **B25**, 101; **B25**, 115.
Schreyer, H., W. Dietsche and H. Kinder, 1984, Laser Induced Nonequilibrium Superconductivity - a Spatially Resolving Phonon Detector, in: LT17, ed. U. Eckern, A. Schmid, W. Weber and H. Wuhl (Elsevier, Amsterdam).
Shields, J.A., J.P. Wolfe and S. Tamura, 1989, Z. Phys. **B76**, 295.
Shields, J.A., S. Tamura and J.P. Wolfe, 1991, Phys. Rev. **B43**, 4966.
Shields, J.A. and J.P Wolfe, 1989, Z. Phys. **B75**, 11.
Taborek, P. and D. Goodstein, 1979, J. Phys. **C12**, 4737.
Taborek, P. and D. Goodstein, 1980, Solid State Commun. **33**, 1191.
Tamura, S., 1986, Propagation of Large-Wave-Vector Phonons, in: Phonon Scattering in Condensed Matter V, ed. A.C. Anderson and J.P. Wolfe (Springer, Berlin).
Tamura, S., D.C. Hurley and J.P. Wolfe, 1988, Phys. Rev. **B38**, 1427.
Tamura, S., 1990, Phonon Transmission Through Periodic, Quasiperiodic and Random Superlattices, in: Phonons89, ed. S. Hunklinger, W. Ludwig and G. Weiss (World Scientific, Singapore).
Tamura, S., 1992, Measurement of Phonon Elastic Scattering Rates by Phonon Imaging and Monte Carlo Simulations, in: 7th Int. Conf. on Phonon Scattering in Condensed Matter.
Wichard, R, and W. Dietsche, 1992, A New Phonon Focusing Phenomenon due to Elastic Mode coversion on Silicon Surfaces, in: Acoustical Imaging 19, ed. H. Ermert and H Harjes (Plenum, New York).
Wolfe, J.P., 1989, Adv. Solid State Phys. **29**, 75.
Wolfe, J.P., 1993, Heat Pulses and Phonon Imaging (Springer, to be published).

PHONON IMAGING AT ULTRASONIC FREQUENCIES: THE DYNAMIC RESPONSE ON ANISOTROPIC SOLIDS

A.G. Every, K.Y. Kim*, and W. Sachse*

Physics Department, University of the Witwatersrand
P.O. Wits 2050, South Africa
*Department of Theoretical and Applied Mechanics
Cornell University, Ithaca, N.Y. 14853, USA

I. INTRODUCTION

Recently there have been a number of ultrasonics experiments performed that elicit detailed information on the elastodynamic response of anisotropic solids to transient point-like forces. In this lecture I will be describing methods for computing the dynamic Green's functions of anisotropic solids, and showing how these response functions are able to account for the observations. In these experiments small-aperture transducers or acoustic lenses are used to achieve wide-angle radiation and detection sensitivity as well as high spatial resolution. Many salient aspects of thermal phonon imaging, such as phonon focusing, polarization selectivity and mode conversion at surfaces also feature in these ultrasonic experiments, but there are neverthless significant differences. Unlike thermal phonons, the ultrasonic waves are coherent, and because their characteristic wavelengths are several orders of magnitude larger than those of phonons, the far field condition is not as well satisfied, and diffraction effects feature prominently. Various modes of excitation and detection exist, both broadband and monochromatic, some of which tend to disguise these diffraction effects, while others reveal them clearly.

The extension of phonon imaging to the ultrasonic domain has practical advantages which derive from the greatly reduced attenuation rate at the lower frequencies. As a consequence, measurements can be performed at room temperature and on much larger specimens, which need not be single crystals. Indeed, much of the impetus for developing ultrasonic phonon imaging arises from its potential value in the aerospace industry and elsewhere for materials characterization and the non-destructive testing of components made from fibre composites, polycrystalline textured metals and other elastically anisotropic solids. There are also a variety of biological materials such as wood and bone which could usefully be studied with these techniques.

There is an extensive body of published data and analytical results on the dynamical response functions of *isotropic* solids and efficient computer codes exist for dealing with a variety of problems (see e.g. Aki and Richards, 1980; Ceranoglu and Pao, 1981). Much less information is, however, available on the dynamical Green's functions of *anisotropic* solids. Formal solutions in terms of integral transforms have been available for some time (see e.g Payton, 1983), but it is only for a few special cases that closed-form solutions are known, and it is only recently, outside the domain of seismology, that extensive comparisons between numerical results and experimental data have begun to appear in the literature.

The first part of this lecture will be devoted to a brief derivation of integral expressions for the frequency and time domain Green's functions of anisotropic solids and the presentation of illustrative numerical results. In the latter part of the lecture I will describe a number of ultrasonic experiments that have recently been performed or are underway at present, and show how they can be interpreted in terms of the dynamical Green's functions. Every et al. (1990, 1991) using laser generation and piezoelectric detection of ultrasound in a Si single crystal have been able to identify various single- and multi-pass wave arrivals and have demonstrated phonon focusing at ultrasonic frequencies. Hauser et al. (1992) and Weaver et al. (1992) using transmission acoustic microscopy have observed phonon focusing diffraction patterns in a number of anisotropic solids. Wesner et al. (1992) have followed a similar approach, except that they retain phase information and are able to generate holograms. The existence of these diffraction effects had earlier been predicted for thermal phonons by Maris (1983) and are in fact a general characteristic of the unfolding of caustics at finite wavelength (Berry, 1976). The feasibility of doing phonon imaging with ultrasonic waves had been suggested by Novikov and Chernozatonskii (1988). Finally, I will be describing some experiments we are carrying out (Kim et al. 1993a, 1993b, 1993c), which involve on the one hand the use of small-aperture shear transducers to observe the focusing of FT waves in Si, and on the other capillary fracture to simulate a suddenly applied force. Mention should be made of the observation of the focusing of surface acoustic waves in silicon by Kolomenskii and Maznev (1991).

II. DYNAMIC GREEN'S FUNCTIONS

In this section we derive integral expressions for the frequency and time domain Green's functions of infinite anisotropic elastic continua. We set out by considering the Green's function $G_{sp}(\mathbf{x}, t)$ given by

$$\left(\rho \delta_{rs} \frac{\partial^2}{\partial t^2} - C_{r\ell sm} \frac{\partial^2}{\partial x_\ell \partial x_m}\right) G_{sp}(\mathbf{x}, t) = \delta_{rp} \delta(\mathbf{x}) F(t). \tag{1}$$

Physically $G_{sp}(\mathbf{x}, t)$ represents the s'th component of the displacement at point $\mathbf{x}$ and time t in response to a force in the p'th direction with time dependence $F(t)$, applied at the origin. The array of $G_{sp}(\mathbf{x}, t)$ form a tensor of second rank. In what follows $F(t)$ will either be sinusoidal or a step function. The formal solution to Eq. (1) in terms of integral transforms has been considered by a number of authors including Buchwald (1959), Duff (1960), Cameron and Eason (1967), Burridge (1967), Yeatts (1984), Tverdokhlebov and Rose (1988) and Tewary and Fortunko (1992), and the methods have been reviewed by Payton (1983), Cottam and Maradudin (1984) and van der Hijden (1987). Carrying out a quadruple space-time Fourier transform on Eq. (1), one obtains

$$L_{rs}(\mathbf{k}, \omega) g_{sp}(\mathbf{k}, \omega) = \frac{1}{(2\pi)^3} \delta_{rp} f(\omega), \tag{2}$$

where
$$f(\omega) = \frac{1}{2\pi} \int F(t) e^{i\omega t} dt, \tag{3}$$
$$L_{rs}(\mathbf{k}, \omega) = C_{r\ell sm} k_\ell k_m - \rho \omega^2 \delta_{rs}, \tag{4}$$

and
$$g_{sp}(\mathbf{k}, \omega) = \frac{1}{(2\pi)^4} \int d^3x dt G_{sp}(\mathbf{x}, t) e^{-i(\mathbf{k}\cdot\mathbf{x}-\omega t)}. \tag{5}$$

From Eq. (2) it follows that
$$g_{sp}(\mathbf{k}, \omega) = \frac{1}{(2\pi)^3} \left(L^{-1}(\mathbf{k}, \omega)\right)_{sp} f(\omega) = \tilde{g}_{sp}(\mathbf{k}, \omega) f(\omega), \tag{6}$$

which defines the Fourier domain Green's function $\tilde{g}_{sp}(\mathbf{k}, \omega)$, whose inverse Fourier transform with respect to $\mathbf{k}$ is

$$\tilde{G}_{sp}(\mathbf{x}, \omega) = \frac{1}{(2\pi)^3} \int d^3\mathbf{k} \left(L^{-1}(\mathbf{k}, \omega)\right)_{sp} e^{i\mathbf{k}\cdot\mathbf{x}}. \tag{7}$$

Combining Eqs. (6) and (7) and inverse Fourier transforming with respect to t we obtain
$$G_{sp}(\mathbf{x}, t) = \int \tilde{G}_{sp}(\mathbf{x}, \omega) f(\omega) e^{-i\omega t} d\omega. \tag{8}$$

Using the spectral resolution theorem (Yeatts, 1984), we can write

$$\left(L^{-1}\right)_{sp} = \sum_n \frac{\Lambda_{sp}^{(n)}}{\rho v^{(n)2} k^2 - \rho \omega^2} = \sum_n \frac{s^{(n)2} \Lambda_{sp}^{(n)}}{\rho(k^2 - \omega^2 s^{(n)2})}, \tag{9}$$

where the sum is taken over the three eigenvalues $\rho v^{(n)2} k^2$ of the tensor $\Gamma_{rs} = C_{r\ell sm} k_\ell k_m$, $v^{(n)}$ is the phase velocity and $s^{(n)} = 1/v^{(n)}$ the slowness, $\Lambda_{sp}^{(n)} = U_s^{(n)} U_p^{(n)}$, and $\mathbf{U}^{(n)}$ is the eigenvector associated with $\rho v^{(n)2} k^2$.

In performing the integration in Eq. (7) it is convenient to orient the k_3-axis in the direction of $\mathbf{x}$ and transform to polar coordinates so that $\mathbf{k} = k(\sin\theta\cos\phi, \sin\theta\sin\phi, \cos\theta)$ and $d^3\mathbf{k} \to d\Omega k^2 dk$, where $d\Omega = d(\cos\theta) d\phi$ is the element of solid angle in k-space. Integration with respect to k is facilitated by taking the limits to be $-\infty$ to ∞ rather than 0 to ∞, and to compensate, the angular integral is taken only over the forward hemisphere such that $\cos\theta \geq 0$, rather than over the entire sphere. Thus one obtains

$$\tilde{G}_{sp}(\mathbf{x}, \omega) = \frac{1}{8\pi^3 \rho} \sum_n \int_\cap d\Omega s^{(n)2} \Lambda_{sp}^{(n)} \int_{-\infty}^{\infty} \frac{e^{ikx\cos\theta} k^2 dk}{(k^2 - \omega^2 s^{(n)2})}, \tag{10}$$

where $x = |\mathbf{x}|$. The poles that are encountered in the k-integral are handled by ascribing a small imaginary part $i\epsilon$ to ω, corresponding to the slow switching on of the force. The path of integration is then completed in the upper or lower half complex plane depending on the sign of $x\cos\theta$. This leads through the identity

$$\int_{-\infty}^{\infty} \frac{e^{iku} k^2 dk}{(k^2 - a^2)} = \pi i a e^{ia|u|} + 2\pi\delta(u), \tag{11}$$

to the following result

$$\tilde{G}_{sp}(\mathbf{x}, \omega) = \sum_n \left\{ \frac{i\omega}{8\pi^2 \rho} \int_\cap d\Omega s^{(n)3} \Lambda_{sp}^{(n)} e^{i\omega s^{(n)} x\cos\theta} + \frac{1}{8\pi^2 \rho x} \int_0^{2\pi} d\phi s^{(n)2} \Lambda_{sp}^{(n)} \right\}. \tag{12}$$

The second integral is taken over the circle for which $\cos\theta = 0$.

For isotropic solids the angular integrals can be performed analytically, leading to Eq. (4.35) of Aki and Richards (1980). In the case of anisotropic solids, except for certain special cases where analytic results are known, the angular integrals have to be performed numerically. The first term in Eq. (12), because it involves 2D integration, is the most CPU intensive, and it is the rapid variation in the phase factor that tends to be the crucial consideration. One chooses a rectangular grid of points in the variables ϕ and $\cos\theta$ to sum over, with the spacing small enough that the variation of $\omega s x \cos\theta$ between neighbouring grid points is much smaller than 2π.

At relatively low frequencies such that $\omega s x < 10\pi$ say, the numerical summation requires a modest amount of CPU time. At higher frequencies, where the grid size needs to be small, the numerical integration is made manageable by invoking the stationary phase approximation, and limiting the summation to small regions around directions where the phase $\omega s x \cos\theta = \omega \mathbf{s}.\mathbf{x}$ is stationary. These correspond to points on the slowness surface where the outgoing normal and hence group velocity $\mathbf{V}$ are parallel to $\mathbf{x}$. At very high frequencies these regions shrink to a very small size, within which the factor $s^3 \Lambda$ can be taken to be a constant, and the equation for the slowness surface approximated by $s_3 = L_1 s_1^2 + L_2 s_2^2$ in a locally oriented coordinate system, where L_1 and L_2 are the local principal curvatures. The integral can now be performed analytically yielding a contribution to $\widetilde{G}_{sp}(\mathbf{x},\omega)$ of the form

$$\widetilde{G} \approx \frac{\Lambda e^{i\omega \mathbf{s}.\mathbf{x}}}{\sqrt{|L_1 L_2|}}, \tag{13}$$

for each such point. The intensity or energy flux associated with each contribution is proportional to $|\widetilde{G}|^2$, and is thus inversely proportional to $|K| = |L_1 L_2|$, the magnitude of the Gaussian curvature of the slowness surface, consistent with the ray approximation.

A. Numerical Results

Figure 1(a) shows the focusing pattern for the ST and FT modes of (001)-oriented Si, the angular range being $\pm 27°$ in the [100] and [010]-directions. The ST and FT modes are degenerate in the [001]-direction due to the 4-fold rotational symmetry, and their polarization patterns are shown in Fig. 1(b). Figure 1(c) shows the spatial variation of the Green's function $\widetilde{G}_{33}(\mathbf{x},\omega)$ in the $x_3 = 10$mm plane for frequency $f = \omega/2\pi = 10$ MHz, with darkness representing the magnitude of $\widetilde{G}$. Starting at a very high frequency, as the frequency is lowered each line caustic unfolds into an Airy diffraction pattern. The fringes broaden and begin to merge becoming fewer in number as the frequency is lowered further. By 10 MHz, although the broad overall features of the original focusing pattern, namely the intense ST square in the centre and the diagonal ridges have survived, the fine structure in the focusing pattern has been obliterated.

In the evaluation of $\widetilde{G}_{33}(\mathbf{x},\omega)$, Λ_{33} for the FT modes is very close to zero throughout, because the polarization vectors for these modes deviate very little from the (001)-plane, and so the FT modes make a negligible contribution to $\widetilde{G}_{33}(\mathbf{x},\omega)$ and there is no discernible relic of the narrow FT focusing ridges. This holds true for any axisymmetric excitation which preserves the symetry about the [001]-axis. A contribution from the L modes has not been included in the calculation of Fig. 1(c) because in experiments that relate to this Green's function and that will be described later, tone bursts are used, and by time gating the detected signal the longitudinal and transverse mode contributions can be isolated.

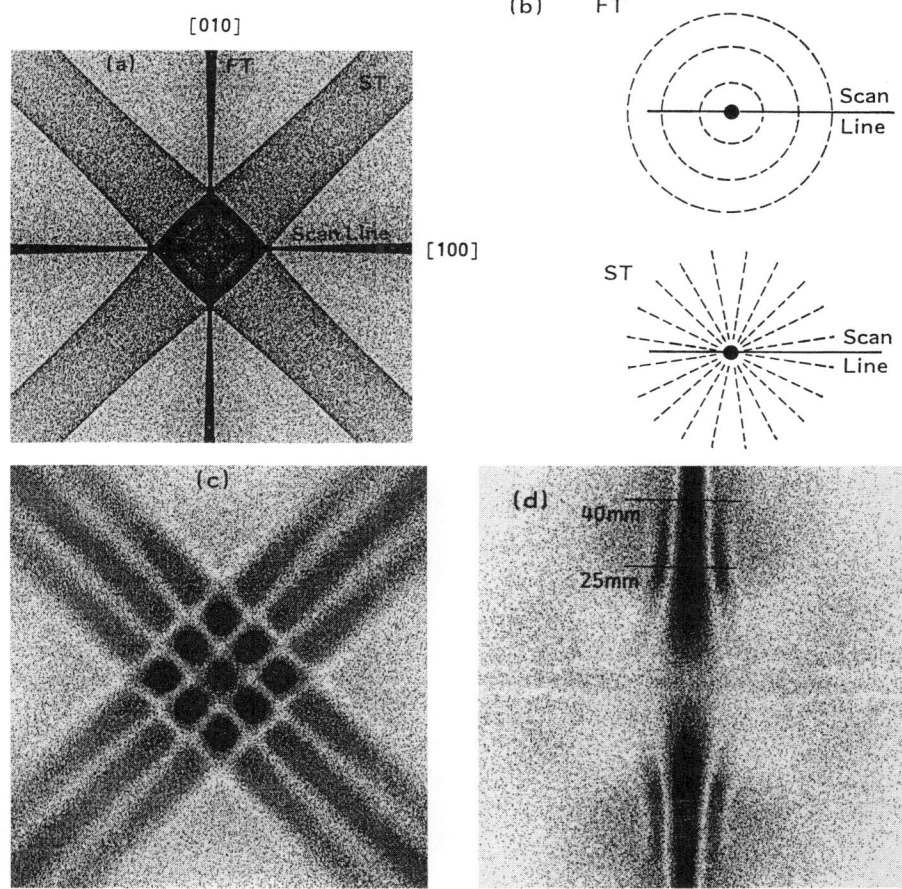

Figure 1. (a) Focusing pattern of Si. (b) Polarisation patterns of the ST and FT modes. (c) Spatial variation of $|\tilde{G}_{33}(\mathbf{x},\omega)|$. (d) Spatial variation of $|\tilde{G}_{11}(\mathbf{x},\omega)|$.

When the directions of force and detection are normal to the [001]-direction, they do couple onto the FT modes. Fig. 1(d) shows the spatial variation of $|\tilde{G}_{11}(\mathbf{x},\omega)|$ in the $x_3 = 49.15$ mm plane (for comparison with experiments to be described later) and frequency $f = 2$MHz. Only the contribution of the FT modes has been included in the calculation. Due to the fact that there is now a preferred axis in the (001)-plane, the 4-fold symmetry about the [001]-axis has been lost. The narrow FT focusing ridges in the [010]-direction have been diffraction broadened and the FT focusing ridges in the [100]-direction have been suppressed.

B. Response to a Suddenly Applied Force

In some experiments concerned with the dynamic response of solids, the force has a step function time dependence, i.e.

$$F(t) = \Theta(t) = \begin{cases} 0, t < 0 \\ 1, t > 0 \end{cases} \tag{14}$$

From a computational point of view the evaluation of the Green's function $G_{sp}(\mathbf{x},t)$ for a step function force is simpler than for other time dependences, and the response to other types of time dependent forces can in any case be derived from $G_{sp}(\mathbf{x},t)$. For instance, in the case of an impulse, i.e. a force with a δ-function time dependence, the response is the time derivative of $G_{sp}(\mathbf{x},t)$.

The Fourier transform of $F(t)$ is $f(\omega) = -1/2\pi i\omega$, and so from Eq. (8) it follows that

$$G_{sp}(\mathbf{x},t) = \frac{1}{2\pi}\int_{-\infty}^{\infty}\frac{\widetilde{G}_{sp}(\mathbf{x},\omega)}{-i\omega}e^{-i\omega t}d\omega. \tag{15}$$

On substituting for $\widetilde{G}_{sp}(\mathbf{x},\omega)$ from Eq. (12) and using the fact that $\frac{1}{2\pi}\int e^{i\omega(\mathbf{s}\cdot\mathbf{x}-t)}d\omega = \delta(t-\mathbf{s}\cdot\mathbf{x})$ and $\frac{1}{2\pi}\int \frac{1}{-i\omega}e^{-i\omega t}d\omega = \Theta(t)$ we obtain

$$G_{sp}(\mathbf{x},t) = \sum_{n}\left\{\frac{-1}{8\pi^2\rho}\int_{\Omega}d\Omega s^{(n)3}\Lambda_{sp}^{(n)}\delta(t-\mathbf{s}\cdot\mathbf{x}) + \frac{\Theta(t)}{8\pi^2\rho x}\int_{0}^{2\pi}d\phi s^{(n)2}\Lambda_{sp}^{(n)}\right\}. \tag{16}$$

To evaluate $G_{sp}(\mathbf{x},t)$ at a particular time, the first term can be reduced to a one dimensional integral summed over directions for which $t - \mathbf{s}\cdot\mathbf{x} = 0$, and which have to be located numerically. However, it is more likely that the entire time dependence of $G_{sp}(\mathbf{x},t)$ is required, in which case it is simpler, and not necessarily more demanding on computer time, to perform the 2D integration as a sorting and counting process. The time interval between $t = 0$ and the arrival of the last wave, i.e. the largest value of $\mathbf{s}\cdot\mathbf{x}$, is divided into N slots, which in our calculations we have taken to be 240. A rectangular grid of $\cos\theta$, ϕ points is chosen, and for each one the value of $\mathbf{s}\cdot\mathbf{x}$ determines the slot in which the corresponding value of $s^3\Lambda_{sp}$ is accumulated. For times exceeding the maximum of $\mathbf{s}\cdot\mathbf{x}$, $G_{sp}(\mathbf{x},t)$ has the constant value given by the second term in Eq. (16).

The procedure described above is akin to determining the frequency distribution function of a 2D lattice, and similar van Hove-type singularities are encountered here. Singularities in the time dependence of $G_{sp}(\mathbf{x},t)$ are associated with points on the slowness surface where $\mathbf{s}\cdot\mathbf{x}$ is stationary, i.e. where the group velocity points in the direction of $\mathbf{x}$. These points in time are called wave arrivals. The behaviour in the immediate vicinity of such points can be determined, as with the stationary phase approximation, by approximating the equation of the slowness surface as a paraboloid and taking $s^3\Lambda_{sp}$ to be constant. The first term in Eq. (16) can then be evaluated analytically (Payton, 1984), and what emerges is that for convex and concave regions of the slowness surface (L_1, L_2 both negative or both positive) $G_{sp}(\mathbf{x},t)$ displays a discontinuity of magnitude proportional to $\Lambda/\sqrt{|K|}$, together with a change in slope, while for saddle shaped regions of the slowness surface (L_1, L_2 opposite in sign) $G_{sp}(\mathbf{x},t)$ displays a logarithmic divergence. In symmetry directions it can happen that a particular Λ_{sp} is zero. In this case that Λ_{sp} is expanded about the point before integrating. This leads to a lower order singularity, which often takes the form of a change in slope.

C. Numerical Results

Figure 2(a) shows $G_{33}(\mathbf{x},t)$ as a function of t for $\mathbf{x} = (0,0,h)$, $h = 10$ cm, in polycrystalline aluminium. This type of waveform is typical for isotropic media. The displacement is zero until the first L wave arrival at $t = h/v_L$, at which point there is a discontinuity followed by a parabolic rise. The T wave arrival at $t = h/v_T$ is accompanied by a change in slope, following which the displacement is constant. There is no discontinuity here because $\Lambda_{33}^T = 0$, but away from the x_3-axis, where $\Lambda_{33}^T \neq 0$, there would be a discontinuity. Likewise, for a point in the (001)-plane $\Lambda_{33}^L = 0$, there is no discontinuity, only a chane of slope at the L wave arrival.

Figure 2(b) shows $G_{33}(\mathbf{x}, t)$ as a function of t for $\mathbf{x} = (0, 0, h)$, $h = 10$ cm, for single crystal silicon. The L wave arrival is not much different from that of Aℓ, but the behaviour for the T wave arrivals is much more complicated. Fig. 2(c) shows a (010)-section of the ray surface of Si, which provides the key. Because of the nature of the polarization pattern of the FT branch, there is no FT singularity. The ST singularities correspond to the three closely spaced points where the ST sheet of the ray surface intersects the x_3-axis. Point 1 corresponds to a pure-T mode for which $\Lambda_{33} = 0$, and there is only a change in slope here. Point 2 is associated with a saddle-shaped region of the slowness surface and so there is a logarithmic divergence. Point 3 corresponds

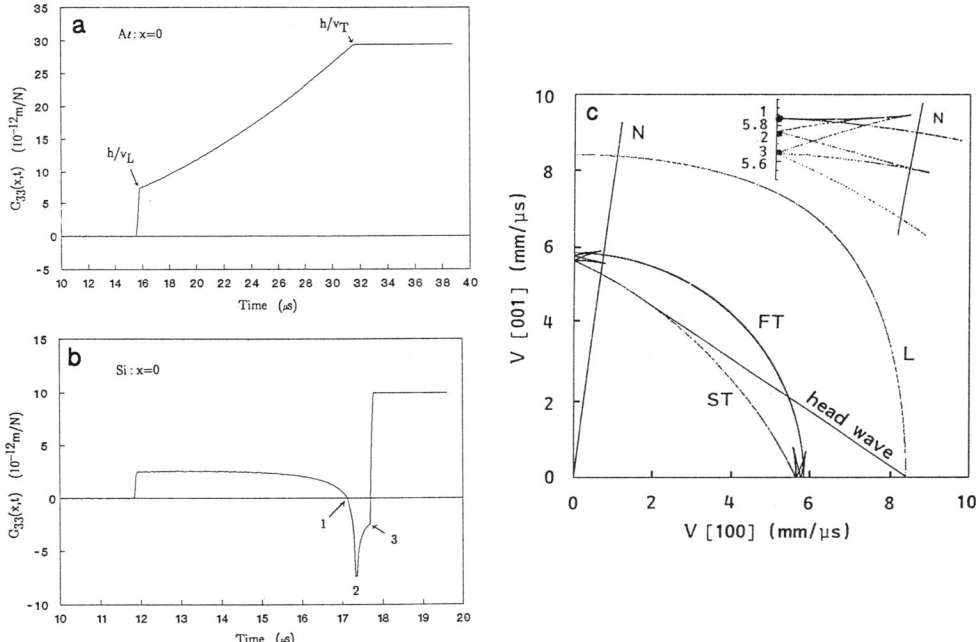

Figure 2. (a) $G_{33}(\mathbf{x}, t)$ for isotropic Aℓ. (b) $G_{33}(\mathbf{x}, t)$ for Si. (c) (010)-section of the ray surface of Si.

to a convex portion of the slowness surface and so there is a discontnuity. This is the last wave arrival, after which the displacement is constant.

For elastic half spaces, there are in addition one or more head waves, the most significant (in the present context) of which is shown in Fig. 2(c). In simple terms the head wave can be thought of as the envelope of T waves which are created in order to satisfy the boundary conditions as the L wavefront proceeds along the surface. The head wave front is the locus of linear filaments, each of which is terminated by phase matched L and T rays. A full discussion of head waves lies outside the scope of this lecture, but the subject is mentioned here because there are head wave features in some of the experimental results to follow.

III. OBSERVATION OF TRANSIENT WAVEFORMS

A number of techniques have been used to study various aspects of the dynamic Green's functions of anisotropic solids, all of which involve the use of small aperture transducers or focusimg lenses to generate and detect transient waveforms. These approximate the conditions of point force and sensing that underly the definition of the Green's functions. A discussion of these techniques and some of the results that have emerged follows.

A. Laser Generated Ultrasound in Anisotropic Solids

Extensive investigations have been carried out of laser generated ultrasound in isotropic solids, mainly with the objective of establishing methods for non-destructive testing (For reviews, see Scruby et al., 1982; Hutchings, 1988; Castagnede and Berthelot, 1988). These measurements are carried out at room temperature, under which conditions thermal phonons propagate by a process of diffusion at a rate much less than that of the speed of sound. We emphasize that in spite of the superficial similarity with thermal phonon imaging, what is observed in these experiments are not the thermal phonons, but thermoacoustically generated sound waves.

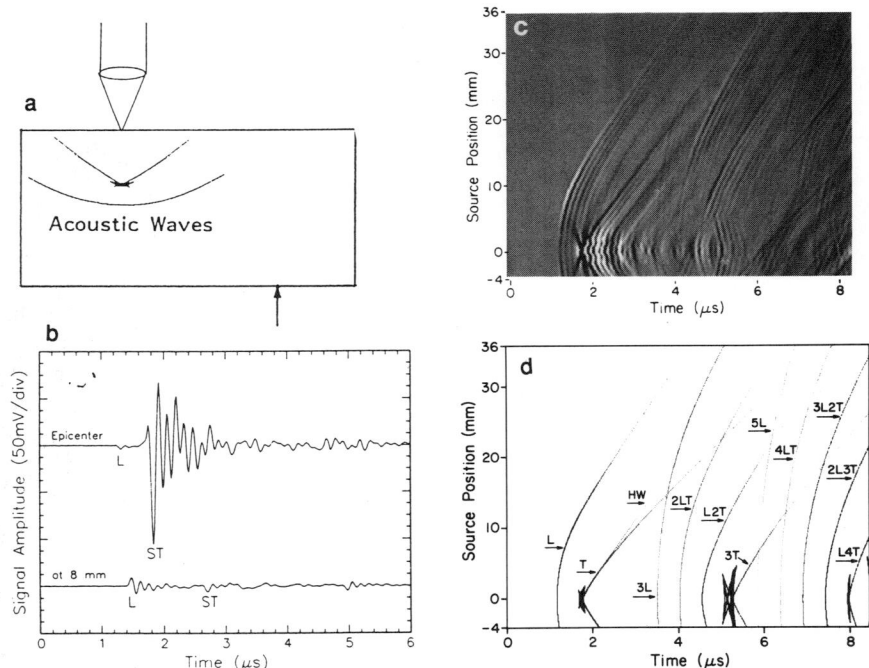

Figure 3. (a) Experimental set-up. (b) Waveforms for Si. (c) Measured scan-image for (001)-oriented Si. (d) Corresponding calculated scan-image.

Recently this technique has been applied to a number of anisotropic solids including silicon and zinc single crystals and fibre composites (Every et al., 1991, Sachse et al. 1990). The approach we have used is depicted in Fig. 3(a). A focused Q-switched laser delivers a heat pulse to a small region on the surface of a specimen, causing a transient acoustic wave to be launched into the specimen. The wave is measured on the opposite face using a small-aperture piezoelectric transducer. The generation and detection processes are both axisymmetric in form, and only waves with a significant sagittal component to their polarizations feature in the observations. At low power densities the laser pulse causes the sudden rise in temperature of the surface. Thermal expansion within the surface is constrained by the underlying material, and this leads to the sudden appearance of a localised radial stress field which causes acoustic waves to be radiated. The Green's function $G_{33}(\mathbf{x}, t)$ for a step function force normal to the surface provides a partial explanation of the measurements.

Figure 3(b) shows typical waveforms we have obtained with a 10mm thick (001)-oriented silicon single crystal for detection at epicenter (i.e., directly opposite the source) and at 8mm off epicenter in the [100] direction. The sensitive transducers we have used exhibit a tendency to ring when triggered by a wave arrival, which can be attributed to the sudden deposition of the high concentration of acoustic energy accompanying the wavefront singularity (discontinuity or logarithmic divergence). Little information can be inferred directly from these signals concerning the continuous portions of the waveforms inbetween the singularities.

In order to more clearly distinguish wave arrivals from noise and ringing, a large number of waveforms for a closely spaced set of excitation points are stacked together, and the resulting (x,t) response represented as a gray-scale image as in Fig. 3(c). Fig. 3(d) shows the theoretical wave arrivals (wavefront singularities) obtained using Monte Carlo methods. It features the first L and ST wave arrivals, the arrivals of L and ST waves which have passed three or five times through the crystal, with or without undergoing mode conversion on reflection, and also the L→ST head wave. These various wave arrivals are all in very good agreement with the signal onsets in the measured scan image. There is no discernable presence of FT single- or multi-pass waves, or any mode conversion sequence or head wave involving FT waves. This is because the FT waves are almost perfectly SH polarized, and are therefore uncoupled from the axisymmetric excitation and detection. Moreover, on reflection there is negligible mode conversion between FT modes and the other two branches. Data such as this provides a means for measuring the elastic constants of anisotropic solids (Aussel and Monchalin, 1989; Castagnede et al., 1990; Every and Sachse, 1990).

B. Phonon Imaging with Scanning Acoustic Microscopy

Scanning transmission acoustic microscopy has been used by Hauser et al. (1992) and Weaver et al. (1992) to study the frequency domain dynamic response of a number of anisotropic solids including metal, insulating and semi-conducting crystals and fibre composites. Their experimental set-up is shown in Fig. 4(a). A pair of water immersion acoustic lenses are used to generate and detect ultrasonic tonebursts which are focused to a diameter of about 200 μm on the surface of the sample. One of the transducers is kept fixed while the other is raster scanned to yield a 2D ultrasonic flux pattern. Fig. 4(b) shows an image they have obtained with a (100)-oriented silicon cube of 2cm side, using 15MHz tonebursts of 500 ns duration. The diagonal fringes and square grid of diffraction spots at the centre are fully accounted for by the dynamic Green's function $\tilde{G}_{33}(\mathbf{x}, \omega)$ corresponding to the experimental parameters. The diffraction fringes are more closely spaced than in Fig. 1(c) because of the higher frequency and greater distance. To explain the remaining details, such as the radial modulation of the intensity towards the four corners, Weaver et al. (1992) in their calculations have taken account of the finite angular width of the incoming and outgoing beams in the water, and the angular dependence of the transmission across the fluid-solid interface.

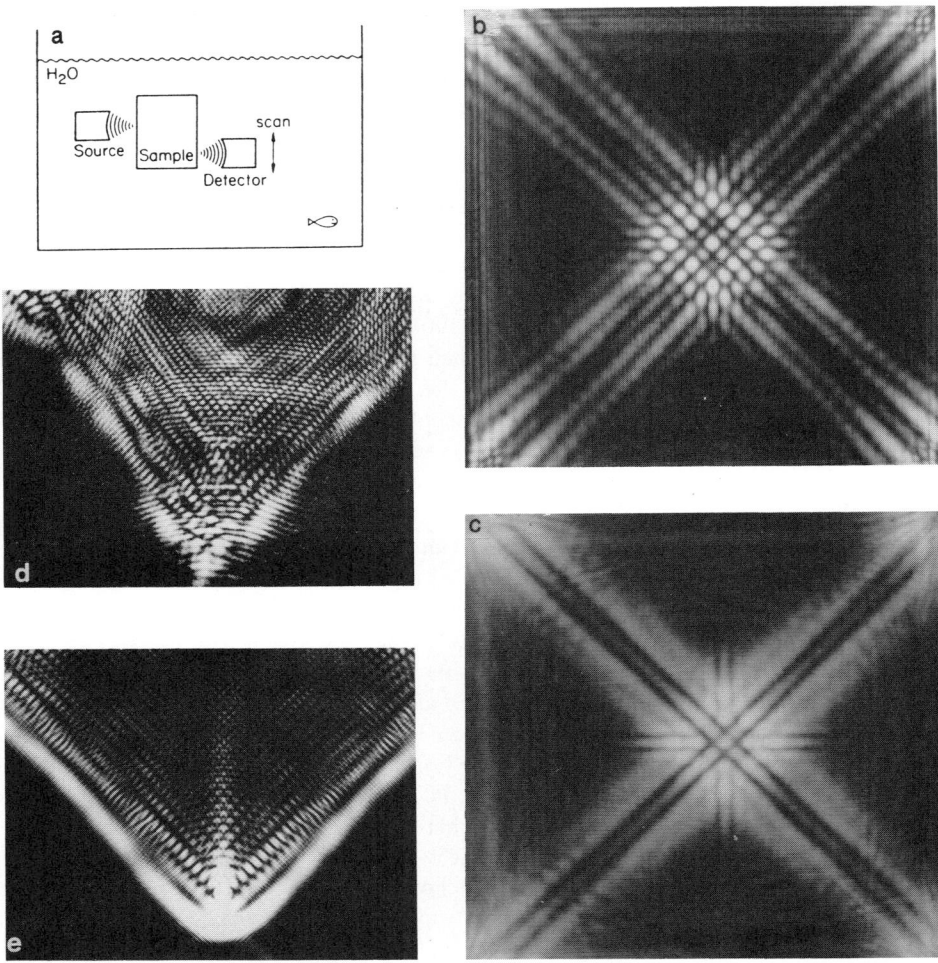

Figure 4. (a) Experimental set-up. (b) Image obtained by Weaver et al. (1992) for (100)-oriented Si. (c) Corresponding calculated image. (d) Image obtained by Wesner et al. (1992) for (100)-oriented GaAs. (e) Corresponding calculated image.

Their calculated image, shown in Fig. 4(c) is in good agreement with their measured one. Because of the axisymmetric nature of the excitation and detection, there is no perceptible presence of SH-polarized FT modes in these images.

Wesner et al. (1992) have used basically the same technique to study a number of crystals, except that they retain phase information and are able to generate both intensity and holographic images. Fig. 4(d) shows an intensity image they have obtained at 392MHz with a 4.8mm thick (100)-oriented GaAs crystal, on a 1.2mm×1.6mm scan area. For this high frequency they use the stationary phase approximation in their calculations, thereby obtaining Fig. 4(e), which is in good agreement with experiment.

C. Observation of Focusing with Shear Transducers

The methods of the previous two sections are limited to the study of modes with a significant sagittal component of polarization and provide no evidence of the FT modes in (001) silicon. Kim et al. (1993a) have sidestepped this limitation by using small aperture piezoelectric shear transducers to generate and detect ultrasonic tonebursts in a 49.15 mm thick (001)-oriented Si crystal. Our experimental setup is shown in Fig. 5(a). Source and detector are both polarized in the [100]-direction and are thus able to couple onto the SH-polarized FT modes. By scanning the detector in the [100]-direction at various distances from the [100]-axis, the diffraction broadened FT focusing pattern can clearly be observed. Some of the detected waveforms obtained when the detector is scanned at a distance of 40 mm from the [100]-axis are shown in Fig. 5(b), in which x indicates the distance from the [010]-axis, where the focusing is most intense.

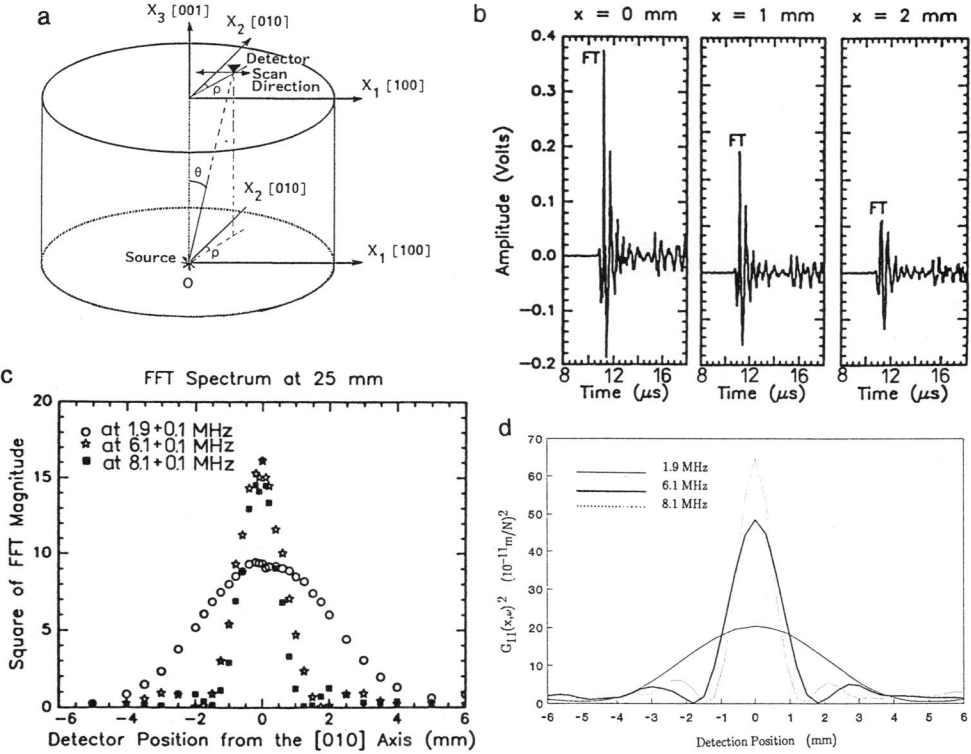

Figure 5. (a) Experimental set-up. (b) Typical waveforms for Si. (c) Dependence of Fourier components of signal on frequency and x. (d) Variation of $|\tilde{G}_{11}|^2$ with frequency and x.

Since the FT modes are almost totally uncoupled from the ST and L modes in reflection, there is no need to consider a head wave contribution to the waveforms, and the infinite elastic continuum Green's function $\tilde{G}_{11}(\mathbf{x}, \omega)$ shown in Fig. 1(d) provides an adequate account of the data. The focusing pattern for $\tilde{G}_{11}(\mathbf{x}, \omega)$ does not have 4-fold symmetry about the [001]-axis. The strongly focused FT modes near the (100)-plane are almost perfectly [100]-polarized and contribute maximally to $\tilde{G}_{11}(\mathbf{x}, \omega)$, while near to the (010)-plane, even though the FT modes there are strongly focused, they contribute negligibly to $\tilde{G}_{11}(\mathbf{x}, \omega)$, since they are very nearly [010]-polarized. This is confirmed by experiment. On scanning the detector in the [010]-direction across the [100]-axis, the FT signal is very small and no significant focusing is observed. From symmetry it is the other way around for $\tilde{G}_{22}(\mathbf{x}, \omega)$.

In the far field limit the FT focusing caustics alongside each (100)-plane are about 1.7° apart at their furthest, giving rise to a narrow focusing ridge. As the frequency is lowered, keeping the distance fixed, diffraction causes the progressive broadening of this ridge, as shown in Fig. 1(d) for f=2 MHz. Fig. 5(c) shows the squared magnitudes of the 8.1, 6.1 and 1.9 MHz Fourier components of the observed waveforms as a function of x for a scan line 25 mm from the [100]-axis. Each of these curves peaks at the centre of the focusing ridge, but they differ in width. The 8.1 MHz curve is the narrowest, having a half-width of about 1 mm, a minimum at 1.5 mm and then a small secondary peak at 2 mm. The 6.1 MHz curve shows the same behaviour, but is slightly broader. The 1.9 MHz curve is the broadest, with a half width of about 3 mm. Only the relative variation of these intensities is significant and not their absolute magnitudes. Fig. 5(d) shows the variation of $|G_{11}|^2$ with x for the same three frequencies. These calculated curves are in very good agreement with the measured data with regard to the widths and positions of the various peaks and minima (Kim et al., 1993b).

D. Capillary Fracture Waveforms

A technique that to a good approximation is able to simulate a normal force on a surface with a step function time dependence is the following (Kim et al., 1989). A thin glass capillary is placed on the surface of the sample and the sharp edge of a razor blade, aligned at right angles, is gradually pressed down on the capillary until it causes it to break suddenly. At this instant the load on the surface drops abruptly to zero. The waveform that is generated is of large amplitude and can be accurately monitored with a small-aperture capacitive transducer. This method has been applied to a variety of solids.

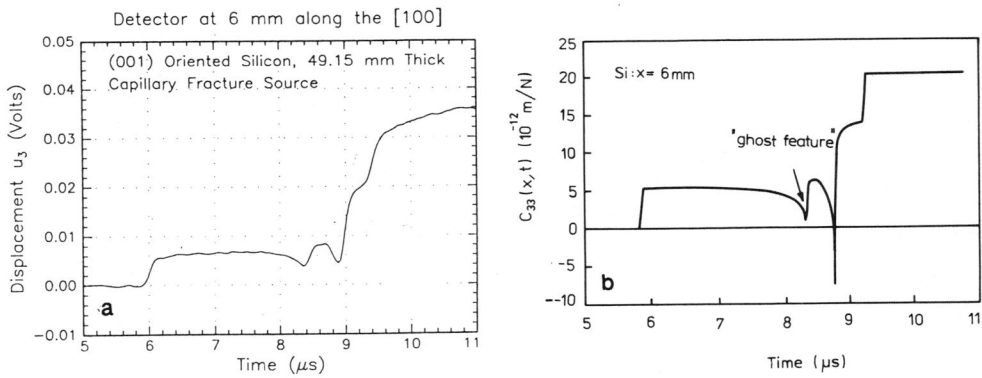

Figure 6. (a) Measured capillary fracture waveform for Si. (b) Corresponding calculated waveform.

Figure 6(a) shows a waveform we have obtained using this technique on a 49.15mm thick (001)-oriented Si crystal, with detection at 6mm from epicentre in the [100]-direction (Kim et al., 1993c). Fig. 6(b) shows $G_{33}(\mathbf{x},t)$ as a function of t corresponding to these experimental parameters. Apart from the smoothing effect of the measured waveform due to the finite response time of the equipment and the finite size of the source and detector, there is good agreement between the measured and calculated waveforms. The line **N** in Fig. 2(c) corresponds to the source-detector direction, and its intersections with the ray surface match up with the singularities in Fig. 6(b). The feature labelled "ghost feature" is actually a slightly rounded deep minimum and not a logarithmic singularity. Line **N** does not intersect the outermost fold of the ST sheet of the ray surface, but passes close to its cuspidal edge, and this accounts for the presence of the minimum. Since detection is near epicenter, there is no head wave, which first appears at about 20° from epicenter.

IV. REFERENCES

Aki, K. and P. Richards, 1980, Quantitative Seismology (Freeman, San Francisco).
Aussel, J.D. and J.P. Monchalin, 1989, Ultrasonics **27**, 165.
Berry, M.V., 1976, Adv. Phys. **25**, 1.
Buchwald, V.T., 1959, Proc. R. Soc. Lond. **A253**, 563.
Burridge, R., 1967, Quart. J. Mech. Appl. Math. **XX**, 42.
Cameron, N. and G. Eason, 1967, Quart J. Mech. Appl. Math. **XX**, 23.
Ceranoglu, A.N. and Y.H. Pao, 1981, J. Appl. Mech. **48**, 125; **48**, 133; **48**, 139.
Castagnede, B., J.T. Jenkins, W. Sachse and S. Baste, 1990, J. Appl. Phys. **67**, 2753.
Castagnede, B. and Y. Berthelot, 1992, J. Acoustique **5**, 417.
Cottam, M.G. and A.A. Maradudin, 1984, Surface Linear Response Functions, in:
  Surface Excitations, ed. V.M. Agranovich and R. Loudin (Elsevier, Amsterdam).
Duff, G.F.D., 1960, Phil. Trans. Roy. Soc. **252**, 249.
Every, A.G. and W. Sachse, 1990, Phys. Rev. **B42**, 8196.
Every, A.G. and W. Sachse, 1991, Phys. Rev. **B44**, 6689.
Hauser, M.R., R.L. Weaver and J.P. Wolfe, 1992, Phys. Rev. Lett. **68**, 2604.
Hutchings, D.A., 1988, Ultrasonic Generation by Pulsed Lasers, in: Physical
  Acoustics, Vol **XVIII**, ed. W.P. Mason and R.N. Thurston (Academic, Boston).
Kim, K.Y., L. Niu, B. Castagnede and W. Sachse, 1989, Rev. Sci. Instrum. **60**, 2785.
Kim, K.Y., A.G. Every and W. Sachse, 1993a, to be published.
Kim, K.Y., A.G. Every and W. Sachse, 1993b, to be published.
Kim, K.Y., A.G. Every and W. Sachse, 1993c, to be published.
Kolomenskii, A.A. and A.A. Maznev, 1991, JETP Lett. **53**, 423.
Maris, H.J., 1983, Phys. Rev. **B28**, 7033.
Novikov, V.V. and L. Chernozatonskii, 1988, Sov. Phys. Acoust. **34**, 215.
Payton, R.G., 1983, Elastic Wave Propagation in Transversely Isotropic Media
  (Martinus Nijhoff, The Hague).
Sachse, W., A.G. Every and M.O. Thompson, 1990, Am. Soc. Mech. Eng. AMD-
  Vol. **116**, 51.
Scruby, C.B., R.J. Dewhurst, D.A. Hutchings and S.B. Palmer, 1982, Laser Generation of Ultrasound in Metals, in: Research Techniques in Nondestructive Testing,
  Vol. **V**, ed. R.S. Sharpe (Academic, New York).
Tewary, V.K. and C.M. Fortunko, 1992, J. Acoust. Soc. Am. **91**, 1888.
Tverdokhlebov, A. and J. Rose, 1988, J. Acoust. Soc. Am. **83**, 118.
van der Hijden, 1987, Propagation of Transient Elastic Waves in Stratified Anisotropic Media (North Holland, Amsterdam).
Weaver, R.L., M.R. Hauser and J.P. Wolfe, 1992, in press.
Wesner, J., K.U. Wurz, K. Hillmann and W. Grill, 1992, Imaging of Coherent
  Phonons, in: 7th Int. Conf. on Phonon Scattering in Condensed Matter.
Yeatts, F.R., 1984, Phys. Rev. **B29**, 1674.

PHONON PATTERNS OF CUBIC CRYSTALS MONTE CARLO SIMULATION PROGRAM

W. M. Gańcza, T. Paszkiewicz

Institute of Theoretical Physics, University of Wrocław
Pl. M. Borna 9, PL–50–204 Wrocław, Poland

1. INTRODUCTION

The images of crystals provide global information on their internal structure and dynamics. Among them we may mention the neutrongrams, roentgenograms, neutron and x-ray topographs or the sound topographs. With the exception of sound beam topography, the experimental techniques supplying the above images were invented long time ago. In the late seventies Wolfe and collaborators (cf. [1]) invented the phonon imaging method which produced a global view of energy flux anisotropy in the crystal, called the energy focusing pattern. This method is discussed in this volume by Arthur Every.

Dietsche and collaborators invented another phonon imaging experiment giving quasimomentum focusing patterns (cf. [2], [3]) (detectors sensitive to phonon quasimomentum are reviewed by Werner Dietsche in this volume).

Phonon focusing patterns characterize the anisotropy of fluxes of energy and quasimomentum of phonons. This information is very important for the proper understanding of the results of many experiments with the use of phonon beams. For this reason it is important to develop fast methods of obtaining the phonon focusing patterns and extracting usable information from them.

Here, we present a simple program which provides both kinds of phonon focusing patterns and permits some measurements to be made on the images produced.

2. CHARACTERISTICS OF PHONONS

Weakly disturbed states of crystals can be described in terms of quasiparticles (cf. [4]). Quasiparticles related to vibrational motions of a whole crystal are called phonons. For lattices containing s particles in each unit cell there are $3s$ different kinds of phonons labeled by polarization and branch index $j(j = 0, 2, .., 3s − 1)$. Here, we shall confine ourselves to acoustic phonons ($j = 0, 1, 2$). A phonon carries the energy $\epsilon_j = \hbar\omega_j$ and quasimomentum $\mathbf{p} = \hbar\mathbf{k}$. Generally energy ϵ_j and, of course the frequency ω_j, depend on the quasimomentum $\hbar\mathbf{k}$ (or the wave vector $\mathbf{k}$; $\mathbf{k} = (k, \theta_k, \phi_k)$)

Die Kunst of Phonons, Edited by T. Paszkiewicz and
K. Rapcewicz, Plenum Press, New York, 1994

$$\epsilon_j = \hbar\omega(\mathbf{k}, j), \qquad (j = 0, 1, 2). \tag{1}$$

For small wave vectors ($ka \ll 1$, where a is the lattice constant) the dispersion law (1) is linear in the magnitude of the wave vector $k = |\mathbf{k}|$

$$\omega = c(\hat{\mathbf{k}}, j)k. \tag{2}$$

The phase velocity c depends on the direction of the wave vector $\hat{\mathbf{k}} = \mathbf{k}/k, (\hat{\mathbf{k}} = (\theta_k, \phi_k))$.

The equation

$$\omega(K) = \omega_0, \tag{3a}$$

where K stands for $(\mathbf{k}, j)$, defines in k-space a surface of constant frequency (ω-surface for short). Each ω-surface has the symmetry of the considered crystal.

Having the phase velocity $c(\hat{K})$ ($\hat{K}$ denotes the pair $(\hat{\mathbf{k}}, j)$, $j = 0, 1, 2$) one can introduce the slowness s

$$s(\hat{K}) = c^{-1}(\hat{K}).$$

The polar plot of the slowness gives the slowness surface. The equation of an ω-surface is equivalent to

$$k = s(\hat{K})\omega. \tag{3b}$$

Thus, an ω-surface, which is in fact a polar plot of k, has the same shape as the slowness surface. The ratio of sizes of the surfaces is simply the frequency ω_0.

The group velocity is given by

$$\mathbf{v}(K) = \nabla_k \omega(K). \tag{4}$$

For long-wavelength acoustic phonons the group velocity depends only on $\hat{K}$. This is a vector which is always in the direction of the maximum rate of change of frequency and is normal to the ω-surface, i.e. is it normal to the slowness surface too. For an anisotropic solid the group velocity $\mathbf{v}(\hat{K})$ for a phonon with the wave vector $\mathbf{k}$ and polarization j is not, in general, in the same direction as $\mathbf{k}$.

When the tip of the wave vector $\mathbf{k}$ moves across the ω-surface direction of the group velocity changes. The rate of this change depends on the curvature of the ω-surface. In a flat region the group velocity vectors are almost parallel. The local geometry of the ω-surfaces are characterized by the Gaussian curvature Γ, which is the product of the two principal curvatures Γ_1, Γ_2

$$\Gamma = \Gamma_1 \Gamma_2.$$

The principal curvatures can be related to the derivatives of the group velocity in two perpendicular directions in the plane tangent to the ω-surface at the given point (cf. [5], [6]).

In the case of experiments with the transmission of pulses of phonons (time-of-flight spectroscopy and phonon focusing) in place of the Gaussian curvature one uses the focusing (amplification) factor $\mathcal{A}$. Consider a solid angle $d\Omega_k^{(j)}$ which subtends an area element of the ω-surface (i.e. the solid angle subtended by the suitable bundle of directions of wave vectors). To this solid angle there corresponds the solid angle $d\Omega_v^{(j)}$

subtended by the bundle of group velocity vectors. The ratio $d\Omega_k^{(j)}/d\Omega_v^{(j)}$ defines the focusing factor $\mathcal{A}$

$$\mathcal{A}_j = d\Omega_k^{(j)}/d\Omega_v^{(j)}. \tag{5}$$

One can show that the focusing factor can be written in terms of local geometric characteristics of the ω-surfaces: the magnitude of the Gaussian curvature $\Gamma_j(\hat{\mathbf{k}})$, the length k of the radius vector tipping the point $\mathbf{k}$ and the angle between the direction of this vector and the direction of the group velocity $\mathbf{v}(\mathbf{k},j)$

$$\mathcal{A}_j(\mathbf{k}) = [\hat{\mathbf{k}}\hat{\mathbf{v}}(\mathbf{k},j)]/[k^2|\Gamma_j(\mathbf{k})|]. \tag{6}$$

At points on which the group velocity is stationary with respect small variations, the Gauss curvature vanishes and the focusing coefficient becomes infinite. If we consider the ω-surface there will be lines along which $\Gamma = 0$. Along these lines $\mathcal{A}$ is infinite. These *parabolic* lines divide the ω-surface into regions of positive and negative curvature. The more detailed classification of regions of an ω-surface is obtained by considering both principal curvatures Γ_1, Γ_2

$\Gamma_1 > 0, \quad \Gamma_2 > 0 \quad$ convex region $\qquad \begin{cases} \Gamma_1 > 0, \quad \Gamma_2 < 0 \\ \Gamma_1 < 0, \quad \Gamma_2 > 0 \end{cases}$ saddle region

$\Gamma_1 < 0, \quad \Gamma_2 < 0 \quad$ concave region

$\Gamma_1 = 0, \quad \Gamma_2 \neq 0; \quad \Gamma_1 \neq 0, \quad \Gamma_2 = 0 \quad$ lines of parabolic points.

The slowness surface can be divided into convex, concave and saddle regions (Fig. 1). Phonon focusing patterns are the Gauss maps of the slowness surfaces. So they provide information about slowness surfaces.

3. MEASUREMENT OF FOCUSING FACTOR

Consider the typical arrangements in a phonon focusing experiment (Fig. 2a) [8]. A phonon (heat) pulse is generated on one face of a crystalline specimen by exciting a small spot on a metal film deposited on the surface. This spot (i.e. the source of phonons) is heated above the ambient temperature with a focused laser beam or a beam of electrons. This creates a highly inhomogeneous distribution of energy and quasimomentum. At low temperatures and low excitation powers in perfect specimens phonons emanating from this heated region undergo negligible bulk scattering and travel ballistically with the group velocity through the crystal to opposite face, where they are detected with a suitable detector. The superconducting bolometers or the tunnel junction detectors are sensitive to the energy transported by phonons. In the presence of heat pulses the electrons in semiconductor heterojunctions undergo phonon drag so such detectors are sensitive to the quasimomentum falling onto them (Fig. 2b) (cf. [9], [10]).

Our program simulates phonon focusing when the source of phonons is fixed and the detector is movable (cf. the paper by Akimov et al. in this volume). Physically there is no difference between the results of experiment with movable source and fixed detectors and vice versa.

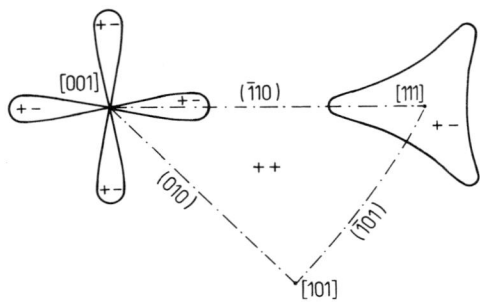

Figure 1. Regions of different curvature of the slowness surface for a slow transverse mode (after Every [7]).

If the heat source is of short duration the detector signal consists of a number of separate pulses. These pulses correspond to the arrival of phonons of different polarization having group velocity vectors in the source-detector direction. The magnitudes of phonon pulses vary with direction. The source is raster scanned over the metallic film and the boxcar integrated detector signal is recorded as an image.

A common assumption in the theoretical description of the phonon imaging is that the phonon can be regarded as a wave packet, of dimension much bigger than the lattice constant a, but much smaller than the mean free path l (which for ballistic phonons is of the order of linear dimension of the specimen). These wave packets move along trajectories. Since the sources of heat pulses are incoherent the phonon wave packets are described by a distribution function $f(K, \mathbf{r}, t)$ which obeys the Boltzmann equation. Assume that the normal to the detector surface is $\hat{\mathbf{n}}$ and that the origin of a Cartesian coordinate system is located at the point source. A small area detector is located at a point $\mathbf{r} = (r, \theta_r, \phi_r)$. Phonons with the group velocity vector $\mathbf{v}(\hat{\mathbf{K}}) = (v(\hat{K}), \theta_v(\hat{K}), \phi_v(\hat{K}))$ parallel to $\mathbf{r}$ reach the detector. We shall enumerate the points of the ω-surface with the normal parallel to $\mathbf{r}$, i.e. solutions $\theta_j^{(i)}, \phi_j^{(i)}$ of the set of two equations

$$\theta_v^{(j)}(\theta_k, \phi_k) = \theta_r, \qquad \phi_v^{(j)}(\theta_k, \phi_k) = \phi_r, \qquad (7)$$

with the index i ($i = 1, 2, \ldots, n$). The density of energy $e(\mathbf{r})$ falling onto the detector surface is [9]

$$e(\mathbf{r}) = \frac{(\hat{\mathbf{n}}\hat{\mathbf{r}})}{2\pi r^2} \frac{1}{3} \sum_{j=0}^{2} \mathcal{A}_j^{(i)}(\theta_r, \phi_r) \hbar\omega_0, \qquad (8a)$$

and a component η of the quasimomentum density (cf. [9])

$$p_\eta(\mathbf{r}) = \frac{(\hat{\mathbf{n}}\hat{\mathbf{r}})}{2\pi r^2} \sum_{j=0}^{2} \{\hat{\eta}\mathbf{s}_j^{(i)}(\theta_r, \phi_r)\} \mathcal{A}_j^{(i)}(\theta_r, \phi_r) \hbar\omega_0, \qquad (8b)$$

where for an arbitrary function F_j of θ_k and ϕ_k

$$F_j^{(i)}(\theta_r, \phi_r) \equiv F_j[\theta_j^{(i)}(\theta_r, \phi_r), \phi_j^{(i)}(\theta_r, \phi_r)].$$

Note that the quasimomentum focusing patterns besides the local geometric characteristics contain also information about the linear dimension of the slowness surface.

Unfortunately, the calculation of the density of energy and the density of quasimomentum is a difficult numerical problem (cf. [9, 10]). Therefore, usually one relies on different variants of Monte Carlo simulations. In Sect. 5 we describe a very simple program performing the Monte Carlo simulations.

4. PHASE AND GROUP VELOCITIES OF LONG–WAVELENGTH ACOUSTIC PHONONS

For long-wavelength acoustic phonons a crystalline medium is equivalent to an elastic continuum. The propagation of acoustic waves in an elastically anisotropic medium is governed by the set of three linear differential equations known as the Christoffel equation (cf. [11]). A disturbance of such a medium is represented by a set of position-and-time dependent displacements

$$\rho \frac{\partial u_\alpha}{\partial t} = C_{\alpha\mu,\beta\nu} \frac{\partial^2 u_\beta}{\partial x_\mu \partial x_\nu}, \tag{9}$$

where ρ is the mass density and the tensor of elastic constants $\mathcal{C}$ has components $C_{\alpha\mu,\beta\nu}$. The set of solutions of Eq. (9) in the form of plane waves

$$\mathbf{u} = \mathbf{e} e^{i(\mathbf{k}\mathbf{r} - \omega t)}, \tag{10}$$

forms a basis. Therefore, an arbitrary solution can be written as a superposition of plane waves. The phase velocity $c = \omega/k$ satisfies an algebraic equation

$$| \Gamma_{\alpha\beta} - \rho c^2 \delta_{\alpha,\beta} | = 0, \tag{11}$$

where the propagation tensor Γ has elements

$$\Gamma_{\alpha\beta} = C_{\alpha\mu,\beta\nu} \hat{k}_\mu \hat{k}_\nu. \tag{12}$$

On making the replacement

$$3\rho c^2 = T + S,$$

where

$$T = Tr\Gamma,$$

Every arrived at the following cubic equation for S [11]

$$S^3 - 3GS - 2H = 0, \tag{13}$$

where

$$3G = \Lambda_{12}^2 + \Lambda_{23}^2 + \Lambda_{31}^2 + \Lambda_{11}\Lambda_{22} + \Lambda_{22}\Lambda_{33} + \Lambda_{33}\Lambda_{11}, \tag{14a}$$

$$2H = \Lambda_{11}\Lambda_{22}\Lambda_{33} + 2\Lambda_{12}\Lambda_{23}\Lambda_{31} + \Lambda_{11}\Lambda_{23}^2 + \Lambda_{22}\Lambda_{31}^2 + \Lambda_{33}\Lambda_{12}^2, \tag{14b}$$

$$\Lambda_{\alpha\beta} = 3\Gamma_{\alpha\beta} - T\delta_{\alpha\beta}. \tag{14c}$$

The parameterization solutions to Eq. (11) proposed by Every depend on characteristic combinations of the elastic constants and two functions of the components of $\hat{\mathbf{k}}$.

Consider the case of a cubic medium. There are three such combinations [11]

$$C_1 = C_{11} + 2C_{44}, \quad C_2 = C_{11} - C_{44}, \quad K = C_{11} - C_{12} - 2C_{44}.$$

We define two dimensionless parameters

$$s_2 = C_2/C_1, \quad s_3 = K/C_1,$$

and rename the coefficient C_1

$$C_1 \equiv s_1.$$

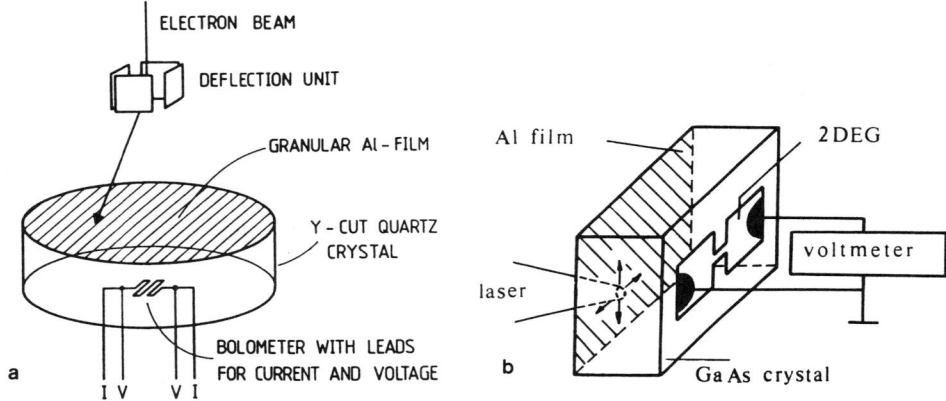

Figure 2. a,b

The phase velocity and components of the group velocity can be expressed in terms of two dimensionless functions of $\hat{K}$, viz.

$$c(\hat{K}) = c\tilde{c}(\hat{K}), \tag{15}$$

and

$$v_\alpha(\hat{K}) = c\tilde{v}_\alpha(\hat{K}) \quad (\alpha = 1, 2, 3), \tag{16}$$

where

$$c = \sqrt{s_1/(3\rho)}. \tag{17}$$

The dimensionless velocities $\tilde{c}(\hat{K}), \tilde{v}_\alpha(\hat{K})$ depend on the components of $\hat{\mathbf{k}}$ and on the two parameters s_2, s_3 via five functions. Three of these functions are obtained by putting $\hat{\mathbf{k}}^2 = 1$ in the Every expressions (19-23) and $(A18 - A19)$ [11], i.e.

$$G(\hat{\mathbf{k}}) = s_2^2 - 3s_3(2s_2 - s_3)\mathcal{P}(\hat{\mathbf{k}}) \equiv s_2^2 G_0(\hat{\mathbf{k}}),$$
$$H(\hat{\mathbf{k}}) = s_2^3 - [s_2 s_3(2s_2 - s_3)\mathcal{P}(\hat{\mathbf{k}})/2] + [27s_3^2(3s_2 - 2s_3)\mathcal{Q}(\hat{\mathbf{k}})/2] \equiv s_2^3 H_0(\hat{\mathbf{k}}),$$

and
$$\psi(\hat{\mathbf{k}}) = \arccos[H(\hat{\mathbf{k}})/G^{3/2}(\hat{\mathbf{k}})]/3 = \arccos[H_0(\hat{\mathbf{k}})/G_0^{3/2}(\hat{\mathbf{k}})]/3,$$
where
$$\mathcal{P}(\hat{\mathbf{k}}) = \hat{k}_1^2\hat{k}_2^2 + \hat{k}_1^2\hat{k}_3^2 + \hat{k}_2^2\hat{k}_3^2, \text{ and } \mathcal{Q}(\hat{\mathbf{k}}) = \hat{k}_1^2\hat{k}_2^2\hat{k}_3^2.$$

The remaining two functions are vector functions with components
$$G_\alpha(\hat{\mathbf{k}}) = 2s_2^2 - 3s_3(2s_2 - s_3)(1 - \hat{k}_\alpha^2),$$
and
$$H_\alpha(\hat{\mathbf{k}}) = 2s_3^2 - 3s_2s_3(2s_2 - s_3)[1 + \mathcal{P}(\hat{\mathbf{k}}) - \hat{k}_\alpha^2] + 9s_3^2(3s_2 - 2s_3)\hat{k}_\gamma^2\hat{k}_\delta^2, (\gamma \neq \delta, \delta \neq \alpha).$$

We can now write the explicit expressions for $\tilde{c}(\hat{K})$ and the $\tilde{v}_\alpha(\hat{K})$, viz.
$$\tilde{c}(\hat{K}) = \sqrt{1 + S(\hat{K})}, \tag{18}$$
and
$$\tilde{v}_\alpha(\hat{K}) = [1 + S_\alpha(\hat{K})]\hat{k}_\alpha/\tilde{c}(\hat{K}), \tag{19}$$
where
$$S(\hat{K}) = 2\sqrt{G(\hat{K})} \cos[\psi(\hat{\mathbf{k}}) + 2\pi j/3], \tag{20}$$
and
$$S_\alpha(\hat{K}) = [S(\hat{K})G_\alpha(\hat{\mathbf{k}}) + H_\alpha(\hat{\mathbf{k}})]/[S^2(\hat{\mathbf{k}}) - G(\hat{\mathbf{k}})]. \tag{21}$$

From (15-18), and (20) it follows that the shape of surfaces of constant frequency (or the slowness surfaces) depend only on s_2 and s_3. The parameter s_1 scales their linear dimensions. For weakly anisotropic cubic media the slowness surfaces and polarization vectors were studied in [12].

5. DESCRIPTION OF PROGRAM OF MONTE CARLO SIMULATIONS

5.1. FUNCTIONS OF THE PROGRAM

The present Monte Carlo simulation program produces the focusing patterns of energy and quasimomentum for long wave-length acoustic phonons for all three polarizations as well as the partial pattern for a chosen polarization (or polarizations) in a cubic crystal. The orientation of the pattern plane as well as the crystal density and elastic constants are defined by the user. A view of the screen output of the simulation is presented in Fig. 3.

One can define a detector reaction for the energy of phonons and all three components of the quasimomentum vector independently.

The program allows also for the geometrical measurement of the pattern. With three movable cursors one can choose three points and the program will find the distance separating them as well as the angles of the corresponding triangle whose vertices are these points (cf. Fig. 3).

It is assumed that the sample has the form of a perpendicular parallelepiped with the squares at the base. The user defines the side length and the height of this parallelepiped.

One can easily switch from one to another experiment. At a given instant of time, one can see the results of only one of the experiments. However, the results of other experiments are stored.

5.2. THE ALGORITHM

We used the Monte Carlo method. Individual phonons are generated sequentially by a source located at a chosen point inside the perpendicular parallelepiped with a randomly chosen direction for its wave vector. For a given frequency ω and polarization j, we obtain the quasimomentum vector $\mathbf{k}$, phase velocity $c(\hat{\mathbf{k}}, j)$ and group velocity $\mathbf{v}(\hat{\mathbf{k}}, j)$ as well as the coordinates of the surface point hit by a phonon $(\hat{\mathbf{k}}, \omega, j)$. A value of a quantity collected by a suitable detector is updated by a value added by an incoming phonon.

The program does not use the symmetry properties, so it can produce the focusing patterns on an arbitrary crystalline plane. Besides the highly symmetric pattern for energy density it additionally gives the less symmetric pattern for quasimomentum density.

The energy density and the density of each component of the quasimomentum vector can be obtained separately or together. This allows to obtain the quasimomentum density vector along any axis.

Leaving the pending experiment the computer stores the updated maps of energy density or density of the quasimomentum component. Hence, one can perform the simulation in steps, during the time when the computer is available. One can stop the calculation and start it again at any instant of time.

The phonon patterns are visible fairly quickly – after several minutes. With a longer lapse of time, the quality of the patterns becomes better – more smooth and with greater contrast.

The values of measured quantities are coded as real numbers on a grid of 480×480 points. Thus for a smaller area the resolution is better.

5.3. USAGE

Once running, on the screen there is a window with entries for ten experiments (some of them may be empty) and, at the bottom of the screen, the data of the specified experiment. When the experiment is running, there is no way to change the data. When you want to run a new experiment, you should supply all the required data. When the data are not correct, you cannot run an experiment, and the cursor moves to the location of any incorrect data. The data are:

Description - a short description of the experiment (indicated in an experiment directory).

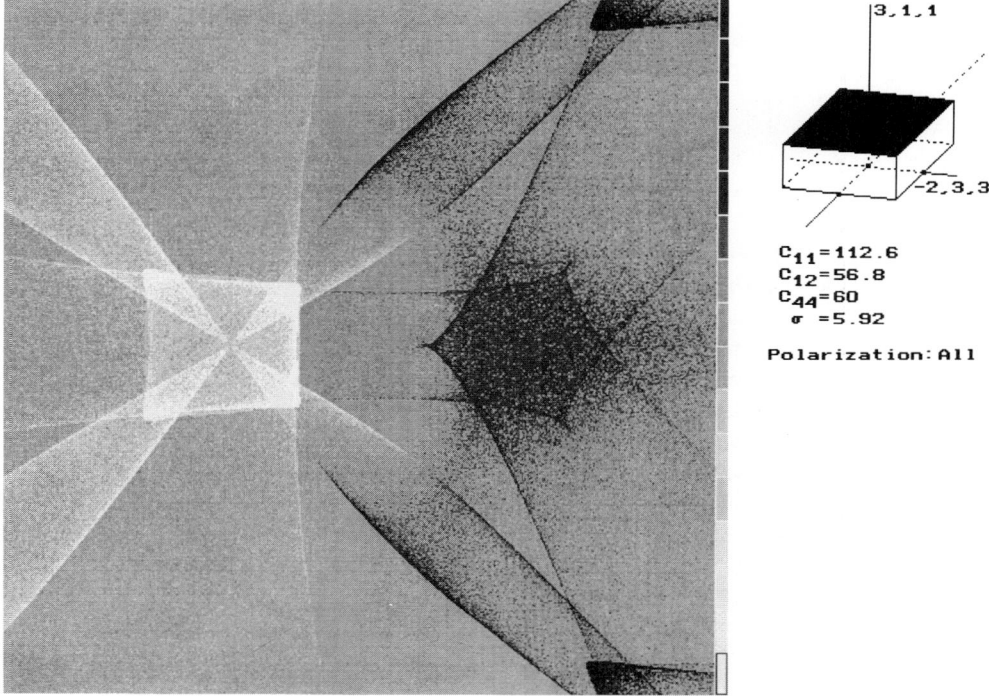

Figure 3. Example of visual output of the simulation for GaAs. In the inset is indicated the normal to the plane of the focusing pattern and one of the axes of the Cartesian coordinate system.

Elastic constants - a list of elastic constants. In this version of the software, one can define only $C11, C12$ and $C44$ (for example: $c11 = 112.6$ $c12 = 56.7$ $c44 = 60'$ units).

Density - density of the crystal.

Z-axis - the direction of the z axis (for example $<1,1,0>$).

X-axis - the direction of the x axis.

Source position - the position of the source of phonons - a point $(0,0,0)$ is in the middle of the bottom of the surface. All surfaces have a coordinate 1 (for example: a point $(0,0,0.5)$ is in the middle of the crystal, irrespective of the size).

Width of the crystal

Length - in this program the specimen surface is a square - you can define only the side length of this square.

Detector sensitive to ... - the detector is sensitive to the energy and each of the components of quasimomentum of the phonons. The recommended value is 1.

Polarization - the polarization of phonons (the numbers 0 for longitudinal, 1 - for fast quasitransverse, 2 for slow quasitransverse).

On this screen one can:

Open an old experiment (pointed by cursor) $<\text{ctrl}-O>$

Create a new experiment $<\text{ctrl}-C>$

Delete a chosen experiment <ctrl−D>
Quit the program <ctrl−Q>

When you choose Open or Create, the screen is switched to the graphics mode and the simulation begins. One can stop the simulation by pressing the $F10$ key - after a while (to save a data to disc) you return to the main screen, which displays a list of experiments.

On the screen of simulation (cf. Fig. 3), you can press the $F1$ key - to measure the pattern. After that, you will see a cursor on the screen (in fact there are three cursors, but in the same position).

The hot keys for this part of the program are:
arrow keys - to move the cursor a pixel up, down, left or right
Home - to move the cursor 5 pixels left
End - to move the cursor 5 pixels right
PgUp - to move the cursor 5 pixels up
PgDn - to move the cursor 5 pixels down
'+' and '−' - to zoom the picture around the active (bright) cursor
Ins - to change an active cursor
Enter - to see the distance separating two chosen points, and the angles of the corresponding triangle.
$F10$ - to quit, and return to the simulation.

5.4 System requirements

IBM $PC386$ or $PC486$ or compatible computer
$2MB$ RAM ($1.5MB$ XMS Memory)
minimum $1.5MB$ of hard disc free space for one experiment, max $15MB$ for 10 experiments
VGA or SVGA video card and monitor (colour or monochrome).

REFERENCES

[1] G.E. Northrop, J.P. Wolfe, "Phonon imaging: theory and application", in: W.E. Bron (ed.), "Nonequilibrium phonons", p. 165, New York: Plenum Press 1985.

[2] W. Dietsche, "Electron phonon interaction in semiconductor heterostructures" in: "Phonons '89", S. Hunklinger, W. Ludwig, G. Weiss (eds.), World Scientific, Singapore, 1990.

[3] T. Paszkiewicz, "Ballistic transport in anisotropic ordered phases of condensed matter: the phonon imaging", in: "Ordering phenomena", Z.M. Galasiewicz, A. Pękalski (eds.), World Scientific, Singapore, 1991.

[4] M.I. Kaganov, I.M. Lifshitz, Quasiparticles, Moscow, Nauka, 1976.

[5] M. Lax, V. Narayanamurti, "Phonon magnification and the Gaussian curvature of the slowness surface in anisotropic: Detector shape effects with application to $GaAs$", Phys. Rev. **B22**, 4876, 1980.

[6] H.J. Maris, "Phonon focusing", in:"Nonequalibrium phonons in nonmetallic crystals", W.Eisenmenger, A.A. Kapylanskii (eds.), p. 51. Amsterdam, Elsevier 1986.

[7] A.G. Every, "Ballistic phonons and shape of the ray surface in cubic crystals", Phys. Rev. **B24**, 3456, 1981.

[8] R. Eichele, R.P. Huebener, H. Seifert, "Phonon focusing in quartz and sapphire imaged by electron beam scanning", Z. Phys. *B*-Condensed Matter **48**, 89, 1982.

[9] Cz. Jasiukiewicz, D. Lehmann, T. Paszkiewicz, "Phonon images of crystals. I Energy and quasimomentum focusing patterns: application to GaAs", Z. Phys *B*-Condensed Matter **84**, 73 1991.

[10] Cz. Jasiukiewicz, D. Lehmann, T. Paszkiewicz, "Phonon images of crystals. II Image of crystalline GaAs obtained by the phonon-drag effect in GaAs-GaAlAs hererostructures", Z. Phys *B*-Condensed Matter **86**, 225, 1992.

[11] A.G. Every, "General closed-form expressions for acoustic waves in elastically anisotropic solids", Phys. Rev. **B22**, 1746, 1980.

[12] T. Paszkiewicz, M. Wilczyński, Scattering of long-wavelength acoustic phonons by isotopic impurities. Spectra of the collision integral and diffusion equation for crystalline media with cubic symmetry, Z. Phys. B–Condensed Matter **88**, 5, 1992.

SURFACE PHONON FOCUSING AT ULTRASONIC FREQUENCIES

Al.A. Kolomenskii and A.A. Maznev

General Physics Institute of the Russian Academy of Sci.
38 Vavilov Str., 117942 Moscow, Russia

Abstract

Surface phonon focusing is examined with laser excitation of pulsed surface acoustic waves (SAWs). $10-ns$ SAW pulses are generated by a focused beam of a Q-switched Nd:YAG laser. To visualize the SAW amplitude distribution we use a "dust patterns" arising as a result of SAW-induced dust particle removal from the surface under investigation. The SAW amplitude measurements with the probe beam deflection technique show that the above patterns reflect adequately the SAW amplitude angular dependence. Strong SAW focusing is observed on cubic crystals such as Si, Ge, GaAS. The results obtained are compare with theory.

1. INTRODUCTION

The distinct anisotropy of acoustic energy flux emanating from a point source embedded in an anisotropic elastic solid is referred to as phonon focusing. There has been considerable work on this effect in the area of ballistic transport of bulk thermal phonons [1]. Recently bulk phonon focusing at ultrasonic frequencies has been observed [2].

Surface acoustic waves can also be focused as it has been shown theoretically by several authors [3-5]. In particular, calculations for some cubic crystals were presented in Ref. 4. As far as we know, however, phonon focusing of SAWs has never been observed directly.

In the present paper we report the experimental results on the phonon focusing effect with ultrasonic SAWs generated coherently by laser irradiation [6,7].

2. EXPERIMENTAL PROCEDURE

To generate SAW pulses a focused beam of a Q-switched Nd:YAG laser (wavelength 1.06 μm, pulse duration 10 ns, pulse energy 10 mJ) was used, with the 1/e radius of the focusing spot being equal to 7 μm. SAW pulses were generated in the ablation regime with a characteristic duration corresponding to that of the laser pulse.

The qualitative examination of the SAW amplitude anisotropy was accomplished using a visualization technique, based on SAW-induced fine particle removal from the surface [8]. The sample surface was dusted with $1-2$ μm Al_2O_3 particles. High local accelerations in the SAW pulse result in an inertial force acting on a particle that can exceed the adhesive force and detach the particle from the surface. The degree of removal of particles from the surface depends on the SAW amplitude. In a

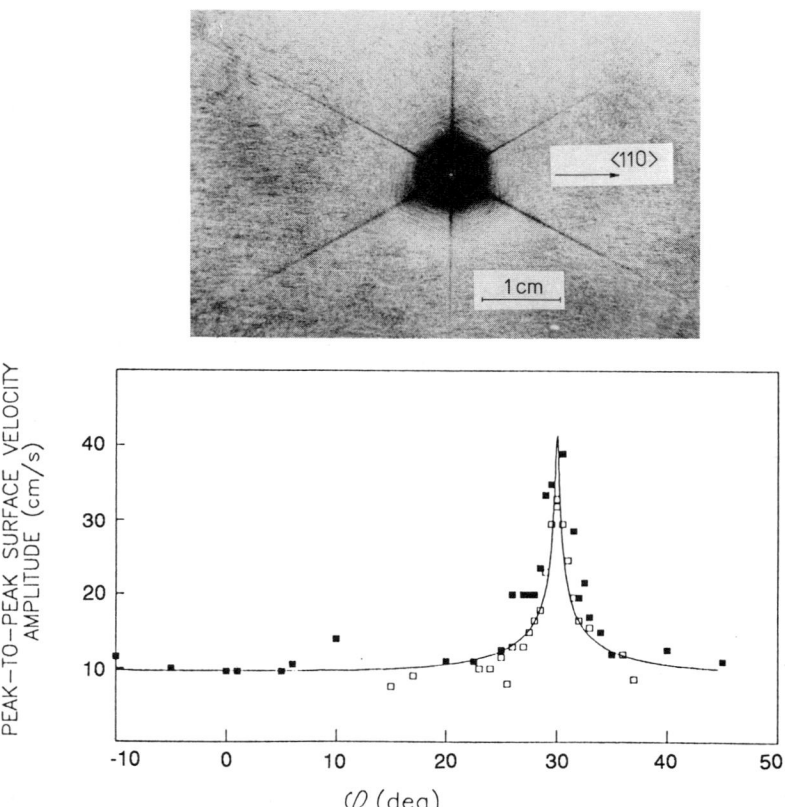

Fig. 1. Surface phonon focusing in the (111) plane of silicon. (a) Focusing pattern: photograph of the dusted surface after laser-acoustic action. (b) Measured and calculated angular dependencies of the amplitude of the normal component of the surface velocity in a SAW pulse. The angle φ is measured from the $<110>$ direction. The black and white circles refer to different series of measurements. For the calculated curve an arbitrary amplitude scale is used.

photograph made with side illumination, a dusty surface looks light and that cleaned of particles looks dark, such a photograph thereby displays qualitatively the SAW amplitude distribution. It should be noted that the surface accelerations in the SAW reached high values, $w_\perp \approx 10^8$ m/s^2, so the removal of micron-size particles was possible [8].

Along with the above visualization method, we measured the SAW amplitudes using a probe beam deflection technique that allowed us to register surface inclinations as small as 2×10^{-6} rad with 10 ns temporal resolution [6,8].

3. RESULTS AND DISCUSSION

Shown in Fig. 1(a) is a photograph of the dusted (111) surface of a silicon wafer after the laser action. Clear-cut dark tracks, making an angle $\varphi = 30°$ with the $<110>$ axis, correspond to those directions in which strong focusing occurs. The pattern reflects the symmetry of the acoustic properties of the (111) surface: the normal to the surface is a threefold axis and there is a symmetry relative to the center and the $<110>$ axis.

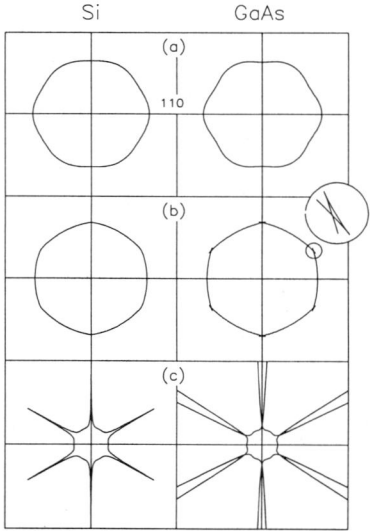

Fig. 2. SAW focusing calculations for the (111) surfaces of Si and GaAs; (a) the SAW slowness curves; (b) the wavefronts of SAWs propagating from a point of source; (c) the polar plots of SAW amplitude.

In Fig. 1(b), the angular dependence of the SAW amplitude measured using the laser probe is presented. Indeed, a sharp maximum can be seen at $\varphi = 30°$. Thus we may conclude that our "dust pattern" reflects correctly the SAW amplitude anisotropy.

To compare the results obtained with theory, calculations have been carried out within the ray approach, in which the measure of the ray density anisotropy is provided by the focusing factor [1,5]

$$A = |d\varphi/d\theta|^{-1}$$

where θ is the wave vector angle and φ is the SAW group velocity angle. The angular dependence of the SAW amplitude a large distance away from a point source is then given by

$$u(\varphi) \approx A^{1/2} \approx \frac{1}{|1 + c''_{\theta\theta}/c|^{1/2}}.$$

The left part of Fig. 2 shows the SAW slowness curve, the SAW wavefront and the polar plot of the SAW amplitude versus the direction of observation for the (111) silicon surface, calculated using $c(\theta)$ dependence from Ref. 9. In the instance under consideration there are no caustics since c''/c never reaches the value (-1); however, in the direction of maximum SAW velocity $\theta = \varphi = 30°$ it is close to this value $(c''_{\theta\theta}/c|_{\theta=30°} = -0.92)$, so that rather strong focusing occurs in this direction. The theoretical angular dependence of the SAW amplitude is also presented in Fig. 1(b) showing good agreement with the experimental data.

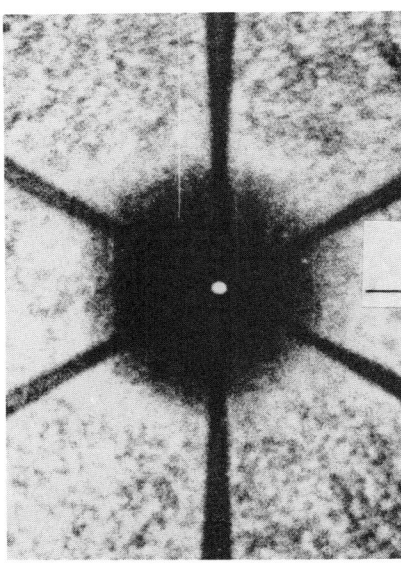

Fig. 3 Focusing pattern on the (111) surface of GaAs.

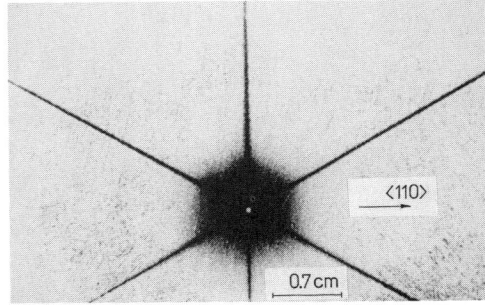

Fig. 4 Focusing patterns on the (111) surface of germanium.

It is of interest, as well, to examine the (111) surface of a cubic crystal with stronger anisotropy, where caustics have to appear. We have chosen GaAs, whose anisotropy factor $\eta = 2c_{44}/(c_{11} - c_{12}) = 1.8$ is considerably larger as compared to that of silicon ($\eta = 1.56$) [9] as an example. In the right part of Fig. 2 the results for the (111) surface of GaAs are presented being calculated again on the basis of the $c(\theta)$ dependence from Ref. 9. Concave sections of the slowness curve which cause cusps of the SAW wavefront can be seen. The singular directions (caustics), at which the focusing factor A becomes infinite, correspond to the cusps and are positioned symmetrically near the direction $\varphi = 30°$. Previously, the singular focusing directions in the (111) cut of GaAs have been calculated by Chernozatonskii and Novikov [5] and our results are in agreement with theirs.

The focusing pattern for the (111) GaAs surface obtained by us experimentally is shown in Fig. 3. The agreement between theory and experiment is seen to be good.

Shown in Fig. 4 is the focusing pattern for the (111) surface of Ge, whose anisotropy factor $\eta = 1.66$ is slightly larger than that of Si. Calculated results [4,5] show that in this case there is also a pair of caustics near $\varphi = 30°$, with the angle between the caustics being equal to $\approx 1°$. The pattern obtained experimentally, however, shows the nonsplitting focusing tracks, so that the angle between caustics can not exceed $0.3°$. The origin of this discrepancy between the theory and experiment is not clear at present.

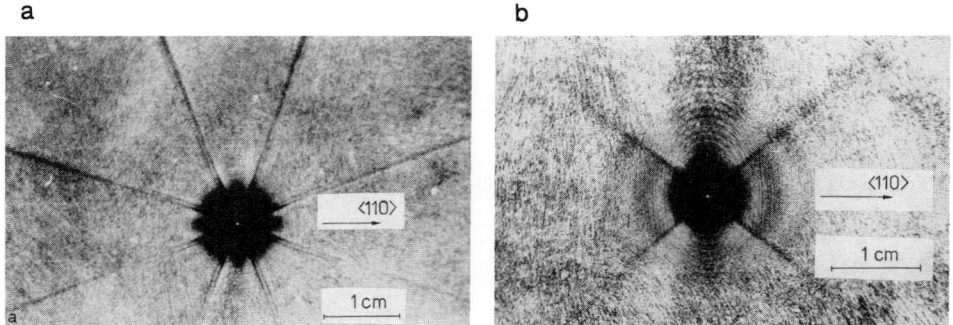

Fig. 5 Focusing patterns on the (a) (100) and (b) (110) surfaces of Si.

We examined as well (110) and (100) cuts of silicon (see Fig. 5). The focusing pattern for the (110) surface of silicon is similar to the calculated one for the (110) surface of Ge [4]. For the (100) surface, however, there is substantial discrepancy between the calculated and experimental patterns [7], this is believed to be caused by Rayleigh wave coupling with the slowest bulk shear mode in the vicinity of the $<110>$ direction.

4. CONCLUSION

Surface phonon focusing has been examined for the first time experimentally using laser generation of SAW pulses. Different focusing patterns have been observed in cubic crystals depending on the elastic constant ratios and surface orientation. As a whole the results obtained agree with the theory, although some discrepancies have been observed.

REFERENCES

[1] See, for a review, G.A. Nortrop and J.P. Wolfe, in *Nonequilibrium Phonon Dynamics*, ed. by W.E. Bron (Plenum, New York, 1985).

[2] A.G. Every, W. Sachse, K.Y. Kim and M.O. Thompson, Phys. Rev. Lett. **65**, 1446 (1990); M.R. Hauser, R.L. Weaver, and J.P. Wolfe, Phys. Rev. Lett. **68**, 2604 (1992).

[3] H. Shirasaki and T. Makimoto, J. Appl. Phys. **49**, 658, 661 (1978); **50**, 2795 (1979).

[4] R.E. Camley and A.A. Maradudin, Phys. Rev. B **B 27**, 1959 (1983);
S. Tamura, K. Honjo, Jap. Journ. Appl. Phys. **20** (Supplement 20-3), 17(1981).

[5] L.A. Chernozatonskii and V.V. Novikov, Solid State Commun. **51**, 643 (1984); V.V. Novikov and L.A. Chernozatonskii, Sov. Phys. Solid State **28**, 233 (1986).

[6] A preliminary report has been published: Al.A. Kolomenskii and A.A. Maznev, JETP Lett. **53**, 423 (1991).

[7] Al.A. Kolomenskii and A.A. Maznev, submitted to Phys. Rev. B.

[8] Al.A. Kolomenskii and A.A. Maznev, Sov. Tech. Phys. Lett. **17**, 483 (1991).

[9] *Akusticheskie kristalli,* edited by M.P. Shaskol'skaya (Nauka, Moscow, 1982) [in Russian].

TIME–DEPENDENT SPECIFIC HEAT OF CRYSTALS AND GLASSES AT LOW TEMPERATURES

N. Sampat and M. Meissner

Hahn–Meitner–Institut Berlin
Neutronenstreuung 2
Glienicker Strasse 100
D–14109 Berlin

INTRODUCTION

At low temperatures heat travels in solids predominantly by phonons. When a heat pulse is applied at the surface of a bulk sample, a hot phonon flux distribution propagates through the sample with the group velocity of sound. Depending on the static and dynamic properties of the solid, the thermal phonon equilibrium system interacts with the traveling phonon distribution. Below 1K, in a perfect dielectric crystal there is pure ballistic transport with no intrinsic scattering. In a glassy system, however, strong scattering processes exist due to anharmonic interactions and diffusive heat transport is observed.

In this contribution we will discuss the question: How does temperature establish itself in a solid shortly after the injection of a heat pulse? This problem of phonon thermalization is closely related to the measurement of the specific heat. Traditionally, the specific heat is determined by the ratio of the initially applied heat Q to the subsequent temperature rise ΔT, which normally is measured on sec–time scales. However, using short–time thermometry on a μsec–time scale it is possible to study the evolution of temperature $\Delta T(t)$ in the solid and to follow the internal thermalization processes in terms of a "time–dependent" specific heat $C(t) = Q/\Delta T(t)$. Presenting heat pulse data on NaF and a–SiO$_2$ below 1K, it will be shown that for both the crystal and the glass a time–dependent specific heat can be physically meaningful.

For dielectric crystals ballistic phonon propagation has been studied extensively [1]. It has been shown that the hot phonon distribution converts to a new equilibrium distribution through inelastic phonon scattering due to specular and diffusive reflection from the internal walls of the solid. For a high–purity NaF single crystal we observe that the thermalization process takes place on a μsec–time scale and that the subsequent temperature profile is totally in agreement with the Debye–model of elastic wave excitations.

In order to study similar thermalization schemes in glasses, we applied the same heat pulse and thermometry techniques as for crystals. At 1K the heat diffusion through a thin disc of vitreous silica with a thickness $L = 1$mm is completed on a

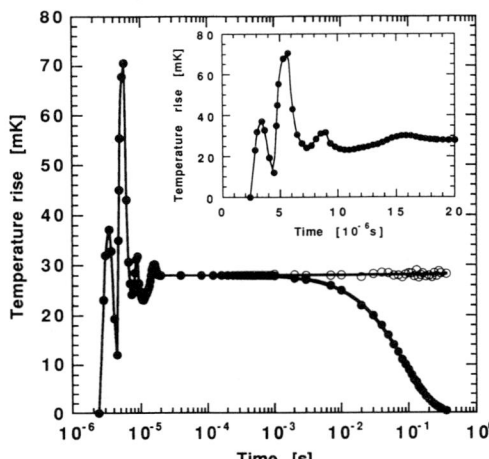

Figure 1: Temperature profile versus time after applying a heat pulse of power P and duration $\Delta t = 1\mu\text{sec}$ with a gold film heater (area $= 1 \times 1\text{mm}^2$) on a high purity NaF single crystal (001–direction, $L = 1\text{cm}$) at temperature $T_0 = 0.4\text{K}$. On μsec–time scale the carbon film thermometer measures the arrival of ballistic longitudinal and transversal phonons (see inset). Shortly afterwards the first reflection peak and a broad maximum at about 10 μsec, the temperature rise $\Delta T(t)$ is constant for $t > 20\mu\text{sec}$ and yields the correct Debye specific heat $C_D = P \cdot \Delta t / \Delta T$ of NaF at 0.4K. For times $t > 1$msec the temperature starts to decrease due to the heat flow from the sample to the thermal bath via the thermal link. This external relaxation can be described by an exponential function (solid line) and can be corrected to the measured data (open circles). By this technique the temperature profile is shown to be constant up to $t = 0.4\text{sec}$.

msec–time scale. Assuming adiabatic boundary conditions, uniform heating on the surface area and 1–dimensional heat flow, the sample should be thermalized on this time scale. However, for times $t > 1$msec, a decrease of the internal temperature of the sample is observed due to the coupling of low energy excitations in the glass. From these experimental results we show that the measured temperature profile $\Delta T(t)$ is in good agreement with the theoretical prediction of the tunneling model [2], where a logarithmic time–dependence of the specific heat is predicted [3].

EXPERIMENTAL

Specific heat measurements were performed with the transient heat pulse technique [4]. In this method the sample is clamped between thin copper rods in order to thermalize it to the base temperature T_0. Given a sample–to–bath time constant $\tau_{SB} = R \cdot C$ (capacity C of the sample and thermal boundary resistance R between the copper/sample contact area) the heat flow from sample to bath is an exponential function of time.

In order to reduce addenda and establish a fast time–response we used a gold film heater (thickness 20nm) and a carbon film thermometer (thickness 500nm, area 1mm^2). The carbon film was produced by rubbing pencil material onto the surface of the sample and preparing a resistance of about 100Ω. Below 1K the carbon thermometer shows increasing sensitivity and a response time in the sub–μsec range [5].

By utilizing both the short heat pulse ($\Delta t < 1\mu$sec) and short–time thermometry technique we were able to observe the temperature profile $\Delta T(t)$ within a quasi–adiabatic time–window ranging from the initial thermalization time τ_i to the external thermalization time τ_{SB}. As this time–window covers many orders of magnitude (from μsec to sec), the temperature profiles were measured in at least three different ranges of time with the help of signal averaging technique.

RESULTS AND DISCUSSION
Display of temperature in a Debye solid: NaF

We have measured time–resolved temperature profiles on a single crystal of NaF ($m = 9.4$g) at $T = 0.4$K. A heat pulse of power P and duration $\Delta t = 1\mu$sec was applied on one side of the crystal. The temperature rise $\Delta T(t)$ at the opposite side ($L = 1$cm) was monitored within the quasi–adiabatic time–window $t = 1$ μsec...0.1sec. For an ideal crystalline solid at low temperature T_0, thermalization takes place only by inelastic diffusive phonon scattering at the surface, leading to a new equilibrium temperature $T = T_0 + \Delta T$ [6]. Depending on the roughness of the surface this thermalization process can establish within a small number of phonon reflections (Casimir limit). However, for perfect surfaces the probability for elastic specular reflections is enhanced and thermalization times can increase up to msec–time scales [7]. Once the temperature equilibrium is established, the specific heat can simply be calculated as $C_p = P \cdot \Delta t / \Delta T$.

In Fig. 1, the measured temperature profile is shown on a linear scale for the time range of the arrival of ballistic phonons. The leading edge times for longitudinal and transversal phonons in (100)–direction are in good agreement with previous measurements of the first and second sound on the same crystal by McNelly et al. [8]. As

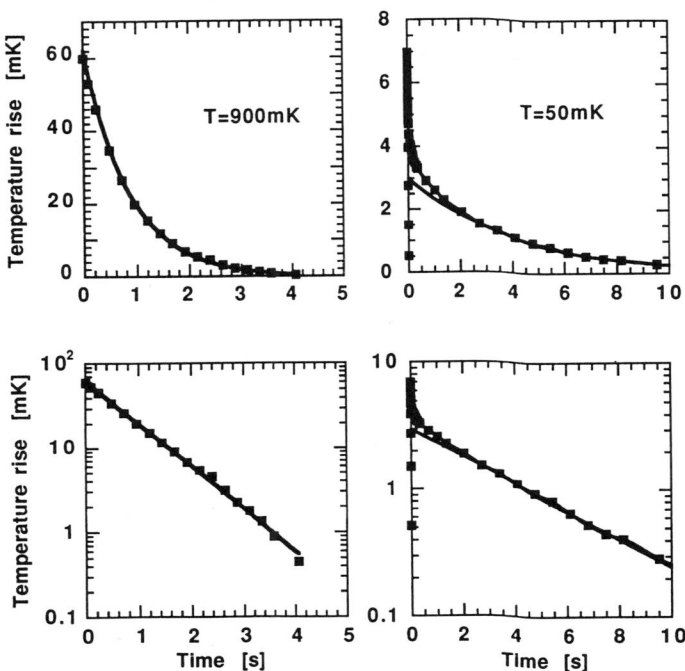

Figure 2: Temperature profiles versus time on a linear time scale for vitreous silica at $T_0 = 0.9$K and $T_0 = 0.05$K plotted with linear (above) and logarithmic (below) scale for the temperature rise. Full lines show the exponential sample–to–bath relaxation. At $T_0 = 0.05$K a strong "overshoot" of the temperature rise is observed indicating a hot phonon temperature around $t = 0$ and a subsequent decrease of the internal temperature of the glass. By using time–resolved thermometry this temperature relaxation can be studied in terms of a time–dependent specific heat (see Fig. 3). At higher temperatures the overshoot effect diminishes rapidly and is no longer observable above 0.5K.

can be seen from both the μsec–range (inset) and the overall logarithmic time–scale display of Fig. 1, the thermalization process is established for times $t > 20\mu\text{sec}$. This means that in the NaF crystal with dimensions of $1 \times 1.8 \times 1.8\text{cm}^3$ about four phonon reflections are sufficient to construct the new phonon equilibrium.

As the temperature profile for $t > 20\mu\text{sec}$ displays a time–independent specific heat in agreement with the Debye value, the experiment demonstrates that thermometry on μsec–time scales is reliable. For shorter times the thermometer reduces to a bolometer measuring the arrival of ballistic phonons and the conversion of phonon flux to a thermal phonon distribution. As the internal temperature can be measured within a quasi–adiabatic time–window any internal relaxation process to the phonon bath can be observed. Using the time–resolved temperature technique on crystals with tunneling impurities, the lattice–impurity coupling in RbBr : OH and KBr : CN has been measured. In case of RbBr, the OH$^-$–phonon coupling times were in good agreement with the spin–lattice times received by optical methods [9].

Time–dependent specific heat of vitreous silica

In disordered solids the transport of heat is a diffusive process. At low temperatures it is assumed that heat is transported by elastic waves, which – due to the strong anharmonicity – have short lifetimes. Assuming the dominant phonon approximation, in vitreous silica at 1K the dominant phonon wavelength is about $\lambda_{dom} = 50\text{nm}$ and the mean free path is about $l_{mfp} = 10\mu\text{m}$ [10]. In order to thermalize a glass on a fast time scale the heat diffusion should take place on a short distance of the order given above. Unfortunately, for very thin samples of vitreous silica (thickness of about 100μm) the addenda capacities of heater and thermometer strongly contribute to the temperature profiles [11].

By this reason we have performed time–dependent specific heat measurements on a disc shaped sample with diameter $d = 15\text{mm}$ and thickness $L = 2\text{mm}$. At 1K a short heat pulse applied uniformly to the total surface of the sample produces a temperature rise to the maximum on the other side at a diffusion time $t_D = L^2/D = 1.3\text{msec}$, where the thermal diffusivity $D = 30\text{cm}^2/\text{sec}$ at this temperature. Adjusting a sample–to–bath time constant for this sample of about 1sec we arranged a quasi–adiabatic time–window of three orders in magnitude.

In Fig. 2 two temperature profiles are shown on the sec–time scale for $T = 0.9\text{K}$ and 0.05K. For both temperatures the sample-to-bath relaxation is characterized by the exponential decrease of the temperature. For $T = 0.05\text{K}$ and for $t < 1\text{sec}$ the temperature profile exhibits a strong overshoot indicating a hot temperature, i.e. a smaller specific heat.

Expanding the time scale to the μsec–range (Fig. 3, upper graph) displays the initial diffusive thermalization which is completed at about 1msec. As the transport equation for the diffusive increase of the temperature (here for $t < 1\text{msec}$) is a rather complicated matter, the time–dependent effects for the increasing profile will not be discussed here. Recently, a complete description of the overall temperature profile using coupled Boltzmann equations has been given [12].

Using the simple picture of a quasi–adiabatic time–window, the maximum temperature rise of about $\Delta T = 7\text{mK}$ can be interpreted as a short–time specific heat. A comparison to the temperature rise, which can be extrapolated from the long–time relaxation $\Delta T_0 = 3\text{mK}$, yields an increased specific heat on a sec–time scale by more than a factor of two. More clearly, the evolution of the time–dependence is shown by plotting $1/\Delta T$, which is a direct measure of the specific heat (Fig. 3, lower graph).

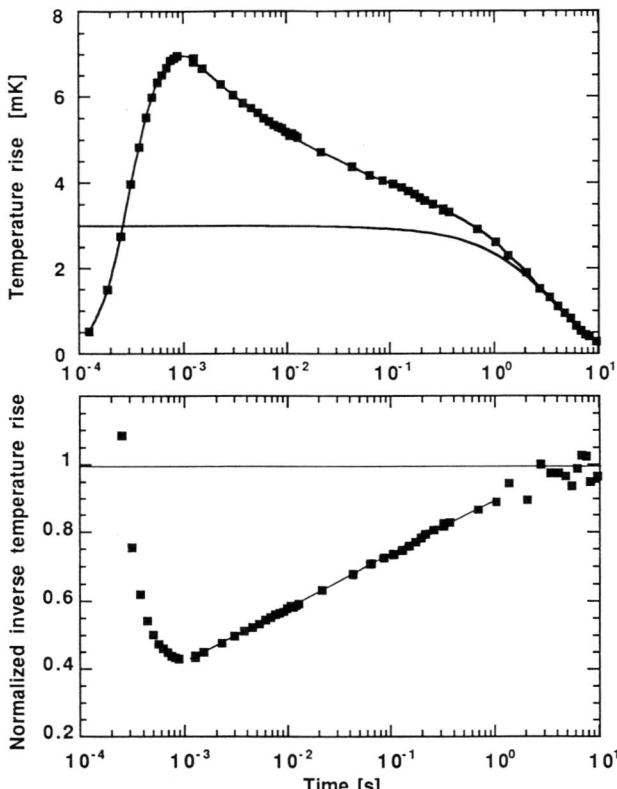

Figure 3: Above: temperature profile versus time after applying a heat pulse with a gold film heater on a thin disc of vitreous silica (thickness 2mm) at temperature $T_0 = 0.05$K. The full line represents the external sample–to–bath relaxation $\Delta T = T_0 \exp(-t/t_{SB})$ which dominates the profile for $t > 2$sec (see Fig. 2). Below: inversed temperature profile taken from above and corrected by the sample–to–bath relaxation. Within the quasi–adiabatic window of about 1msec to 1sec a logarithmic time–dependence of the specific heat is shown by the straight line.

Within the quasi–adiabatic time–window the time–dependence is truly logarithmic as shown by the straight line (within an relative accuracy of 10% for the different data sets).

Earlier measurements on vitreous silica [4, 13] and on the orientational glass KBr : KCN [14] have shown that the observed time–dependences are in good agreement to the tunneling model. According to theory [3], the contribution of two–level–systems (TLS) to the specific heat results in $C_p = const.\ PT \ln[t/(4 \cdot \tau_{min})]$, where P is the spectral density of TLS and τ_{min} is the shortest relaxation time of the coupling between phonons and TLS. The spectral density P can be evaluated from the slope of the logarithmic increase and is a direct measure of this quantity. It has been pointed out that the tunneling model is strongly supported by this direct observation of the TLS coupling to the phonon bath as the experimental time scale expands. However, the problem of the universal microscopic structure of the TLS still remains unsolved and cannot be answered by these experiments.

CONCLUSIONS

By utilizing the heat pulse technique and fast–time thermometry we have measured the evolution of temperature *vs.* time for a perfect crystal and for a glass at temperatures below 1K. For a crystal, the initial arrival of heat is by ballistic phonon transport and the evolution of temperature can be observed on a time scale of the velocity of sound arrival times. The subsequent temperature profile is a direct measure of the specific heat contributions in the crystal. While for a pure phonon system the specific heat is independent of time, low energy contributions from subsystems (impurities, defects) show up with a characteristic time–dependence. For example, on a NaF crystal we have demonstrated how the concept of a time–dependent specific heat can be applied for a sample with a given geometry and experimental time–scale.

In a glass the transport of heat is a diffusive process. As the diffusion length is strongly dependent on temperature (or phonon frequency) a fast thermalization of a glass is possible only for a thin sample at very low temperatures. For a thin disc of vitreous silica with thickness of 1mm the initial thermalization is on msec–time scale (below 1K). In order to measure the subsequent temperature profile due to internal relaxation processes in the glass, the boundary conditions must be strictly adiabatic for orders of magnitude of the diffusion time. Arranging this time–window for the experimental set-up, a slow increase of the specific heat is observable at temperatures below 0.5K. According to the tunneling model for TLS a broad spectrum of relaxation times exists, which results in a weak logarithmic time–dependence of the specifc heat of glasses. Within the experimental window, ranging from 1msec to 1sec, the measured temperature profile for vitreous silica is in agreement to the theoretical prediction.

REFERENCES

[1] For a review, see G.A. Northrop and J.P. Wolfe, in "Non Equilibrium Phonon Dynamics", ed. by W. E. Bron, Plenum, New York (1985).

[2] For a review, see S. Hunklinger and A. K. Raychaudhuri, Prog. in Low Temp. Phys. **9**, 265 (1986).

[3] J.L. Black, Phys. Rev. **B 17**, 2740 (1978).

[4] A.K. Raychaudhury and R.O. Pohl, Phys. Rev. **B 25**, 1310 (1982).

[5] W. Knaak, Thesis, Technische Universität Berlin, unpublished (1986).

[6] T. Klitsner and R.O. Pohl, Phys. Rev. **B 36**, 6551 (1987).

[7] W. Knaak, T. Hauss, M. Kummrow, and M. Meissner, in "Phonon Scattering in Condensed Matter V", ed. by A.C. Anderson and J.P. Wolfe, Springer-Verlag, Berlin, p. 174 (1986).

[8] T.F. McNelly, S.J. Rogers, D.J. Channin, R.J. Rollefson, W.M. Goubau, G.E. Schmidt, J.A. Krummhansl, and R.O. Pohl, Phys. Rev. Lett. **24**, 100 (1970).

[9] S. Kapphan, J. Phys. Chem. Sol. **35**, 621 (1973).

[10] A.C. Anderson, in "Amorphous Solids", ed. by W.A. Phillips, Springer-Verlag, Berlin, p. 65 (1981).

[11] M.T. Loponen, R.C. Dynes, V. Narayanamurti, and J.P. Garno, Phys. Rev. Lett. **42**, 457 (1980); Phys. Rev. **B25**, 1161 (1982).

[12] P. Strehlow, W. Dreyer, and M. Meissner, in "Phonon Scattering in Condensed Matter VII", ed. by M. Meissner and R.O. Pohl, Springer-Verlag, Berlin, p. 229 (1993).

[13] W. Knaak and M. Meissner, in "Proc. 17th Int. Conf. Low Temp. Physics, LT-17", ed. by U. Eckern, A. Schmid, W. Weber, and H. Wuhl, North-Holland, Amsterdam, p. 667 (1984).

[14] J.J. DeYoreo, W. Knaak, M. Meissner, and R.O. Pohl, Phys. Rev. **B 32**, 6091 (1986).

OPTICAL STUDIES OF NONEQUILIBRIUM PHONONS IN SEMICONDUCTORS

A.V. Akimov, A.A. Kaplyanskii, and E.S. Moskalenko
A.F. Ioffe Physical–Technical Institute,
194021 St. Petersburg, Russia

1. INTRODUCTION

Over the years luminescence has became a powerful technique for the detection of terahertz acoustic phonons in crystals [1]. Used mainly in insulators the method is based on the interaction of nonequilibrium phonons with electronic emitting states of "probe" impurities in photoexcited crystals (such as ruby, doped flourite, etc.). The technique has an advantage over the methods using superconducting devices because of a much wider spectral range of phonon frequencies and the lack of the boundary between the crystal and detector.

In this paper we present the results of our carried out during last 6 years studies of high–frequency phonons in semiconductor crystals by measuring the effect of nonequilibrium phonons on the edge luminescence in semiconductors at liquid helium temperatures. In comparison with insulators, semiconductors have two specific features which are significant for luminescence detection of phonons.

The first feature is the existence in an optically excited semiconductor of "free" quasiparticles - free electrons or/and holes and excitons. Radiative transitions are due to different types of electron - hole recombination (band–band, exciton, band–impurity luminescence). Nonequilibrium phonons, injected into the crystal may change the concentration of free quasiparticles, as well as the energy and momentum distribution of them which may reveal in luminescence in different ways. The heating of electron (exciton) gas leads to the broadening of corresponding distribution function in the energy band and of corresponding luminescence lines. The phonon heating of electrons often results in the quenching of band–band and band–impurity luminescence due to competing processes of thermally activated carrier capture by nonradiative recombination centres. Phonons also often induce the drag effect [2,3,4] which is accompanied by the spatial shift of luminescence region in the sample.

The second specific feature is the small binding energy of localized electronic states in semiconductors - levels of shallow impurities and of bound excitons. Therefore THz acoustic phonons may induce ionization of shallow impurities and dissociation of bound excitons. These processes lead to the increasing of free quasiparticle luminescence and, correspondingly, to the quenching of luminescence lines due to localized states (donor - acceptor recombination, bound exciton luminescence, etc.).

The luminescence technique in semiconductors often posesses frequency selectivity for phonons. Indeed, in the case of localized states phonon–induced ionization (dissociation) may only take place if phonon energy is larger than threshold being equal to the binding energy of shallow impurity or bound exciton. In the case of phonon interaction with free quasiparticles, the phonon modes that appear active in the interaction, may be found from the well known energy and momentum conservation rules. Only phonons with energies

$$\hbar\omega = 2m^*s^2 + 2\left(2m^*s^2 E\right)^{1/2} \qquad (1)$$

are active in electron (exciton) - phonon interaction, m^* is the quasiparticle effective mass, s is a sound velocity.

In the present paper experimental studies of nonequilibrium phonons using luminescence of semiconductors are reviewed. In Sect. 2 the exciton–phonon luminescence of cuprous oxide crystals is used to study the phonon spectrum of hot spot. In Sect. 3 we describe the exciton drag effect in silicon and the application of this effect to the studies of the different properties of terahertz phonons - reflection from interface, isotope scattering, phonons in amorphous materials. In Sect. 4 we present two experiments where luminescence technique is used for the study of electron–phonon interaction in two–dimensional electron gas. In the Conclusion we briefly give review of some other experiments where optical detection of phonons in semiconductors is used.

2. PHONON HOT SPOT AND EXCITON LUMINESCENCE OF CUPROUS OXIDE CRYSTALS

In the case of optical excitation of semiconductors the major part of absorbed energy is transformed into phonons which are emitted into the lattice in the processes of energy relaxation and recombination of photoexcited carriers and excitons. The intense interband excitation of crystal at $T = 2K$, when light penetration is small ($\sim 1\mu m$), results in the formation of so–called phonon "hot spot" * (HS) [5], which is a small local region near the surface strongly overheated relative to the cold bulk. Such HS may be formed in crystal also under injection of enough strong heat pulses from the metal film. A long ($10^{-6} - 10^{-5}$s) time of cooling which depends on the power injected is characteristic for the HS [6]. It is due to processes of inelastic phonon–phonon scattering and of elastic isotope scattering which prevent the ballistic escape of phonons, especially of high–frequency ones, from the HS.

The question arises whether the frequency distribution of phonons in HS is a local equilibrium one and may be described by a definite temperature. It is possible in principle to get the answer by measuring the energy distribution of excitons interacting with the phonons in HS region.

The study [7] was performed on 1s–ortoexcitons in cuprous oxide crystals. The spectral shape $I(E)$ of luminescence band in 614nm region which is due to radiative

---

*For detailed description of HS cf. D.V. Kozakovtsev, Y.B. Levinson, Theory of formation dynamics and explosion of a hot spot, in "Physics of Phonons", T. Paszkiewicz, ed., Springer, Berlin - Heidelberg, 1987

annihilation of 1s–excitons with emission of optical phonon 109cm^{-1} was studied. As was shown earlier [8], the shape of $I(E)$ directly reflects the kinetic energy distribution $f(E)$ of excitons in 1s–band

$$I(E) \sim \rho(E) \, f(E) \qquad (2)$$

where $\rho(E) \sim \sqrt{E}$ is the density of states in exciton band. The value of E is measured as the distance from the spectral frequency $\nu_0 - 109\text{cm}^{-1}$, where $\nu_0 = 16400\text{cm}^{-1}$ is the frequency of pure electronic quadrupole 1s–exciton transition. The lifetimes of 1s–ortoexcitons in Cu_2O ranges from 10^{-9} to 10^{-8}s [9].

The Cu_2O single crystal is immersed in pumped liquid helium ($T_0 = 1.7$K). Strong pulses of $2nd$ harmonic of YAG : Nd laser ($\lambda = 530$nm, pulse duration 200ns, maximal pulse power $P = 100\text{W/mm}^2$, repetition rate 5kHz) pump $0.1 \times 0.1 \times 0.01\text{mm}^3$ volume of the sample, exciting simultaneously both phonon HS and luminescence

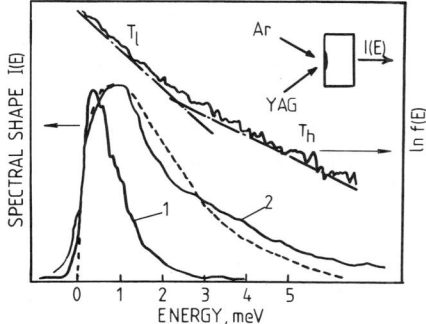
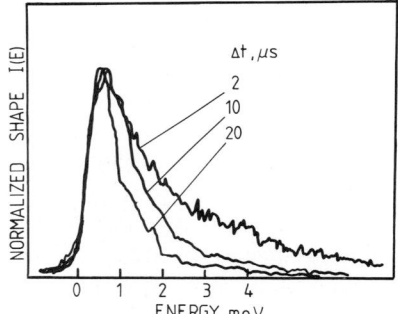

Fig. 1. Experimental shape $I(E)$ and $\ln(I(E)/E)$ in the exciton luminescence spectra of Cu_2O (see text). Theoretical distributions: MB spectrum for $T = 16$K (dashed line) and $\ln f(E)$ for $T = 14$K, $T = 22$K (dashed–doted).

Fig. 2. Exciton luminescence spectra of Cu_2O measured with time delay after intense pulse excitation.

of 1s–excitons. By using a weak probe (inset Fig. 1) cw – Ar–laser beam (514nm, 0.2W/mm^2) which excites the luminescence (but not the HS!) we could measure the luminescence spectrum $I(E)$ with various time delays Δt after action of main YAG–pulse.

Figure 1 shows typical shapes $I(E)$ in the luminescence spectrum measured during the time of YAG–pulse action ($\Delta t = 0$) at low ($P = 0.4\text{W/mm}^2$, curve 1) and high ($P = 100\text{W/mm}^2$, curve 2) levels of pumping density. A broadening of $I(E)$ shape with increase of P is seen, which indicates the heating of exciton gas. Fig. 2 shows exciton luminescence spectra measured with different time delays after YAG–pulse action. It is seen that shape $I(E)$ remains broadened during long times $\sim 10\mu$s after YAG–pulse action. These times strongly exceed the lifetimes of ortoexcitons ($\tau_0 \sim 10^{-9}$s) and of all other possible electronic states in Cu_2O, excited by YAG–laser.

Hence the broadening of $I(E)$ under strong optical excitation may be due only to the heating of phonon system in the excited volume ($HS!$), which they may remain during microsecond time scales. The heating of the exciton gas reflected in $I(E)$ broadening is produced obviously by interaction of 1s–excitons with acoustic phonons of HS.

It is important to stress, that the shape of $I(E)$ broadened by strong optical pumping cannot be approximated by a single Maxwell–Boltzmann (MB) distribution with any fixed temperature (see Fig. 1, dotted line). Hence the experimental exciton energy distribution function $f(E)$ is not Boltzmann exponent ($e^{-E/kT}$). In Fig. 1 $f(E)$ is approximated by two Boltzmann distributions with different temperatures for the low energy ($T_l = 14K$) and high energy ($T_h = 22K$) parts, the value of T_l being significantly smaller than T_h. The crossover energy between the two distributions with T_l and T_h is nearly 2meV.

The observed absence of equilibrium MB distribution in exciton gas interacting with HS phonon gas means that frequency distribution of phonons in HS is also nonequilibrium (non–Planckian). The fact that the temperature for the "low energy" part of the exciton distribution $I(E)$ is lower than for "high energy" part ($T_l < T_h$) shows that in a phonon gas interacting with excitons the effective temperature for low frequency phonons is lower than for high frequency ones. Indeed, the expression (1) for Cu_2O parameters ($m^* = 3m_0$, $s = 4.8 \; 10^5 cm/s$) shows, that excitons with energy $E \approx 2meV$ interact inelastically with low frequency ($\omega < 3.5meV$) phonons, whereas excitons with $E \approx 6meV$ interact predominantly with high frequency phonons ($\omega \approx 6meV$). Of course the description of phonon spectrum by means of two temperatures is very crude and serves only as a semiquantitative illustration of the nonequilibrium non–Planckian nature of the phonon spectrum in an HS, which has the deficit of low frequency phonons[†]. This deficit originates from the frequency dependence of anharmonic and defect phonon scattering processes, which permits low frequency phonons to escape the HS region more easily [10].

3. THE EFFECT OF HEAT PHONON PULSES ON EXCITON LUMINESCENCE IN SILICON

Silicon serves as a model crystal in the physics of nonequilibrium phonons in solids due to the possibility of having pure and perfect single crystals with a large mean free path of subterahertz phonons at low temperatures. Many experiments in Si with ballistic phonon beams generated and detected mostly by superconducting tunnel junction technique were done [11]. The first studies of nonequilibrium phonons in Si by means of optical technique were reported in Ref. [12]

3.1. Experimental Technique

In typical experiments (inset Fig. 3) an ultrapure Si single crystal ($\rho \sim 10^4 \; \Omega cm$) immersed in liquid helium (LH) ($T = 1.8K$) is illuminated by cw − Ar–laser beam, which produces $e - h$ pairs near the surface, which bind into free excitons (FE). The majority part of the FE annihilate near the surface due to strong surface recombination. Other FE diffuse into the sample forming an excitonic cloud (EC), whose

---

[†]cf. the paper by Maksimov et al. this volume

thickness is $r \sim \sqrt{D\tau_0} \sim 1\mu m$, where $D = 100 cm^2/s$ [13] is the FE diffusion constant, and $\tau_0 \sim 1\mu s$ [14] - is the FE bulk lifetime. The latter is determined mainly by the capture of FE by residual impurities, $\tau_0 \sim -w_{FB}^{-1}$, which results in the formation of bound excitons (BE), where w_{FB} is the $FE \to BE$ capture probability.

A thin metal film "h" on the opposite face of the sample is heated by short (200ns) current pulses, phonon heat pulses being injected into the sample. The effect of these nonequilibrium phonon pulses on the EC is studied by measuring the phonon–induced temporal changes $\Delta I(t)$ in the intensity of the lines of the EC luminescence spectrum. The spectral emission of FE (LO, TO assisted transitions 1.099, 1.097eV) [15]) and of BE (LO assisted transitions 1.093eV [16]) was measured.

3.2. Observation of Phonon–Induced Drag of Excitons

It was established that phonon induced relative changes of FE luminescence intensity are of several tens percent and drastically depend on the boundary conditions on the surface with EC, namely on the nature of media in contact with this surface (superfluid LH, He gas, vacuum) and on the perfectness of Si surface.

Figure 3 shows the phonon–induced luminescence pulses $\Delta I^F(t)$ for FE and $\Delta I^B(t)$ for BE for the case of standard ("ordinary") Si surfaces with the thin (70 Å) oxide film usually formed by oxidation of surfaces in the open air atmosphere. In the case of surface–vacuum boundary the phonon–induced enhancement of FE-luminescence is observed (Fig. 3a). This "positive" pulse has a sharp leading edge, coinciding in time with time of ballistic flight of phonons from the heater (h) to the detector (EC). In the case of surface - LH boundary the sign of pulse $\Delta I^F(t)$ is negative indicating the phonon–induced quenching of FE-luminescence (Fig. 3b). It can be seen (Fig. 3b) that the amplitude of negative pulses increases with increase of heat pulses power. Above some injected energy threshold, when the LH around the sample boils and a helium bubble is formed, the sign of $\Delta I^F(t)$ is reversed and becomes positive (Fig. 3b).

Finally, the sign of the phonon induced pulses of BE luminescence $\Delta I^B(t)$ observed in all experiments is negative, irrespective to the above boundary conditions near the Si surface with EC (Fig. 3c).

While discussing the observed effect of phonon pulses on EC luminescence we have to consider the interaction of acoustic phonons coming from the heater "h" both with FE (i) and BE (ii).

(i) The inelastic FE–phonon scattering processes result in redistribution of energy and momenta between the phonon and exciton gases. The energy transfer and FE heating by phonon pulse does not change the total number of FE in EC and thus must not change the intensity of FE luminescence. If, however, the distribution of momenta in the phonon pulse remains anisotropic in the EC region the transfer of momenta from phonons to FE may lead to the FE drag towards the surface where they effectively recombine, the FE luminescence being quenched. In Si only the low frequency acoustic phonons with $\omega < 1.8$meV (0.4THz) are active in FE drag, because of the energy and momentum conservation conditions (1). Quantitatively, the phonon flux induced FE drag may be characterized by exciton drift velocity $v = s\Delta\bar{n}/\bar{n}$, where s is the sound velocity, $\bar{n}$ is the mean occupation number for low frequency phonons (< 0.4THz), $\Delta\bar{n} = n^+ - n^-$ describes the anisotropic distribution of phonon occupation numbers in a flux with phonon momentum projections, directed normally towards the surface (n^+) and back (n^-) [3].

(ii) Dissociation of bound excitons $BE \to FE$ which may be induced by "high–frequency" acoustic phonons with frequencies $\Omega > \Delta E$, where $\Delta E = 3.8$meV (0.9THz) is the binding energy of BE in Si, results obviously in enhancement of FE–luminescence and in quenching of BE luminescence. The efficiency of $BE \to FE$ processes may be characterized by the value of the real bulk lifetime τ of FE in EC in the presence of high–frequency phonons. This time appears to be increased in comparison with the previously considered bulk lifetime τ_0 in the absence of phonons due

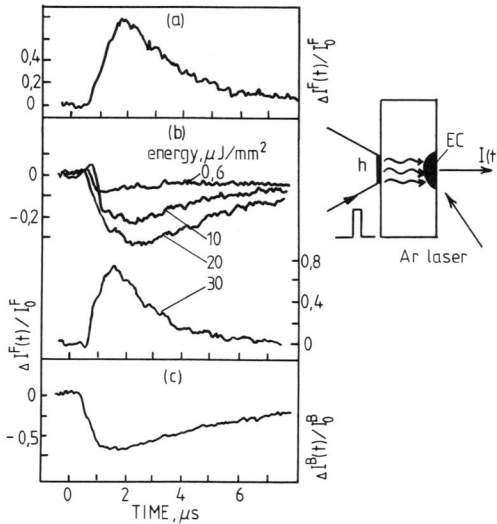

Fig. 3. Scheme of experiments and phonon induced luminescence pulses for Si sample with an ordinary surface in contact with vacuum (a), liquid helium (b,c); $a,b - FE$, c - BE.

to repetitive $BE \to FE$ processes of liberation of captured ($FE \to BE$) excitons $\tau = \tau_0(1 + w_{BF}/w_0)$. Here w_0 ($w_0^{-1} = 1\mu s$ [14]) is the probability of Auger annihilation of BE, while $w_{BF} \sim N$ is the probability of phonon-induced dissociation $BE \to FE$, with N being the occupation number of active phonons with $\Omega > 0.9$THz.

Thus, there are two phonon frequency depending contributions of the opposite sign to the total phonon induced FE–luminescence signal $\Delta I^F(t)$. In the "quasistationary" approximation (drift velocity v=const, bulk FE lifetime τ=const) the ratio of the intensities of FE–luminescence in the presence (I) and absence (I_0) of phonons is [12]

$$I/I_0 = (\tau/\tau_0)^{1/2} \left[\beta + \left(\beta^2 + 1\right)^{1/2}\right]^{-1} \quad (3)$$

the coefficient $\beta = 0.5\, v\sqrt{\tau/D}$ - characterizing the relative contribution of the processes of FE drag (v) and of dissociation $BE \rightarrow FE$ (τ). Depending on the relative value of τ and v both the increase $(I/I_0 > 1)$ and quenching $(I/I_0 < 1)$ of the intensity of FE–luminescence is possible.

The relative role of the FE drag and $BE \rightarrow FE$ dissociation processes strongly depends on experimental conditions, which determine the transmission and reflection of acoustic phonons at the crystal surface with EC. Indeed, if the reflection of acoustic phonons from the crystal boundary is high enough, the back phonon flux compensates the incident one and the net directed phonon flux in the EC region becomes negligible. Hence, no phonon flux induced FE drag towards the surface occurs in this case and the resulting signal $\Delta I^F(t)$ of FE–emission must be positive since it is due exclusively to the $BE \rightarrow FE$ dissociation induced by phonons with $\Omega > 0.9$THz. This explains the positive sign of $\Delta I^F(t)$ observed for Si crystal–vacuum (Fig. 3a) and crystal–He gas boundaries, where 100% reflection of phonons takes place. The negative sign of the signals $\Delta I^F(t)$ observed for all oxidized Si surfaces in contact with LH (Fig. 3b) can be attributed to the predominant role of luminescence quenching due to FE drag accompanied by a strong surface recombination of FE. The FE drag occurs here because of the known strong phonon transmission through such boundaries (Kapitza anomaly [11]) resulting in the emergence of a uncompensated phonon flux towards the crystal surface. Finally the universal negative sign of BE–luminescence pulses $\Delta I^B(t)$ (Fig. 3c) is explained by phonon–induced $BE \rightarrow FE$ dissociation processes which occur in all situations.

Thus for the first time the phonon flux induced drag of FE in Si crystals was observed. Erlier the drag of FE by phonons was observed in Ge at 2K [16] and in CdS at 77K [3].

3.3. Reflection of Phonons from Crystal–Liquid Helium Boundary and Kapitza Anomaly

As was mentioned in Section 3.2, the sign of the phonon induced FE–luminescence signal is sensitive to the behaviour of phonons on the crystal surface with EC. It allows the use of the luminescence technique described above for studying the nature of anomalously low heat resistance of the solid–liquid helium boundary which is not consistent with the predictions of the acoustic mismatch theory (Kapitza anomaly). As it was shown earlier in the experiments on the reflection of nonequilibrium phonons from surfaces performed for some crystals by the tunnel junction technique [11] the anomalously high phonon conductance through crystal–LH boundary is due to imperfections of different kind of real crystal surfaces. Indeed, the atomically pure surfaces of alkali fluorides in LH give practically 100% reflectance of phonons (290GHz) which indicates the absence of the Kapitza phonon conductance for such surfaces [17]. It seemed to be interesting to study phonon reflection from "ideal" as–cleaved Si surfaces in LH.

Fig. 4a shows phonon induced FE–luminescence pulses $\Delta I^F(t)$ measured for an atomically pure (111)–surface of Si, which was prepared by cleavage of the sample in situ in LH. A positive $\Delta I^F(t)$ pulse is observed with a steep leading edge corresponding to the ballistic arrival of phonons in the EC region. Fig. 4b shows the pulse $\Delta I^F(t)$ measured under the same conditions but after 20 hours oxidation of this cleaved (111)–surface in the open air at room temperature. It is seen that oxidation of a freshly cleaved surface results in the inversion of sign of the main part of the signal which becomes negative. These results (Fig. 4) directly indicate the existence of strong phonon reflection from a freshly cleaved atomically pure Si surface in contact with LH, the phonon reflectance decreasing significantly with the oxidation of this surface. These results correlate with those obtained for alkali fluorides [17] and correspond fully to the above mentioned conclusions about the close relation between the anomalous Kapitza phonon conductance and irregularities of the structure of real surfaces. It is interesting to note that strong phonon reflection from high quality Si surface prepared by laser annealing was observed in [18].

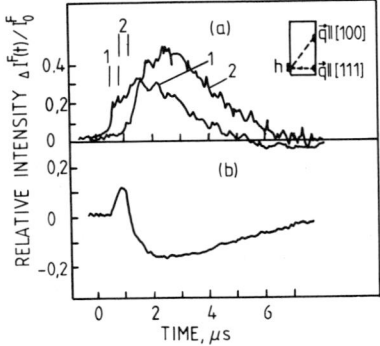
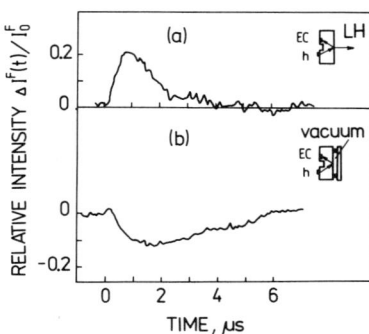

Fig. 4. Phonon induced FE luminescence pulses for a freshly cleaved (a) and oxidized (b) Si surface (111) in contact with LH. The $h-EC$ direction [111] (a–curve 1, b) and [100] (a–curve 2). The arrival times for LA and TA ballistic phonon are indicated by vertical lines.

Fig. 5. Phonon induced FE luminescence pulses for Si in back geometry of experiment.

3.4. Isotopic Scattering of Terahertz Phonons

The propagation of acoustic phonons with frequencies as high as 1 THz and greater was studied in Si. The fact was used that under the conditions of compensation of phonon flux, when the FE drag is small, the positive $\Delta I^F(t)$ signal is due to bound exciton dissociation $BE \rightarrow FE$ induced by high frequency phonons $\Omega > 0.9$ THz (3.8meV).

Figure 4a shows FE–luminescence pulses $\Delta I^F(t)$ observed in the sample with a freshly cleaved (111)–surface for two directions of phonon propagation from the heater

"h" to EC - along [111] and along [001], in the latter case "h" and EC on opposite faces of sample are shifted and not situated on the same normal to the faces. The leading edges of the $\Delta I^F(t)$ pulses correspond to the shortest times of phonon arrival from "h" to EC. This indicates ballistic phonon propagation, as well as the observation of focussing effect for TA–phonons propagating along [001]; indeed, the pulse with **q** [001] is stronger than with **q** [111] (cf. ref. [4], although the distance "$h - EC$" is twice as long in the former case (Fig. 4a, inset).

In addition to the ballistic component in propagation, the prominent bulk scattering was observed for high frequency phonons with $\Omega > 0.9$THz. This scattering reveals itself in the geometry of the experiment, in which both the heater (h) and detector (EC) are on the same face of the sample, with deep cut between them (Fig. 5, inset), preventing the direct ballistic flight of phonons from "h" to EC. In such a geometry the luminescence pulses from EC are induced by phonons which are scattered in the bulk or reflected from the opposite face of the sample. The observed sign $\Delta I^F(t)$ is negative in the case of full reflection at the crystal–vacuum boundary of the low frequency phonons, participating in FE drag (Fig. 5b). But when effective escape of such phonons in LH occurs, a positive pulse is observed (Fig. 5a), which is induced by high frequency phonons, which posess strong ($\tau^{-1} \sim \Omega^4$) bulk Rayleigh scattering.

The first observation of a prominent ballistic component for $\Omega = 0.9$THz phonons, as well the observation of strong bulk scattering of phonons with $\Omega > 0.9$THz show that the mean free path l for THz phonons in pure silicon is of the order of several mm. This value coincides well with the theoretical estimate [19] $l = 4$mm for the Rayleigh scattering of 1THz phonons from Si isotopes in natural material (4.7% ^{29}Si, 3.05% ^{30}Si).

3.5. Luminescence Study of the Scattering of High Frequency Phonons in Amorphous Semiconductor Films

The luminescent method is used to study the phonon transmittance through thin amorphous semiconductor (Si, Ge) films [20].

The samples used were parallel–side plates consisting of crystalline (111) silicon (c-Si) substrate $10 \times 5 \times 2$mm^3 and amorphous film (AF) deposited on it (Fig. 6a). Main results were obtained on hydrogenated amorphous silicon (a – Si : H) films and also on amorphous germanium (a – Ge) films. The samples were immersed in pumped liquid helium ($T = 1.7$K). Heat pulses were injected into AF by means of pulse (duration $2\ 10^{-7}$s) current heating of thin (100 Å) constantan film "h", deposited on AF. From "h" into AF a broad phonon spectrum is injected. Phonons transmitted through AF reach c – Si substrate and propagate ballistically to its opposite side, where phonon detection with time resolution 50ns is performed.

Opposite to AF side of c-Si is excited by cw – Ar–laser. As a result EC is formed near the surface of c-Si and phonon induced changes in the intensities of FE and BE lines $I^F(t)$ and $I^B(t)$ respectively were measured.

The observed $\Delta I^F(t)$ signal has negative sign (Fig. 6b) and its amplitude increases with AF thickness d. This result definitely shows that transmission of nonequilibrium phonons through a – Si : H film is accompanied by effective phonon down–conversion from high–frequency to low–frequency part of phonon spectrum, caused by inelastic scattering of phonons. Actually such conversion leads to the enhancement of FE drag by phonons with $\omega < 0.4$THz and to the reduction of dissociation rate $BE \to FE$ induced by phonons with $\Omega > 0.9$THz.

As a result the $\Delta I^F(t)$ negative signal amplitude (Fig. 6b) has to increase while the thickness increases. At the same time frequency redistribution in phonon spectrum must not have a strong influence on BE luminescence (Fig. 6c), which is effected by phonons with $\Omega > 0.9$THz and $\omega < 0.4$THz in the same way. Fig. 6d shows luminescence pulses measured for $a-\text{Si}:\text{H}$ films with different hydrogen concentration N_H(6%, 10%, 16%). It is seen that amplitude of negative $\Delta I^F(t)$ signal increases with N_H. These results show that centers responsible for inelastic scattering includes H. Our conclusion about effective phonon down-conversion in $a-\text{Si}:\text{H}$ is in agreement with experiments [21].

4. LUMINESCENCE STUDY OF PHONON ABSORBTION AND EMISSION BY 2DEG

The study of phonon absorbtion and emission by a two-dimensional electron gas ($2DEG$) provides detailed information on the nature of electron-phonon interaction [22]. In previous experiments these processes has been studied using electrical methods of phonon detection. In the present section we present the experiments where luminescence detection of noneqilibrium phonons is used.

4.1. Effect of Acoustic Phonons on Impurity Luminescence of Quantum-Well GaAs/AlGaAs structure

We investigated a MBE structure of $5 \times 5\text{mm}^2$ composed of (100) layer of GaAs (layer thickness $d = 100\text{Å}$) with $\text{Al}_{0.3}\text{Ga}_{0.7}\text{As}$ barries (80Å). This structure was deposited on a (001)-oriented semiinsulating GaAs substrate (of thickness 0.4mm) with buffer layer of slightly p-GaAs. The AlGaAs barries were practically impermeable to carriers in the GaAs layers, so that the motion of carriers in the narrow gap material sould be regarded as quasi two-dimensional. The effects considered below were determined by the properties of the individual layer and the large number of the layers affected only the additive enhancement of the luminescence signal emitted by the whole structure.

In experiments [23] carried out at $T = 1.8$K we used $He-Ne$ laser to excite the photoluminescence spectrum of GaAs layers. The spectrum is well known [24] and consists of a line due to free excitons in quantum wells (QW) in GaAs with a maximum at $E = 1.544$eV and a wide inhomogeneosly broadened band in the range $E = 1.51 - 1.54$eV due to conduction band-acceptor (carbon) transition in GaAs quantum wells. In addition to the luminescence from the GaAs quantum wells, the spectrum includes a band $E = 1.49$eV representing conduction band - acceptor transition in bulk GaAs buffer layer.

The side of GaAs substrate opposite to the QW structure was excited (inset Fig. 7) by pulses of the 2nd harmonic $YAG:\text{Nd}$ laser ($\lambda = 530$nm, pulse duration $2 \cdot 10^{-7}$s, energy density per pulse less than 100mJ/cm^2), which created HS (see Sec. 2). This HS acted as a source of nonequilibrium phonons which reached the QW structure after passing through the 0.4mm substrate. We investigated the influence of these phonon pulses on the spectrum of luminescence due to the QW structure and bulk GaAs in the buffer layer (BL).

We found that phonon pulses reduced cosiderably the intensities of the GaAs impurity luminescence bands conduction band - acceptor both in QW and in BL. The "negative" going luminescence pulses $I(t)$ are shown in Fig. 7a. Figure 7b shows the results of our measurements of steady state thermal quenching of the QW and BL luminescence bands observe under the same excitation conditions by He–Ne laser as in the case of phonon pulse experiment.

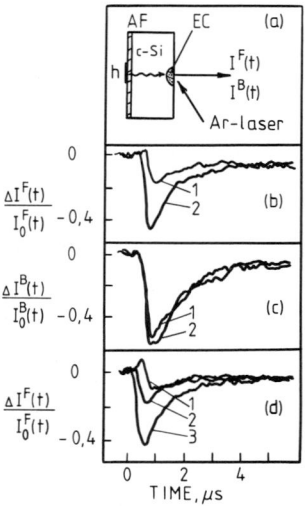

Fig. 6. Experimental scheme (a) and the relative intensities of FE (b,d) and BE (c) luminescence pulses for two thicknesses of AF $a-$Si : 16%H (b,c) ($1-d=0.2$mm; $2-d=1.0$mm) and for different concentrations $N_H(d)$ in AF with $d=1.0$mm (N_H : 6% – 1; 10% – 2; 16% – 3).

A comparison of the amplitude of phonon–induced "negative" luminescence pulse $I(t)$ in QW and BL (Fig. 7a) with measured (Fig. 7b) curves of the equilibrium temperature quenching of these bands has yieled "effective" temperature T_{eff} in the QW and BL under HS conditions. It has been found that T_{eff} in QW always exceeds by a few degrees T_{eff} in the bulk BL.

The temperature quenching of the impurity luminescence in our samples is due to the heating of electron gas in material [23], which enhances the role of competitive processes of temperature–activated electron trapping at the nonradiative recombination centers. Hence the experimentally observed inequality $T_{eff}^{2D} > T_{eff}^{3D}$ implies that electron heating by phonon pulse in QW is stronger than that in the bulk BL.

The parameters of phonon pulses in QW and BL are very close. Therefore we attribute the difference in the degree of electron heating in QW and BL to the difference of phonon absorbtion and correspondingly electron–phonon interaction of the $2DEG$ in QW and in a 3–dimensional electron gas ($3DEG$). In $3DEG$ the requirement of energy and momentum conservation in the electron–phonon interaction event places a constraint on the phonon frequency at 0.3meV. For $2DEG$ [22,25] in QW where the phonon momentum across the layer is not conserved in an interaction event, the region of allowable phonon frequencies broadens to 1meV [23]. Thus while in a bulk BL only relatively low–frequency acoustic phonons contribute to the T_{eff}^{3D}, in the $2DEG$ in QW the role of high frequency phonons is also essential.

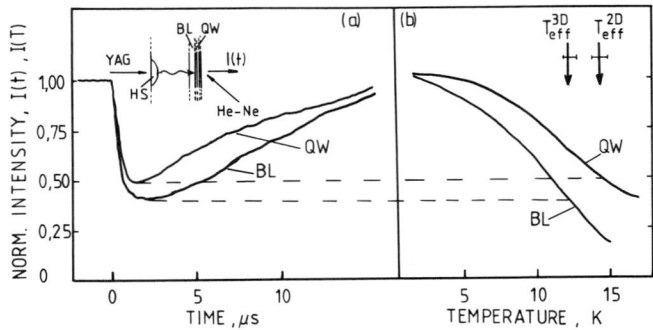

Fig. 7. Luminescence pulses $I(t)$ (a) and curves representing thermal qeunching of the luminescence under steady–state heating conditions (b).

It is known (see Sec. 2) that the hot spot phonon spectrum is non–Planckian and is characterized by a relative deficiency in low–frequency phonons. So the lack of low-frequency phonons in the hot spot spectrum results in the effective temperature in the bulk material, where T_{eff}^{3D} is determined by low–frequency phonons, being lower than that in the QW, where high–frequency phonons contribute to T_{eff}^{2D}.

In general note that this T_{eff} difference can manifest itself in experiment only under the conditions of a nonequilibrium phonon spectrum.

4.2. Luminescence Detection of Phonons Emitted from the First and Second Sub–Bands of a $2DEG$ in Silicon

An exciton cloud detector has been used [26] to measure the frequency distribution of the phonons emitted by a heated $2DEG$ in a Si-$MOSFET$ at power densities at which there is significant thermal population of an excited subband.

Studies of acoustic phonon emission from a heated two–dimensional electron gas ($2DEG$) have largely been carried out at sheet densities n_s and electron temperatures T_e for which $E_F < E_{ex}$ (E_{ex} - the energy of the first excited subband) and $k_B T_e \ll (E_{ex} - E_F)$ so that the emission is very largely from transitions within the ground state subband. At higher values of E_F and T_e thermal population of an excited subband becomes significant and changes are expected in the nature of the emission. In the present work measurements have been made on (001) Si $MOSFET$ samples for which it is known that E_F enters the first excited subband when $n_s = 4.9\ 10^{16}\text{m}^{-2}$ [27]. Furthermore, the values of power input P_m needed to produce significant population of E_{ex} had been determined as a function of n_s by observing the fall in resistivity for $P > P_m$ caused by the presence of higher mobility electrons in the excited subband [28].

The measurements used an exciton cloud as a phonon detector (see Sec. 3). The cloud was formed at the substrate face opposite the 2DEG using a 0.3mm diameter $cw-argon$ - laser beam. The substrate was immersed in liquid helium at 2K at which temperature all the excitons were bound (BE) by the acceptor impurities as could be seen from the luminescence arising from exciton annihilation which showed no trace of the free exciton (FE) spectrum. The $2DEG$ was then heated by 100ns pulses and the phonons incident on the exciton cloud created a pulse of FE luminescence and a decrease (negative–going pulse $I^B(t)$) of BE luminescence caused by the dissociation of excitons from acceptor impurities. Since the binding energy is known to be 3.8meV, $I^B(t)$ reflects the intensity of phonons of frequency > 920GHz. The signal was integrated over a time of 500ns.

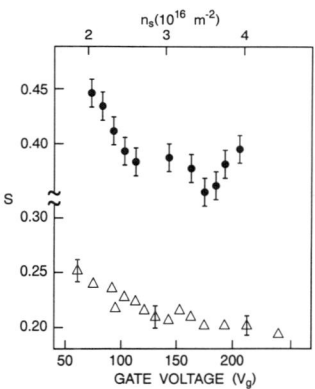

Fig. 8. The relative decrease in BE luminescence, induced by phonon pulse as a function of n_s for two values of injected power density. Triangles $-P = 20\text{mW/mm}^2$; dots $-P = 60\text{mW/mm}^2$.

Fig. 8 shows that for two values of the input power $P > P_m$ the relative change S in BE luminescence fell significantly as n_s was increased (P was kept constant as n_s was varied). The area of the $2DEG$ was large compared with the exciton cloud so we discard the possibility that this decrease is due to changes in angular distribution and attribute it to a shift in spectrum to lower frequencies (< 920GHz). This contrasts with the shift to higher frequencies expected from the emission from a single subband as a result of the increase in the cut–off values of both the in–plane ($2k_F$) and out–of–plane (a^{-1}) wavenumbers (k_F is the Fermi wavenumber and a the thickness of the $2DEG$). However, it would appear to be consistent with part of the emission arising from intrasubband transitions from an excited subband where k_F is lower and the electrons less strongly confined.

5. OTHER OPTICAL STUDIES OF NONEQUILIBRIUM PHONONS

In the final section we shall briefly review other studies of nonequilibrium phonons in semiconductors using optical technique. The kinetics of optical nonequilibrium phonons was studied in several experiments [29] using Raman scattering. Luminescence technique and nonequilibrium phonons were used to study Auger–processes [30]. The phonon–induced scattering of FE between different exciton bands in Si was discussed [31]. The phonon–induced anti-Stokes acoustic phonon wing of zero–phonon lines ("vibronic spectrometer") was used to analize the phonon spectrum of metal film injectors [32].

ACKNOWLEDGEMENTS

We wish to thank all our coworkers in A.F. Ioffe Institute, Nottingham University in England and Institute of Physics in Prague. Particularly we are very greateful to Prof. L.J. Challis (Nottingham University) for and Dr. J. Kočka (Institute of Physics, Prague) for cooperation in the last years. Experiments which are described above were supported by Academy of Sciences of the USSR (now Russian Academy of Sciences). The experiment described in Sec. 4.2 is partly financed by The Royal Society, United Kingdom.

REFERENCES

[1] W.E. Bron, Phonon generation, transport and detection through electronic states in solids, in "Nonequilibrium Phonons in Nonmetalic Crystals", W. Eisenmenger and A.A. Kaplyanskii, eds., North–Holland, Amsterdam (1986), Ultrashort transient dynamics of phonons and electrons in semiconductors, in "Physics of Phonons", T. Paszkiewicz, ed., Springer, Berlin–Heidelberg, 1987.

[2] L.V. Keldysh and N.N. Sibeldin, Phonon wind in higly excited semiconductors, in "Nonequilibrium Phonons in Nonmetalic Crystals", W. Eisenmenger and A.A. Kaplyanskii, eds., North–Holland, Amsterdam (1986).

[3] N.N. Zinov'ev, I.P. Ivanov, V.I. Kozub, and Y.D. Yaroshetskii, Exciton transport by nonequilibrium phonons and its influence on the recombinational irradiation of semiconductors at high density excitation, Sov. Phys. JETP **57**,1027 (1983).

[4] Cz. Jasiukiewicz, D. Lehmann, T. Paszkiewicz, Phonon images of crystals I. Energy and quasimomentum focusing patterns: application to GaAs, Z. Phys. B - Condensed Matter **84**, 73(1991).

[5] J.C. Hensel and R.C. Dynes, Interaction of electron–hole drops with ballistic phonons in heat pulses: the phonon wind, Phys. Rev. Lett. **39**, 969 (1977).

[6] M. Greenstein, M.A. Tamor and J.P. Wolfe, Propagation of laser generated heat pulses in crystals at low temperatures: spatial filtering of ballistic phonons, Phys. Rev. **B 26**, 5604 (1982).

[7] A.V. Akimov, A.A. Kaplyanskii and E.S. Moskalenko, Phonon hot spot in cuprous oxide crystals, Sov. Phys. Sol. State **29**, 288 (1987).

[8] F.I. Kreingold and B.S. Kulinkin, Temperature dependence of exciton luminescence and the phonon spectrum in the Cu_2O and Ag_2O, Optika i Spektroskopiya **33**, 706 (1972); A. Compaan and H.Z. Cummins, Raman scattering, luminescence, and exciton–phonon coupling in Cu_2O, Phys. Rev. **B 6**, 4753 (1972).

[9] J.S. Weiner, N. Caswell and P.Y. Yu, Ortho to para–exciton conversion in Cu_2O a subnanosecond time–resolved photoluminescence study, Sol. St. Comm. **46**, 105 (1983).

[10] Y.B. Levinson, Phonon propagation with frequency down conversion, in "Nonequilibrium Phonons in Nonmetalic Crystals", W. Eisenmenger and A.A. Kaplyanskii, eds., North–Holland, Amsterdam (1986).

[11] D. Marx and W. Eisenmenger, Phonon scattering at crystal surfaces, Z. Physik **B**–Cond. Matt., **48**, 277 (1982).

[12] A.V. Akimov, A.A. Kaplyanskii, E.S. Moskalenko and R.A. Titov, Drag of excitons by heat generated phonon pulses in silicon, Sov. Phys. JETP **67**, 2348 (1988).

[13] M.A. Tamor and J.P. Wolfe, Drift and diffusion of free excitons in Si, Phys. Rev. Lett. **44**, 1703 (1980).

[14] R.B. Hammond and R.N. Silver, Temperature dependence of the exciton lifetime in high–purity silicon, Appl. Phys. Lett. **36**, 68 (1980).

[15] P.J. Dean, J.P. Haynes and W. Flood, New radiative recombination processes involving neutral donors and acceptors in silicon and germanium, Phys. Rev. **161**, 711 (1967).

[16] B. Etienne, M. Voos and C. Benoit a la Guillaume, Exciton and droplet intrainment by nonequilibrium phonons in pure Ge, in Proc. 14th Int. Conf. on Physics of Semiconductors, Edinburgh, B.L.H. Wilson ed., Inst. Phys. Conf. Ser. No 43, Inst. of Physics, Bristol (1978), p.387.

[17] J. Weber, W. Sandman, W. Dietsche and H. Kinder, Absence of anomalous Kapitza conductance on freshly cleaved surfaces, Phys. Rev. Lett., **40**, 1469(1978). H. Kinder, Superconducting tunnel junctions, very high frequency phonons and the Kapitza resistance, in "Physics of Phonons", T. Paszkiewicz, ed., Springer, Berlin–Heidelberg (1987).

[18] W. Dietsche, H. Kinder and P. Leiderer, Kapitza resistance of laser–annealed surfaces, in "Phonon Scattering in Condensed Matter", W. Eisenmenger, K. Lassmann and S. Dottinger, Springer-Verlag, Berlin, Heidelberg, N.Y., Tokyo (1984).

[19] D.V. Kazakovtsev and Y.B. Levinson, The effect of phonon scattering in the substrate on temperature dynamics of a phonon film injector, Phys. Stat. Sol. b **136**, 425 (1986).

[20] A.V. Akimov, A.A. Kaplyanskii, J. Kočka, E.S. Moskalenko and J. Stuchlik, Scattering of Terahertz phonons in amorphous Si and Ge, Sov. Phys. JETP **73**, 742 (1991).

[21] W. Dietsche and H. Kinder, Spectroscopy of phonon scattering in glass, Phys. Rev. Lett. **43**, 1413 (1979); T.I. Galkina, A.Yu. Blinov, M.M. Bonch–Osmolovskii, O. Koblinger, K. Lassman and W. Eisenmenger, Down–conversion of high–frequency acoustic phonons, Phys. Stat. Sol. **b 144**, K87 (1987);
J. Mebert, B. Maile and W. Eisenmenger, High frequency phonon transmission through amorphous films, in Phonons 89, S. Hunklinger, W. Ludwig and G. Weiss, eds., World Scientific, Singapure (1990).

[22] L.J. Challis, G.A. Toombs and F.W. Sheard, Acoustic phonon interaction with a two–dimensional electron gas, T. Paszkiewicz, ed., Lecture Notes in Physics vol.205, Springer, Berlin (1987).

[23] A.V. Akimov, A.A. Kaplyanskii, V.I. Kozub, P.S. Kop'ev and B.Ya. Mel'tser, Effects of acoustic phonon pulses on impurity luminescence of quantum–well semiconductor structures, Sov. Phys. Solid State **29**, 1058 (1987).

[24] Zh.I. Alferov, P.S. Kop'ev, B.Ya. Mel'tser, A.M. Vasil'ev, S.V. Ivanov, N.N. Ledentsov, I.N. Uraltsev and D.R. Yakovlev, Intrinsic and impurity luminescence in GaAs–AlGaAs structures with quantum wells, Sov. Phys. Semicond. **19**, 439 (1985).

[25] V. Karpus, Energy and momentum relaxation of two–dimensional charge carriers interacting with deformation acoustic phonons, Sov. Phys. Semicond. **20**, 6 (1986).

[26] A.V. Akimov, L.J. Challis, J. Cooper, C.J. Mellor and E.S. Moskalenko, Phonon emission from the first and second subbands of a two–dimensional electron gas in silicon detected by exciton luminescence, Phys. Rev. **B 45**, 1137 (1992).

[27] N.P. Hewett, P.A. Russell, L.J. Challis, F. Ouali, V.W. Rampton, A.J. Kent and A.G. Every, Hot electron effects and phonon emission from a two–dimensional electron gas (2DEG), Semicond. Sci. Technol. **4**, 955 (1989).

[28] J. Cooper, F. Ouali and L.J. Challis, Semicond. Sci. Technol. **7**, B570 (1992).

[29] D. von der Linder, J. Kuhl and H. Klingenberg, Phys. Rev. Lett. **44**, 1505 (1980);
J.A. Kash, J.C. Tsang and J.M. Hvam, Phys. Rev. Lett. **54**, 2151 (1985);
W.E. Bron, T. Juhasz and S. Mehta, Phys. Rev. Lett. **62**, 1655 (1989).

[30] B.I. Gelmont, N.N. Zinov'ev, D.I. Kovalev, V.D. Kharchenko, Y.D. Yaroshetskii and I.N. Yassievich, Auger recombination of bound excitons induced by acoustic phonons, Sov. Phys. JETP **67**, 613 (1988).

[31] N.N. Zinov'ev, D.I. Kovalev, Y.D. Yaroshetskii and A. Yu. Blank, Phonon induced transitions between exciton subband in silicon, JETP Lett. **53**, 154 (1991).

[32] N.N. Zinov'ev, D.I. Kovalev, V.I. Kozub and Y.D. Yaroshetskii, Kinetics of nonequilibrium acoustic phonons in a thin semiconducting specimen, Sov. Phys. JETP **65**, 746 (1987).

MONTE-CARLO CALCULATED NONEQUILIBRIUM PHONON PULSES IN GaAs

D.V. Kazakovtsev*, B.A. Danilchenko, and I.A. Obukhov
*Institute of Solid State Physics, Russian Academy of Sciences,
142432, Chernogolovka, Moscow Region,
Russia,
Institute of Physics, Ukrainian Academy of Sciences,
252008, Kiev,
Ukraine

Abstract

Time-of-flight spectra for nonequilibrium phonons in GaAs crystals are obtained by Monte-Carlo simulation procedure. The calculations use the quasi-isotropic Tamura model for anharmonic phonon decay and the Rayleigh ω^{-4} law for isotope scattering. The obtained spectra demonstrate a transition from the ballistic regime to quasidiffusive propagation regime via quasiballistic one as the anharmonic lifetime of phonons is changed. In spite of the intermode conversion time being short, the pulse shapes are different for the starting phonon of LA and TA mode.

1. Introduction

A number of experiments use phonon generation by optically excited electron system of the crystal as a source of nonequilibrium phonon pulses. An especially wide set of such experiments was carried out on GaAs crystals (see [1,2] and references therein). A typical feature of this kind of phonon source is a broadband energy distribution of nonequilibrium acoustic phonons, initially generated directly by excited carriers or via fast intermediate stage of optical phonon creation. The energy distribution of the initial acoustic phonons is not well known, but it is common to assume that their frequencies extend up to the boundary ones for acoustic modes. Meanwhile the propagation of high frequency acoustic phonons has a rather complicated character, as they have short mean free paths due to both isotope scattering and anharmonic three phonon processes. Phonon generation occurs on one side of a crystalline specimen, and detector is usually placed on the opposite side at a distance of the order of centimeter. So most of the phonons contributing to the pulse should experience a certain number of elastic and anharmonic scattering events. Thus, there exists a problem of calculation of time dependency of the shape of the phonon pulse, observed in such experiments.

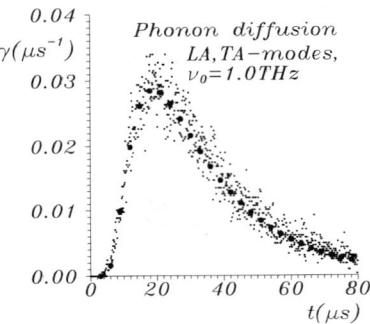

Fig. 1. Normalized phonon energy rate γ as a function of time for initial LA phonons with $\nu_0 = 1.5$ THz for the case of purely diffusive motion diffusion $\tau = \infty$.

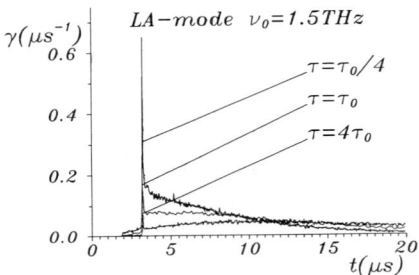

Fig. 2. Normalized phonon energy rate γ as functions of time for initial LA phonons with frequency $\nu_0 = 1.5$ THz for different values of the anharmonic decay time τ.

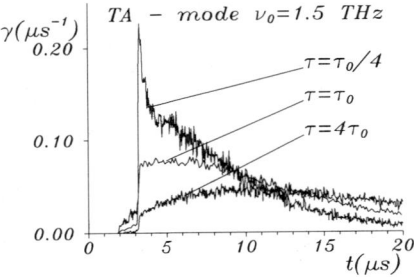

Fig. 3. Normalized phonon energy rate γ as functions of time for initial TA phonons with frequency $\nu_0 = 1.5$ THz for different values of the anharmonic decay time τ.

2. Physical Model of the Propagation Process

We calculated the nonequlilibrium phonon energy rate at the detector as a function of time for a set of frequencies of initially generated phonons. We restrict ourselves to the case of bolometer registration, when the electrical signal at any instant of time is proportional to the incident nonequlilibrium phonon energy rate at the detector.

The acoustic phonon spectrum was assumed to be that of a certain isotropic medium, i.e. to consist of one longitudinal and two degenerate transverse nondispersive isotropic branches. Parameters were taken from [3]. Namely, the averaged phase velocities are $v_L = 5.13 \cdot 10^5$ cm/s for the LA branch and $v_T = 3.02 \cdot 10^5$ cm/s for the TA one. For the isotope scattering time the well established value $\tau_I(\nu) = 1.35 \cdot 10^{41} \nu^{-4} s$ was taken.

We assumed the nonequilibrium phonon occupation numbers during the whole process to be much smaller than unity, so only anharmonic decay events are relevant. An essential feature of our model is that all possible ratios of frequencies of phonons, created by three-phonon decay of the initial one, are taken into account in a straightforward manner. These are restricted only by the energy and quasimomentum conservation laws in an elementary three-phonon anharmonic decay event. The TA phonon branch is not subject to decay, but mode conversion due to isotope scattering is present. For the LA mode the anharmonic decay time is $\tau(\nu) = A\tau_0(\nu)$, where $\tau_0(\nu) = 1 \cdot 10^{54} \nu^{-5} s$ is the value calculated in [3]. The constant A for different calculations is changed to reveal the effect of inelastic scattering and is a set of powers of two.

We followed the motion of an initial LA or TA with frequency ν_0, starting from the centre of a sphere at the time $t = 0$. In a cascade of scattering events with possible multiplication of number of phonons due to decays all the phonons were tracked until they reached an $R = 1$ cm sphere. Summation of a large number of histograms obtained by the above outlined procedure yield the resulting time-of flight spectra.

3. Pure Diffusion without Anharmonic Processes

To check the program we exclude decay processes by assuming $\tau = \infty$. Then, the propagation of phonons in the crystal is purely diffusive with the appropriate diffusion constant averaged over polarizations. We are interested in the the energy rate through the $R = 1$ sphere, normalized by the total input phonon energy E_0, $\gamma = E_0^{-1} \cdot dE/dt$. Using the diffusion equation Green function inside the sphere with a zero boundary condition at $R = 1$, we obtain for γ the following analytical expression

$$\gamma(t) = 2\pi^2 D \sum_{n=1}^{\infty} (-1)^n n^2 \exp\left(-\pi^2 n^2 Dt\right).$$

The $\gamma(t)$ values, calculated for a set of different times t are depicted in Fig. 1. with the closed circles alongside the Monte-Carlo results for the same starting phonon frequency, represented here by dots. One can hardly expect a better coincidence of analytical and numerical simulation results. This fact encourages us to use our program taking into account anharmonic processes in the parameter domain, where one cannot make use of any analytical expressions.

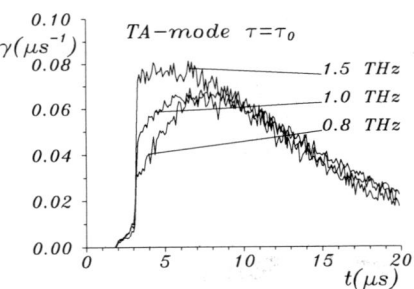

Fig. 4. The same as in Fig. 2, for $\tau = \tau_0$ and different frequencies of an initial phonon ν_0.

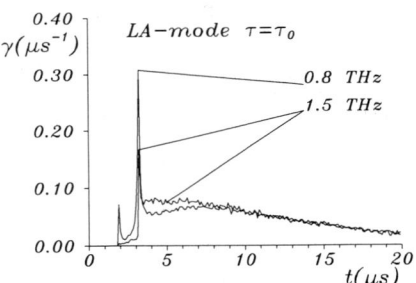

Fig. 5. The same as in Fig. 3, for $\tau = \tau_0$ and different frequencies of an initial phonon ν_0.

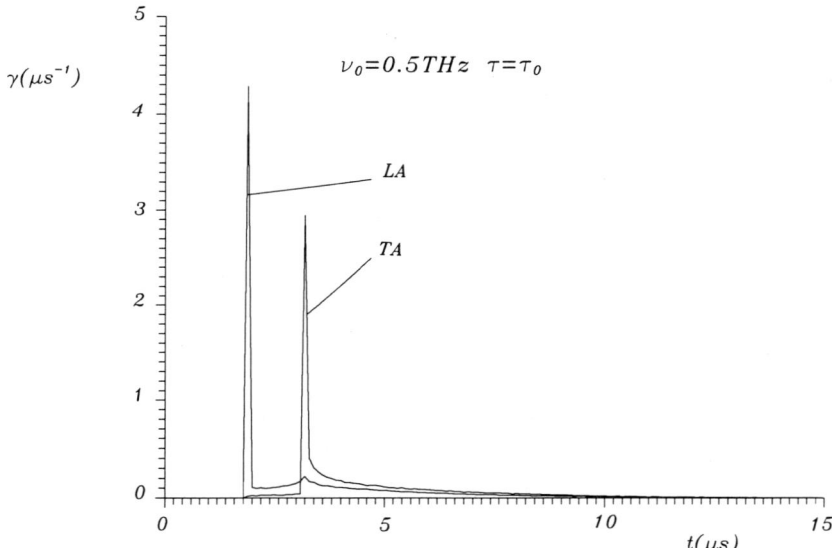

Fig. 6. Typically ballistic pulse shapes, obtained for $\nu_0 = 0.5$ THz and $\tau = \tau_0$.

4. Results and Discussion

The calculated normalized energy rate γ for the initial phonon frequency $\nu_0 = 1.5$ THz with the account of both elastic and inelastic scattering events are represented in Fig. 2, 3. For an initial phonon of LA mode we see a sharp ballistic pulse at the time corresponding to TA pnonon time of flight accompanied by a quasiballistic tail at $\tau = \tau_0/4$ and a quasidiffusive pulse at $\tau = 4\tau_0$. For an initial TA phonon at the time of the ballistic TA pulse, an abrupt increase of phonon rate appears, the step being larger for quasiballistic case $\tau = \tau_0/4$ than for quasidiffusive one $\tau = 4\tau_0$.

The changes of the pulse shape with starting frequency ν_0 change (Fig. 4,5) are somewhat similar to those due to the variation of the anharmonic time τ. For $\nu_0 = 0.5$ Thz the pulses (Fig. 6) are typically ballistic. With the increase of ν_0 for the initial phonon of LA mode the magnitude of the ballistic pulses diminishes, and a quasidiffusive tail appears. For an initial TA phonon, the changes with the increase of ν_0 cannot be explained so easily.

REFERENCES

[1] Ulbrich R.G., in "Nonequilibrium Phonon Dynamics", Ed. by W.E. Bron, Plenum Press, New York (1985), pp. 3-26.

[2] B.A. Danilchenko, V.N. Poroshin, M.I. Slutskii, M. Ashe, phys. stat. sol. (b) **136**, 63 (1986).

[3] M.T. Ramsbey, J.P. Wolfe, S. Tamura, Z. Phys. B – Condensed Matter **77**, 209 (1989)

[4] S.Tamura, Phys. Rev. **B.31**, 2574 (1985).

INFLUENCE OF SAMPLE TEMPERATURE AND PUMPING INTENSITY ON THE PROCESSES OF HOT SPOT FORMATION AND DEGRADATION

A.A. Maksimov, D.A. Pronin, I.I. Tartakovskii
Institute of Solid State Physics
Russian Academy of Sciences
Chernogolovka, Moscow, dist., 142432
Russia

There is an important problem in the low temperature physics of solid state connected with the relaxation and propagation of nonequilibrium phonons. One of the most interesting phenomena, concerning phonon propagation, is the so-called "hot spot", that is a small part of a sample where the temperature significantly exceeds the bulk value. It appears, for instance, as a result of pulsed laser irradiation of the substance when a considerable part of absorbed energy is transformed to acoustic phonons.

As a rule, after optical excitation acoustic phonons appear as a result of optical phonon decay (which may be emitted by the relaxing electronic excitations). In this case the spectral distribution of acoustic phonons in a hot spot may differ drastically from the equilibrium Planck distribution. In particular, the mode temperatures of the high frequency phonons exceed those of the low-frequency phonons. This fact complicates significantly theoretical consideration of hot spot degradation making it necessary to take into account both the processes of phonon interaction (their coalescence and decay to lower-frequency phonons) and spatial phonon diffusion. This problem is rather difficult and has not been solved in general. Nevertheless there is a simple way to discuss this problem, which is based on the following speculations.

Since the mean free path λ of phonons strongly increases as the frequency Ω decreases ($\lambda \propto \Omega^{-4}$ for scattering by impurities), cooling of a heated region is initiated by the diffusion of low frequency phonons. But at low temperatures cooling may be caused by phonons ballistically propagating with the sound velocity. At the same time the high-frequency phonons remain in the hot spot region and their occupation numbers decrease via anharmonic decays more effectively than due to their own diffusion. The characteristic frequency of the phonons decreases with time, due to the spectral evolution of the phonon distribution in a hot spot caused by decay processes and spatial expansion. This this spatial expansion of the heated region may be accelerated in time up to its "explosion". The latter occurs when a considerable part of the hot spot energy is concentrated in phonons, whose diffusion length are comparable with the hot spot dimension. If subsequent expansion is possible it occurs in the quasidiffusive regime.

We investigated the degradation of a hot spot obtained near one of the antracene crystal surfaces at helium temperatures. We used a nitrogen laser pulse to produce a hot spot and another one delayed in time as a temperature probe.

The time dependence of the front and rear surface temperature after excitation by the laser pulse are presented on Fig. 1. Three relative steps of the absorbed energy relaxation inside a crystal can be distinguished. The first step is completed within 100ns. It is associated with ballistically propagating phonons. The second step is characterized by the spreading of the heated region into the cold bulk of the crystal. The third step is associated with the process of the slow cooling of a uniformly heated crystal in a helium vapour atmosphere during the time interval between the laser pulses. This process lasts for about 100 μs.

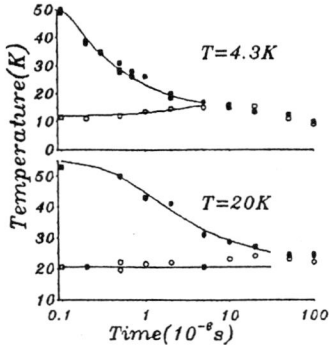

Fig. 1. Temperature temporal dependencies on the front (solid circles) and rear (open circles) surfaces after laser excitation at different initial temperatures T. The solid curves are the results of the calculations.

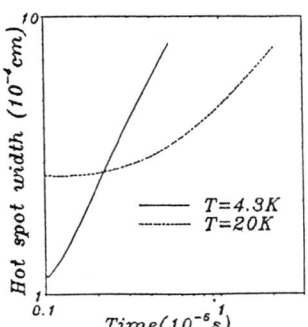

Fig. 2. The calculated hot spot width.

The experimental results were discussed in terms of the local thermal conductivity with both specific heat and thermal conductivity as well-known functions of temperature. The plane experimental geometry validates the use of a $1D$-equation with the x-axis perpendicular to the sample surface (depth). The equation of thermal conductivity used for the numerical calculations is

$$C(T(x,t))\frac{\partial T(x,t)}{\partial t} = \frac{\partial}{\partial x}\left(K(T(x,t))\frac{\partial T(x,t)}{\partial x}\right),$$

$$\frac{\partial T(x,t)}{\partial x}\bigg|_{x=0} = 0, \quad \frac{\partial T(x,t)}{\partial x}\bigg|_{x=d} = 0, \quad T(x,t)\big|_{t=0} = T_0(x),$$

where $C(T) = \alpha T^3$ is the specific heat capacity and the thermal conductivity coefficient $K(T) = K_0 \exp(\theta/T)$, θ is Debye temperature. We have supposed $T_0(x)$ in the form of Gaussian distribution:

$$T_0(x) = \exp\left(-\frac{x^2}{a^2}\right)(T_F - T_R) + T_R$$

Good agreement between experimental results and model calculations was achieved (Fig. 1). Analyzing the results of calculations we draw a conclusion on the strong sample temperature influence on the process of hot spot formation immediately after the laser pumping. A temperature decrease leads to the fact that a significant part of the absorbed energy leaves the region of hot spot localization in the form of ballistically propagating phonons. As a result the degradation of the hot spot is accelerated.

The solution of the equation used to describe the experimental results has been analyzed. It was found that the initial conditions strongly affect the time evolution of a hot spot width (Fig. 2). This non-linear equation (one has more reasons to employ it at low temperatures) may show both heat propagation with almost constant speed (as is expected for a quasidiffusive mechanism) and a square root as in the usual thermal conductivity model. Thus, it should be noted that heat propagation with the constant speed is not a good criterion for detecting quasidiffusion, because it may be also explained by nonlinear thermal conductivity [1].

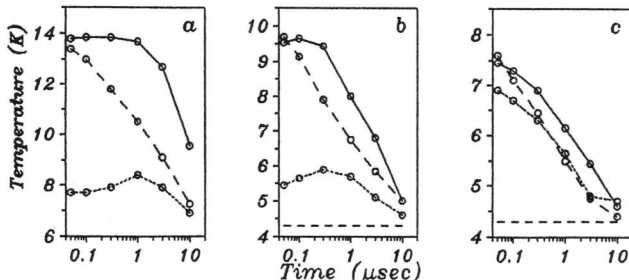

Fig. 3. The temporal dependencies of the temperature after laser excitation at the different distances from the excitation spot edge: solid curve -120 μm inside, dashed curve -30 μm inside and dotted curve -30 μm outside the excited region. Pumping power: $a - 11$ kW/cm^2, $b - 5.6$ kW/cm^2, $c - 2.8$ kW/cm^2.

The kinetics of hot spot formation and degradation have a strong dependence on the initial temperature of the sample. Another important parameter is the intensity of the laser pumping. To investigate this problem we used a frequency sensitive phonon detector [2]. We produced a hot spot in the semi-surface on the thin crystal plates, the probe laser pulse was formed as a narrow strip parallel to the edge of the excited region and was placed at the different distances from it. This permitted us to measure the temporal dependence of the phonon mode temperature around the edge and thus to investigate the kinetics of hot spot edge degradation [3].

At a high level of the laser pumping (Fig. 3a) the energy absorbed in the excited area might be able to provide the own "cloud" of the temperature immediately around the region of excitation. This results in a considerable decrease of phonon flow and thus locks the hot spot in the excitation area for a long period of time ("self-locking"). In Fig. 3a the hot spot keeps its edge sharp up to 1 μs. At a low excitation level (Fig. 3c) overheating appears to be too small to produce self-locking. In this case the energy rapidly flows out of the excited region. At intermediate pumping power (Fig. 3b) the situation seems to be very close to the hot spot formation, but the hot spot in this case does not persit for more than 300ns.

REFERENCES

[1] D.A. Pronin, Phys. Stat. Sol. (b) **173**, 533 1992

[2] D.V. Kazakovtsev, A.A. Maksimov, D.A. Pronin and I.I. Tartakovskii, Pis'ma Zh. Eksp. Teor. Fiz., **49**, 52, 1989, (Sov. Phys. JETP Lett. **49**, 61, 1989)

[3] To be published

BALLISTIC PHONON PROPAGATION IN AT-CUT QUARTZ

B. Sujak-Cyrul, J. Szczepański, and T. Tyc

Institute of Low Temperature and Structure Research
Polish Acedemy of Sciences
50-950 Wroclaw 2, P.O. 937, Poland

INTRODUCTION

The AT-cut is a cut distinctly showing piezoelectrical properties of quartz in practical applications (transducers), but phonon propagation in AT-cut quartz has not yet been investigated in detail. Our aim was to measure the phonon pulses propagating in the direction perpendicular to the AT-cut surfaces, to identify the observed signals on the basis of the theoretical pure ballistic phonon propagation model, to determine if the influence of piezoelectricity on phonon propagation in this direction is detectable and also to observe if our experimental results obtained for a very thin phonon detector depend on the value of the current supplying the detector.

Koss and Woolf [1] have investigated the influence of the inherent piezoelectricity on phonon focusing patterns (centered on the z and $-y$ directions) in quartz, taking into account the "stiffened" elastic constants in their calculations. The comparison between their experimental data and calculations shows that the influence is rather small. However some features observed in the phonon focusing pattern are theoretically incomprehensible [1,2] so further analysis and time dependent measurements are needed [2].

Superconducting thin films with a thickness smaller than the magnetic penetration depth $\lambda(T)$ are treated as two-dimensional or even quasi-one-dimensional [3]. For such superconductors the complexity of effects induced by current is reduced in comparison with three-dimensional superconductors, but still these effects can appear and influence the measured time-dependent bolometer signal.

EXPERIMENT

We have measured the phonon pulse propagation in AT-cut quartz using the conventional time-of-flight method. Samples with thickness $d = 4.1\ mm$, cut from high quality synthetic α-SiO_2, had mechanically and chemically polished surfaces $16.5 \times 20.5\ mm$. A radiator (phonon generator) and a bolometer (phonon detector) were evapourated on the two opposite surfaces and the radiator-bolometer direction was perpendicular to the polished surfaces with an accuracy not worse than 5°. The radiator, a thin-film of silver with resistance of about 10 to 60 Ω in 4.2 K and an active area S_R smaller than 0.01 mm^2, was heated by electric rectangular pulses (duration 25 ns, amplitude 5 to 10 V, repetition rate 5 kHz). The bolometer, a thin-film of superconducting aluminium with thickness of about 20 $\overset{\circ}{A}$, resistance of about 70 to 150 Ω in 4.2 K and an active area S_B smaller than 0.01 mm^2 or 0.1 mm^2, was prepared by vacuum deposition of Al in the presence of O_2. For bolometers made in this way the $N-S$ transition temperature T_c was within the range from 1.6 to 2.1 K and the sensitivity was about 1000 Ω/K. For measurements of the time-dependent bolometer signal, caused by phonon pulses, the bolometer was supplied by a direct current (I_B from 0.05 to 1.1 mA) and its working point was stabilised with an accuracy better than 0.001 K.

CALCULATIONS

We have estimated the time-dependent bolometer signal using a numerical program written on the basis of the general closed-form expressions for the phase velocities and polarization vectors of acoustic waves in elastically anisotropic solids (based on Every's paper [4] and assuming that the generated phonons have an isotropic distribution in wavevector space and can propagate ballistically without any interaction (giving ideal patterns of phonon focusing). The transmission of phonons across interfaces radiator-substrate and substrate-bolometer is assumed to be without reflection, refraction and change of polarization, whereas the geometric dimensions and orientation of the radiator and detector areas as well as the rectangular shape of the heat pulse are explicitly taken into account. Because the proper inclusion of piezoelectricity greatly complicates the calculations, we used the same constant-field elastic constants, taken from Bechmann [5], that Rösch and Weis [6] and also Müller and Weis [7] used in their calculations for quartz (but for different directions).

RESULTS

The results of our measurements of the ballistic phonon propagation, for different bolometer currents I_B from 0.05 to 1.1 mA and two different sizes of bolometer active area S_B are presented in Fig. 1 and Table 1. For S_B =0.01 mm^2 (1st case) for the lower I_B five singularities have been observed whereas for the higher I_B four singularities have been observed (two of the modes indicated in Table 1 by stars, separated for lower I_B, are now observed as one singularity with the maxima positions in the middle of both of them), all with stable positions of the maxima (± 0.01 μs) for the different currents I_B (Fig. 1a, b). For $S_B = 0.1$ mm^2 (2nd case) all singularities observed in the 1st case plus one more very weak singularity have been observed with stable positions of the maxima for different currents I_B(Fig. 1c). On the basis of our calculations all the observed singularities have been identified with phonon modes and the values of the group velocities have been calculated (see Table 1 and Fig. 2). The agreement between the experimentally observed and theoretically predicted positions of the maxima shows that the piezoelectricity does not influence strongly the phonon propagation in this direction.

Table 1. Values of the group velocities v_g of different modes of phonons in AT-cut synthetic quartz.

	Volume of v_g $[km/s]$			
Mode	$S_B = 0.01\ mm^2$		$S_B = 0.1\ mm^2$	
	calc.	exp.	calc.	exp.
L	5.4	5.5	5.5	5.6
T_1	-	-	4.9	5.1
T_1	4.5	4.6*	4.4	4.6
T_1	4.3	4.4*		
T_2	4.0	4.1	4.1	4.0
T_2	3.8	3.6	3.8	3.6

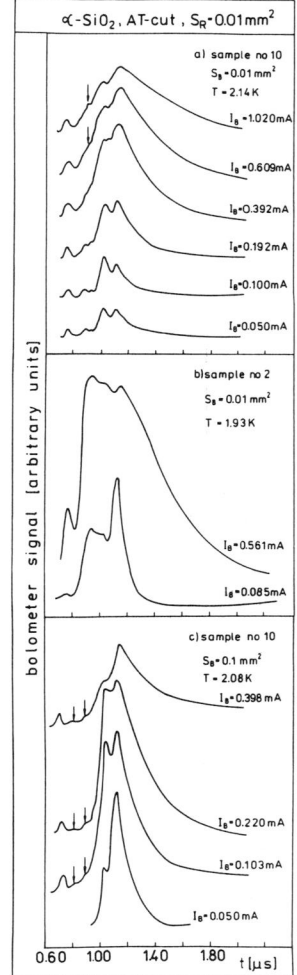

Figure 1. Measured time dependent bolometer signal. Weak maxima are indicated by arrows.

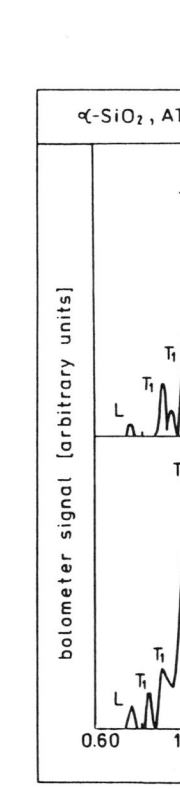

Figure 2. Calculated time dependent bolometer signal.

A different ratio of the amplitudes of the maxima, not always being in agreement with the theoretically predicted ratio, has been observed for different samples and different currents I_B. The variability of this ratio, observed with an increase of I_B, can be explained in some cases (see Fig. 1a) by an increase of the bolometer relaxation time τ, whereas other changes of the ratio (see Fig. 1b, c) could suggest that the absorption of different phonon modes by the bolometer depends on the I_B value.

More detailed analysis of experimental results will be continued [8].

REFERENCES

[1] G. Koos, J. Wolfe, Phys. Rev. **B 30**, 3470 (1984)

[2] A. G. Every, Phys. Rev. **B 36**, 1448 (1987)

[3] R. Tidecks, "Current Induced Non-equilibrium Phenomena in Quasi-One-Dimensional Superconductor", Springer-Verlag, Berlin (1990)

[4] A.G. Every, Phys. Rev. **B 22**, 1746 (1980)

[5] R. Bechmann, Phys. Rev. **110**, 1060 (1958)

[6] F. Rösch, O. Weis, Z. Phys. **B 25**, 101 (1976)

[7] G. Müller, O. Weis, Z. Phys. B - Condensed Matter **80**, 15 (1990)

[8] B. Sujak-Cyrul, J. Szczepański, T. Tyc, to be published

EFFECT OF RESONANT SCATTERING BY PARAMAGNETIC CENTERS ON THE PROPAGATION OF NONEQUILIBRIUM PHONONS

K.L. Aminov

Kazan Physical-Technical Institute
Kazan, Russia

INTRODUCTION

At low temperatures phonon anharmonic decay processes decrease enough to make it possible to get ballistic phonon propagation at frequencies up to $10^{11} - 10^{12}$ Hz. On the other hand, phonon scattering at resonant centres, if present, becomes more effective since the difference between the population of centre levels increases. It is known from experiments that the heat conductivity at low temperatures is reduced mainly by resonant scattering. And although resonant phonons in a crystal find themselves in a narrow band compared to the total spectra of acoustic vibrations, they play a crucial role in such process as paramagnetic relaxation. Even for low concentrations of resonant centres the lifetimes of the resonant phonons is reduced enough to make phonon propagation more diffusive than ballistic. It may result in phonon trapping being an example of the general problem of radiation trapping in resonant media (see [1,2]).

Here the effects of phonon resonant scattering are considered theoretically on the basis of a simple model. Particular attention is paid to nonlocal transport properties. Also the effects of nonresonant scattering and resonant combinational scattering are discussed.

1. SIMPLE RESONANT SCATTERING MODEL

To consider the effects of strong resonant phonon scattering on phonon propagation, a simple resonant scattering model may be applied. Assume that the resonant centres have two levels and that the scattering of phonons may be represented as two elementary processes:

a) a centre can absorb a phonon q with probability $W_q N_q$, where N_q is the total number of phonons q and W_q is the probability of transition;

b) a centre can emit a phonon q with probability $W_q(N_q + 1)$. While these two processes take place in some volume, the kinetic equations may be obtained for the nonequilibrium phonon density $n_q(\mathbf{r},t)$ and for the density of nonequilibrium excitations of a centre $N(\mathbf{r},t)$, supposing these nonequilibrium densities to be small compared with the equilibrium ones:

$$(\partial/\partial t + \gamma) N(\mathbf{r},t) = \sum_q \gamma_q n_q(\mathbf{r},t), \qquad (1)$$

$$(\partial/\partial t + \mathbf{v}_q \nabla + \gamma_q) n_q(\mathbf{r},t) = \chi \gamma_q N(\mathbf{r},t).$$

Here $\mathbf{v}_q$ is the group velocity of a phonon q; $\gamma_q = W_q(N_1^0 - N_2^0)$ is its inverse lifetime, $(N_1^0 - N_2^0)$ is the equilibrium difference between the population of the centre levels; $\gamma = \Sigma_q W_q(2N_q^0 + 1)$ is inverse relaxation time of a centre, where N_q^0 is the equilibrium population of the phonon mode q; $\chi = \gamma/\Sigma_q \gamma_q$ so that the conservation law for the total number of excitations is fulfilled:

$$(\partial/\partial t) \int d\mathbf{r}\, P(\mathbf{r},t) = 0, \qquad (2)$$

where $P(\mathbf{r},t) = N(\mathbf{r},t) + \Sigma_q n_q(\mathbf{r},t)$.

An exact solution of (1) can be obtained taking a time Laplace and a spatial Fourier transformation:

$$f(\mathbf{k},\lambda) = \int_0^\infty dt \int d\mathbf{r}\, f(\mathbf{r},t)\, \exp(i\mathbf{k}\mathbf{r} - \lambda t).$$

Then the solution of (1) can be presented in the form:

$$N(\mathbf{k},\lambda) = F(\mathbf{k},\lambda)^{-1} \left\{ N(\mathbf{k},t=0) + \sum_q \gamma_q n_q(\mathbf{k},t=0)/(\lambda + \beta_q) \right\},$$

$$n_q(\mathbf{k},\lambda) = \{n_q(\mathbf{k},t=0) + \gamma_q \chi\, N(\mathbf{k},\lambda)\} \Big/ (\lambda + \beta_q), \qquad (3)$$

$$F(\mathbf{k},\lambda) = \lambda + \gamma - \chi \sum_q \gamma_q^2/(\lambda + \beta_q), \qquad \beta_q = \gamma_q + i\mathbf{k}\mathbf{v}_q.$$

As we are interested in long time (hydrodynamic) behaviour of the system, we may neglect fast fluctuations and use a coarse grain approximation, i.e. we are interested in the averaged values of some variable $f(t)$:

$$\tilde{f}(t) = \int dt'\, f(t')\, S(t-t'), \qquad (4)$$

where $S(t)$ is an averaging function normalized to obey: $\int dt\, S(t) = 1$. The appropriate time scale is determined by the τ - characteristic width of $S(t)$. In the frequency domain, the expression (4) looks like

$$\tilde{f}(\omega) = f(\omega)\, S(\omega), \qquad (4a)$$

i.e. averaging in the time domain is filtering in the frequency domain. The respective filter width $\Delta\omega$ is determined by $\Delta\omega\tau \sim 1$. In the case of our model the characteristic time is the time of the phonon's ballistic escape from an excited volume. Since the volume size L is related to the interval of the wave numbers ΔK by $\Delta K L \sim 1$, it is reasonable to choose $\tau \gg (\Delta K v)^{-1}$, where v is the mean phonon velocity. So on this time scale, we need only solutions from (3) with $\lambda \ll \Delta K v$, i.e. we may neglect λ in comparison with β_q. Hence we have:

$$P(\mathbf{k}, \lambda) = P_0(\mathbf{k})/(\lambda + \lambda(\mathbf{k})), \qquad (5)$$

where

$$P_0(\mathbf{k}) = N(\mathbf{k}, t=0) + \sum_q \gamma_q n_q(\mathbf{k}, t=0)/\beta_q, \qquad \lambda(\mathbf{k}) = B(\mathbf{k})/A(\mathbf{k}),$$

$$A(\mathbf{k}) = \sum_q \left\{ \gamma_q/\gamma + \gamma_q^2/\left((\mathbf{k}\mathbf{v}_q)^2 + \gamma_q^2\right) \right\}, \qquad B(\mathbf{k}) = \sum_q \gamma_q (\mathbf{k}\mathbf{v}_q)^2 / \left((\mathbf{k}\mathbf{v}_q)^2 + \gamma_q^2\right).$$

Also for centre excitations and strongly scattered phonons (i.e. when $\gamma_q > \tau^{-1}$) we have:

$$N(\mathbf{k}, \lambda) = P(\mathbf{k}, \lambda)/A(\mathbf{k}), \qquad n_q(\mathbf{k}, \lambda) = \gamma_q \chi \, P(\mathbf{k}, \lambda)/\beta_q A(\mathbf{k}). \qquad (6)$$

Since $\lambda(\mathbf{k}) \to 0$ with $\mathbf{k} \to 0$, equation (5) is just another form of the conservation law for the total number of excitations (2). The solutions (5), (6) give the following picture of transport of nonequilibrium excitations with spatial size L: All phonons can be divided into two groups - resonant phonons with $\gamma_q > v/L$ and nonresonant phonons with $\gamma_q < v/L$. During the time $\tau > L/v$ the nonresonant phonons escape from the excited volume and a local equilibrium is established between resonant phonons and centres as defined by (6). On a large time scale given by τ this resonant excitation propagates from the excited volume as defined by (5).

After a back Fourier transformation, (5) looks like

$$(\partial/\partial t) \, P(\mathbf{r}, t) + \nabla^2 \int d\mathbf{r}' \, K(\mathbf{r} - \mathbf{r}') \, P(\mathbf{r}', t) = 0, \qquad (5a)$$

$$K(\mathbf{r}) = (1/2\pi)^3 \int d\mathbf{k} \, \exp(-i\mathbf{k}\mathbf{r}) \, \lambda(\mathbf{k})/k^2.$$

Upon comparison with γ_q^2, we may neglect $(\mathbf{k}\mathbf{v}_q)^2$ in (5) for small k, and we have $\lambda(\mathbf{k}) \simeq Dk^2$ and (5a) becomes a diffusion equation with D as the diffusion coefficient (it may generally be anisotropic). But there are cases when the diffusion coefficient D becomes divergent since the main part of $\lambda(\mathbf{k})$ is contributed by phonons with $\mathbf{k}\mathbf{v}_q \sim \gamma_q$. The equation (5a) remains then nonlocal. This means that there is a channel of fast transport of excitations via such phonons. Possible physical realizations of nonlocal transport for resonantly scattered phonons are discussed in following sections.

2. QUASIDIFFUSIVE PROPAGATION OF PHONONS SCATTERED BY ANISOTROPIC PARAMAGENTIC CENTERS

It is known that different types of scattering may occur depending on the detailed microscopic interaction between centres and phonons and also on interactions between centres. In [3,4] was shown that in the absence of phase relaxation, phonons scatter elastically due to energy conservation. To describe properly the state of the centre subsystem we need the spectral density of excitations, $N(\mathbf{r}, t, \omega)$, so that $\int d\omega \, N(\mathbf{r}, t, \omega) = N(\mathbf{r}, t)$

is the full centre excitation density. The kinetic equations (1) may be rewritten in this case in the form:

$$(\partial/\partial t + \gamma) \, N(\mathbf{r}, t, \omega) = <\gamma_q n_q(\mathbf{r}, t)>_\omega, \qquad q = (\mathbf{q}s), \tag{7}$$

$$(\partial/\partial t + \mathbf{v}_q \nabla + \gamma_q) \, n_q(\mathbf{r}, t) = \chi_\omega \, \gamma_q \, N(\mathbf{r}, t, \omega), \qquad \chi_\omega = \gamma/<\gamma_q>_\omega,$$

$$<f_\mathbf{q}>_\omega = \left(\sum_s \int d\mathbf{q} \, f_\mathbf{q} \, \delta(\omega_q - \omega)\right) \bigg/ \left(\sum_s \int d\mathbf{q} \, \delta(\omega_q - \omega)\right),$$

where s is the wave number index, ω_q is the frequency of the phonon q. Since (7) is a particular form of (1), we can readily use the solution (5):

$$P(\mathbf{k}, \lambda, \omega) = P_0(\mathbf{k}, \omega) / (\lambda + \lambda(\mathbf{k}, \omega)), \tag{8}$$

with

$$\lambda(\mathbf{k}, \omega) = \left\langle \gamma_q (\mathbf{k}\mathbf{v}_q)^2 / ((\mathbf{k}\mathbf{v}_q)^2 + \gamma_q^2) \right\rangle_\omega / (1 + <\gamma>_\omega / \gamma).$$

For $\mathbf{k}$ small enough, in the isotropic case, supposing $\gamma_q = \gamma_{ph}$ and $\mathbf{v}_q = v(\mathbf{q}/q)$ we get a diffusive type of phonon propagation[3,4] with the diffusion coefficient

$$D = (1/3)(v^2/\gamma_{ph})/(1 + \gamma_{ph}/\gamma). \tag{9}$$

Another situation can take place in an anisotropic resonant medium[5,6] and a paramagnetic crystal is a good example. Resonant phonon scattering is caused by the electron-phonon interaction

$$H_{e-ph} = \sum_{\lambda \Gamma} V(\lambda \Gamma) \, \epsilon(\lambda \Gamma),$$

where $V(\lambda \Gamma)$ are the electron operators and $\epsilon(\lambda \Gamma)$ is the lattice strain both transforming as the λ component of the Γ representation of the point group at the ion site. The inverse lifetime of a phonon with wave vector $\mathbf{q}$ and wave number s in the case of small concentration of resonant centres is

$$\gamma_{qs} = \left(\pi \omega_{qs} n_c g(\omega_{qs})/\rho h v_{qs}^2\right) \left| \sum_{\lambda \Gamma} \langle 1 | V(\lambda \Gamma) | 2 \rangle \, [\epsilon(\lambda \Gamma)]_{qs} \right|^2, \tag{10}$$

where ρ is the mass density; n_c is the concentration of resonant centres; the resonant line form $g(\omega)$ is normalized to obey $\int d\omega \, g(\omega) = 1$;

$$[\epsilon(\lambda \Gamma)]_{qs} = \sum_{ij} S(\lambda \Gamma \mid ij) \, [\epsilon_{ij}]_{qs}, \qquad [\epsilon_{ij}]_{qs} = [q_i e_j(\mathbf{q}s) + q_j e_i(\mathbf{q}s)]/2q,$$

$S(\lambda \Gamma \mid ij)$ is the transformation from symmetric lattice strains to distortions ones; $e_i(\mathbf{q}s)$ is the i-th component of phonon $\mathbf{q}s$ polarization vector. Let us consider some symmetry cases. We shall use the isotropic elastic medium approximation for lattice vibrations with

$$\omega_{q1} = \omega_{q2} = \omega_{q3} = vq,$$
$$\mathbf{q} = q(\sin \vartheta \cos \varphi, \sin \vartheta \sin \varphi, \cos \vartheta),$$
$$\mathbf{e}(\mathbf{q}1) = (\sin \vartheta \cos \varphi, \sin \vartheta \sin \varphi, \cos \vartheta),$$

$$\mathbf{e}(\mathbf{q}2) = (\cos\vartheta\cos\varphi, \cos\vartheta\sin\varphi, -\sin\vartheta),$$
$$\mathbf{e}(\mathbf{q}3) = (-\sin\varphi, \cos\varphi, 0), \tag{11}$$

choosing the Z axis to be in some symmetry direction.

a) $<1 \mid H_{e-ph} \mid 2> = V(\epsilon_{xz} + i\epsilon_{yz})$. One example of such a symmetry is an Ho^{3+} ion in an LiRF$_4$ crystal (point symmetry group S_4). In this case, the averaged inverse phonon lifetime is

$$\langle\gamma_q\rangle_\omega = \left(\pi\omega/(9\rho h v^2)\right) n_c g(\omega) \mid V\mid^2,$$

and the asymptotic value for $\lambda(\mathbf{k},\omega)$ is

$$\lambda(\mathbf{k},\omega) = (1/36)(kv/\langle\gamma_q\rangle_\omega)^{3/2}\,\gamma\,\langle\gamma_q\rangle_\omega/\left(\langle\gamma_q\rangle_\omega + \gamma\right)$$

$$\left\{\pi^{-1/2}\,\Gamma(1/4)^2\,|\sin\theta|^{3/2} + 2^{1/2}\int_0^{2\pi} d\phi\,|\cos\theta + \sin\theta\cos\phi|^{3/2}\right\},$$

where $\mathbf{k} = k(\sin\theta\cos\Phi, \sin\theta\sin\Phi, \cos\theta)$.

b) $\langle 1 \mid H_{e-ph} \mid 2\rangle = V(\epsilon_{xx} - \epsilon_{yy})$. In this case

$$\langle\gamma_q\rangle_\omega = \left(2\pi\omega/(9\rho h v^2)\right) n_c g(\omega) \mid V\mid^2,$$

$$\lambda(\mathbf{k},\omega) = B(\theta,\Phi)(kv/<\gamma_q>_\omega)^{3/2}\ln(<\gamma_q>_\omega/kv)\gamma<\gamma_q>_\omega/(<\gamma_q>_\omega + \gamma),$$

$$B(\theta,\Phi) = (5^{1/2}/36)\Big\{|\cos\theta|^{3/2}$$

$$+ |\sin\theta|^{3/2}\left(|\cos(\phi - \pi/4)|^{3/2} + |\cos(\phi + \pi/4)|^{3/2}\right)\Big\}.$$

c) $\langle 1 \mid H_{e-ph} \mid 2\rangle = V\theta_z$, where $\theta_z = (\partial U_x/\partial y - \partial U_y/\partial x)/2$ - rotation tensor Z component. In this case

$$<\gamma_q>_\omega = \left(2\pi\omega/(9\rho h v^2)\right) n_c g(\omega) \mid V\mid^2,$$

$$\lambda(\mathbf{k},\omega) = (2/9)(kv/<\gamma_q>_\omega)^2\ln(<\gamma_q>_\omega/kv)(\cos\theta)^2$$

$$\gamma<\gamma_q>_\omega/(<\gamma_q>_\omega + \gamma).$$

The three particular cases given above demonstrate an empirical rule that can be used to obtain the asymptotic behaviour of $\lambda(\mathbf{k},\omega)$ for different types of anisotropy. Let's call "peculiar", a point situated on a sphere centreed at the origin of $\mathbf{q}$ space, if for $\mathbf{q}$ directed through this point, $\gamma_q = 0$. Then the cases considered above represent examples of three types of the geometry of peculiar points: a) A line of peculiar points: $\lambda \propto k^{3/2}$;
b) Crossed lines of peculiar points: $\lambda \propto k^{3/2}\ln(1/k)$;
c) Individual peculiar points: $\lambda \propto k^2 \ln(1/k)$.

Different types of nonlocal phonon propagation with specific time and size dependencies may be observed in each of this cases.

3. QUASIBALLISTIC PHONON TRANSPORT IN A MEDIUM WITH HOLSTEIN TYPE RESONANT PHONON SCATTERING

In a resonant scattering mechanism in which the strong phase relaxation of resonant centres is implied was proposed by Holstein for resonant light scattering in gases[1]. This means that after absorbing a phonon the centre "forgets" about its excitation conditions before emitting another phonon and the emitted phonon is not correlated with the absorbed one. Equations (1) imply exactly such a mechanism. Therefore we can directly use the results of Sect. 1. It is known that nonlocal excitation transport is possible for Holstein type of resonant scattering due to excitations escaping via resonant line wings. To the consider the asymptotic behaviour of damping factor $\lambda(\mathbf{k})$ for small $\mathbf{k}$ for different line forms, let us make the isotropic approximation with $\gamma_q = \gamma_{ph} g(\omega_q)$, $\mathbf{v}_q = v\mathbf{q}/q$, where $g(\omega)$ is the resonant line form with $g(\omega_0) = 1$ at the resonant frequency ω_0, and the line width γ_0 is determined by $\gamma_0 = \int g(\omega) d\omega$ and γ_{ph} is inverse lifetime of phonons with resonant frequency. Then we have from (5):

$$\lambda(k) = \gamma_{ph} J_1(k)/(J_2(k) + \gamma_0 \gamma_{ph}/\gamma), \tag{12}$$

$$J_1(k) = \int g(\omega)\{1 - S(\alpha,\omega)\} d\omega, \qquad J_2(k) = \int g(\omega) S(\alpha,\omega) d\omega,$$

$$S(\alpha,\omega) = (g(\omega)/\alpha)\,\mathrm{arctg}(\alpha/g(\omega)), \qquad \alpha = kv/\gamma_{ph}.$$

For $\alpha \ll 1$ it is possible to get asymptotic expressions for J_1, J_2 in some particular cases of $g(\omega)$: a) Rectangular line : $g(-\gamma_0/2 < \omega < \gamma_0/2) = 1$, $g(\omega) = 0$ otherwise;

$$J_1(k) \simeq (1/3)\gamma_0 \alpha^2; \qquad J_2(k) \simeq \gamma_0.$$

b) Lorenz line: $g(\omega) = \gamma_0^2/(\gamma_0^2 + \pi^2(\omega - \omega_0)^2)$;

$$J_1(k) \simeq \left(\sqrt{2/3}\right)\gamma_0\sqrt{\alpha}; \qquad J_2(k) \simeq \sqrt{2\gamma_0}/\sqrt{\alpha}.$$

c) Gauss line: $g(\omega) = \exp(-\pi(\omega-\omega_0)^2/\gamma_0^2)$;

$$J_1(k) \simeq \left(\sqrt{\pi/4}\right)\gamma_0\alpha/\left(\ln(1/\alpha)\right)^{1/2}; \qquad J_2(k) \simeq \left(2/\sqrt{\pi}\right)\gamma_0\left(\ln(1/\alpha)\right)^{1/2}.$$

In the case of a rectangular line shape, we have diffusive transport with the same diffusion coefficient as in (9). With non zero line wings (cases b, c) the propagation becomes nonlocal as reflected in the non quadratic asymptotic behaviour of $\lambda(k)$. Another feature also has to be noted. According to (12), two limiting cases of resonant media can be considered. The first one with $\gamma \ll \gamma_{ph}$ may be regarded as analogous to photon propagation since in the situation of resonant photon trapping the condition $\lambda_{ph} = v/\gamma_{ph} \ll L$ has to be satisfied. In this case

$$\lambda(k) = (\gamma/\gamma_0) J_1(k). \tag{13}$$

Estimating the time of resonant excitation trapping in a volume with size L as $\tau \sim 1/\lambda(1/L)$ for Lorenz and Gauss lines we have respectively:

$$\tau_L \sim \left(3/\sqrt{2}\right)\gamma^{-1}\left(\gamma_{ph}L/v\right)^{1/2} \tag{14}$$

$$\tau_G \sim \left(4/\sqrt{\pi}\right)\gamma^{-1}\left(\gamma_{ph}L/v\right)\left(\ln(\gamma_{ph}L/v)\right)^{1/2} \qquad (15)$$

These estimates agree up to a numerical factor with those obtained for the Holstein model in [2]. Another situation with $\gamma \gg \gamma_{ph}$ is more common for phonons. In this case

$$\lambda(k) = \gamma_{ph}J_1(k)/J_2(k), \qquad (16)$$

and respective trapping time estimates are:

$$\tau_L \sim 3L/v \qquad (17)$$

$$\tau_G \sim (8/\pi)(L/v)\ln\left(\gamma_{ph}L/v\right) \qquad (18)$$

The qualitative difference between phonon and photon propagation features as seen from (13-18) can be considered as retardation effect caused by the finite sound velocity[7]. It should be cautioned that the result (17) for Lorenz line has to be accepted with care since initially it was implied that $\lambda(k) \ll kv$. Nevertheless it may be concluded from (17,18) and also (15) that nonlocal propagation of a resonant excitation via resonant line wings is rather quasiballistic than quasidiffusive. In view of this principal difference between this result and the result for the system with no spectral migration considered in section 2 and since the assumption was made in the Holstein scattering mechanism that an excitation can jump over the whole spectrum, we are encouraged to check the validity of this assumption for each particular case. Therefore investigations of the details of spectral migration are significant to a proper understanding of propagation picture. See, e.g., [3,8].

4. NONRESONANT SCATTERING

For high energy phonons nonresonant scattering by defects and anharmonic decay may predominate over resonant scattering. To consider these effects let us modify equations (1) by including nonresonant scattering terms:

$$(\partial/\partial t + \gamma)N(\mathbf{r},t) = \sum_q \gamma_q^R n_q(\mathbf{r},t), \qquad (19)$$

$$(\partial/\partial t + \mathbf{v}_q\nabla + \gamma_q)n_q(\mathbf{r},t) = \chi\,\gamma_q^R\,N(\mathbf{r},t) + \gamma_q^D\,\langle n_q(\mathbf{r},t)\rangle,$$

$$\gamma_q = \gamma_q^R + \gamma_q^D + \gamma_q^A, \gamma = \gamma_R + \gamma_L, \qquad \langle\ldots\rangle = \sum_q \ldots \Big/ \sum_q 1.$$

Here $\gamma_q^R, \gamma_q^D, \gamma_q^A$ are the phonon damping due to resonant scattering, elastic defect scattering and anharmonic decay respectively; γ_R is the centre excitation damping caused by interaction with phonons, γ_L is the damping caused by another factors (e.g., the fluorescent decay rate in the case of optically created resonant centres). Taking, for simplicity, the isotropic model for anharmonic decay and defect scattering, i.e., $\gamma_q^A = \gamma_{ph}^A, \gamma_q^D = \gamma_{ph}^D$ for all q, we can obtain in the way similar to that of section 1 hydrodynamic equations similar to (5) with

$$\lambda(\mathbf{k}) = G(\mathbf{k}) + \gamma^*, \qquad (20)$$

$$G(\mathbf{k}) = \beta\left\langle(\gamma_{ph}^D + \gamma_q^R)(\mathbf{k}\mathbf{v}_q)^2/((\mathbf{k}\mathbf{v}_q)^2 + \gamma_q^2)\right\rangle,$$

$$\gamma^* = \beta \left(\gamma_L \gamma_{ph}^R / \gamma_R + \gamma_{ph}^A \left(1 - \gamma_{ph}^A / \gamma_{ph}\right)\right),$$

$$\beta^{-1} = 1 - \gamma_{ph}^A / \gamma_{ph} + \gamma_{ph}^R / \gamma_R, \gamma_{ph} = \langle 1/\gamma_q \rangle, \qquad \gamma_{ph}^R = \langle \gamma_q^R \rangle.$$

The non transport damping γ^* appears because conservation law (2) is violated by phonon anharmonic decay ($\gamma_{ph}^A \neq 0$) and nonresonant centre damping ($\gamma_L \neq 0$). It is seen from (20) that nonlocal transport effects are removed by anharmonic decay and defect scattering at the lapse of time when $\gamma_{ph}^A + \gamma_{ph}^D > L/v$ since there is no longer fast escape channel. Propagation becomes diffusive after that lapse of time. The interesting point is that different mechanisms play a role at different phonon energies because of their different frequency dependence. The characteristic frequency dependence of the trapping time was directly observed for phonons in $CaF_2 : Eu^{2+}$ in [9] and interpreted with above theory in [10].

5. RESONANT PHONON SCATTERING AT MULTILEVEL CENTERS

In the previous sections we have dealt with phonon scattering at two-level resonant centres. But in a paramagnetic crystal paramagnetic ions are multilevel systems and there can be cases when more than one electron transition frequency finds itself in the spectra of the lattice vibrations. The prominent role of resonant combinational scattering (RCS) in phonon trapping was revealed in [11] where the effect of magnetic field on phonon trapping in a ruby was investigated. It is possible to extend the model equations (1) to consider phonon scattering at several electronic levels. The relaxation of nonequilibrium excitations in that case is similar to that in the case of a two-level system. A local equilibrium is established between each electronic transition and the resonant phonons. On a long time scale $\tau \gg L/v$ the evolution can be presented as a relaxation of the different resonant packets as the whole resonant excitation. But now there is a coupling between packets due to RCS. The relaxation of a hydrodynamic wave now becomes a multi-exponential process as determined by the set of equations:

$$\sum_k (\lambda V_{ik} + W_{ik}) P_k = P_i^0, \qquad (21)$$

where P_k is effective density of excitation of electronic transition k. With the assumption of not overlapping transitions we have:

$$\mathbf{V} = \mathbf{1} + s\mathbf{A}, \qquad \mathbf{W} = s\mathbf{B}, \qquad (22)$$

where

$$A_{ik} = \delta_{ik} A_k, B_{ik} = \delta_{ik} B_k,$$

$$A_k = b_k \left\langle \gamma_{kq}^2 / ((\mathbf{k}\mathbf{v}_q)^2 + \gamma_{kq}^2) \right\rangle, \qquad B_k = b_k \left\langle \gamma_{kq} (\mathbf{k}\mathbf{v}_q)^2 / ((\mathbf{k}\mathbf{v}_q)^2 + \gamma_{kq}^2) \right\rangle,$$

$$b_k = \gamma_k / \langle \gamma_{kq} \rangle, \qquad s_{\alpha\beta,\gamma\epsilon} = p_\alpha (\delta_{\beta\epsilon} - \delta_{\beta\gamma}) + p_\beta (\delta_{\alpha\gamma} - \delta_{\alpha\epsilon}).$$

Here p_α is the population of α-th electronic level; γ_k is the inverse relaxation time of the electronic transition k; γ_{kq} is the inverse lifetime of the phonon q resonant with the transition k. The damping factors λ are determined by the equation:

$$\det(\lambda \hat{V} + \hat{W}) = 0. \qquad (23)$$

For N electronic levels this equation has $N-1$ nonzero solutions since only N transitions from $N(N-1)/2$ are independent. One more zero solution reflects the fact that the electronic excitation does not leave some of the ion levels.

Acknowledgment

The author wishes to thank Professor B.Z. Malkin for constant attention to this problem and very useful discussions.

REFERENCES

[1] Holstein T.: Phys. Rev. **72**, 1212–1233 (1947)

[2] Kaplyanskii A.A., Basoon S.A.: in Mod. probl. in Cond. Matter Scien. **16**, 373-453 (1988)

[3] Levinson Y.B. JETF **75**, 234–248 (1978)

[4] Malyshev V.A., Shekhtman V.L. Fizika Tv. Tela **20**, 2915–2928 (1978)

[5] Malyshev V.A., Shekhtman V.L. Opt. i Spectr. **46**, 800–808 (1979)

[6] Aminov K.L., Malkin B.Z. Pis'ma JETF **48**, 508–510 (1988)

[7] Aminov K.L. Fizika Tv. Tela **32**, 2234–2239 (1990)

[8] Solovyev A.E. Fizika Tv. Tela **32**, 2198–2204 (1990)

[9] Akimov A.V., Kaplyanskii A.A., Syrkin A.L. Pis'ma JETF **33**, 410–414 (1981)

[10] Aminov K.L. Fizika Tv. Tela **33**, 581–587 (1991)

[11] Basoon S.A., Kaplyanskii A.A., Shekhtman V.L. JETF **82**, 1945–1963 (1982)

SINGULARITIES OF THE HEAT CONDUCTIVITY IN THIN DIELECTRIC SLABS

J. Czerwonko,[1] and M. L. Kaganov[2]

[1]Institute of Physics, Technical University of Wrocław,
50–370 Wrocław, Poland
[2]P. L. Kapitza Institute for Physical Problems, Moscow, Russia

Usually charge transport in thin metallic slabs (i.e., in macroscopic samples much thinner than the electron mean free path (m.f.p.)) is considered in the relaxation time approximation[1] However, for thin slabs we have shown[2] that accounting for the gain term of the collision integral for impurity scattering does not alter the two leading terms of the asymptotic expansion[3,4] of the conductivity coefficient. It seems interesting to consider a similar question for the low–temperature heat conductivity of a thin slab of a dielectric. In this case, phonons are the main heat carriers.

Generally, for small temperature gradients, the energy current obeys the Fick law, i.e., it is proportional to the temperature gradient $\nabla T(r)$

$$Q_i = -i\kappa_{ik}(\partial T/\partial x_k) . \qquad (1)$$

Here Q is the density of the energy current, T is the temperature, κ_{ik} a component of the heat conductivity tensor. The summation convention over repeated vector indices is assumed. The number of independent components of κ is determined by the crystallographic class.

Let us consider a monocrystalline slab of the thickness $2d$, with the z–axis perpendicular to the slab ($|z| < d$), and with the temperature gradient directed along the slab. Let us assume that:

(i) $T \ll \Theta_D$, where Θ_D is the Debye temperature,

(ii) $T \ll \hbar v/d$, where v is the sound velocity. Hence, one can disregard the quantization of the phonon energy spectrum of the slab,

(iii) $l_T \gg d$, where l_T is the m.f.p. of the thermal acoustic phonons, ie., phonons with energy $\varepsilon \sim T$, ($k_B \equiv 1$).

Under such conditions the stationary transport in dielectric solids is described by the time–independent Boltzmann equation. As follows from the condition (ii), we shall restrict ourselves to the contribution of nondispersive acoustic phonons for which the energy is proportional to the length of the quasi–momentum p

$$\varepsilon = v(\mathbf{n})p,$$

where $v(\mathbf{n})$ is a generally anisotropic phase velocity. The deviation of the distribution function from equilibrium (described by the Planck function $F(p) = [exp(\varepsilon/T) - 1]^{-1}$) obeys the Boltzmann equation

$$v_z(\partial f/\partial z) + f/\tau = -v_x(\partial F/\partial T)(\partial T/\partial x). \qquad (2)$$

The group velocity v_a, $(a = x, z)$ depends on the propagation direction $\mathbf{n}$ and the

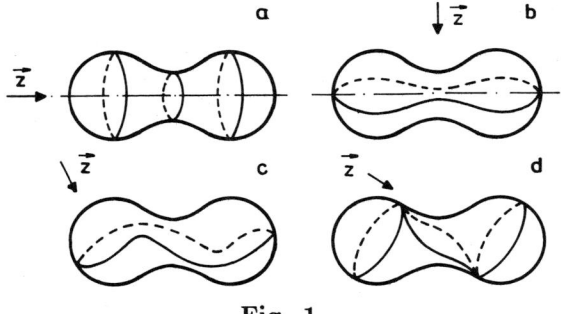

Fig. 1

relaxation time τ, generally, depends on both $\mathbf{n}$ and energy ε. The application of the relaxation time approximation is justified by the condition (iii) if phonon surface scattering is, at least, partially diffuse. For simplicity, we will assume in the following diffuse boundary conditions, namely

$$f|_{v_z>0, z=-d} = f|_{v_z<0, z=d} = 0 . \qquad (3)$$

The index, indicating one of the three acoustic branches of phonons, has been omitted. In the final expression, all polarizations will be taken into account. The restriction to the p–linear part of the energy spectrum is justified by the assumption (i), provided that only the main terms of the low-temperature κ–expansion are considered.

It is easy to verify that the solution of eq. (2), fulfilling the boundary conditions (3) has the form

$$f(z) = -\tau v_x(\partial F/\partial T)(\partial T/\partial x) \{1 - exp[-(z/v_z + d/|v_z|)/\tau]\} . \qquad (4)$$

As a result, the density of the heat current along the slab is given by

$$Q_x = \sum_{s=1}^{3} \int d^3 p v_x^{(s)} \varepsilon_s f_s(z)/h^3 \equiv -\kappa_{xx}(z)(\partial T/\partial x). \qquad (5)$$

Substituting the polarization dependent function $f_s(z)$ in the form (4) and introducing the average heat conductivity over the slab thickness,

$$\bar\kappa_{xx} = \int_{-d}^{d} dz \kappa_{xx}(z)/2d,$$

one finds

$$\bar{\kappa}_{xx} = 2\sum_{s=1}^{3}\int_{(+)} d^3p\,\left(v_x^{(s)}\right)^2 \varepsilon_s \tau_s (\partial F_s/\partial T) \times$$

$$\times \left\{1 - \tau_s v_z^{(s)}\left[1 - exp\left(-2d/\tau_s v_z^{(s)}\right)\right]/2d\right\}(2\pi\hbar)^{-3}. \tag{6}$$

Here, the subscript $(+)$ of the integral denotes the restriction to the half–space in reciprocal space such that $v_z^{(s)} > 0$. Taking into account the assumption (i) leading to $\varepsilon = v_s(\mathbf{n})p$ and introducing the dimensionless integration variable $x = \varepsilon_s/T$ we get from eq. (6) our central result

$$\bar{\kappa} = 2T^3\int_0^\infty dx\,x^4 e^x\,(e^x-1)^{-2}\int_{(+)} d\Omega \sum_s \tau_s(x,\mathbf{n})\left[v_x^{(s)}(\mathbf{n})\right]^2$$

$$\times \left\{1 - \tau_s(x,\mathbf{n})v_z^{(s)}(\mathbf{n})\left[1 - \exp\left(-2d/\tau_s(x,\mathbf{n})v_z^{(s)}(\mathbf{n})\right)\right]/2d\right\}\left[2\pi\hbar v^{(s)}(\mathbf{n})\right]^{-3}. \tag{7}$$

If impurity scattering plays the predominant role, then $\tau_s(\varepsilon,\mathbf{n}) = R_s(\mathbf{n})\hbar\theta_D^3/\varepsilon^4 \equiv R_s(\mathbf{n})\hbar\theta_D^3/T^4 x^4$, where $R_s(\mathbf{n})$ is a dimensionless positive function.[5] In this case, in agreement with the ideas of Peierls and Pomeranchuk[6,7], the integral (7) diverges if $d \to \infty$. The umklapp–processes or the scattering on the slab boundaries remove this divergence. For isotropic systems, eq. (7) gives

$$\bar{\kappa} = \sum_s \left[\theta_D^3 R_s/Tv_s(2\pi\hbar)^3\right] \times$$

$$\times \int_0^\infty dx\,\{2/3 + [S_1(q_0(x)) - S_3(q_s(x)) - 1/4]/q_s(x)\}\,e^x\,(e^x-1)^{-2}, \tag{8}$$

where

$$S_n(q) \equiv \int_0^1 dc\,c^n exp(-q/c), \qquad q_s(x) \equiv 2dT^4 x^4/\hbar v_s R_s\theta_D^3. \tag{9}$$

In the summation over the polarization index s, we deal with longitudinal acoustic phonons and, as we consider an isotropic medium, two kinds of transverse acoustic phonons[5] having the same energy. The asymptotic expansion of eq. (9) has the following form:[2,3]

$$S_1(q) = 1/2 - q - q^2(\ln q + C - 3/2)/2 + q^3/6 + 0(q^4),$$

$$S_3(q) = 1/4 - 2q/3 - q^3/6 + 0(q^4 \ln q), \tag{10}$$

where C is the Euler constant. Substituting (10) into (8) and calculating the typical integral containing the Planck function, we find

$$\bar{\kappa}_{xx} = \sum_s \left(dT^3\pi^2/15\hbar^3 v_s^2\right)\left[-\ln q_s - 22/3 + 3C + (360/\pi^4) \times\right.$$

$$\left.\times \sum_{k=2}^\infty k^{-4}\ln k + 32\pi^4 q_s/3 + O(q_s^2 \ln q_s)\right] \approx \tag{11}$$

$$\approx 0.65797 \sum_s k_{T_s}^3 v_s d\,[-\ln q_s - 5.3470 + 1039.03 q_s].$$

Here $q_s \equiv q_s(1)$ is the ratio of the slab thickness and the m.f.p. for a thermal acoustic phonon of polarization s, $\varepsilon_s = T$ and $k_{T_s} = T/\hbar v_s$. The sum of the series appearing in eq. (10) is equal to 0.068911. Note that the familiar estimate of the low temperature bulk heat conductivity is $k_T^3 v^2 \tau_u$, where τ_u is the characteristic time for a resistive process (leading to the finite κ for the bulk sample).[5] The formula (11) is valid only for $q_s \gg 1$. The third term in the square bracket of the rhs of eq. (11) becomes of the order of five percent of the main term if $q_s \sim 10^{-4}$. Similarly, if $q_s \sim 10^{-47}$ the second term is of the same order of the main term. It is worth to emphasizing that only for $q_s \lesssim 0.047$ the sum of the two first term in the square brackets of the rhs of eq. (11) is positive. The large coefficients appearing in the correction terms in eq. (11) are the price paid for the very sharp dependence of $\tau_s(x)$ at small values of x.

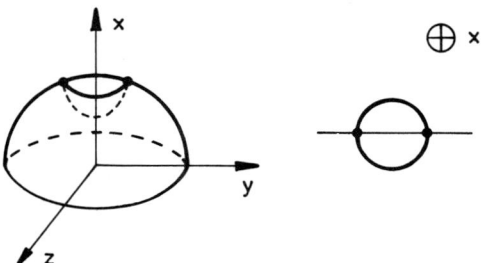

Fig. 2

As a result of the appearance of the lines of parabolic points on the isoenergetic phonon surface‡ responsible for caustics in the phonon fluxes,[8] the heat conductivity should have topological singularities. For example, let us consider the isoenergetic surface of the form of a dog's bone, well known as one of possible Fermi surfaces of metals.[9] If the z-axis is directed along the axis of the bone, we have three belts $v_z = 0$, whereas if z is perpendicular to this axis, we have only one belt $v_z = 0$. Hence, for some intermediate direction, the belt $v_z = 0$ should be self–intersecting, (cf. Fig. 1). This leads to a stronger singularity compared to the logarithmic one in the electric conductivity.[10] This singularity has its source in the integrals over the isoenergetic surface.[10] Because $d^3p = dS_s d\varepsilon/v$, with dS_s being an area element of the isoenergetic surface of phonons with polarization s, the same integral appears also in eq. (7) and, hence, κ_{xx} contains also a singularity stronger than a logarithmic one. In such a case the heat conductivity is determined mainly by the phonons from the vicinity of the point of self-intersection, analogously as for electric transport in metals.[10] Hence, if the slab surface is so oriented that the belt $v_z^{(s)} = 0$ of the isoenergetic surface of phonons of polarization s contains self–intersection points then the heat conductivity can be estimated to be

$$\kappa \sim k_{T_s}^3 v_s d \left[(\ln q_s)^2 + 0(\ln q_s) \right], \qquad (12)$$

phonon of polarization s exhibits self-intersection points. The contribution of other polarizations as well as the correction connected with the neighborhood of self-intersection points are contained in the $0(\ln q_s)$ term. All coefficients of eq. (12) are calculated at the self-intersection point. It is worth emphasizing that the mechanism leading to the large coefficients in the correction terms, indicated for the isotropic systems, works

---

‡it is convenient for us to integrate over the unit sphere and to use image of the isoenergetic surface.

also for anisotropic ones because of x^{-4} dependence of $\tau_s(x,\mathbf{n})$. Hence, the proportion of the coefficient of the main term to the coefficient of the correction term should be of the same order of magnitude for the isotropic and anisotropic systems, cf. (11) and (12), respectively.

The Miller indices of the slab plane in the case indicated in Fig. 1 are usually not small if the belt $v_z^{(s)} = 0$ contains self–intersection points. This can be a source of troubles, namely, the preparation of a suitable sample may be difficult. On the other hand, if the isoenergetic surface has, e.g., a crater with its centre lying on the x–axis, being a symmetry axis of the crystal, then the curve defined by conditions $\varepsilon_s = const$, $v_z^{(s)} = 0$ has, at least, two self–intersection points. Each of them contributes to the heat conductivity in the manner of (12). In the case, the preparation of samples may be easier, cf. Fig 2.

In conclusion, it is worth emphasizing the close analogy of the phenomena discussed here and the properties of thin metallic slabs.[2,3,10] For the electric conductivity of thin slabs of metals, the values of q_s for which the logarithmic term is important, are not as small as for heat conductivity of thin slabs of dielectrics. In the later case, $q_s \to 0$ if $T \to 0$, i.e., we deal with a strong temperature dependence, in contrast to metals.

For anisotropic systems, we can get neither the correction term nor the prefactor of $(\ln q_s)^2$ using the technique of reference.[10] Because of difficulties with the asymptotic expansion of the integral in eq. (7), which is two dimensional for anisotropic systems, (not to mention the presence of an additional integral over x), it is rather necessary to calculate them numerically.

It is worth to emphasize that the logarithmic factor in the familiar expression of the heat conductivity coefficient has been usually disregarded, cf. for example paper by Furzhi and Maximer.[11] However, this term will be clearly visible in differential measurements of the heat conductivity.

A preliminary version of this paper was prepared during the 20-th *Winter School of Theoretical Physics* organized by the University of Wrocław. The authors are greatly indebted to organizers of the School for the creative atmosphere. The support of the grant No. 20949101 of the Committee of Scientific Research of Poland is gratefully acknowledged by the first of authors.

REFERENCES

[1] K. Fuchs, Proc. Camb. Phil. Soc. **34**, 100 (1938).

[2] J. Czerwonko, Z. Phys. **B80**, 225 (1990).

[3] J. Czerwonko, Physica **A174**, 438 (1991).

[4] J. Czerwonko, M. I. Kaganov and G. Ya. Lyubarskii, in this volume,

[5] V. L Gurevich, Transport in Phonon Systems", North-Holland, Amsterdam (1986).

[6] R. E. Peierls, Ann. d. Phys. **3**, 1055 (1929).

[7] I. Ya. Pomeranchuk, J. Phys. USSR **4**, 259 (1941).

[8] H. J. Maris, Phonon Focusing, in: "Nonequilibrium Phonons in Nonmetallic Crystals", W. Eisenmenger and A. A. Kaplyanskii, eds., Elsevier, Amsterdam (1986).

[9] I. M. Lifshitz, M. Ya. Azbel and M. I. Kaganov, "Electron Theory of Metals", Consultants Bureau, New York (1973).

[10] M. I. Kaganov, D. V. Kamshilin and A. A. Nurmagambetov, Fiz. Niz. Temp. **15**, 289 (1989). [Sov. J. Temp. Phys. **15**, 162 (1989)].

[11] R. N. Gurzhi and A. O. Maksimov, Fiz. Nizk. Temp. **3**, 356 (1977).

ACOUSTIC PHONON INTERACTION WITH TWO-DIMENSIONAL ELECTRON AND HOLE SYSTEMS

Lawrence J. Challis and Anthony J. Kent

Physics Department
University of Nottingham
Nottingham NG7 2RD
United Kingdom

1. INTRODUCTION

Phonons play a significant part in the behaviour of electrons and holes in two-dimensional gases, 2DEGs and 2DHGs (for convenience in this paper we often use the abbreviation 2DEGs to cover both 2D electron and hole gases). Indeed phonon techniques have proved effective not only in studying electron-phonon interactions but also in obtaining other information on the properties of 2DEGs such as the location of dissipation and the magnitude of the energy gaps in the fractional quantum Hall state. In these lectures we aim first to provide an introductory review to the techniques being used in the field with illustrations of the type of information they provide. We then describe a number of recent experiments in which we have participated. As 2D systems form an important part of this winterschool, the review is prefaced by a short introduction to their properties and the way in which they interact with acoustic phonons. Theoretical aspects of this latter topic are covered in more detail in the lectures in this volume by Professor Shik.

1.1. Two-dimensional Electron Gases

Two-dimensional electron gases can be formed in semiconductors by electrostatic confinement. Frequently the confinement is achieved at a single interface between two different materials, a heterojunction, but can also be obtained by sandwiching one material between two plates of another. The potential well that is formed in these ways has discrete one-dimensional states/energy levels combined with in-plane states of energy $\hbar^2 k^2/2m$ which together form a series of overlapping subbands as shown in figure 1. These can filled from donors placed outside the well (2DHGs can be produced in a similar way) and the number of electrons n_s per unit area of well (the sheet density) determines the Fermi energy E_F. If E_F lies well below the energy of the first excited subband and $k_B T_e \ll E_F$ then essentially all the electrons are in the first subband and the effect of modest electric or magnetic fields or changes in T_e can only affect the in-plane motion of the electrons. So in many respects the system behaves two-dimensionally. Further confinement can be achieved in one or both of the in-plane directions resulting in one- and zero-dimensional systems, quantum wires and quantum dots. In quantum wires the energy levels lie in one-dimensional subbands and the electrons can only move parallel to the wire while in quantum dots, the electrons are wholly confined and have discrete energy levels like an atom.

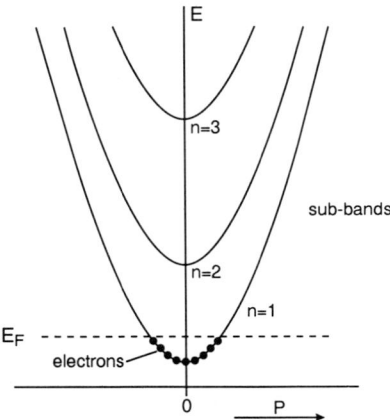

Fig. 1. The energy levels of a two-dimensional system showing the overlapping subbands.

The density of in-plane states of a 2DEG is independent of energy in zero magnetic field but in a field, the spectrum separates into a series of highly degenerate Landau levels separated by $\hbar\omega_c$ where the cyclotron frequency $\omega_c = Be/m^*$ (the additional splitting that can occur as a result of spin is neglected here). The degeneracy of each spin-split level is Be/h m^{-2}. This description neglects disorder which broadens the Landau levels, localises the levels in the wings and extends them into the gaps between the levels. The eigenstates can conveniently be described in the Landau or asymmetric gauge in which the localisation due to the magnetic field is confined to one direction only which we take to be the x-direction. So the wavefunction in the y-direction is unchanged and for free electrons in a 2DEG of sides L_x, L_y is $\exp(ik_y y)$ with $k_y = 2b\pi/L_y$ where $b = 1, 2..$ (periodic boundary conditions). In the x-direction, the wavefunction consists of a harmonic oscillator function $(\exp(-\frac{1}{2}(x-x_0)^2/l_B^2)$ in the ground state) whose length scale is the magnetic length, $l_B = (\hbar/eB)^{1/2} = 26/B^{1/2}$ nm. So the wavefunction has the form of a narrow strip $\sim L_y \times l_B$ (a Landau strip) and since the harmonic oscillator wavefunction for the eigenstate of a particular k_y is centred at the coordinate $x_0 = -l_B^2 k_y$, the centre of each of the Be/h m^{-2} Landau strips is displaced from its neighbour although they are heavily overlapping in samples of typical size.

It was stated earlier that, in the absence of disorder, the Be/h states m^{-2} in a Landau level are all degenerate. This is true for the large majority of the states in wide samples but not for those at the very edge of the sample. The confinement of the electrons to the 2DEG region evidently implies that the potential energy must rise at the edges of

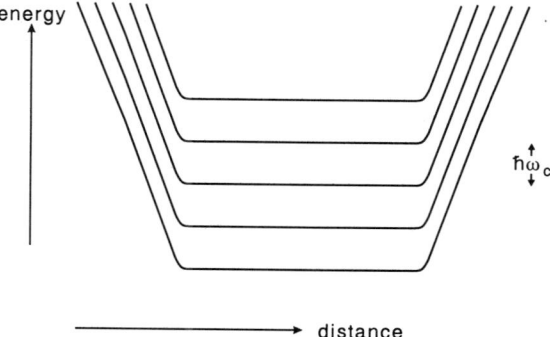

Fig. 2. The variation in the Landau level energy across the width of a 2DEG showing the increase at the edges of the samples due to the confinement potential.

the sample; the 2DEG is contained within a wide potential well. So the energies of the Landau levels also rise as shown in figure 2. The states in the sloping regions of potential are referred to as edge states and despite their relatively small number they appear to play an important role in the properties of 2DEGs. Their significance in phonon experiments will be discussed later in these lectures.

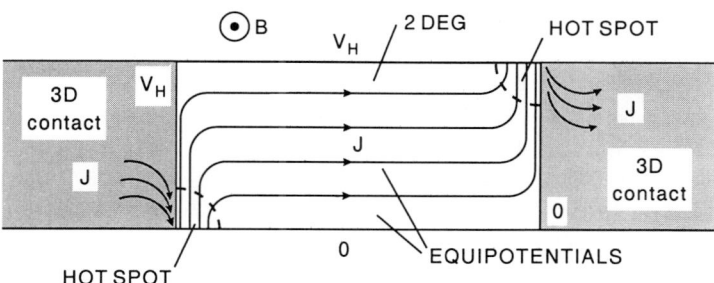

Fig. 3. Equipotentials in the quantum Hall regime (schematic).

The integer quantum Hall effect occurs when E_F is midgap (between two Landau levels) so that, at low temperatures ($k_B T_e \ll E_F$), if we neglect edge states, essentially all extended levels above E_F are empty and those below E_F are filled with electrons and there are no empty levels to scatter into. If now an electric field is applied across the width of the sample, the combination of crossed electric and magnetic fields causes the electrons to drift with velocity E/B in the direction $\underline{E} \times \underline{B}$. Since no scattering is possible, the current is dissipationless at $T_e=0K$ and remains very small at non-zero temperatures provided $k_B T_e \ll \hbar\omega_c$ for currents $I < I_c$ where I_c is a critical current. In Hall bar geometry, $\underline{E}$ is the Hall field caused by the current passing through the sample. The other remarkable feature is that the Hall resistance R_H (= Hall voltage /current) equals h/ne^2 where n, an integer, is the number of filled Landau levels below E_F ($n=n_s h/Be$). This can be explained either in terms of the quantisation of the phase-integral round a closed electron orbit or in terms of edge states[1] but it is not clear that there is yet a full description.

A further property of a Hall bar in the quantum Hall state of relevance to the dissipation that occurs is that the resistance R between the source and drain contacts is equal to the Hall resistance R_H. This situation arises because the electron concentration in the 3D contacts is very much larger than that in the 2DEG. So the Hall voltage across the width of the 2DEG is very much larger than that across the contacts which therefore act effectively as equipotentials. Since there is no voltage drop along the length of the sample ($\rho_{xx} = 0$), the equipotentials run parallel to the length with a value of V_H on one side and 0 on the other for most of the sample. However, at the ends they have to curve round to become parallel to the two contacts as shown schematically in figure 3. (In practice the equipotentials are much more closely bunched together at the edges and wider apart in the middle because of the increased charge at the edges needed to set up the Hall voltage.[2]) The resulting field distribution forces the current to enter and leave at two diagonally opposite corners. The current has to cross the width of the sample falling through a potential difference V_H and so the source-drain resistance $R=R_H$. The location of the dissipation or Joule-heat $I^2 R = I^2 R_H$ must evidently occur at the current entry and exit corners where the current density and electric field is very large. This presumably leads to quantum Hall breakdown through the same mechanisms that occur in the bulk when $I = I_c$.

The fractional quantum Hall and Wigner crystal states of a 2DEG are also of considerable current interest. Both owe their existence to the repulsive Coulomb potential energy between the electrons. This can often be neglected in considering the properties of 2DEGs. But at low temperatures and sheet densities this is no longer true in high quality samples. The signature of the fractional state, a minimum in ρ_{xx} and a plateau in R_H, is similar to that of the integer state but in this case it occurs when the lowest Landau level is fractionally filled to an extent $\nu=1/3$, $2/5$, $3/7$ etc (ν is called the filling factor) rather than wholly filled with E_F in the middle of a gap between two Landau levels. The similarity of the signature suggests that E_F must also lie in the middle of a gap in the fractional state so we conclude that there are also gaps within a Landau level at particular values of ν. Information on the size of these gaps obtained from measurements of $\rho_{xx}(T_e)$ show that their sizes are comparable to the Coulomb energy between the electrons rather than the cyclotron energy as in the integer state.

The charge carriers in the fractional state appear to be fractionally charged quasiparticles with, for example, $e^* = 1/3 e$ for $\nu=1/3$ and we can picture this as arising because the electrons distribute themselves over all the degenerate states in the Landau level so that each state has an average charge $1/3 e$. The quasiparticle can be pictured as a vortex in the dissipationless electron fluid. The requirement for phase quantisation $\oint \underline{k} \cdot \underline{dl} = \lambda 2\pi$ ($\lambda=1,2..$) means that λ flux quanta (flux=$\lambda h/e$) have to pass through the vortex. Now at filling factor ν, the charge associated with a flux quantum is νe so the charge associated with a vortex is $e^* = \nu \lambda e = 1/3 e$ for $\lambda=1$. The axes of the Be/h m^{-2} vortices are parallel to the magnetic field and so normal to plane of the 2DEG and in the ground state the vortices all rotate in the same direction. An excited state is obtained by raising a vortex to a higher energy state above the gap. This leaves a 'hole' below the gap which behaves like a vortex of opposite rotation and positive charge 1/3e : an antivortex. The lowest energy state is a bound exciton-like state of the excited vortex with the antivortex which is called a magnetoroton.

In the absence of kinetic energy (and disorder), the lowest energy state of a 2DEG is evidently a solid with the minimum repulsive potential energy occurring when the electrons are arranged in a hexagonal/triangular lattice. This Wigner solid has been seen very clearly for a 2D gas of electrons over helium and for ions below the surface of helium but is more difficult to form and observe inside a semiconductor. The main difficulty is to reduce the kinetic energy sufficiently in comparison with the potential energy. At low temperatures the kinetic energy is essentially the zero-point motion $(3/5)E_F$ which is proportional to the sheet density n_s while the potential energy varies inversely with the electron separation and so varies as $n_s^{1/2}$. So to form a solid, n_s has to be reduced: the 2DEG has to be expanded, and calculations suggest that in GaAs solidification should occur at values well below those presently attainable. The situation is much more promising in Si and 2DHGs in GaAs since the kinetic energies are substantially less because of the larger effective masses and there is evidence of possible solid formation in Si.[3] Solidification can also be obtained in a 2DEG in GaAs if a large magnetic field is applied. The localisation produced by the field assists the formation of the solid and there is experimental evidence of solidification using a number of different techniques including interaction with surface acoustic waves. The interpretation is still controversial however. The experiments clearly show that the electrons are localised but it is difficult to be certain that the localised system has a degree of long range order. Observation of this order would demonstrate that the localisation is indeed mainly caused by the potential energy between the electrons as in a Wigner solid and not by the potential energy associated with the disorder.

Further details of these systems can be found in the book by Weisbuch and Vinter[4] and there is also an introductory review written by one of us[5] which contains references to other introductory material though not to a very recent article on Wigner solids.[6]

1.2. Acoustic Phonon Interaction with Two-dimensional Electron Gases

Acoustic phonon interaction with 2DEGS and even lower-dimensional systems is believed to proceed via the same coupling mechanisms that exist in 3D systems (deformation potential and piezoelectric) and there is no clear evidence for any difference in coupling strength. (The strength increases with effective mass so is appreciably stronger in Si and 2DHGs in GaAs than in 2DEGs in GaAs). The effective strength is modified however by differences in screening associated with the dimensionality and further differences arise, particularly in the angular distribution of the phonons emitted or absorbed, because of differences in the requirements of momentum conservation. The localisation of the electrons into a region of effective thickness a means that phonons can be emitted or absorbed with non-zero wavevector components $q_\perp$ normal to the 2D plane even though the electron wavevector in this direction is unchanged during the process. The limit on $q_\perp$ set by the uncertainty principle is $q_\perp < a^{-1}$ and this is reflected in the electron-phonon matrix elements which roll-off at higher values of $q_\perp$. At the lower frequencies, phonon emission and absorption is only constrained by energy conservation and in-plane momentum conservation. In zero-field, the maximum change that can occur in the in-plane momentum of an electron is around $2\hbar k_F$ so that $q_\parallel < 2k_F$. The details of these interactions have been discussed by a number of authors in both 2D[7] and lower-dimensional systems.[8]

In a magnetically quantised system, phonons can be emitted/absorbed either as a result of interLandau level transitions with $\omega = j\omega_c$, j=1,2,3.. (cyclotron phonon emission)[9] or as a result of intraLandau level transitions. In principle these could occur either within the bulk of the sample or at the edges. The limit on $q_\parallel$ is also changed by magnetic field. Since the x-coordinate of a Landau strip $=-l_B^2 k_y$, an electron has to be displaced $l_B^2 q_y$ in the x-direction if it emits or absorbs phonon momentum $\hbar q_y$. Now the Landau strips have widths $\sim l_B$ so when $l_B^2 q_y > l_B$ or $q_y > l_B^{-1}$ there is a rapid fall in the overlap between the initial and final states and so the matrix element. There are also constraints on q_x and, when these are combined, the in-plane constraint becomes $q_\parallel < l_B^{-1}$. At low magnetic fields, the transitions take place from Landau levels with indices given by $E_F = (n+1/2)\hbar\omega_c$ which have widths $\sim (2n+1)^{1/2} l_B$. So the momentum conservation rule becomes $q_\parallel l_B^2 < 2(2n+1)^{1/2} l_B$. Now $2(2n+1)^{1/2} = 2(2E_F/\hbar\omega_c)^{1/2} = 2(\hbar/Be)^{1/2} k_F = 2 l_B k_F$ so, in the low field limit, the conservation rule moves smootnly over to the zero field form $q_\parallel < 2k_F$.

As the magnetic field is increased, the wavenumber at the cyclotron frequency $q_c = \omega_c/v_s$, where v_s is the sound velocity, becomes larger than the inverse thickness a^{-1}. For transverse modes in GaAs ($a \sim 5$nm for a typical heterojunction) this occurs at a cyclotron frequency ~ 200GHz corresponding to B~ 0.5T so that the cyclotron phonon emission process falls rapidly as the field is increased beyond this value. Inter-Landau processes should however still contribute to hot electron relaxation in the bulk of the samples, though less efficiently than at lower fields, since the constraints of momentum conservation can be avoided by the emission of two oppositely directed phonons q, $-q$, each of frequency $\omega_c/2$[10] and edge-state transitions are also likely to be significant.[11] Momentum conservation is also not a constraint in the emission of low q optical phonons although this normally only becomes significant at relatively high electron temperatures.

2. EXPERIMENTAL TECHNIQUES

The main types of technique are illustrated schematically in figure 4. Figure 4(a) shows a 2DEG heated by passing a current through it and the detection of the phonon emission on the further face of the substrate. Figure 4(b) illustrates experiments for measuring phonon absorption, transmission or reflection from an incident phonon beam generated by a heater or ultrasonic transducer. The intensity observed in such experiments is a convolution of the angular dependence of the phonon emission etc into q-space with the phonon focussing effects associated with the elastic anisotropy of the substrate.

2.1. Phonon Emission

Phonons emitted from a 2DEG travel ballistically across the substrate if their frequency is ≤1000GHz. So information on features such as their angular distribution, the intensity and location of the emission and occasionally their frequency distribution can be obtained from detectors at various points on the two substrate surfaces. Both CW and pulse techniques have been used. In both cases, the intensity I_s at the surface is usually determined from the local rise in temperature, $\Delta T \propto I_s$, and is measured by some form of bolometer although frequency sensitive detectors have also been used to obtain spectroscopic information.

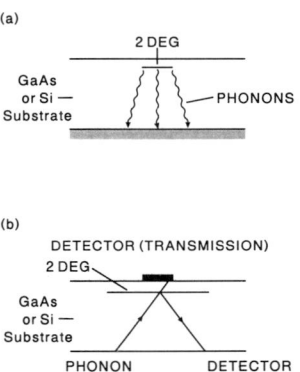

Fig. 4. (a) Phonon emission. The 2DEG is heated electrically ($P=I^2R$) and the angular distribution of the phonons emitted are detected from detectors on the opposite face of the substrate. (b) Phonon Scattering. The phonons incident on the 2DEG are generated by a heater or an ultrasonic transducer. Absorption can be measured by using the 2DEG as a bolometer and transmission and reflection, from detectors placed as shown.

In our own laboratory, for CW measurements, ΔT is recorded on a 250 μm square contact of silver foil which is glued to the surface and also connected to a copper wire leading to a carbon thermometer. The thermometer is supported in a cage 15cm above the sample to minimise magnetoresistance effects. ΔT is usually measured with respect to a reference point on the sample involving the use of a second contact and thermometer. Temperature changes of a few μK can be detected allowing measurements to be made down to input powers of $\leq 1\mu$W. The technique has been used in zero magnetic field to observe sheet density changes of the angular dependence of the phonon emission implied by the selection rule $q_\parallel < 2k_F$ and also the change in angular distribution that occurs when E_F enters the first excited subband.[12] In magnetic fields we have used it to observe the dissipation at the current entry and exit points of a Hall bar in Si and GaAs[13,14] and observed oscillations in the cyclotron phonon emission from a narrow strip of 2DEG[15]. In Stuttgart the helium fountain pressure has been used to obtain information on ΔT[16]. A local temperature rise produces a small rise in the thickness of the helium film on the surface of the substrate which can readily be detected optically. Further details of this elegant technique and its use in exploring dissipation in the quantum Hall regime are given in the lectures by Dr Dietsche.

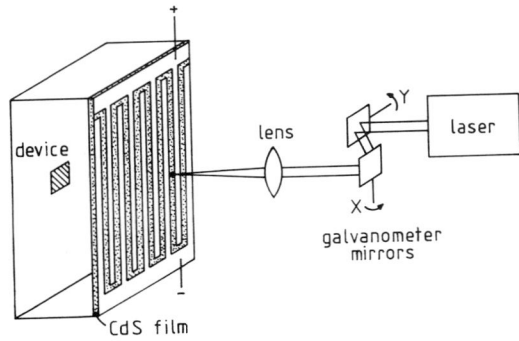

Fig. 5. Phonon imaging system based on an extended CdS bolometer.[24]

Pulse techniques have the advantage that from time-of-flight measurements, information can be obtained on the relative intensities of longitudinal and transverse phonons. Measurements have been made using discrete detectors of various types including bolometers and also superconducting tunnel junctions and exciton cloud detectors which provide spectroscopic information on the phonons emitted. Using these techniques a wide range of information has now been obtained on both 2DEGs and more recently 2DHGs. The first emission experiments on 2DEGs were on GaAs[17] and showed that the emission was dominated by transverse modes and indeed very weak LA mode emission was only reported rather recently[18]. These experiments were followed by detailed studies on Si[19] which showed the marked change in angular dependence that occurs with sheet density reflecting the requirements of in-plane momentum conservation. In this case both TA and LA modes were present in the emission. Further studies on GaAs include observations of the change in pulse shape that takes place when the input power is raised to the point where crossover occurs from predominantly acoustic to optic phonon emission[20]. There have also been a number of studies of phonon emission in magnetic fields showing evidence of both cyclotron phonon[21] and optic phonon[22] emission, the latter being seen through the appearance of magnetophonon resonances in the detected signal. Studies of phonon emission from 2DHGs are comparatively recent and are discussed later in this paper. The pulse technique has been extended to obtain an image of the phonon intensity on the opposite face - a "phonograph". The first such imaging system used a large area superconducting tunnel junction[23] but the interest in producing images in magnetic fields resulted in a system based on CdS[24]. In the CdS system the image is typically formed by a 3x3mm^2 array of 100μm strips of evaporated CdS film whose resistance is measured using Cu strips as electrodes as shown in figure 5. CdS is a semi-insulator at low temperatures because the electrons from the shallow donors become trapped by deep levels. However if it is briefly illuminated by a 100μm diameter laser beam, some of the electrons excited to the conduction band become trapped by the shallow donors and at low temperatures are unable to return to the deep traps since this requires the involvement of large q

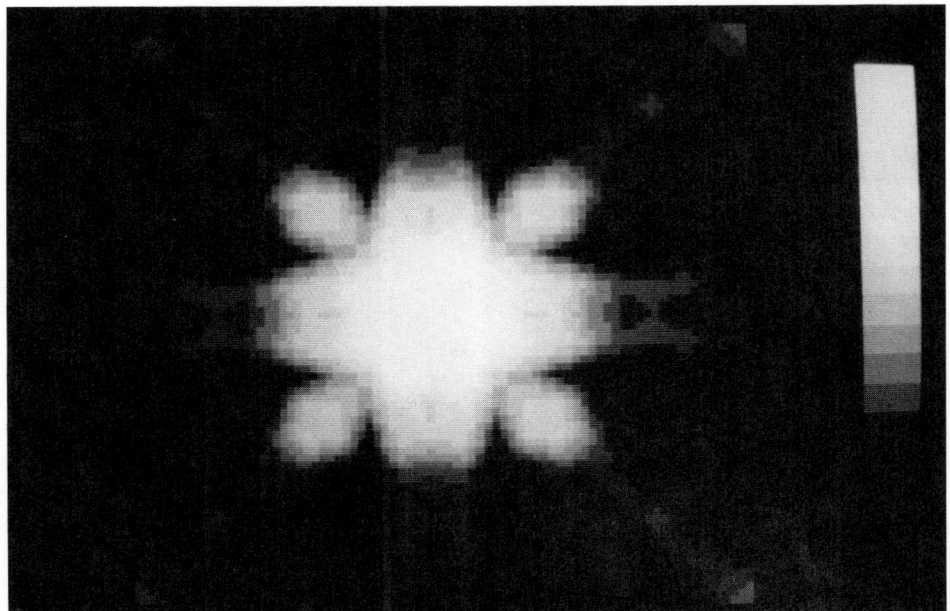

Fig. 6. An image of the phonons emitted from a (100) Si MOSFET using the system shown in figure 5.[25]

phonons. So the laser beam creates a 100μm-wide conducting path between two Cu strips whose resistance is very sensitive to small increases in temperature caused by an incident phonon pulse. When the phonon intensity on this conducting path has been measured, the area is desensitized by putting a large voltage between the Cu strips. This heats the CdS locally creating the phonons necessary to allow the electrons to return to deep levels and so restores the resistance to its original large value. The laser beam can then be moved to the next position and by raster-scanning the beam, an image of I_s can be built up as in the previous technique. A phonon image of the emission from a 2DEG in a Si MOSFET produced in this way[25] is shown in figure 6. The intensity is seen to be largely confined to a cone around the normal to the 2DEG as expected. It falls away at larger angles but in certain directions - those lying in the (100) and (110) planes - it is strongly enhanced by phonon focussing. This can also be seen in figure 7 which shows a line scan across the image. For comparison we also show a scan across an image from a 3D metal film.

2.2. Phonon Scattering

The majority of experiments of this type have been made using heat pulses from current or laserheated metal films though there have also been lower frequency experiments using coherent ultrasonic sources. These latter include both bulk ultrasonics at 10GHz[26] and surface acoustic waves up to 2GHz.[27] The first experiments to show phonon scattering by a 2DEG was by Hensel and Dynes.[28] They observed the fall in specular reflection from the Si surface of a MOSFET when the 2DEG was formed by

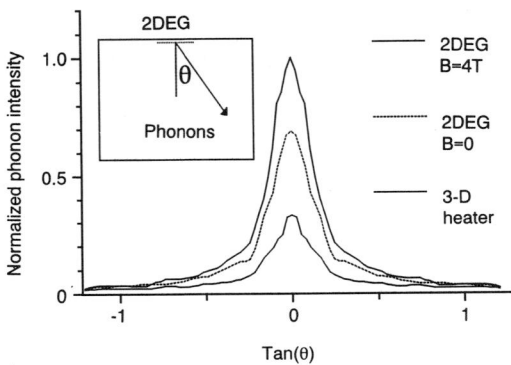

Fig. 7. Line scans of the intensity for phonons propagating in the (110) plane, i.e. taken horizontally through the centre of the image in figure 6. In each case the power input to the 1×1 mm² 2DEG or heater is the same. The increased height of the θ=0 peak in the case of the 2DEG is due to the emission being concentrated in a narrow cone of angles close to the normal. Application of a magnetic field to the 2DEG restricts the emission to an even smaller range of angles.

the gate potential. They originally attributed this to absorption but later concluded that the effect was too large to be explained in this way and was more likely to be due to interference between the reflections from the Si surface and from the 2DEG. These experiments have been followed by many others both in zero and high magnetic fields and in the latter case show oscillations attributable to both inter and intra Landau level transitions; we refer to a recent review for descriptions and references of some of this work.[29] For illustration here we describe two techniques which we are using in Nottingham to obtain information on phonon absorption in the integer and fractional quantum Hall regimes.

One of the methods is based on a fixed electrically heated phonon source and is able to provide some limited spectroscopic information while the second uses a movable phonon source to probe the spatial dependence of the interaction. Both techniques use the resistance of the 2DEG or 2DHG as the phonon detector and for this reason they are sometimes referred to as phonoconductivity experiments.

Figure 8 shows the experimental arrangement using the fixed heater. The back face of the substrate is polished and a small heater consisting of a 50Ω strip of metal is vacuum evaporated onto it. Short bursts of nonequilibrium phonons can be generated by passing electrical current pulses, which are typically a few tens of nanoseconds long, through the heater. The spectrum of the phonons is approximately Planckian with a peak at $\hbar\omega_p \approx 2.8 k_B T_h$, where T_h is the temperature of the heater. These phonons propagate ballistically across the substrate which is at liquid helium temperatures and are incident on the 2DEG. The interaction of the phonons with the electron gas produces small changes in its mobility which can be detected by passing a constant bias current and using a two- or four-terminal voltage measurement. Using acoustic mismatch theory [30] we are able to determine T_h as a function of the electrical power, P_h, supplied to the heater:

$$P_h = \Gamma A(T_h^4 - T_l^4) ,$$

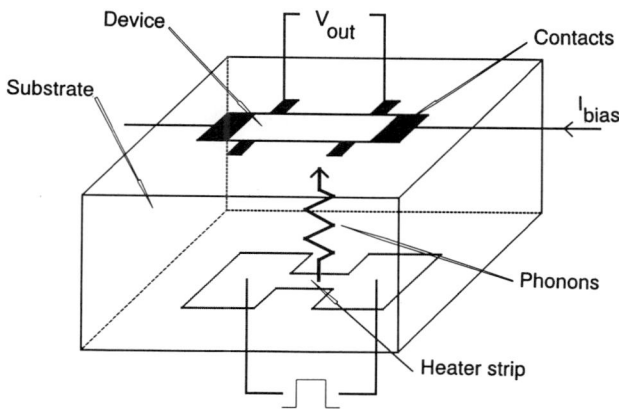

Fig. 8. Sample geometry for the phonoconductivity experiments. Phonons are generated by passing a current pulse through the 50 Ω heater strip and detected through the change in resistance of the device.

where A is the area of the heater, T_l is the substrate temperature and Γ is the acoustic mismatch constant: for constantan on GaAs, $\Gamma=524$ Wm^{-2}K^{-4}. Therefore by varying P_h we can obtain some limited spectral information regarding the electron-phonon interaction. Using this technique we have been able to observe evidence of cyclotron phonon absorption by electrons in the silicon inversion layer[31] and probe the size of the energy gap in the fractional quantum Hall state (see later).

A development of this technique replaces the fixed heater with a movable phonon source. The experimental arrangement, shown in figure 9, is similar to the phonon imaging method developed by Wolfe and co-workers.[32] A metal film is deposited on the polished back face of the substrate and locally heated by thermalising a pulsed laser beam. The angle of incidence of the phonons on the 2DEG can be varied by moving the laser beam over the metal film by means of a pair of computer-controlled galvanometer mirrors. By raster scanning the laser a 2D map of the angular dependence of the electron phonon interaction can be created to a resolution of about 20μm. Figure 10 shows some examples of images obtained using this method. The images are of the phonon absorption by a 2DEG in a silicon MOSFET at different 2DEG sheet densities, n_s. It is clear from these that increasing the sheet density enables phonons incident at larger angles to the 2DEG normal to interact more strongly with the electrons because of the increase in $2k_F$. Most of the detail of the images is due to phonon focussing in the substrate which gives rise to a highly anisotropic flux of phonons on the 2DEG. In principle it is possible to deconvolve these effects and extract the angular dependence of the 2D electron-phonon interaction to compare with theory. In practice, however, this is rather difficult, especially in the case of the transverse modes for which phonons having different wavevector directions can have the same group velocity direction. For this reason the opposite approach is used. It is fairly straightforward to calculate numerically the angular dependence of the phonon absorption corresponding to various n_s and heater temperatures etc. These results can be fed as initial distributions into monte-carlo phonon focussing simulation programs[33] and the results compared with experiment. Figure 11 shows the results of such a calculation compared with linescans through the experimental images. Good agreement is obtained using just one adjustable parameter - the heater temperature.

The phonon focussing effects have been put to good use in some recent experiments. Because of the strong focussing of transverse modes close to the (100) direction, the phonon flux within about 12 degrees to the normal from the source is at least an order of magnitude stronger than elsewhere. On a standard 0.4mm thick substrate this results in an area of high intensity on the opposite face which is only about 100μm across. So phonons can be used to probe small areas of an extended 2DEG. The resolution is ensured by setting the boxcar detector gate to exclude any phonons that have not travelled the shortest path between the phonon source and the 2DEG. Figure 12 shows two images made using a 3×1mm^2 active area silicon MOSFET in a strong magnetic field.[34] At integer filling factor, in the quantum Hall regime, the only response, a de-

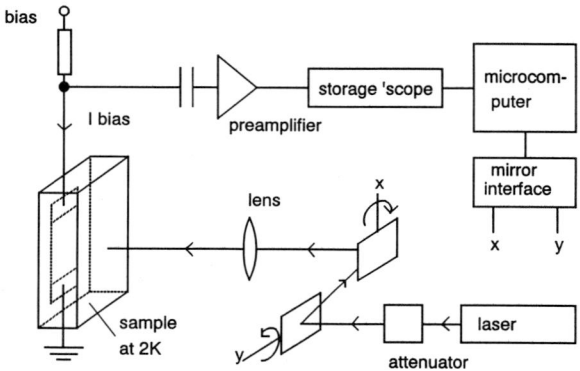

Fig. 9. Experimental imaging system. The sample is mounted in a optical access 7T cryomagnetic system. A pulsed Nd-YAG laser is thermalised in a metal film to generate phonon pulses. The laser spot position is controlled by a pair of computer controlled galvanometer mirrors. Phonoconductivity signals are detected using a high-speed digital storage oscilloscope or boxcar averager and recorded as a function of mirror position in the computer.

crease in two-terminal resistance, is from the contact regions near the current entry and exit points. This is due to phonon activated conduction in the disordered contact regions and is strongest at regions of highest current density. There is no response from the bulk because the phonons have insufficient energy to excite electrons from the filled levels below E_F to the empty states above and so disrupt the dissipationless state. On the other hand at half integer filling factor intraLandau level transitions are possible and the bulk response, an increase in resistance, is dominant.

The main drawback of the scanning technique is that it is hard to determine accurately the laser power absorbed in the metal film and hence the phonon frequency spectrum. For this reason the fixed heater and scanning methods are complementary.

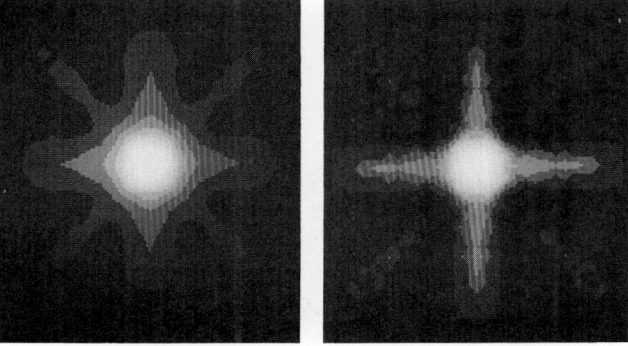

Fig. 10. Phonoconductivity images of the absorption of phonons by a (100) Si MOSFET in zero magnetic field. (a) $n_s = 2.2 \times 10^{15}$ m^{-2}; (b) $n_s = 4.6 \times 10^{15}$ m^{-2}.

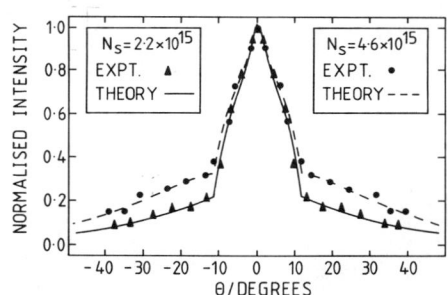

Fig. 11. Comparison of linescans taken horizontally through figures 10 (a) and (b) with theoretical calculations of the phonon absorption including focussing effects.

3. RECENT INVESTIGATIONS

We next describe some recent experiments with which we have been associated.

3.1. Dissipation in the Integer Quantum Hall regime

It has already been noted that when a current is passed through a Hall bar with the 2DEG in the integer quantum Hall state, the Joule heating $I^2 R_H$ is expected to occur in regions of the 2DEG at two diagonally opposite corners where the electrons enter and leave. The positions of the corners should depend on the direction of the magnetic field but not on the direction of the current.

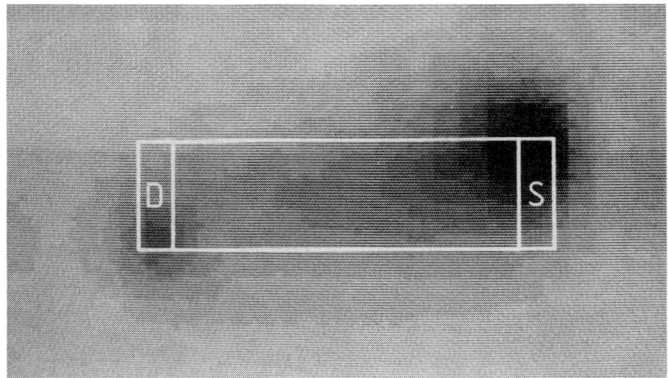

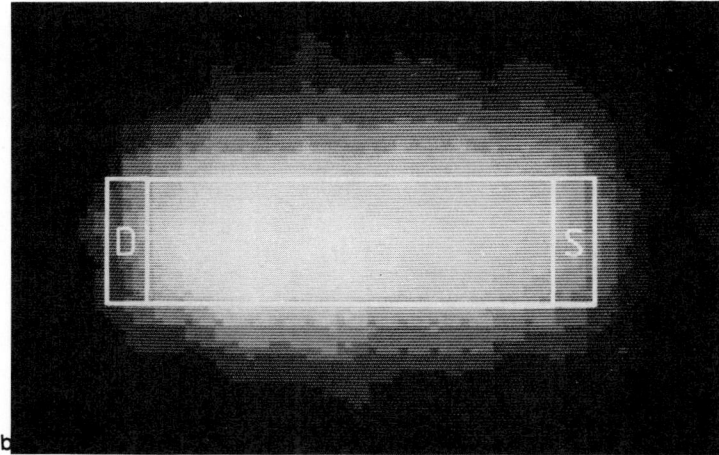

Fig. 12. Phonoconductivity image of extended 2DEG in a Si MOSFET and in a strong magnetic field. (a) In the quantum Hall regime the only response, a decrease in resistance, occurs when the phonons are incident near the current entry and exit points. (b) At half integer filling factor an increase in resistance is observed when the phonons are incident on the bulk. The response in (b) is due to intra-Landau level scattering of electrons by phonons.

A direct demonstration of the location of the dissipation was first made by observing the phonon emission from a corner and the centre of a Hall bar in a Si MOSFET.[13] The changes that occurred in these as E_F was moved through the Landau level spectrum were consistent with the description given above as were the changes with field direction and it was also found that no detectable change occurred when the current was reversed showing that the dissipation was equally divided between the electron entry and exit corners. These experiments were made using the CW technique described earlier but they were confirmed on similar Si MOSFET samples using pulsed currents and an imaging bolometer.[34] Measurements carried out by Dietsche's group using steady currents on 2DEGs in GaAs/(AlGa)As heterostructures showed interestingly different behaviour however.[16] Below a particular value of current ($I=110\mu A$ for i=2 which was an order of magnitude smaller than the likely critical current for quantum Hall breakdown in the bulk sample) the behaviour was identical to that we had observed in Si but for higher currents the dissipation appeared to decrease substantially at the electron exit point compared with that at the entry point.

These experiments and some electrical transport experiments by van Son et al[34]

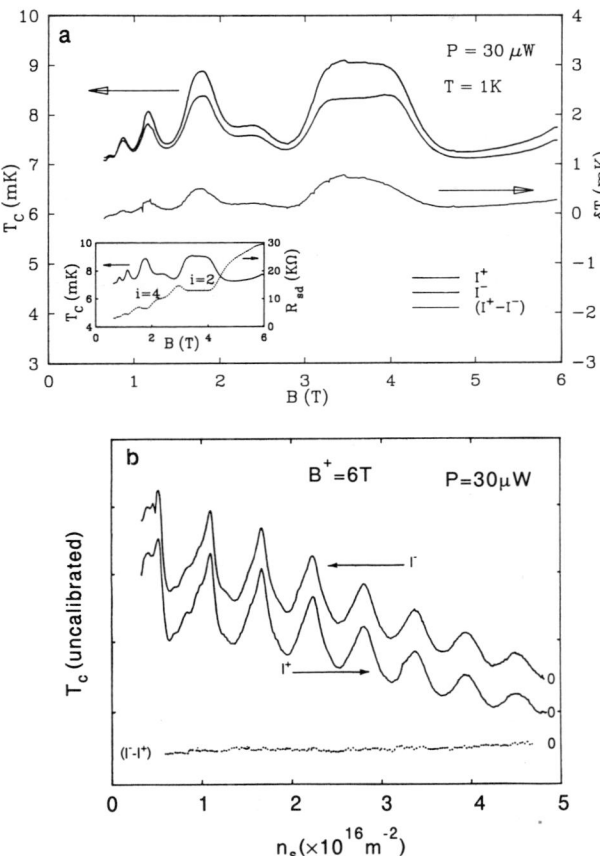

Fig. 13. The variation of phonon intensity, T_c, opposite a corner of a magnetically quantised 2DEG for two current directions I^+ and I^-. The difference between the two intensities is shown by $I^+ - I^-$. The intensities are shown for (a) GaAs as a function of B[13] and (b) Si as a function of n_s[13].

raise the question as to whether a small degree of asymmetry also exists in Si above a certain current. This might have been too small for us to have detected in our first experiments so we carried out a second series[13] using an improved technique capable of detecting differences in phonon intensity when the current was reversed of around ∼1 per cent. The technique included a calibration procedure which allowd us to set an upper limit to the change. We also explored a wider range of powers, 0.3-2000μW, than in the earlier work but were unable to detect any differences between the dissi-

173

pation at the current entry and exit points though they may of course exist below our detection level. Another possibility, that the difference is associated with the differences in the two techniques being used, appears to be ruled out since measurements in Stuttgart on a similar Si sample to that used in Nottingham showed equal dissipation at the two corners over the whole range[36] and measurements in Nottingham on GaAs samples found unequal dissipation as shown in figure 13; data for Si are also shown for comparison. The behaviour observed for GaAs was however somewhat different from that observed in Stuttgart.[14] We found an increase in asymmetry with power but in our case it was not as abrupt as that observed in Stuttgart. The source-drain resistance was found to be rather accurately equal to R_H showing that the asymmetry was not associated with the breakdown of quantization.

The difference in asymmetry between Si and GaAs is intriguing but we need to look rather carefully at what we are measuring in the GaAs experiments. What we observe is that there is a change in phonon signal on a small detector opposite a corner when the current direction is reversed. Now while this may indicate that there is a change in magnitude of the dissipation, it could also indicate a change in the position or size of the heated area or a change in the angular distribution of the emitted phonons. For example, suppose the magnitude of the dissipation is the same at both corners but, for some reason, the heated area is larger at the electron exit than at the electron entry. If one switches the corner being observed from entry to exit by reversing the current, a smaller proportion of the phonons will hit the detector resulting in a fall in signal. We believe we have ruled out this particular possibility by recent experiments in which we used two detectors.[37] One of these was immediately opposite the corner and the other was 250μm away along the edge of the Hall bar. So if the hot spot increases in size, the signal on the detector opposite the corner should fall as observed but the signal on the second detector should rise. We observed in fact that the signal falls on *both* detectors.

So what is happening in the corners? Presumably the dissipation heats up the electrons exciting some of them to higher Landau levels. At the electron entry corner, the hot electrons are drifting away from the contact so must largely relax within the 2DEG itself. However, at the exit corner, the hot electrons are drifting into the 3D contact so now some of them may relax within the contact itself. This emission is likely to be more isotropic than that from the 2DEG leading to a fall in intensity directed towards the detector on the opposite face. Now if this is the reason behind the asymmetry in GaAs why is it not seen in Si? The explanation may lie in the very different relaxation rates in Si and GaAs. In Si, at the magnetic fields used in the experiments, the electrons can readily relax by emitting cyclotron phonons because the wavevector corresponding to the cyclotron frequency is less than or comparable to the inverse thickness of the 2DEG. However in GaAs, for the same magnetic field, the cyclotron frequency is three times larger than in Si and since the 2DEG is also appreciably thicker (a^{-1} is smaller), cyclotron phonon emission is strongly suppressed at fields above $\sim$1T as noted earlier. So it is possible that in Si most of the emission from both corners occurs from the 2DEG regions so that rather few hot electrons contribute to the current entering the 3D contact while in GaAs there is a significant hot electron current. It would be very interesting to examine whether there are any changes in the phonon spectrum from the two corners since the phonon emission from the 3D contact is likely to occur at lower frequencies, ($3k_B T/\hbar$) than that from the magnetically quantised 2DEG (ω_c).

There have in fact been a number of quantitative discussions on the distribution of the magnitude of the dissipation between the two corners. The first by Büttiker[1] predicted that the dissipation should be greater at the electron exit point though this was later retracted in favour of equal dissipation at each[38] and this was also the conclusion of Komiyama and Hirai.[39] Two other groups[40] have predicted unequal dissipation could occur though in both cases this was through the presence of additional resistance at one of the corners resulting in a total source-drain resistance $R > R_H$ which was not observed in the present work.

Information on the electron temperature of heated electrons in quantising magnetic

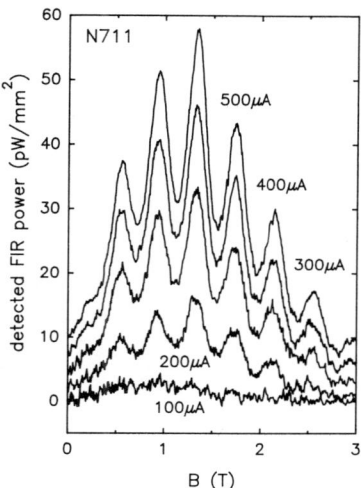

Fig. 14. FIR power detected per unit area of 2DEG as a function of magnetic field for various currents I.[41]

fields and so on the phonon relaxation is also being obtained by measuring the far-infrared (FIR) intensity emitted by a 2DEG as a function of input power and magnetic field[41]. The 2DEG in a GaAs heterostructure is placed at the centre of the magnetic field at one end of an evacuated lightpipe and heated by current pulses. The back surface of the GaAs substrate is prepared with a mirror finish to form an etalon which should give rise to interference maxima and minima for particular values of the FIR wavelength emitted by the hot electrons. The total intensity of the FIR emission is measured by a broad-band InSb detector located at the other end of the lightpipe in an essentially field free region. The useful sensitivity of the detector covers the range 0-900GHz and examples of the magnetic field dependence of the detector output for constant voltage input are shown in figure 14. From the thickness of the substrate, the pronounced oscillations in detected intensity are readily attributed to narrow-band FIR emission at the cyclotron frequency and this has been confirmed by including a low-pass filter in the light path and observing the fall in emission when the cyclotron frequency is tuned beyond the cutoff frequency of the filter. The intensity falls by around 70%, figure 15, showing that while most of the FIR is at the cyclotron frequency around 30% occurs at lower frequencies. The nature of the electronic transitions responsible for this is not yet known but intra-Landau level transitions within edge states seem a possible source.

We hope next to image the variation in CR intensity and so T_e by moving a slit across the surface of the Hall bar. At present the measured intensity is essentially independent of position presumably as a result of the strong reflection from the back surface.

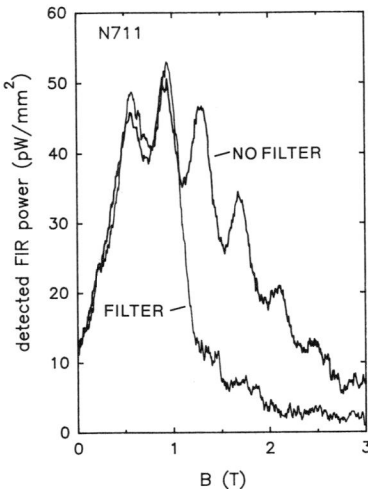

Fig. 15. The effect of a low pass filter on the FIR detected.[41]

Work is in progress to reduce this reflection so that the intensity from a small area of surface should indicate the **local** temperature of the 2DEG underneath. However at present the values of T_e of 5-30K for the powers shown in the figure are averages over the Hall bar obtained from the average CR intensity by making use of Drude theory. Preliminary analysis suggests that the magnetic field dependence of these average T_e values is weaker than that calculated for relaxation by inter-Landau level transitions which may imply that edge states are also contributing to phonon relaxation.[11]

3.2 Quantum Oscillations in Cyclotron Phonon Emission

For much higher input power densities ($k_B T >_e \hbar \omega_c$) than those used in 3.1, Joule heating and phonon emission occur throughout the whole of the 2DEG sample. The emission should be very largely cyclotron phonon emission from inter-Landau level transitions provided the frequency is low enough to satisfy the constraints of momentum conservation which is the case for Si at modest magnetic fields, B < 6T. The efficiency of the cyclotron emission is proportional to the product of the number of electrons in the upper state and holes in the lower which is greatest when E_F is midgap and the decrease in efficiency as E_F moves away is expected to be enhanced by increases in the screening of the electron-phonon interaction. So the efficiency should oscillate as E_F moves through the Landau level spectrum.

CW measurements to examine these changes were made by observing the intensity of phonons opposite a narrow strip or bridge of 2DEG between two large 2DEG areas of a Si MOSFET.[42] Because of its large aspect ratio, most of the resistance between the contacts to the 2DEG areas occurs within the bridge so that quite high power densities can be achieved for modest power inputs. The phonon intensity in the forward direction is shown in figure 16 as a function of filling factor for two powers. The power and so the total phonon intensity was kept constant during the sweep so any changes in signal are attributed either to the location or the angular dependence of the emission. At the lower power the forward intensity is seen to be least when E_F is midgap as expected from the work reported in the previous section : most of the emission is occurring at the current entry and exit corners which are well removed from the detector and the signal observed varies as ρ_{xx}. However, at the higher power, the signal is greatest when E_F is midgap where the cyclotron emission is greatest. We conclude that when the efficiency of cyclotron emission falls as we move away from midgap a small but increasing part of the emission is taking place by a parallel channel with a wider angular distribution so that fewer phonons are emitted in the forward direction. We are presently analysing possible alternatives such as intraLandau level transitions, edge-state transitions and optic phonon emission to see which is the dominant.

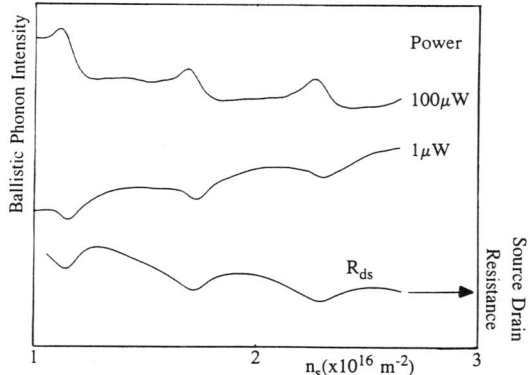

Fig. 16. The variation of the phonon intensity emitted normal to a heated 2DEG in a magnetic field of 6 T at low (1 μW) and high (100 μW) powers.[15] The broken line shows the source-drain resistance with minima when E_F is midgap.

3.3 Phonon Induced Breakdown in the Quantum Hall Regime

The phonoconductivity technique described in 2.2 and used to investigate the positional dependence of phonon scattering in silicon quantum Hall devices has also been used to study GaAs quantum Hall devices with rather different results.[43] The devices in this case were based on a GaAs/(AlGa)As heterojunction having a 2DEG sheet density of $7.8 \times 10^{15} \text{m}^{-2}$ and a 4.2K mobility of 70 $\text{m}^2\text{V}^{-1}\text{s}^{-1}$. The $3 \times 1 \text{mm}^2$ active area was defined by etching and ohmic contacts formed at each end. Figure 17 shows a set of images taken at integer filling factor near $B=1$T. It shows quite clearly that the response, an increase in the two-terminal resistance by 10^{-5} to 10^{-6}%, is strongest when phonons are incident close to one edge of the 2DEG. Furthermore this edge changed sides upon reversal of magnetic field or current direction. This behaviour suggests that the response might be due to phonon assisted tunnelling between edge states or edge and bulk states[43]. A decrease in the device resistance was observed when the phonons were incident near the current entry and exit points; the explanation for this is the same as in the case of the silicon device (see Section 2.2).

Classically the edge electrons can be visualised as performing skipping orbits around

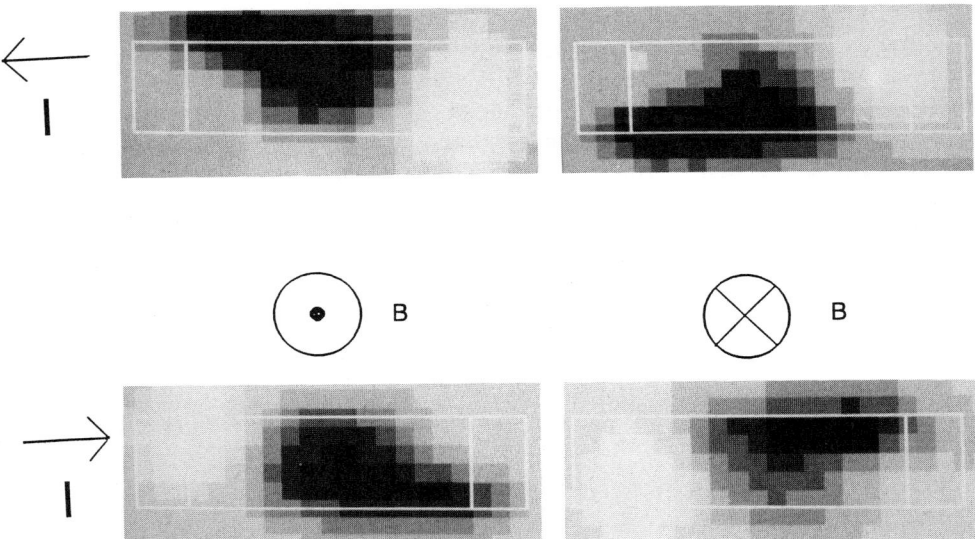

Fig. 17. Phonoconductivity images of an extended 2DEG in a GaAs/(AlGa)As heterojunction in the quantum Hall regime. The main response, an increase in two-terminal resistance by just 0.01%, occurs at the sample edge which is at the higher potential. This changes sides with reversal of the magnetic field or current polarity.

the periphery of the device. The direction of propagation of the edge electrons depends on the magnetic field direction and which edge of the device they are on. According to the edge state picture of the quantum Hall effect each edge state at the Fermi level acts as a pseudo-one-dimensional channel of resistance h/e^2 and the Hall resistance depends on how many such channels there are connected in parallel. The propagation of an edge electron is unaffected by a moderate amount of disorder. However, processes that cause backscattering, that is transfer of an electron from one edge of the sample to the other, can lead to the loss of quantization as pointed out by Büttiker[1]. In this experiment, we believe backscattering is triggered by the intense beam of nonequilibrium phonons which assist tunnelling of electrons from the higher potential edge states near the Fermi level to empty bulk states above it. They then diffuse through the empty level to the other edge which is at a lower potential. Changing the magnetic field or current direction changes the relative potential of the two edges and so reverses the images. We note, however, another recent interpretation, due to Shik[43], attributes the loss of quantization to a combination of inter and intra-edge state transitions.

The question remains as to why these edge effects are observable in GaAs but not in Si. We believe the reason is due to the different nature of the edges in the two 2DEG samples. In Si the edges are defined by the fringing field of the gate and the potential rises less steeply than at the edges of a GaAs sample which are defined by etching. In silicon samples, therefore, much higher energy phonons are required to assist tunnelling between adjacent edge levels and these are not normally present in the spectrum from the heater.

3.4 Phonon Emission by 2DHGs

For nearly a decade phonons have been used to good effect as a probe of two-dimensional electron gases but have not been extended to 2DHGs until recently since it has been hard to produce high quality 2D hole gases in GaAs. With recent MBE growth developments, including the use of novel substrate orientations such as (311)A, it has been possible to produce 2D hole gases which are good enough exhibit fractional quantised Hall effects and provide evidence of Wigner solidification.[44]

There is good evidence that the coupling of acoustic phonons to 2D holes is much

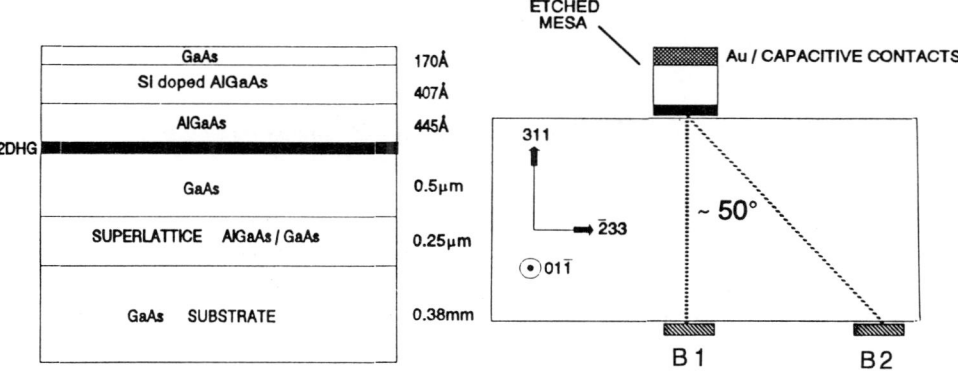

Fig. 18. MBE layer structure of p-type (311)A GaAs/(AlGa)As heterojunction and sample geometry for phonon emission experiments. B1 and B2 are superconducting Al bolometers.

stronger than that to 2D electrons. For example, it is found that the mobility remains strongly temperature dependent to below 1K. This in itself is not overwhelming evidence because such behaviour could be due to temperature dependent screening of impurity scattering as observed in Si devices. However, it is also found that the hole temperature is not easily raised in a transport experiment, suggesting that there is efficient energy relaxation at low hole temperatures. The first direct phonon experiments on 2D hole gases[45] also indicated strong coupling to acoustic modes but the data were limited and the quality of the samples available at that time was not comparable to that of modern devices. In the present work we have carried out a systematic study of the phonon emission from warm holes and the phonon drag of cold holes in a GaAs/(AlGa)As heterojunction grown by MBE on a (311)A semi-insulating (chromiumfree) GaAs substrate. A preliminary report appears elsewhere.[46]

The samples, see figure 18, had a hole sheet density of $1.6 \times 10^{15} \mathrm{m}^{-2}$ and the mobility at 1.5K was $10 \mathrm{m}^2 \mathrm{V}^{-1} \mathrm{s}^{-1}$. A device of active area $0.1 \times 0.1 \mathrm{mm}^2$ was defined by etching and $1 \times 1 \mathrm{mm}^2$ capacitative contacts were fabricated at each end. Capacitative contacts were used because ohmic contacts to (311) material generally have rather high resistances and we wish to avoid unwanted phonon emission from these. The channel resistance was 1.2kΩ at 1.5K and the contact capacitance 1.5nF. The holes were heated by applying 20ns current pulses and the emitted phonons detected using superconducting Al bolometers on the opposite side of the 380μm substrate to the device. The power delivered to the device and the device resistance were then both determined from the amplitude of the forward and reflected pulses on the 50Ω line from the pulse generator. The resistance is used to estimate the hole temperature T_h from comparison with steady state mobility vs temperature measurements.

Figure 19 shows a series of heat pulses recorded on the bolometer directly oppo-

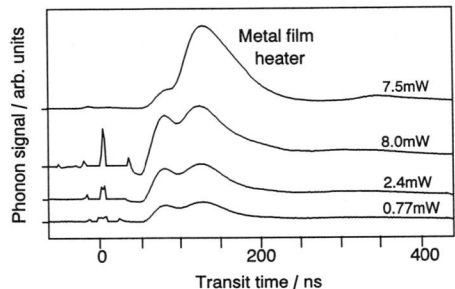

Fig. 19. Heat pulses detected by bolometer B1 at different excitation levels. The longitudinal (LA) pulse arrives 70 ns after the excitation pulse and the transverse (TA) after 110 ns. Also shown, for comparison, is the heat pulse obtained when using a metal thin film heater in place of the 2DHG. Otherwise the geometry is the same.

site the 2DHG for a range of excitation powers. No signal was detected on a second bolometer which measured the intensity emitted at about 50 degrees to the normal. The heat pulses in the normal direction consist primarily of ballistic phonons, both longitudinal and transverse and for comparison we also show the heat pulse from a metal film heater. It can be seen that coupling to the LA mode is much stronger in the 2DHG than in the heater and it is also of interest that the angular dependences are very different since when the film heater was used a signal was observed on the second bolometer. The strong LA signal from the 2DHG emission is also in marked contrast with the absence of any LA signal from a 2DEG at least in the normal direction[17,18,20] as has already been noted. The absence of signal on the second bolometer from the 2DHG can be attributed to the $2k_F$ cutoff due to in-plane momentum conservation. At such a low n_h this would restrict emission to angles less than 30 degrees at even the lowest hole temperatures used in this experiment.

Because of their higher velocity, the q-vector of LA modes is smaller than that of TA modes of the same frequency. We would therefore expect that the momentum conservation cutoffs would affect the LA modes less severely. This is evident from

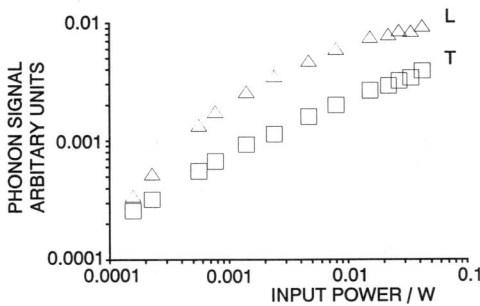

Fig. 20. Amplitude of LA and TA phonon signals as a function of 2DHG excitation power. In this device a total power input of 1 mW corresponds to 67 pW per hole.

figure 20 which shows the increase of ballistic pulse height with power dissipation. Initially the height of the LA pulse increases more rapidly than that of the TA because it is less affected by the $q_\perp \leq a^{-1}$ cutoff. Eventually however, at about 10mW, the LA also becomes affected and both pulse heights increase at the same rate. At low hole temperatures the phonon spectrum should be approximately Planckian with a peak determined by T_h. However, as the temperature increases, the peak in the thermal distribution exceeds the cutoff in $q_\perp$ so that only subthermal phonons can be emitted resulting in a decrease in the temperature dependence of the power emitted. This means that we can determine the "thickness" a of the 2D hole gas from the temperature at which the rate of increase of the LA mode height slows. We obtain $a \approx 2$nm which is in fair agreement with variational calculations[47] and considerably less than that of a 2DEG of similar sheet density.

Figure 21 shows the hot hole energy relaxation rate as a function of T_h. The approximate T_h^5 dependence is characteristic of energy relaxation by acoustic phonon

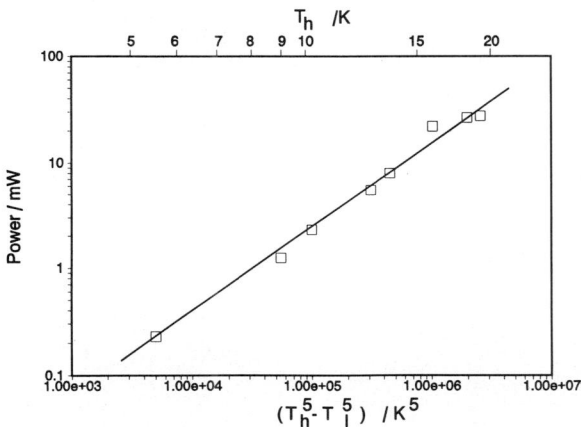

Fig. 21. Energy relaxation rate as a function of hole temperature.

emission by 2D carriers[7] and it continues up to the highest powers used corresponding to 120pW/hole and T_h= 20K. This is in marked contrast with 2DEGs in GaAs in which crossover from predominantly acoustic phonon emission to optic phonon emission occurs at about 5pW/electron and T_e =50K[20].

3.5 Phonon Measurements of the Energy Gap in the Fractional Quantum Hall Regime

Phonons appear to provide a valuable spectroscopic technique for determining energy gaps in the fractional quantum Hall regime. For bulk states, phonon absorption should not be possible at frequencies less than that of the gap corresponding to the excitation of a magnetoroton but should be possible at higher frequencies. The first determinations using this technique[48] lead to gap energies comparable to those obtained for high quality samples from the exponential dependence of the longitudinal magnetoresistance[49] and from optical luminescence.[50]

For the ρ_{xx} minimum corresponding to a particular fractional state measurements were made of the increase in ρ_{xx} produced by an incident phonon pulse from a constantan heater. The increase is assumed to be proportional to the phonon absorption and its size was monitored as the blackbody phonon spectrum from the heater was shifted to higher frequencies by increasing the power. The samples had mobilities $\sim 1 \times 10^6$ cm^{2}V^{-1}s^{-1} and the gaps obtained from the temperature dependence of their resistivities had values comparable to those obtained in this way by other workers for samples of similar mobility. The values we obtained this way were however 2-3 times smaller than those we obtained from phonon measurements on the same samples. We attribute this to the reduction in minimum excitation energy relative to E_F caused by disorder. This lowers the gap measured from $\rho_{xx}(T)$ but should not affect the gap measured by phonons. Further details of these experiments and analysis appear elsewhere in this volume.[51]

The next step in these experiments is to look for changes in the interaction when the electron fluid solidifies to form a Wigner crystal. In particular we hope to see interference effects, Bragg scattering, associated with the long range order.

3.6 Phonon Emission from Excited Subbands

The last experiment to be described is a St. Petersburg-Nottingham collaboration on phonon emission from a Si MOSFET[52]. The phonons are detected by an exciton cloud about 0.3 mm in diameter formed opposite the 2DEG by an incident laser beam. At helium temperatures and in the absence of incident phonons, the weak p-type doping of the substrate is sufficient to bind all the excitons in the cloud to acceptor atoms. The luminescence is therefore wholly characteristic of bound excitons (BE) and no free exciton (FE) luminescence occurs. However, when a current pulse is passed through the 2DEG, the ballistic phonon pulse incident on the exciton cloud creates both a positive-going pulse in the FE luminescence and a negative-going pulse in the BE luminescence as shown in figure 22. Now since the energy required to free a bound exciton is 3.8 meV corresponding to a frequency of 920 GHz, the exciton cloud detector is only sensitive to phonons of frequency >920 GHz and so can be used as a spectrometer.

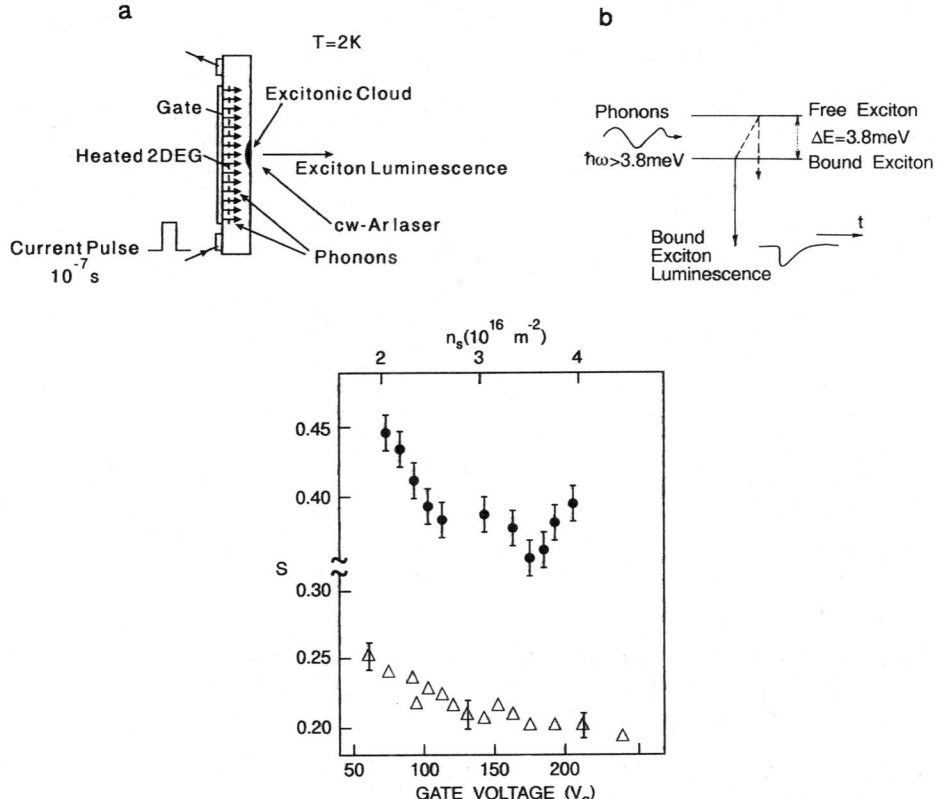

Fig. 22. Phonon detection using an exiton cloud luminescence detector (a) the experimental arrangement, (b) the exciton energy levels and (c) the size of the negative going BE luminescence pulse, S, following pulsed heating of a 2DEG as a function of n_s.[52]

The detector has been used in this way to examine changes in the phonon emission spectrum with sheet density. The power was maintained constant as n_s was varied and high enough to create significant excited subband population. Figure 22(c) shows the fall in BE luminescence and so in high frequency intensity as n_s is increased for two values of input power. This decrease contrasts with the **increase** in high frequency intensity that should occur because of the increase in k_F when only the lowest subband is occupied. But the observed decrease is consistent with an increasing population of excited electrons since these have lower values of both k_F and inverse confinement widths so the momentum selection rules restrict their phonon emission to be of lower frequency than that from the ground state subband. So increasing n_s shifts the spectrum to lower frequencies as observed.

ACKNOWLEDGMENTS

We wish to acknowledge strong support from all our colleagues in the Nottingham NUMBERS group on low-dimensional structures and mention particularly V W Rampton and F W Sheard and also our collaborators in the recent work reported here: A V Akimov, K A Benedict, S Chapman, J Cooper, R Eyles, R Fletcher, R George, P Hawker, M Henini, A Jezierski, S Kravchenko, D J McKitterick, C J Mellor, E S Moskalenko, F F Ouali, A Y Shik, K R Strickland, B Sujak-Cyrul, Xin Zhijun and N N Zinov'ev. We are very grateful to the Science and Engineering Research Council, the Royal Society, the European Commission and the Russian Academy for financial support.

REFERENCES

[1] R. B. Laughlin, *Phys. Rev.* **B23**, 5632 (1981); M. Büttiker, *Phys. Rev.* **B38**, 9375 (1988).

[2] P. F. Fontein, J. A. Kleinen, P. Hendriks, F. A. P. Blom, J. H. Wolter, H. G. M. Lochs, F. A. J. M. Driessen, L. J. Giling and C. W. J. Beenakker, *Phys. Rev.* **B43**, 12090 (1991).

[3] S. V. Kravchenko, V. M. Pudalov, J. Campbell and M. D'Ioria, *Pis'ma Zh. Eksp. Teor.* **54**, 528 (1991) [*JETP Letts*, **54**, 532 (1991)]

[4] C. Weisbuch and B. Vinter "Quantum Semiconductor Structures", Academic Press, San Diego (1991).

[5] L. J. Challis, *Contemp. Phys.*, **33**, 111 (1992).

[6] C. J. Mellor, *New Scientist* **135**, 36 (1992).

[7] V. Karpus, *Fiz. Tekh. Poluprovodn* **20**, 12 (1986) [*Sov. Phys. Semicond* **20**, 6 (1986)]; V. Karpus *Fiz. Tekh. Poluprovodn* **22**, 439 (1988) [*Sov. Phys. Semicond* **22**, 268 (1988)]; L.J. Challis, F. W. Sheard and G. A. Toombs "Physics of Phonons, Lecture Notes in Physics" ed. T. Paszkiewicz, Vol. **285**, p.348, Springer, Berlin (1987) and references therein.

[8] U. Bockelman and G. Bastard, *Phys. Rev.* **42**, 8947 (1990); A. Y. Shik and L. J. Challis *Phys. Rev.* **47**, 2082 (1993).

[9] G. A. Toombs, F. W. Sheard, D. Neilson and L. J. Challis, *Solid State Commun.* **64**, 577 (1987); K. A. Benedict, *J. Phys. Condens. Matter* **3**, 1279 (1991).

[10] V, I, Fal'ko and L, J, Challis, *J. Phys. Condens. Matter* , in press.

[11] A, Y, Shik, *Fiz. Tekh. Poluprovodn* **26**, 855 (1992) [*Sov. Phys. Semicond* **26**, 481 (1992)].

[12] N. P. Hewett, P. A. Russell, L. J. Challis, F. F. Ouali, V. W. Rampton, A. J. Kent and A. G. Every, *Semicond. Sci. Technol.* **4**, 955 (1989).

[13] P. A. Russell, F. F. Ouali, N. P. Hewett and L. J. Challis, *Surf. Sci.* **229**, 54 (1990); F. F. Ouali, L. J. Challis and J. Cooper, *Semicond. Sci. and Technol.* **7**, 608 (1992).

[14] F. F. Ouali, Z. Xin, L. J. Challis, J. Cooper, A. F. Jezierski, and M. Henini, "Phonon Scattering in Condensed Matter" eds. R.O. Pohl. and M. Meissner, Springer, Berlin (1993) in press.

[15] J. Cooper, L. J. Challis, F. F. Ouali, K. A. Benedict and C. J. Mellor, "Phonon Scattering in Condensed Matter" eds. R. O. Pohl and M. Meissner, Springer, Berlin (1993) in press.

[16] U. Klass, W. Dietsche, K. von Klitzing and K. Ploog, *Z. Phys.* **B82**, 351 (1991); U Klass, W. Dietsche, K. von Klitzing and K. Ploog, *Surf. Science* **263**, 97 (1992).

[17] M. A. Chin, V. Narayanamurti, H. L. Stormer and J. C. M. Hwang "Phonon Scattering in Condensed Matter IV" eds. W. Eisenmenger, K. Lassmann and S. Döttinger, Springer, Berlin, p.328 (1984).

[18] J. K. Wigmore, M. Erol, M. Sahraoui-Tahar, C. D. W. Wilkinson, J. H. Davies and C. Stanley, *Semicond. Sci. and Technol.* **6**, 837 (1991).

[19] M. Rothenfusser, L. Koster and W. Dietsche, *Phys. Rev.* **B34**, 5518 (1986)

[20] P. Hawker, A. J. Kent, O. H. Hughes and L. J. Challis, *Semicond. Sci. and Technol.* **7**, **B29**, (1992).

[21] A. J. Kent, V. W. Rampton, M. I. Newton, P. J. A. Carter, G. A. Hardy, P. Hawker, P. Russell and L. J. Challis, *Surf. Sci.* **196**, 410 (1988).

[22] P. Hawker, A. J. Kent, L. J. Challis, M. Henini and O. H. Hughes, *J. Phys. Condens. Matter.* **1**, 1153 (1989).

[23] W. Dietsche, "Phonon Scattering in Condensed Matter" eds. A. C. Anderson and J. P. Wolfe, Springer, Berlin, p.366 (1986).

[24] L. J. Challis and V. W. Rampton SERC Application GR/D/10657 (1984); D. C. Hurley, G. A. Hardy, P. Hawker and A. J. Kent, *J. Phys. E. Sci. Instrum.* **24**, 824 (1989).

[25] A. J. Kent, G. A. Hardy, P. Hawker and D. C. Hurley "Phonons 89" eds. S. Hunklinger, W. Ludwig and G. Weiss, World Scientific, Singapore, p.1010 (1990).

[26] P. J. A. Carter, V. W. Rampton, M. I. Newton, K. McEnaney, MHenini and O. H. Hughes "Phonons 89" eds. S. Hunklinger, W. Ludwig and G. Weiss, World Scientific, Singapore, p.998 (1990).

[27] V. W. Rampton, K. McEnaney, A. G. Kozorezov, P. J. A. Carter, C. D. W. Wilkinson, M. Henini and O. H. Hughes, *Semicond. Sci. and Technol.* **7**, 641 (1992), this paper includes references to earlier work by several groups.

[28] J. C. Hensel, R. C. Dynes and D. C. Tsui, *Phys. Rev.* **B28**, 1124 (1983); J. C. Hensel, B. I. Halperin and R. C. Dynes, *Phys. Rev. Letts.* **44**, 341 (1980).

[29] L. J. Challis, Physics of Low Dimensional Structures eds. P. N. Butcher et. al., Plenum, New York, p.441, (1993).

[30] F. Rösch and O. Weis, *Z. Physik* **B27**, 33 (1977).

[31] A. J. Kent, G. A. Hardy, P. Hawker, V. W. Rampton, M. I. Newton, P. A. Russell and L. J. Challis, *Phys. Rev. Lett.* **61**, 180 (1988).

[32] G. A. Northrop and J. P. Wolfe, *Phys. Rev.* **B22**, 6196 (1980).

[33] G. A. Northrop, *Computer Physics Commun.* **28**, 103 (1982).

[34] A. J. Kent, *Physica,* **B169**, 356 (1991).

[35] P. C. Van Son, P. C. Kruithof and T. M. Klapwijk, *Surf. Sci.*, **229**, 57 (1990).

[36] R. Knott, U. Klass and W. Dietsche, private communication (1991).

[37] X. Zin, F. F. Ouali, L. J. Challis, J. Cooper, A. F. Jezierski and MHenini to be published.

[38] M. Büttiker, private communication (1989).

[39] S. Komiyama and H. Hirai, *Phys. Rev.* **B40**, 7767 (1989).

[40] P. C. Van Son and T. M. Klapwijk, *Europhysics Letters,* **12**, 429 (1990); S. Komiyama, H. Hirai, M. Ohsawa, Y. Matsuda, S. Sasa and T. Fujii, *Phys. Rev.*, **B45**, 11085 (1992).

[41] L. J. Challis, N. N. Zinov'ev, R. Fletcher, B. Sujak-Cyrul, A. V. Akimov and A. F. Jezierski "Phonon Scattering in Condensed Matter" eds. R. O. Pohl and M. Meissner, Springer, Berlin (1993) in press; N. N. Zinov'ev, R. Fletcher, L. J. Challis, B. Sujak-Cyrul, A. V. Akimov and A. F. Jezierski, submitted for publication.

[42] J. Cooper L. J. Challis F. F. Ouali K. A. Benedict and C. J. Mellor, "Phonon Scattering in Condensed Matter" eds. R. O. Pohl and M. Meissner, Springer, Berlin (1993), in press.

[43] A. J. Kent, D. J. McKitterick, L. J. Challis, P. Hawker, C. J. Mellor and M. Henini, *Phys. Rev. Letts.*,**69**, 1684 (1992); A. Y. Shik, private communication (1992).

[44] M. B. Santos, Y. W. Suen, M. Shayegan, Y. P. Li, L. W. Engel and D. C. Tsui, *Phys. Rev. Lett.* **68**, 1188 (1992).

[45] M. A. Chin, V. Narayanamurti, H. L. Stormer and A. C. Gossard , "Proc 17th Int Conf on the Physics of Semiconductors", Eds. J. D. Cadhi and W.A. Harrison, Springer, Berlin, 333 (1985).

[46] A. J. Kent, K. R. Strickland and M. Henini, "Phonon Scattering in Condensed Matter" eds. R. O. Pohl and M. Meissner, Springer, Berlin (1993), in press.

[47] G. Landwehr, "Electronic Properties of Multilayers and Low-Dimensional Semiconductor Structures 2, Eds. J. M. Chamberlain, L. Eaves and J. C. Portal, NATO ASI Series B: Physics Vol. 231, Plenum. New York, 33 (1990).

[48] R. H. Eyles, C. J. Mellor, A. J. Kent, K. A. Benedict, L. J. Challis, S. Kravchenko, N. N. Zinov'ev and M. Henini submitted for publication.

[49] R. L. Willett, H. L. Stormer, D. C. Tsui, A. C. Gossard and J. H. English, *Phys. Rev.* **B37** 8476 (1988).

[50] H. Buhmann, W. Joss, K. von Klitzing, I. V. Kukushkin, G. Martinez, A. S. Plaut, K. Ploog and Z. B. Timofeev, *Phys. Rev. Lett.*, **65**, 1056 (1990).

[51] R. H. Eyles, C. J. Mellor, A. J. Kent, L. J. Challis, S. Kravchenko, N. N. Zinov'ev and M. Henini, this volume.

[52] A. V. Akimov, L. J. Challis, J. Cooper, C. J. Mellor and E. S. Moskalenko, *Phys. Rev.* **45**, 11387 (1992); C. J. Mellor, J. Cooper, L. J. Challis, A. V. Akimov and E. S. Moskalenko "Phonon Scattering in Condensed Matter" eds. R. O. Pohl and M. Meissner, Springer, Berlin (1993) in press.

PHONON EMISSION AND ABSORPTION EXPERIMENTS IN THE QUANTUM–HALL REGIME

F. Dietzel, U. Klass and W. Dietsche
Max-Planck-Institut für
Festkörperforschung, Heisenbergstraße 1
W-7000 Stuttgart 80, Federal Republic of Germany

K. Ploog
Paul-Drude-Institut, Hausvogteiplatz 5-7, O-1086 Berlin,
Federal Republic of Germany

Abstract

The absorption and emission of phonons by a two–dimensional electron gas (2DEG) has been studied in a quantizing magnetic field. The 2DEG was formed at the interface of a $GaAs/Al_xGa_{1-x}As$ heterojunction. Absorption experiments were done by creating acoustic phonons by heating the substrate locally with a focused laser beam. The phonons travelled ballistically through the crystal and were partially absorbed by the 2DEG. This led to a transfer of momentum into the 2DEG (phonon–drag effect) resulting in phonon–induced voltages and currents. These quantities gave detailed information about the interaction between acoustic phonons and the 2DEG as a function of both the incident angle of the absorbed phonons and the magnetic field. We observed that the absolute intensity of the phonon–drag signal was oscillating in phase with the Shubnikov de–Haas oscillations while the angular dependence did not change in the field. These results could be explained with a simple microscopic theory of the electron–phonon–interaction (EPI) together with a macroscopic model for the response of the 2DEG on the absorption of ballistic phonons.

The spatial distribution of the emission of phonons was studied using the fountain-pressure technique. It was found that 2DEG samples in a strong magnetic field dissipate phonons only in areas where a strong electric field is present. This is the case at the current entry and exit points of the contacts and, in the case of the break-down of the quantum-Hall effect, at filaments in the interior of the samples.

1. INTRODUCTION

A two-dimensional electron gas (2DEG) can be formed at the interface between two different semiconductors. This electronic system interacts with the bulk phonons of the substrate material. The matrix element of this interaction contains an overlap integral of the respective wavefunctions of the electron and the phonons. In zero magnetic field this leads to the conservation of the in-plane momentum and of the energy. The interaction strength follows from the symmetry of the phonon strains and from parameters like the deformation potentials and the piezoelectric interactions.

The situation is less clear if magnetic field is applied which is strong enough to quantize the electron-energy spectrum in Landau-levels. The electron wave functions then become localized. Thus the in-plane momentum conservation rule no longer exists and the matrix element will take a more complicated form. A further complication arises because the current and electrostatic-potential distributions are no longer homogeneous over the sample surface.

In this contribution we report on our recent results about the phonon absorption and emission by a 2DEG in a strong magnetic field. The first ones were done by exposing a 2DEG sample to a flux of ballistic phonons. The phonon absorption strength was measured by observing the phonon-drag effect.[1] In the case of the phonon emission we concentrated on the question of the spatial distribution of the emission processes. This was done by utilizing the fountain-pressure of superfluid helium. [2]

Alternative approaches to study the elctron phonon interaction are being used by the group in Nottingham. Their work is being reviewed by Challis and Kent.[3]

2. PHONON ABSORPTION

In this section we briefly review the basic relations describing the electron-phonon interaction (EPI) between 3D bulk acoustic phonons and a 2DEG. In a ballistic phonon absorption experiment where phonon emission of the 2DEG can be neglected the average momentum per unit time $\vec{F}^p$ that is transfered from the phonon system to one electron of the 2DEG is given by[1]:

$$\vec{F}^p = \sum_\beta \frac{I(\beta)}{v_g(\beta)} \hbar \vec{q}_\parallel(\beta) \frac{(\tau^\beta_{e-p})^{-1}}{N_e} . \tag{1}$$

$I(\beta)$ specifies the flux density of ballistic phonons that are charcterised by the index β which labels their wave vector and polarization, $v_g(\beta)$ is the group velocity of these phonons, N_e is the number of electrons in the 2DEG and $\vec{q}_\parallel(\beta)$ is the component of the phonon wave vector in the plane of the 2DEG (x-y-plane). The quantity $(\tau^\beta_{e-p})^{-1}$ is the absorption rate for these phonons which can be calculated by Fermi's Golden rule.[4, 5, 6]

In τ^{-1}, the magnetic field dependence of the EPI enters via the electron wave function Ψ and the density of states $D(E, B)$. Because of the missing translational invariance of the 2DEG its momentum is well defined only in the x–y–plane and there exists momentum conservation in the case of phonon absorption only for this component. Therefore it is natural to distinguish between the phonon wave vector components $\vec{q}_\parallel(\beta)$ and k_z in the z–direction. They both have a maximum value beyond which the absorption probability of the phonons drastically decreases. For k_z this is the condition

$$k_{z,max} a_0 \approx 1 , \tag{2}$$

where a_0 is the 'width' of the 2DEG and for $k_\parallel$ one finds [4, 5]:

$$k_{\parallel,max} \approx 2k_f , \qquad (3)$$

in the absence of a magnetic field or

$$k_{\parallel,max} l_B \approx 1 , \qquad (4)$$

in its presence [6]. k_f is the Fermi wave vector of the 2DEG, l_B is called the magnetic length with $l_B = \sqrt{\hbar/eB}$. These three length scales can be varied by changing the magnetic field (l_B), the carrier density of the 2DEG (k_F) or, in case of a quantum well, width a_0. Aside from these considerations $D(E, B)$ also enters into the expression for τ^{-1} leading to an oscillatory behaviour of the relaxation rate.

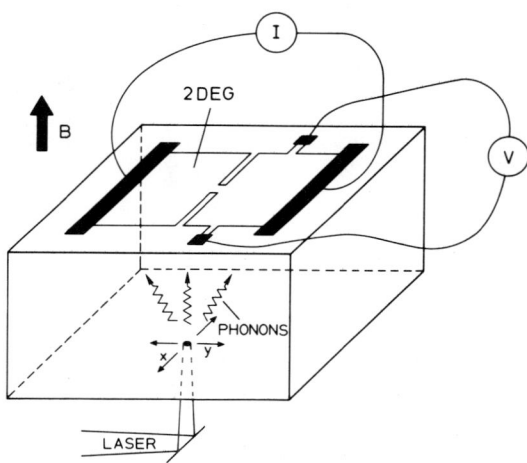

Fig. 1. Schematic view of the experimental setup. The phonons are generated by laser heating an Al film on the bottom of the crystal. The phonon drag is caused by phonons absorbed by the narrow bridge in the centre of the 2DEG structure. Both the phonon induced currents and the perpendicular Hall voltage are recorded as function of laser position.

To get a relation between the momentum transfer rate $\vec{F}^p$ and the resulting electric current density $\vec{j}$ and electric field $\vec{E}$ we treat the phonon induced current in the same way as if it were induced by an external electric current source:

$$\vec{j} = \hat{\sigma}(\vec{E} - \frac{\vec{F}^p}{e}) , \qquad (5)$$

$\hat{\sigma}$ is the conductivity tensor and e the electron charge. In the presence of a strong magnetic field and for $\vec{F}^p$ having a component only in x-direction this tensor equation can be further reduced in the presence of the boundary conditions [7]: $E_x = 0$ and $j_y = 0$. The result in first order in σ_{xx}/σ_{xy} is:

$$E_y = \frac{1}{B\sigma_{xx}n_e} F_x^p , \qquad (6)$$

and
$$j_x = \frac{\sigma_{xy}}{B\sigma_{xx}n_e}F_x^p \ . \tag{7}$$

Eq. 5 implies that heating effects of the 2DEG are neglible in ballistic phonon experiments since the thermoelectric tensor does not enter into this equation. On the other hand the only unknown into this equation is the conductivity tensor that can be experimentally determined. Thus there exists a direct way to determine the momentum transfer rate to the 2DEG.

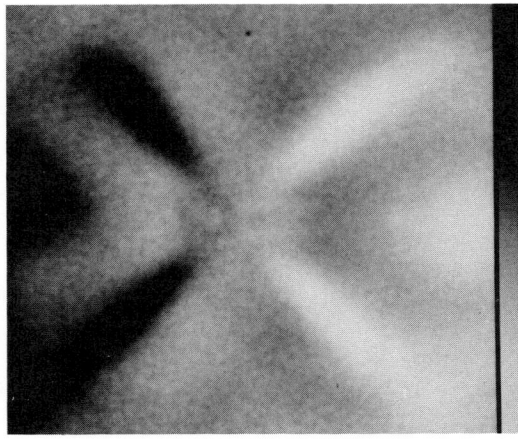

Fig. 2. Phonon-drag image taken at zero magnetic field.

For the experimental investigation we used samples which were standard GaAs/Al$_x$ Ga$_{1-x}$As heterostructures grown on [001] oriented GaAs substrates of about 350 μm thickness. Typical mobilities μ at 4.2 K were $1.6 \cdot 10^6 cm^2/Vsec$ while the carrier densities n_e were $3.76 \cdot 10^{11} cm^{-2}$. The samples were selected so that they showed only negligible parallel conduction under optical illumination.

The 2DEG was structured by wet etching. Ohmic contacts were made by alloying AuGeNi. Onto the substrate side opposite to the 2DEG an aluminum film of 2000 Å thickness was evaporated. The experimental set-up is sketched in Fig. 1. It is similar to that used by Karl et al. [1] except that the contacts were moved at least 3.5 mm from the narrow bridge to avoid thermal contact voltages [9, 10]. We utilized the phonon imaging technique [8]. The sample was placed into a ^{4}He–cryostat operating at 1.1 K and was surrounded by a superconducting solenoid capable of producing fields up to 12 T. Optical access to the sample was provided by a window at the bottom of the cryostat. An argon laser beam ($\lambda = 514.5 \ nm$) was focused on the Al film on the bottom of the substrate. The diameter of the focus was about 10 μm. The absorbed laser light raised the film temperature to typically 15 K [11] depending on the absorbed laser power P_{Lq} which was estimated to vary between 0.7 mW and 12.5 mW. High frequency phonons were emitted by the heated spot which travelled ballistically to the top side where they were partially absorbed by the 2DEG.

Only the narrow bridge with an area of $50\mu m \cdot 50\mu m$ acted as the sensitive area for the ballistic phonon–drag signal. In the larger contact areas the locally induced electric fields cancel each other or are screened due to the finite conductivity of the 2DEG. Both the current through the main contact and the quasi Hall voltage was

Fig. 3. Image of the phonon induced Hall voltage. The large diffuse pattern is due to the Nernst-Ettinghaus effect. The ballistic pattern is atop of this diffuse one and hardly visible in the reproduction.

measured using a Lock-in technique. The signal was recorded by a computer which also controlled the position of the laser focus by two galvanometer-driven mirrors. In order to obtain an 'image' of the phonon–drag signal, the signal was displayed on a monitor as grey tone while the laser focus was raster scanned.

In the experiments, we took images of the phonon drag voltage at different magnetic fields and laser powers. This was done initially by using the longitudinal contacts in Fig. 1. In Fig. 2 the result can be seen for $B = 0\,T$ and an absorbed laser power of about $3\,mW$.

An average grey tone indicates a phonon-drag voltage of 0 V. A light (dark) tone corresponds to a positive (negative) signal. The voltages were always less than $500\,nV$. The anisotropy of the image is a result of the phonon focusing [8] in the GaAs. The sharp ridges are due to fast transverse (FTA) phonons propagating in or near the (100) planes. The two larger round areas are due to longitudinal phonons which are focused near the $\langle 111 \rangle$ directions. The sign reversal (left/right) is due to the different directions of the phonon drag in the two halves of the picture. This image is similar to the one of Karl et al. [1] except that thermal voltages caused by the contacts are now absent. Thus, our image is in better agreement with the one predicted theoretically [1, 12].

By changing the laser power we changed the phonon spectrum of the ballistic phonons impinging on the 2DEG. Although we observed an overall rise in the phonon-drag voltage there is no systematic change in the shape of the pattern. In particular, there was no transition to the image expected from a deformation potential coupling expected at high energies and no indication of the $2k_F$–cutoff in the images.

We attribute this negative result to the low-pass behaviour of the EPI: Even if the phonon intensity is dramatically increased, only a very small fraction at the low-frequency end of the Planck-spectrum is absorbed. In the spectroscopy results of Lega et al. [13], however, both the cutoff and the deformation-potential coupling [12] was observed. Possibly, the thickness parameter a_0 of their 2DEG sample was smaller than ours.

The images change drastically after high magnetic fields had been turned on. In addition, there was not just one type of pattern but four qualitatively different ones that had to be distinguished:

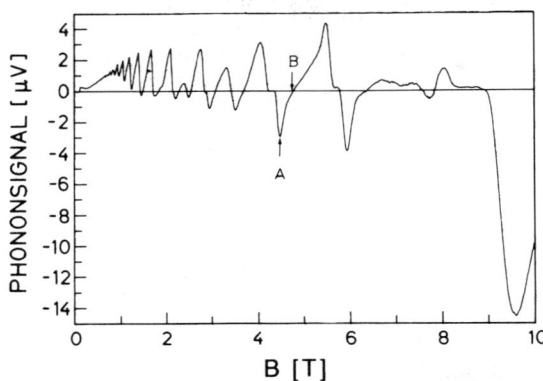

Fig. 4. Magnetic field dependence of the Nernst-Ettinghausen signal. It behaves like S_{xy} in a thermoelectric experiment.

(i) if the Fermi energy lies in the region of localized states, i.e. if we are in the quantum-Hall regime then we do not observe any signal at all.

(ii) if we are just above a QHE plateau, i.e. if the Fermi energy is in the lower half of a region of the extended states then we observe strong signals leading to images like the one shown in Fig. 3. Only a weak part of this image is caused by ballistic phonons. Much more striking are the large diffuse areas of positive and negative voltages in the upper and lower part of the image, respectively.

(iii) if the Fermi energy is in the upper half of the extended states region then the image is qualitatively the same as (ii), but the voltages leading to the diffuse parts of the images reversed sign. Thus the image was dark in the upper part and bright in the lower one. The faint ballistic part in the image remained unchanged.

(iv) consequently, the magnetic field could be set such that the diffuse part of the image vanished. In this situation the ballistic part only became visible. Now the Fermi energy is probably in the centre of the extended states region. The image taken under these conditions were almost the same as the ones obtained at $B = 0$.

We attribute the diffuse signal of thermal voltages of Fig. 3 to a temperature gradient across the sample. The reason is probably a residual heating of the 2DEG by phonons, most likely by the very intense beam of highly focussed phonons along

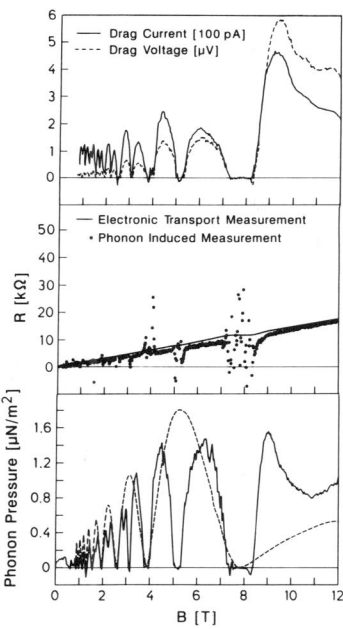

Fig. 5. Top: Ballistic Phonon-drag voltage and current as function of magnetic field, centre: Phonon-Hall resistance, bottom: Phonon force calculated from the data in the top panel. The dashed line is the theoretical model scaled down by a factor of 20.

the [001] direction. This gradient causes a thermal voltage (Nernst-Ettingshausen effect). This interpretation was verified by focusing the laser beam to a position where no ballistic phonon signal is observed and recording the voltage as a function of the magnetic field Fig. 5. This field dependence is qualitatively the same as that of the transverse thermal power S_{xy} observed by [16, 17]. Another feature of this voltage was that it changed sign on reversing the magnetic field direction. When adding two images at two fields $\pm B$ a ballistic phonon-drag image like the one in Fig. 2 was obtained.

The ballistic phonon–drag images have two interesting qualitative features which should be discussed here. One of them was surprising at first:

It seemed as if the electric field induced by the force (1) of the absorbed phonon momenta pointed in the same direction as the momenta themselves. One would have expected that the phonon-drag images in very high fields were rotated with respect to the one in zero field by 90° due to the Lorentz force. Actually, the fields *are rotated* by this value.

However, the spatial current and electric-field distribution are always deformed near the contacts in QHE samples. Thus, the *direction* of the internal electric fields cannot be determined using only the longitudinal contacts.

If, however, a low impedance ampere meter was connected to these contacts and if the 'phonon-drag Hall-voltage' at the the perpendicular contacts was recorded then the internal potential distribution was forced to be homogenous in the regions of both the voltage probes and the imaging range. In this case we found that both the phonon-drag pattern and that of Fig. 3 were rotated by 90°. Thus the electric phonon-drag field was now indeed perpendicular to the phonon momenta.

The second interesting feature is that the shape of the phonon–drag image was not dependent on the magnetic field. Its intensity varied dramatically as will be discussed in the next section but the pattern remained essentially the same at all magnetic fields. This insensitivity to magnetic field is expected because the images are mainly determined by the piezoelectric part of the matrix element which is not dependent on $\vec{B}$. The form factors stemming from the overlap term $\langle i \mid e^{i\vec{q}_s\vec{r}} \mid f \rangle$ are not readily observable because the form factor cutoff in z–direction is the same with and without field while the $2k_F$ cutoff at zero field and the magnetic-length cutoff in high fields are of the same order of magnitude.

For the quantitative analysis of the dependence of the phonon–drag voltage on magnetic field both the phonon-drag induced currents and the corresponding phonon Hall voltages were measured. Separation of the ballistic part of the image from the diffusive background was made by using the sharp focusing features of the ballistic pattern. The intensity of the diffuse background in the images was interpolated in the region of the sharp ridges and then subtracted out. The resulting ballistic voltages and currents are plotted in the top of Fig. 6 as a function of the magnetic field. The phonons which were studied in this way travelled along the [120] crystallographic direction.

It is immediately obvious from the traces that the signal follows roughly the longitudinal resistance in a Shubnikov-de Haas measurement. In particular one sees that it disappears in regions where the conductivity vanishes, i.e. in the regions of the QHE. The amplitude of the signal in the other regions increases strongly with magnetic field. We will show that this behaviour is due to the increase of the density of states at the Fermi edge.

Fig. 6. View of a GaAs Hall-bar sample at about 7 T (filling factor 2). Current= $90\mu A$. The helium drops are clearly visible in the corners. Sample width was 2.5 mm

A first test of the theoretical model is provided by dividing the measured phonon drag Hall voltage by the phonon drag current. According to Eqs. 6 and 7 this should just give the Hall resistance. In Fig. 6, centre, the result is shown together with the Hall resistance measured by standard transport measurements. In the QHE range the result is ill defined, of course, but otherwise the phonon Hall resistance agrees well with the transport Hall resistance. One can now calculate the phonon force $\vec{F}^p$ because σ_{xx} is known from the transport measurements. The result is shown as the solid line in the bottom panel of Fig. 6. Note that the phonon force can be given in absolute units $(\mu N/m^2)$.

In order to compare this experimental result with the theory, we calculated the phonon force quantitatively assuming that the scattering is quasielastic [10]. The phonon spectrum was assumed to be Planck-like as described by [11].

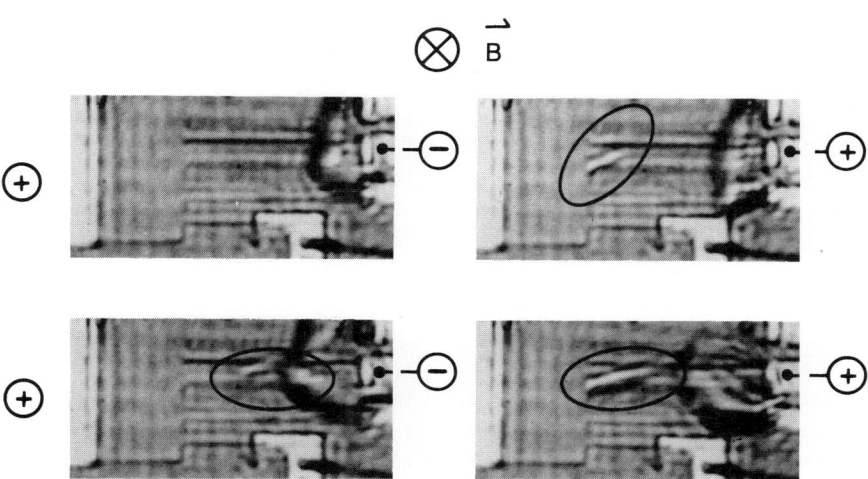

Fig. 7. Breakdown of the QHE in standard samples. Current flows from a wide 2DEG are into a narrow (200μm) strip. Dissipation is visible as diagonal strips. The effect was not symmetric upon current reversal.

The calculated phonon force is shown as a dashed line in the bottom panel of Fig. 6. In view of the large uncertainties in (i) the actual density of extended states, (ii) the phonon power input and loss mechanisms, and (iii) the geometry factors, it is not surprising that the theoretical values had to be scaled down by a factor of 20 to fit approximately the experimental data.

The two main discrepancies in the magnetic field dependence between theory and experiment were actually expected. One is the absence of the spin splitting in the theoretical trace which was not contained in the theoretical DOS. Thus only every other minimum in the experimental trace is reproduced by the theory. The other one is due to the existence of the localized states which are not contained in the model DOS. Thus the wide zero regions (of the QHE) are not found in the theoretical trace.

Apart from these two points the field dependence is well reproduced. Particularly the amplitude of the phonon-force oscillations follow the expected behaviour. This is very well visible at small field where both the spin splitting and the localized states are not very relevant. The pronounced deviations at the highest fields are probably the first indications of the development of the fractional quantum Hall effect.

3. PHONON EMISSION

The electronic transitions which lead to the emission of phonons by a 2DEG in a strong magnetic field take place either between states within a Landau level or between two different ones. The latter would lead to phonon energies of the order of 10 meV at typical fields and are not very likely because of the form factors involved. If, however, a sufficiently strong electric field is applied then the Landau levels become sufficiently tilted to change the overlap of the electronic wave functions. This allows both transitions between neighbouring levels by tunneling and within one level under phonon emission. Theoretical descriptions of these processes are very scarce [18]. The electric field distribution is very inhomogeneous in the 2DEG samples, particularly near the current contacts. Only very recently there have been the first studies of its spatial distribution by electro optical techniques [19].

In the experiments reported here we studied the spatial distribution of the phonon emission by the fountain-effect of superfluid helium. The samples were partly immersed in superfluid helium in such a way that a thin helium film formed on the surface. The emission of phonons led to an increase of the film temperature and consequently to an easily visible helium drop due to the fountain-pressure effect. In the first experiments we studied the behaviour near the current contacts. It has always been assumed that the electric field distribution in a high electron mobility sample has two maxima in opposite corners of the current contacts. This leads indeed to strong phonon emission in those corners. In Fig. 6 we show as an example a GaAs sample. The helium drops in the two opposite corners are easily visible. With reversed field the drops were in the other two corners. One can estimate the electric fields in the corners by assuming that the total Hall voltage is the maximum potential drop over a length equal to the drop diameter. This leads to an electric field of the order of 10 kV/m. This value is about the same as the one which is necessary to cause dissipation in the bulk of a QHE sample. It corresponds to drift velocities E/B of the order of 1000 m/sec. This is close to the sound velocity.

The spatial behaviour of the phonon emission in the bulk of the samples has been studied in GaAs/AlGaAs heterostructures. In circular samples (Corbino geometry) it was found that the dissipation takes place along filaments [2]. Filaments were also observed in standard samples if the drift velocities exceeded the critical values. The filament first formed either at a sample imperfection or if the sample width changed at the transition from wide to narrow sample regions. An example is shown as Fig. 7. The filaments had angles of about 30° with the sample edges. Interestingly these angles became smaller with increasing current. It is rather tempting to interpret this angle θ as a Hall angle. Since $\tan\theta = \omega\tau$ one would then interpret τ as the scattering time between two Landau levels. With increasing current τ and θ would become smaller as it was indeed observed.

4. CONCLUSION

In the absorption experiment, we studied the interaction of acoustic phonons of frequency $f \leq 1000 GHz$ with a 2DEG at a GaAs/Al$_x$Ga$_{1-x}$As interface using the phonon-drag effect and the phonon-imaging technique. The interaction was theoretically described using a free electron approach, thus no screening effects were considered. The macroscopic response of the 2DEG samples was modelled following Efros and [7]. The essential assumption was that the phonon induced currents behave in the same way as those induced by an external current source.

The phonon-drag images in high magnetic fields looked qualitatively the same as in zero magnetic field except for the rotation by 90°. The intensity of the phonon-drag signal oscillated with the density of extended states at the Fermi edge. It was completely zero if the Fermi energy was in the range of the localized states.

The maximum signal in our experiments increased with magnetic field as calculated theoretically using a model density of states by Ando. It turned out to be 20 times smaller than expected theoretically. A reduction of the extended states maxima by a factor of 4 to 5 would be sufficient to reconcile this disagreement. However, there are more uncertainties in the theoretical procedure which are difficult to assess. On the other hand, the overall qualitative behaviour was well reproduced by the calculated results.

Phonon emission was studied experimentally by using the fountain-effect of superfluid helium. It was found that phonon emission (power dissipation) takes place if there exists a sufficiently large local electric field in the 2DEG sample. Such fields always exist near the current entry and exit points. If the applied current (or Hall voltage) exceeds a critical value then local dissipation in the form of filaments was observed. Such filaments have not been predicted.

Unfortunately, there is no accepted model for the onset of dissipation. For further understanding it would be of great interest to perform spectroscopic experiments of the emitted phonons. There are, however, no suitable techniques available in the magnetic fields.

Acknowledgements

We profited greatly from discussions with and support from L.J. Challis, V. Falko, A. Kent, K. v. Klitzing, R. Knott, R. Nötzel, and R. Wichard. This work was partly supported by the SCIENCE program of the European Community.

REFERENCES

[1] H. Karl, W. Dietsche, A. Fischer and K. Ploog, Phys. Rev. Lett. **61**, 2360 (1988).

[2] U. Klaß, W. Dietsche, K. von Klitzing, and K. Ploog, Z. Phys. **82**, 351 (1991).

[3] L.J. Challis, this volume

[4] J. C. Hensel, R. C. Dynes and D. C. Tsui, Phys. Rev. B. **28**, 1124 (1983).

[5] M. Rothenfusser, L. Köster and W. Dietsche, Phys. Rev. B. **34**, 5518 (1986).

[6] S. Tamura and H. Kitagawa, Phys. Rev. B **40**, 8485 (1989).

[7] A. L. Efros and Yu. M. Galperin, Phys. Rev. Lett. **64**, 1959 (1990).

[8] G.A. Northrop and J.P. Wolfe, Phys. Rev. Lett **43**, 1424 (1979).

[9] F. Dietzel, W. Dietsche and K. Ploog, Physica B **165** & **166**, 877 (1991).

[10] F. Dietzel, W. Dietsche and K. Ploog, to be published.

[11] F. Rösch and O. Weis, Z. Phys. B **46**, 33 (1977).

[12] Cz. Jasiukiewicz, D. Lehmann and T. Paszkiewicz, Z. Phys. B - Condensed Matter **86**, 225 (1992).

[13] A. Lega, H. Karl, W. Dietsche, A. Fischer and K. Ploog, Surf. Sci. **229**, 116 (1990).

[14] R. Fletcher, J. C. Maan, K. Ploog and G. Weimann, Phys. Rev. B **33**, 7122 (1986).

[15] C. Ruf, H. Obloh, B. Jung, E. Gmelin K. Ploog and G. Weimann, Phys. Rev. B **37**, 6377 (1988).

[16] R. Fletcher, J. C. Maan, K. Ploog and G. Weimann, Phys. Rev. B **33**, 7122 (1986).

[17] C. Ruf, H. Obloh, B. Jung, E. Gmelin K. Ploog and G. Weimann, Phys. Rev. B **37**, 6377 (1988).

[18] L. Eaves and F.W. Sheard, Semicond. Sci. Techno. **1**, 346 (1986).

[19] P.F. Fontein, P. Hendriks, F.A.P. Blom, J.H. Wolter, L.J. Giling and C.W.J. Beenakker, Surf.Sci. **263**, 91 (1992).

PHONON MEASUREMENTS OF THE ENERGY GAP IN THE FRACTIONAL QUANTUM HALL STATE

R.H. Eyles, C.J. Mellor, A.J. Kent, L.J. Challis
S. Kravchenko*, N. Zinov'ev[+], M. Henini
Physics Department, Nottingham University,
Nottingham, NG7 2RD, UK
*Institute for High Pressure Studies, Troitsk, Moscow, Russia.
[+] Ioffe Institute, St. Petersburg, Russia

The excitation spectrum $\Delta(k)$ of the magnetoroton excitations of the fractional quantum Hall (FQH) state is believed to fall from a maximum $\Delta(0)$ at $k = 0$ to a minimum of value $\Delta_{min} \sim 0.5\Delta(0)$ at $k \sim l_B^{-1}$ (l_B=magnetic length) and ultimately approaches $\Delta(\infty)=0.67\Delta(0)$ (Girvin et al, 1986). Experimental measurements of the excitation spectrum have previously been limited to transport and optical luminescence techniques. We have made initial phonon absorption studies of the FQHE which we believe give a direct measurement of the energy gap near to the magnetoroton minimum.

Figure 1 is a schematic illustration of the experiment. Acoustic phonons are generated on the back face of a GaAs/(AlGa)As heterojunction by electrical pulses in a constantan heater. The phonons travel ballistically across the wafer to interact with the two dimensional system and cause excitations of the incompressible FQH ground state. The phonon absorption is detected by changes in the longitudinal resistance of the electron gas.

The 3x2 mm^2 Hall bar has a carrier density after illumination of 1.4×10^{11} cm^{-2} and a mobility of 1×10^6 cm^{2}V^{-1}s^{-1}. Two 50Ω constantan heaters (60μm x 600μm) were evaporated on to the polished back face of the 0.4mm thick wafer. One of the heaters is positioned over the edge of the Hall bar; the other is over the bulk of the sample (Figure 1).

The sample is mounted on a dilution refrigerator with a base temperature of approximately 40mK in magnetic fields up to 17.4T. A DC current of 50-200nA is passed through the Hall bar with the current source floating with respect to ground potential and the central voltage probe connected to ground. Another voltage probe is connected via a capacitor to the input of a gated amplifier with a gate width of 8μs. When an electri-

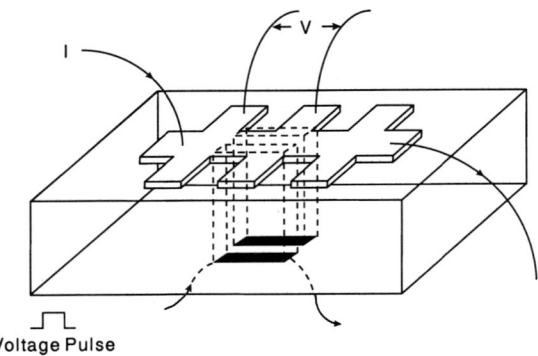

Fig. 1. Experimental arrangement (schematic).

cal pulse of 100ns duration is applied to one of the heaters the ballistic phonon pulse traverses the wafer in approximately 100ns. The change in resistance is detected as a voltage pulse by a gated amplifier.

The energy spectrum of the phonons emitted by the heaters can be assumed to be an approximately black-body spectrum characterised by the temperature of the heater. Using acoustic mismatch theory the heater temperature can be estimated from the power dissipated in it and from a knowledge of the elastic constants of constantan and gallium arsenide.

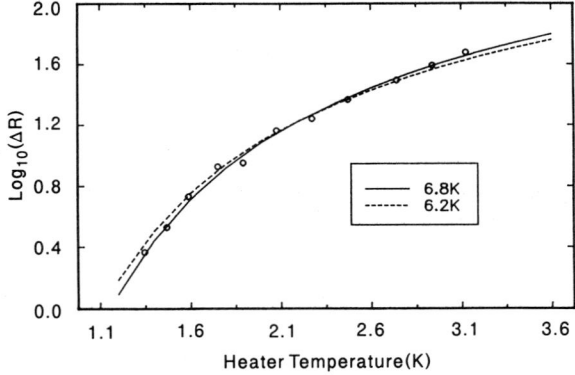

Fig. 2. Increase in resistance produced by phonon pulse from heater at temperature T. The lines are theoretical curves for two values of Δ.

Since the phonons absorption must fall to zero as the in-plane component of the phonons wavevector, $q_\parallel \to 0$, the absorption is thought to occur around the density of states peak at Δ_{min}. So as the heater temperature T_h is varied, the resistance change, which we assume to be proportional to the number of phonons absorbed, should vary as

$$[exp(\Delta_{min}/k_B T_h) - 1]^{-1} ,$$

the phonon intensity at Δ_{min}. This is consistent with the data shown for $\nu=2/3$ at 9.0 T in figure 2 using a value of Δ_{min}=6.8K to within an uncertainty in the fit of about 0.8K. This value is much greater than the activated energy gap of 2.2K, measured by the temperature variation of the minimum of R_{xx}. We attribute this to long range disorder. This should produce variations in the position of the minimum relative to E_F so that the value of Δ_{min} obtained from $R_{xx}(T)$ will be less than the true value by an amount which increases with disorder. However, phonon absorption is an essentially vertical process on this length scale so should measure the full value of Δ_{min}. This is consistent with the fact that our value is in approximate agreement with the gap obtained from transport measurements in very high mobility heterojunctions [Willett et al (1988)]. It is also in quite good agreement with optical measurements of the energy gap [Kukushkin et al (1992)].

A good fit is also obtained to data at $\nu=1/3$, B=13T, for a different sheet density with a value of $\Delta_{min} = 7.0 \pm 1.0K$ which is broadly consistent with the $\nu = 2/3$ value. It also provides a fit to data for $\nu = 3/5$ at 9.7 T but with a gap of 0.5K which is less than the transport gap, 0.8K, for this sample. We suggest that the technique is not applicable in this case since the minimum in R_{xx} is non-zero even at the lowest temperature in contrast to $\nu = 1/3$ and 2/3 so that phonons are likely to be absorbed by thermally excited quasiparticles and holes.

We find no qualitative difference in behaviour between the two heaters.

In conclusion, pulsed phonon measurements can contribute to the understanding of the FQHE in a quantitative manner that is complementary to data obtained by other means. When the minimum in R_{xx} is very small we suggest the technique provides a measurement of the FQHE energy gap close to the magnetoroton minimum.

Acknowledgements

We are grateful to Dr. K.A. Benedict for very helpful discussions and to the Science and Engineering Research Council, the Royal Society and the European Community for financial support.

REFERENCES

S.M. Girvin, A.H. MacDonald, P.M. Platzman, Phys. Rev. **B 33**, 2481 (1986)

I.V. Kukushkin, N.J. Pulsford, K. von Klitzing, K. Ploog, R.J. Haug, S. Koch and V.B. Timofeev, Europhys. Lett. **18**, 63 (1992)

A.H. Macdonald, K.L. Liu, S.M. Girvin and P.M. Platzman, Phys. Rev. **B 33**, 4014 (1986)

R.L. Willett, H.L. Stormer, D.C. Tsui, A.C. Gossard and J.H. English, Phys. Rev. **B 37**, 8476 (1988)

RESPONSE OF TWO-DIMENSIONAL ELECTRON GAS TO PULSES OF A PHONON FIELD

R.N. Gurzhi[1], A.I. Kopeliovich[1], T. Paszkiewicz[2]

[1]Institute for Low Temperature Physics and Engineering
Academy of Sciences of Ukraine, Kharkov, Ukraine
[2]Institute of Theoretical Physics, University of Wrocław
Pl. M. Borna 9, 50–204 Wrocław, Poland

INTRODUCTION

Previously [1] it was shown that in perfect layered conducting systems (of the intercalated graphite type), at low temperatures, qualitatively new relaxation mechanisms exist which are significantly different from those in three-dimensional metals. The reason for this is that a system of interacting two-dimensional electrons and phonons (in general, three-dimensional one) breaks down in quasi-isolated groups. So, instead of one law of conservation of the total momentum of the electrons and phonons, there is an infinite number of approximate laws of conservation. As a result, the Bloch diffusion along Fermi surface caused by electrons colliding with phonons is blocked and far slower processes of superdiffusion come into action. The relaxation due to interelectron collisions in two dimensions also becomes much slower because the collisions are small angle in character. It is difficult to observe these effects in massive samples of layered metals because the mean free path of the electron due to collisions with lattice defects is insufficiently large.

The situation may turn out to be more favourable for $2D$ electron systems in heterojunctions. These have been intensively studied [2], especially, in systems based on GaAs for which a rather high carrier mobility (up to 10^7 cm^2 $nV^{-1}s^{-1}$) is possible. Interest in heterojunctions is explained both by their application in micro electronics and their wide use in studies of a number of fundamental phenomena including the electron-phonon interaction [3]. Examples of this latter type are studies of phonon focusing observed using $2D$ electron gases (2DEGs) in heterojunctions [4,5]. Experiments with phonon beams provide information on the dynamics of ballistic phonons and the electron drag current in a 2DEG also gives details of the properties of the electron system. The purpose of the present work is to show that experiments of this kind can provide detailed information about the dynamic characteristics of electrons at arbitrary local points of the Fermi surface and also details of the relaxation processes in 2DEGs.

THE RELATION BETWEEN THE DRAG CURRENT AND THE LOCAL CHARACTERISTICS OF ELECTRONS ON FERMI–SURFACE

In earlier work [4,5] expressions for the electron drag current produced by a phonon beam were obtained using concrete models of the energy spectrum and the interaction of the electrons with phonons in 2DEGs. However, attention was not paid to the following circumstances. From the laws of conservation of energy and momentum we have

$$\epsilon(\mathbf{p} + \mathbf{q}_\|) = \epsilon(\mathbf{p}) + \hbar\omega(\mathbf{q}) , \qquad (1)$$

where $\epsilon, \mathbf{p}$ are the energy and two-dimensional quasi-momentum of the electrons, $\hbar\omega, \mathbf{q}$ are the energy and momentum of the phonons and $\mathbf{q}_\|$ is the projection of $\mathbf{q}$ on the plane of the 2DEG. From (1), when $\hbar\omega \ll \epsilon_F$, it follows that a phonon with fixed $\mathbf{q}$ can be absorbed only by a small group of electrons near two points on the Fermi surface (it is also assumed that the temperature $T \ll \hbar\omega$, where the Boltzmann constant $k_B = 1$). The inequality $s \ll v$ typical of good conductors (where s, v are phonon and electron velocities, respectively) is also fulfilled in the experiment [4]. If we also have $|\mathbf{q}| \ll p_F$ (p_F, ϵ_F are the Fermi momentum and Fermi energy) and $|\mathbf{q}_\|| / |\mathbf{q}| \gg s/v$, then the above mentioned two points on the Fermi surface obey the condition $\mathbf{v}\mathbf{q}_\| = 0$ and due to symmetry are equivalent. This suggests that it should be possible to scan the Fermi surface of the electron system by changing the direction of the phonon beam. We write down the kinetic equation for a nonequilibrium addition f to the electron distribution function in the following form

$$\frac{\partial f}{\partial t} = -\frac{f - <f>}{\tau_i} + I_{ep}^{abs} \qquad (2)$$

where

$$I_{ep}^{abs} = \frac{2\pi}{\hbar} \sum_{\mathbf{q},\mathbf{p}'} (n_{\mathbf{p}'} - n_{\mathbf{p}}) (\Gamma_{\mathbf{p}'\mathbf{p}\mathbf{q}} + \Gamma_{\mathbf{p}\mathbf{p}'\mathbf{q}}) N(\mathbf{q}, \mathbf{r}, t) ,$$

, and

$$\Gamma_{\mathbf{p}\mathbf{p}'\mathbf{q}} = |B_{\mathbf{p}\mathbf{q}}|^2 \delta_{\mathbf{p}',\mathbf{p}+\mathbf{q}_\|} \delta[\epsilon(\mathbf{p}) - \epsilon(\mathbf{p}') + \hbar\omega(\mathbf{q})] .$$

Here τ_i is the relaxation time of electrons with respect to collisions with impurities; $<...>$ means and average over the isoenergetic surface in $\mathbf{p}$–space corresponding to a given $\mathbf{p}$; I_{ep}^{abs} is the collision integral taking into account the absorption of nonequilibrium phonons by electrons; N is the distribution function of phonons at the point $\mathbf{r}$ where the 2DEG is located; $n_\mathbf{p}$ is the Fermi distribution function and $B_{\mathbf{p}\mathbf{q}}$ is the matrix element of the electron-phonon interaction. Solving equation (2) for the conditions mentioned, we obtain the following expression for the density of the drag current

$$\int_0^\infty \mathbf{j}\, dt = h^{-3} \int \mathbf{J_q}\, N(\mathbf{q}, \mathbf{r}, t)\, d^3q\, dt , \qquad (3)$$

$$\mathbf{J_q} = \frac{2e\omega}{\hbar^2 \pi} |B_{\mathbf{p}\mathbf{q}}|^2 \mathbf{c}\, v^{-1} |\mathbf{ec}|^{-1} \tau_i , \qquad (4)$$

where
$$c_i = \frac{\partial^2 \epsilon}{\partial p_i \partial p_k} e_k, \qquad \mathbf{e} = \mathbf{q}_{\|}/|\mathbf{q}_{\|}|.$$

This expression is true for arbitrary spectra of quasiparticles and their interaction. All electron characteristics entering into (4) are taken at the point $\mathbf{p}$ on the Fermi surface where $\mathbf{vq}_{\|} = 0$. If the Fermi surface consists of several valleys then one has to sum over all groups of carriers in equation (4). Integration over $\mathbf{q}$ in (3) also implies summation over phonon polarization. Since phonons are separated by their velocities on their path to the 2DEG then in (3) only those directions $\mathbf{q}$ are present which correspond to the fixed direction of phonon velocity [5]. The time-of-flight mode separation allows us to deal with one fixed direction $\mathbf{q}$ and, hence, with one electron on the Fermi surface. Note, too that the expression $\mathbf{J_q}$ contains only one characteristic of the electron spectrum, that is, v.

We now consider in more detail the region in which the incident beam makes a small angle to the normal to the 2DEG: $\sin\theta = |\mathbf{q}_{\|}|/|\mathbf{q}| \ll 1$. Note that according to (4), J does not tend to zero at $\theta \to 0$. This region is especially informative as the scanning of the Fermi surface can be performed with almost constant magnitude of $N(\mathbf{q})$. So, it is possible to separate the problem of the study of phonon focusing (this reduces to the determination of the dependence $N(\mathbf{q},\mathbf{r})$) from the study of the properties of $2D$ electrons.

Expression (4) is valid for $\theta \gg s/v$. For $\theta \cong s/v$, the phonon phase velocity in the $2D$ plane $\hbar\omega/q_{\|}$ is comparable to the Fermi velocity. So from (1), the condition $\mathbf{q}_{\|}\mathbf{v} = 0$ should be changed to the form $\mathbf{q}_{\|}\mathbf{v} = \hbar\omega$. Therefore, at $\theta < \theta_c(\mathbf{q}) \cong \hbar\omega/q\,|\mathbf{ve}|_{\max}$, the absorption of phonons by electrons is impossible ("max" means the greatest value in the limits of the Fermi surface). In heterojunctions the existence of this absorption threshold at an angle connected essentially with two-dimensionality was first considered by Vass [6] (a similar problem had been considered in layered metals [7]). The maximum electron velocity, in the fixed direction, may now be determined from the magnitude of $\theta_c(\mathbf{e})$. As the threshold is approached, the drag current increases according to the law

$$J \approx \left[1 - \left(\frac{\theta_c}{\theta}\right)^2\right]^{-1/2}, \qquad \text{at } (\theta - \theta_c) \gg \theta_c \hbar\omega/\epsilon_F, \qquad \theta \ll 1;$$

and reaches the maximum $\sqrt{4\epsilon_F/\hbar\omega}$.

THE STUDY OF SECONDARY PHONONS

Further information on relaxation processes in 2DEGs can be obtained from experiments of the following type. An additional detector is placed on the face of the substratum opposite the 2DEG and used to detect phonons emitted by electrons in the 2DEG following excitation by the primary phonon beam. The registered secondary phonons, as well as the primary ones, appear to be separated by their velocity direction. This should make it possible to study both the anisotropy of the phonon absorption and the anisotropy of the distribution of the phonons emitted in the relaxation process.

Now we will show that the intensity and angular distribution of the secondary phonons are determined largely by the relaxation mechanism in the 2DEG. Consider a kinetic situation resulting from the competition between the processes of phonon emission by nonequilibrium electrons and the processes of scattering of nonequilibrium electrons by impurities. We need to add to the right hand side of kinetic equation (2) an additional term describing the processes of emission

$$I^{em}_{ep} = \frac{2\pi}{\hbar} \sum_{p'q} \left(\tilde{\Gamma}_{pp'q} - \tilde{\Gamma}_{p'pq} \right),$$

$$\tilde{\Gamma}_{pp'q} = \Gamma_{pp'q} [(1-n_p)f_{p'}{}^{-n}{}_{p'}f_p].$$

The result of the solution of this equation corresponds to the following physical picture of the process. The nonequilibrium electrons lose almost completely their extra energy $\delta\epsilon \cong \hbar\omega$ through several processes of phonon emission. (As a result of n acts of phonon emission, the extra energy decreases 2^n times). Each successive phonon emission process requires considerably larger time as the emission time $\tau_{ep} \cong (\delta\epsilon)^{-3}$. So, the relaxation effectively comes to a halt after only a few emission processes.

Note that after an arbitrary number of emission processes, an electron shifts from its initial point on the Fermi surface by a distance of not more than $\hbar\omega/s$. Since $\hbar\omega/sp_F \ll 1$, the motion of a nonequilibrium electron around the Fermi surface arising from phonon emission occurs on both sides with equal probability, so that the mean nonequilibrium momentum and, hence, the electric current are both conserved in this approximation. Therefore, when $\hbar\omega \ll sp_F$ expressions (3) and (4) for the drag current in the 2DEG remains valid even when electron-phonon scattering dominates impurity scattering as the relaxation of nonequilibrium momentum occurs in the same time τ_i. If $\hbar\omega \cong sp_F$, then the dominant electron-phonon scattering leads to a decrease in the numerical coefficient in expression (4). So in spite of the substantial difference of the discussed scattering processes the drag current is the same.

On the other hand, the distribution of secondary phonons contains an essentially anisotropic part connected with electron-phonon scattering. Indeed, until impurity scattering occurs, secondary phonons are emitted by electrons with a momentum of which is close to the momentum $\mathbf{p}$ of the electron that has absorbed the primary phonon. This is why for small $\hbar\omega/sp_F$ and s/v, the momenta of secondary phonons $\mathbf{q}_2$ approximately satisfy the condition $\mathbf{v_p q}_{2\parallel} = 0$. In other words, their momenta are in the same plane as the momenta of the primary phonons and the normal to the plane of the 2DEG[1].

Phonons emitted after the impurity scattering of the nonequilibrium electrons (at $\tau_{ep} \ll \tau_i$, the energy of these phonons is of the order of $\hbar\omega(\tau_{ep}/\tau_i)^{1/3}$) form an almost isotropic background. The anisotropy, A, defined as the ratio of the energy flow of the secondary phonons in the wedge of angles about $\mathbf{q}_{2\parallel} \parallel \mathbf{q}$ to that in the background region is equal, to an order of magnitude,

$$A \cong (\Delta\varphi)^{-1} \left[\frac{\tau_{ep}}{\tau_i} + \left(\frac{\tau_{ep}}{\tau_i}\right)^{-1/3} \right], \qquad \tau_{ep} \equiv \tau_{ep}(\hbar\omega),$$

where $\Delta\varphi$ is the angular width of the wedge. The angular distribution of the emission has a complex structure, i.e. consists of (i) a narrow central part formed by the first

---

[1] At $\hbar\omega \cong sp_F$, the distribution of the secondary phonons is also anisotropic but of a different form.

portion of emitted phonons of width

$$\Delta\varphi_1 = \frac{\omega}{sp_F}\left(\sin\theta + \frac{\hbar\omega}{\epsilon_F}\right), \tag{5}$$

(ii) an outer part due to the all remaining phonons emitted before impurity scattering which has a width

$$\Delta\varphi_2 = \frac{\hbar\omega}{sp_F}\left(3\sin\theta + \frac{\hbar\omega}{\epsilon_F}\right). \tag{6}$$

The ratio of intensities of the central part to the outer part is the order of $6(\tau_i + 8\tau_{ep})\tau_i^{-1}$. We stress that the anisotropy of the secondary phononssecondary phonons is a direct consequence of the two-dimensional character of the electron system.

We next estimate the size of the secondary phonon effect. It is not difficult to show that if the detector lies within the narrow wedge of the angular distribution, the ratio of phonon intensities at the detector to that at the 2DEG is the order of

$$\gamma(S/r_{21}^2)(\Delta\varphi)^{-1}\tau_i(\tau_{ep}+\tau_i)^{-1},$$

where γ is the absorption coefficient of phonon flow impinging on to the 2DEG; S is the area of the 2DEG; r_{21} is the distance between 2DEG and detector. If the 2DEG is of the type used in work [4], then $\gamma \leq 10^{-4}$ and the signal assumed to be registered is extremely small although it could be increased considerably by using a multiple quantum well.

A STUDY OF THE SUPERDIFFUSION PROCESS

The considerable reduction of the momentum relaxation in $2D$ systems compared with $3D$ systems [1] in the case of electron-phonon collisions is related to the phenomena of phonon drag. Therefore, these effects of reduction can be observed only for a thick enought conducting layer, i.e., when the thickness d obeys the inequality $d \gg l_{pe}$, where l_{pe} is the mean free path of phonons with respect to collisions with electrons. In contrast, two–dimensional effects in electron-electron scattering do not depend on the thickness of the 2DEG. As shown in [1] not all relaxation processes are blocked but only these related to the relaxation of the odd part of the electron momentum distribution function. So since the electron contribution to the thermal conductivity of a 2DEG is determined by the energy relaxation rate. The "three-dimensional" lifetime for interelectron collisions ($\tau_{ee} \cong \tau_0(\epsilon_F/T_e)^2, \tau_0 \cong 10^{-15}s$, where T_e is the electron temperature) is the value appropriate for the electron contribution to the thermal conductivity of a heterojunction. In the electron absorption of sound, both even and odd components of the distribution function are important (their ratio depends on the sound wavelength) and thus both "three-dimensional" and superdiffusion times could be inferred [1].

In the case of the secondary phononssecondary phonons discussed above, both energy relaxation and superdiffusion may turn out to be essential. When $\tau_{ee}(\hbar\omega) \equiv \tau_0(\epsilon_F/\hbar\omega)^2 \ll \tau_{ep}$, the electron temperature T_e is established in a time $\tau_{ee}(T_e)$ determined by the energy balance

$$\gamma Q = d\,\tau_{pe}^{-1}(T_e)T_e\,N(T_e), \qquad T_e \approx (Q)^{1/5}.$$

Here, Q is the phonon energy flow of primary phonons; $N(T_e)$ is the equilibrium density of phonons; d is the 2DEG thickness. The secondary phonons give rise to contributions of two types. The first one is due to the high-energy electrons ($\delta\epsilon \approx \hbar\omega$), which have already emitted phonons before interelectron collisions. The angular width of this contribution coincides with that in (5) and (6) but its intensity differs from that considered above in section 2 (weak electron-electron scattering) by a factor of $(1 + \tau_i/\tau_{ep})[1 + \tau_i/\tau_{ee}(\hbar\omega)]^{-1}$.

The second contribution is due to phonon emission after interelectron collisions. When $\tau_{ee}(T_e) \ll \tau_i$, the energy of the emitted phonons is of the order of T_e. The ratio of intensity of this contribution to that for weak scattering is $(\Delta\varphi/\Delta\tilde{\varphi})(T_e/\hbar\omega)$ and its angular width $\Delta\tilde{\varphi}$ is determined by the path length δp_{sd} of electron-electron superdiffusion in a time τ_i, i.e.

$$\Delta\tilde{\varphi} = \Delta\varphi + \Delta\varphi_{sd}, \qquad \Delta\varphi_{sd} = \frac{\delta p_{sd}}{p_F}, \qquad \delta p_{sd} \cong \sqrt{\frac{\tau_i}{\tau_{ee}(T_e)}} \frac{T_e}{v} \cong \sqrt{\frac{\tau_i}{\tau_0}} \frac{T_e^2}{v\epsilon_F}.$$

Frequent interelectron collisions have only a small effect on the background intensity.

Finally we note, that the electron temperature T_e may be used as an independent external parameter if the 2DEG is heated by some additional heat source. However, one is to keep in mind that the increase of T_e rises considerably with the background intensity as it varies as T_e^5.

REFERENCES

[1] R.N. Gurzhi, A.I. Kopeliovich and S.B. Rutkevich, "Kinetic properties of two-dimensional metal systems", Adv. Phys. **36**, 221(1987).

[2] T. Ando, A. Fowler, F. Stern, "Electron properties of $2D$ systems", Rev. Mod. Phys. **54** (1982),
C.Weisbuch and B. Vinter, Quantum Semiconductor Structures, Academic Press, Boston, 1991

[3] L.J. Challis, A.J. Kent, V.W. Rampton, "Magnetic field dependence of acoustic phonon emission and scattering in $2D$ electron systems", Springer Series in Solid State Sciences, Springer-Verlag Berlin Heidelberg, **87**(1989).

[4] H. Karl, W. Dietsche, A. Fischer, K. Ploog, "Imaging of the phonon drag effect in GaAs-AlGaAs heterostructures", Phys.Rev.Lett., **61**,2360(1988).

[5] Cz. Jasiukiewicz, D. Lehmann, T. Paszkiewicz, "Phonon images of crystals II. Image of crystalline GaAs obtained by the phonon–drag–effect in GaAs–GaAlAs heterostructures", Z. Phys. B-Cond. Matt., **86**, 225(1992).

[6] E. Vass, "Theory of the spectral acoustic phonon emission intensity of hot $2D$ electrons in quantized n–inversion layers", Sol. St. Comm., **61**, 127(1987).

[7] Yu.A. Klimenko, A.I. Kopeliovich, "Sound propagation in $2D$ metals",
Fiz. Nisk. Temp., **11**, 1062(1985).

PHONON-DRAG EFFECT IN 1-DIMENSIONAL ELECTRON GASES

Dietmar Lehmann

Technische Universität Dresden, Institut für Theoretische Physik
D-8027 Dresden, Mommsenstraße 13, Germany

1. INTRODUCTION

Phonon images obtained by the phonon-drag effect in two-dimensional electron systems have been investigated both theoretically and experimentally beginning with the experiments of Dietsche and co-workers.[1] But there is also growing interest in the study of the electronic structure and properties of (quasi-) one-dimensional electron gases (1DEG) in which the electron gas is free to move in only one direction and is confined in the other two.[2,3]

In this paper a detailed kinetic picture of the phonon-drag current induced in a 1DEG is first presented. Because the phonon-drag images are a convolution of the phonon quasimomentum focusing (information about the geometry of the slowness surface of the crystal) and the probability of phonon absorption by the 1DEG, a pair of Boltzmann kinetic equations is used in Sect. 2 to describe the two subsystems (a kinetic description of the 3D acoustic bulk phonons and of the (quasi-) 1DEG, respectively). The result is an explicit expression for the instantaneous electric current induced in the 1DEG by a phonon pulse which may be applied to a 1DEG in different geometries (Sect. 3).

This formula is then used to calculate the time integrated current as a function of the phonon propagation direction for the case of a 1DEG formed in a GaAs/AlGaAs heterostructure. To include all possible inter and intra sub-band transitions, all occupied sub-bands of the electron system are taken into account. Comparing the results with the pure phonon quasimomentum focusing image of a GaAs crystal it will be obvious that the influence of the electron-phonon interaction is much stronger than in the case of phonon-drag in two-dimensional systems.[4] This means that measurements of the phonon-drag in 1DEG can give detailed information about the interaction of acoustic phonons with a 1DEG as well as about the 1-dimensional electron system itself.

2. KINETIC DESCRIPTION OF BULK PHONONS AND 1DEG

Under typical experimental conditions the phonons (e.g. created by heating the bottom of the substrate locally with a focused laser) move ballistically through the crystal and the influence of boundary scattering is negligible. As a natural idealisation of such conditions a half-space filled with an anisotropic medium is considered. The z-axis of the Cartesian coordinate frame is perpendicular to the medium boundary and directed towards the medium. At an arbitrary point of the medium boundary, say at $\vec{r} = 0$, there is a point source generating short monochromatic bursts of acoustic long-wavelength phonons of frequency ν_0. It is assumed that the directions of their wave vectors are distributed uniformly over the body angle 2π. It appears from the above description that the deviation δn of the phonon distribution function from the Planck function obeys the Boltzmann equation with a drift and a source term (the collision integral is absent)

$$\frac{\partial \delta n(Q;\vec{r},t)}{\partial t} = -\vec{v}(Q)\frac{\partial \delta n(Q;\vec{r},t)}{\partial \vec{r}} + Z(Q;\vec{r},t),$$

where Q stands for a pair $(\vec{q},j)$, $\vec{q}$ is the phonon wave vector, j enumerates the three polarizations, and $\vec{v}(Q)$ and $\vec{c}(Q)$ are the group and phase velocities of the phonon Q, respectively. The velocities are numerically calculated.[5] According to our conditions the source term Z is equal to

$$Z(Q;\vec{r},t) = \frac{4\pi^2}{3}\frac{c^3(Q)}{\omega^2(Q)}\delta(\vec{r})\delta(t)\delta(\omega(Q) - \omega_0)\Theta(v_z(Q)),$$

where $\delta(x)$ denotes the Dirac delta function, $\Theta(x)$ the Heaviside step function and $\omega(Q) = 2\pi\nu(Q)$ is the angular frequency of the phonon Q. The solution of the Boltzmann equation reads

$$\delta n(Q;\vec{r},t) = \frac{4\pi^2}{3}\frac{c^3(Q)}{\omega^2(Q)}\delta(\omega(Q) - \omega_0)\delta(\vec{r} - \vec{v}(Q)t)\Theta(v_z(Q))\Theta(t).$$

It is supposed that the 1DEG is located at a point $\vec{r}_0$ and that it is initially in equilibrium. The 1DEG is modelled as an electron system with periodic boundary conditions in one direction (here in the x-direction) and confined in the other two. This means, that the electrons are trapped within a 2-dimensional confining potential $V(y,z)$. The one-electron wave function describing the system can be written as

$$\psi_{kmn}(x,y,z) = \frac{1}{\sqrt{l}}e^{ikx}\phi_{mn}(y,z)$$

where l is the length of the 1DEG channel and k is the wave number in the x-direction. The positive integers m and n specify the (mn)th 1D sub-band and $\phi_{mn}(y,z)$ is the normalized sub-band wave function. The corresponding eigenvalue is

$$\varepsilon_{mn}(k) = \varepsilon_{mn} + \varepsilon_k,$$

where ε_{mn} are the energies of the sub-band minima and ε_k is the free-electron energy in the x-direction with effective mass m^*. As a realistic model of the confining potential, the formation of a nearly triangular potential along the z-axis and a square quantum well of infinite height along the y-direction have been assumed, i.e.

$$V(y,z) = \begin{cases} \alpha z & \text{for } 0 \le z \text{ and } 0 \le y \le b \\ \infty & \text{else} \end{cases}$$

where b is the width of the 1DEG channel. Typically $b \gg a$ $(= (12m^*\alpha/\hbar^2)^{1/3})$, so that the solution of the corresponding Schrödinger equation can be approximated for the lower sub-bands by

$$\phi_{1n}(y,z) = \sqrt{\frac{a^3}{b}} \sin(\pi y n/b) z e^{-az/2} .$$

Since the 1DEG is only weakly disturbed by the phonon pulse and thermal phonons are frozen out, the process of reemission of phonons by an excited 1DEG is much slower than the process of thermalization of electrons on impurities. Therefore the response of the electron gas is described by a kinetic equation with a collision integral containing the absorption of phonons $[\partial f/\partial t]_{e-p}^{abs}$ and the scattering of electrons on impurities $[\partial f/\partial t]_{e-i}$. Thus, the electron distribution function of the electrons in the $(1n)$th sub-band $f_{1n}(k; \vec{r}_0, t)$ obeys the following equation

$$\frac{\partial f_{1n}(k; \vec{r}_0, t)}{\partial t} = -v_{1n}(k) \frac{\partial f_{1n}(k; \vec{r}_0, t)}{\partial x} + \left[\frac{\partial f}{\partial t}\right]_{e-i} + \left[\frac{\partial f}{\partial t}\right]_{e-p}^{abs} .$$

Supposing the spatial dimension of the 1DEG is small the dependence of $f_{1n}(k; \vec{r}_0, t)$ on x can be neglected.

It is assumed that collisions of electrons with impurities are elastic. In the relaxation time approximation this term becomes

$$[\partial f/\partial t]_{e-i} = -\left[f_{1n}(k; \vec{r}_0, t) - f_{1n}^0(k)\right]/\tau ,$$

where τ is the relaxation time and $f_{1n}^0(k)$ denotes the equilibrium (Fermi) distribution function. The absorption term is

$$\left[\frac{\partial f}{\partial t}\right]_{e-p}^{abs} = \frac{2\pi}{\hbar} \sum_{n'} \sum_{k'} \sum_Q \{\Gamma_Q(n', k'; n, k) f_{1n'}(k'; \vec{r}_0, t)[1 - f_{1n}(k; \vec{r}_0, t)] \\ -\Gamma_Q(n, k; n', k') f_{1n}(k; \vec{r}_0, t)[1 - f_{1n'}(k'; \vec{r}_0, t)]\}$$

where

$$\Gamma_Q(n, k; n', k') = \frac{\hbar}{2\rho V \omega(Q)} \delta_{k', k+q_x} |\mathcal{F}_{1n}^{1n'}(q_y, q_z)|^2 \delta n(Q; \vec{r}_0, t) \\ |\mathcal{M}(Q)|^2 \delta(\varepsilon_{1n}(k) + \hbar\omega(Q) - \varepsilon_{1n'}(k')) ,$$

and ρ and V are the density and the volume of the specimen, respectively. $\mathcal{M}(Q)$ describes the electron-phonon interaction potential

$$|\mathcal{M}(Q)|^2 = \left|\frac{1}{\epsilon(q_x)} \Xi_D \vec{q} \vec{e}(Q)\right|^2 + \left|\frac{2e\,h_{14}}{\epsilon(q_x)} \frac{e_x(Q)q_y q_z + e_y(Q)q_x q_z + e_z(Q)q_x q_y}{q^2 + q_0^2}\right|^2 .$$

The first term corresponds to the deformation potential, the second one to the piezoelectric coupling. Ξ_D is the scalar deformation potential coupling constant, $\vec{e}(Q)$ is the polarisation vector, h_{14} is the piezoelectric constant, q_0 is the Debye screening length and $\epsilon(q_x)$ is the dielectric function of the 1DEG. At temperatures of about $1K$, both of the two processes can contribute and it is an open question which of the two mechanisms is the dominant one. The form factor $|\mathcal{F}_{1n}^{1n'}(q_y, q_z)|$ which cuts off the absorption for large q_y and q_z is given by

$$|\mathcal{F}_{1n}^{1n'}(q_y,q_z)|^2 = \left|\int dz \int dy\, \phi_{1n'}^*(y,z)\, e^{i(q_y y + q_z z)}\, \phi_{1n}(y,z)\right|^2$$

$$= \left(\frac{1}{1+(q_z/a)^2}\right)^3 \left(\frac{\pi^2\, n\, n'\, \sin\left(\frac{bq_y+(n+n')\pi}{2}\right)}{\frac{bq_y}{2}\left[\pi^2 - \left(\frac{q_y b}{n+n'}\right)^2\right]\left[\left(\frac{\pi(n-n')}{bq_y}\right)^2 - 1\right]\left(\frac{n+n'}{2}\right)^2}\right)^2.$$

By substitution of δn into $[\partial f/\partial t]_{e-p}^{abs}$ only those phonons give rise to the collision integral which move with a group velocity $\vec{v}(Q)$ parallel to the direction $\vec{r}_0$. Thus, the δ-function in δn can be transformed to,[54]

$$\delta(\vec{r}_0 - \vec{v}(Q)t) = \frac{|\mathcal{A}(\vartheta_q,\varphi_q)|}{v^2(\hat{Q})\, t^2\, \sin\vartheta_q} \delta(r_0 - v(Q)t) \sum_i \delta(\vartheta_q - \vartheta_i^j)\delta(\varphi_q - \varphi_i^j)$$

where ϑ_i^j and φ_i^j are solutions of the eqations $\hat{\vec{v}}(Q) = \hat{\vec{v}}(\vartheta_q, \varphi_q, j) = \hat{\vec{r}}_0$ for each polarisation j and the two angles ϑ_q and φ_q describe the direction of $\vec{q}$. $\mathcal{A}(\vartheta_q, \varphi_q)$ is the focusing factor defined as the ratio of the solid angle $\delta\Omega_q$ in the $\vec{q}$-space to the coresponding angle $\delta\Omega_v$ in the $\vec{v}$-space.

3. THE PHONON DRAG EFFECT

The electric current induced by phonons in a 1DEG is now obtained by evaluating

$$J(\vec{r}_0, t) = -\frac{2|e|}{l} \sum_k \sum_n \frac{\hbar\, k}{m^*} \left(f_{1n}(k; \vec{r}_0, t) - f_{1n}^0(k)\right).$$

Upon substitution of the electron distribution function $f_{1n}(k; \vec{r}_0, t)$ and summation over the k-space and all sub-bands, the final result is

$$J(\vec{r}_0, t; \omega_0) = \frac{-|e|}{6\pi\hbar\rho\omega_0 r_0^2} \sum_{i,j} e^{-(t-\frac{r_0}{v(i,j)})/\tau}\, \Theta\left(t - \frac{r_0}{v(i,j)}\right) \Theta(v_z(Q))$$

$$\frac{\cos(\varphi_i^j)}{|\cos(\varphi_i^j)|} \frac{1}{v(i,j)}\, |\mathcal{A}(\vartheta_i^j, \varphi_i^j, j)|\, |\mathcal{M}(\omega_0, \vartheta_i^j, \varphi_i^j, j)|^2$$

$$\sum_{n,n'} |\mathcal{F}_{1n}^{1n'}(\omega_0, \vartheta_i^j, \varphi_i^j, j)|^2\, \mathcal{L}_{n'n}(\omega_0, \vartheta_i^j, \varphi_i^j, j)$$

with $\mathcal{L}_{n'n}(\omega_0, \vartheta_i^j, \varphi_i^j, j)$ defined by

$$\mathcal{L}_{n'n}(\omega_0, \vartheta_i^j, \varphi_i^j, j) = \begin{cases} 1 & \text{for } |l_-^{n'n}| < k_F^{(n')},\ |l_+^{nn'}| > k_F^{(n)}, \\ 0 & \text{else} \end{cases}$$

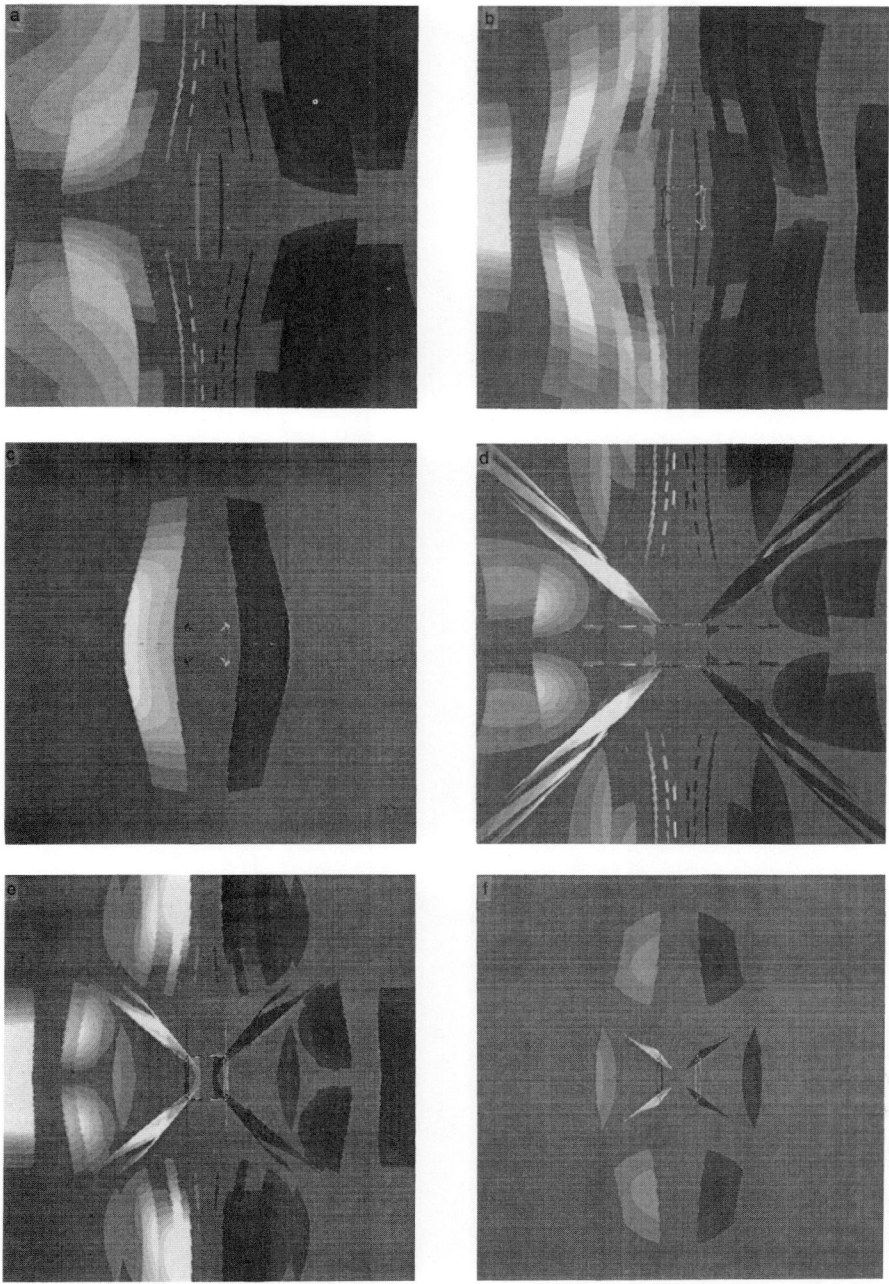

Fig. 1a-f. Patterns of the time-integrated current induced in a 1DEG by beams of monochromatic phonons of frequency $\nu_0 = 120$ GHz. The images are calculated as a function of the phonon propagation direction for different electron densities n_{1D}: 1a-c the contribution of the interaction by the deformation potential and 1d-f of the piezoelectric coupling. 1a,d - $n_{1D} = 1 \times 10^8$ m^{-1}; 1b,e - $n_{1D} = 5 \times 10^7$ m^{-1}; 1c,f - $n_{1D} = 1 \times 10^7$ m^{-1}.

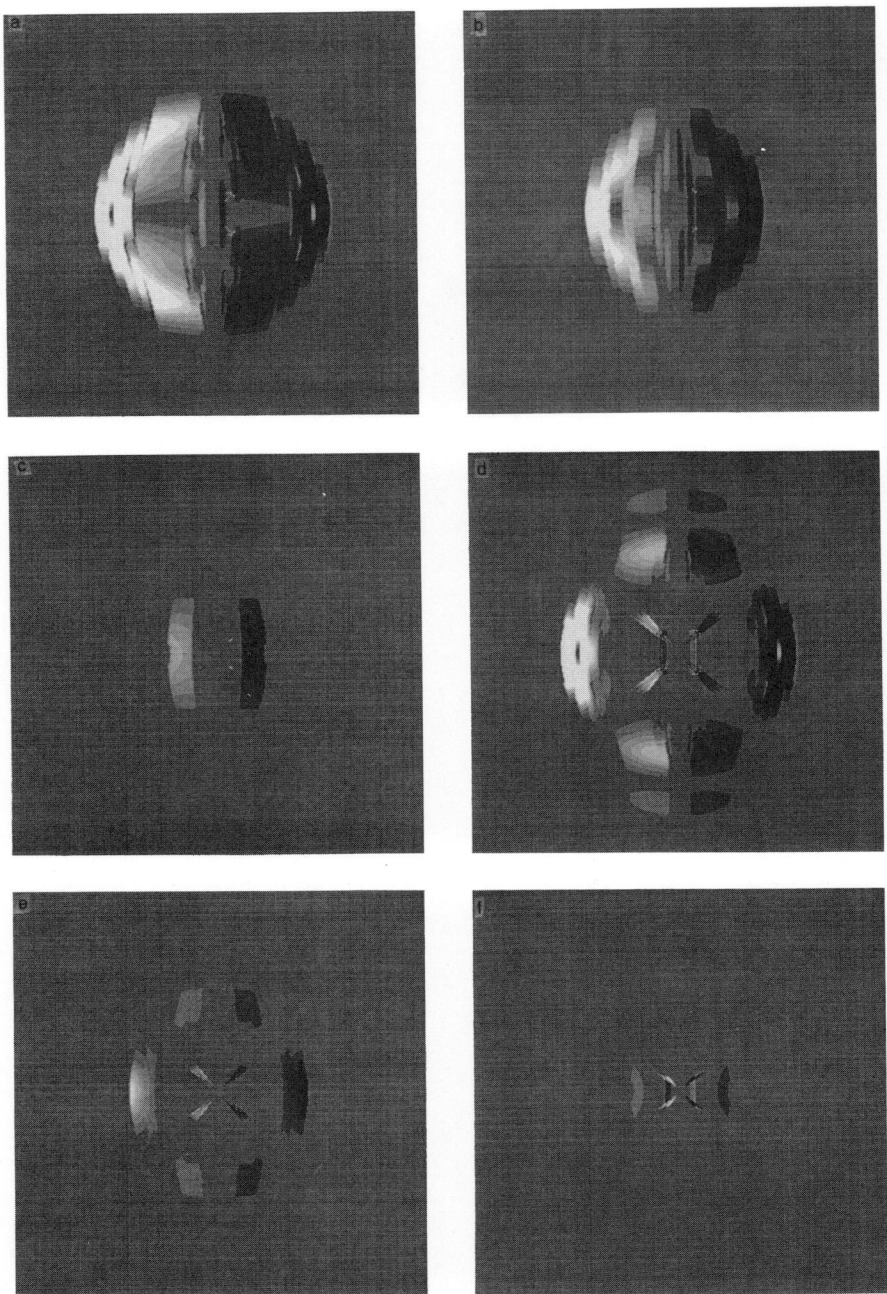

Fig. 2a-f. Patterns of the time-integrated current induced in 1DEG by beams of monochromatic phonons of frequency $\nu_0 = 300$ GHz. The images are calculated as a function of the phonon propagation direction for different electron densities n_{1D}: 2a-c the contribution of the interaction by the deformation potential and 2d-f of the piezoelectric coupling. 2a,d - $n_{1D} = 1 \times 10^8$ m^{-1}; 2b,e - $n_{1D} = 5 \times 10^7$ m^{-1}; 2c,f - $n_{1D} = 1 \times 10^7$ m^{-1}.

where

$$l_\pm^{nn'} = \frac{q_x}{2} \pm \frac{m^*\omega_0}{\hbar q_x} - \frac{1}{2}\left(\frac{\pi}{b}\right)^2 \frac{n^2 - n'^2}{q_x} \quad \text{and} \quad k_F^{(n)} = \sqrt{\frac{2m^*}{\hbar^2}(\varepsilon_F - \varepsilon_{1n})}\,.$$

ε_F is the Fermi energy of the 1DEG determined by the electron density n_{1D}.

Concentrating on the problem of phonon images, numerical calculations were done for the time integrated current $\overline{J(\vec{r}_0, t; \omega_0)}$, in which the current was obtained as a function of the phonon propagation direction (i.e. as a function of the relative position of the phonon source (laser position) and phonon detector (1DEG)). To this end a 1DEG formed in a GaAs/AlGaAs-heterostructure and lying in a plane (001) was simulated. The 1DEG was aligned along the [110] direction and its 'width' was = 100 nm. All possible inter and intra sub-band transitions of electrons were included in these calculations.

The contributions of the two coupling mechanisms between the 3D-bulk phonons and the (quasi-) 1D-electrons (deformation potential and piezoelectric coupling) are presented separately (Fig. 1a-c, 2a-c and Fig. 1d-f, 2d-f, respectively). In each case the results are shown for two different phonon frequencies ν_0 and different electron densities n_{1D}. An average grey tone in the figures corresponds to a vanishing phonon-drag current of zero. Light (dark) tones indicate a positive (negative) current. Comparing the patterns with theoretical results for quasimomentum focusing[5] it is obvious that the electron-phonon interaction strongly distorts the focusing image. This means, that the phonon-drag patterns are very sensitive to changes of the electron density of the 1DEG (typical values of the Fermi wave vector in 1DEG are comparable to the phonon wave vectors in phonon imaging experiments). For electron densities $n_{1D} \geq 5 \times 10^7$ m^{-1}, the influence of higher sub-bands and of inter sub-band transitions results in marked changes of the phonon-drag current. Finally, one can conclude, that measurements of the phonon-drag current in a 1DEG will be in combination with the corresponding model calculations a very helpful tool to study the kinetic properties of 1-dimensional electron systems.

Acknowledgments

I wish to acknowledge the strong support of Tadeusz Paszkiewicz and Czesław Jasiukiewicz from the University of Wrocław in Poland, particularly for assistance in numerical calculations (Cz.J.).

4. REFERENCES

1. H. Karl, W. Dietsche, A. Fischer, and K. Ploog, Phys. Rev. Lett. 61, 2360 (1988)
2. A. Y. Shik, this volume; A. Y. Shik and L. J. Challis, Phys. Rev. B (1993)
3. S. S. Kubakaddi and P. N. Butcher, J. Phys.: Condens. Matter 1, 3939 (1989)
4. Cz. Jasiukiewicz, D. Lehmann, and T. Paszkiewicz, Z. Phys. B 86, 225 (1992)
5. Cz. Jasiukiewicz, D. Lehmann, and T. Paszkiewicz, Z. Phys. B 84, 73 (1991)

THE THERMOELECTRIC BEHAVIOUR OF TWO DIMENSIONAL ELECTRON AND HOLE GASES AND QUANTUM POINT CONTACTS

P N Butcher[1], T M Fromhold[2], R J Barraclough[2], P J Rogers[2], B L Gallagher[2], J P Oxley[2] and M Henini[2]

[1] Department of Physics, University of Warwick, Coventry CV4 7AL, U.K.
[2] Department of Physics, University of Nottingham, Nottingham NG7 2RD, U.K.

1. INTRODUCTION

Considerable progress has been made over the last ten years in understanding the thermoelectric behaviour of low dimensional transport systems. There are two regimes to consider. At low temperatures (below about 1K) the effect of phonons is negligible. This is the "diffusion" thermopower regime which we discuss briefly for 2D electron gases (2DEGS) in Section 2. In Section 3 we discuss a particularly interesting aspect of the formalism for 2DEGS which is that it may be modified easily to apply to the quantisation effects observed recently in quantum point contacts (van Wees et al, 1988; Wharam et al, 1988; van Houten et al, 1992; Molenkamp et al 1992).

Above $T \simeq 1K$ phonon drag effects begin to dominate thermopower behaviour. We outline the semiclassical theory for 2DEGS in Section 4 when the magnetic field $B = 0$. It is in good agreement with experimental data. When the magnetic field B is large the magnetothermopower data for 2DEGS show oscillatory behaviour because of the Landau quantisation of the electron energy levels. The phonon drag theory of these oscillations is outlined in Section 5 and compared with experimental data. Experimental interest has recently shifted from 2DEGS to high mobility 2D hole gases (2DHGS) in GaAs/AlGaAs heterojunctions grown on (311) interfaces. We discuss new experimental results for 2DHGS in Section 6. The data are interpreted by using the phonon drag theory developed in Sections 4 and 5 for 2DEGS with appropriate modifications of the parameters (Barraclough et al, 1993).

2. THE DIFFUSION THERMOPOWER OF A 2DEG

We confine our attention to isotropic 2DEGS with a scalar effective mass m^* and assume that the electron scattering is elastic. We describe the scattering by a relaxation time $\tau(\epsilon)$ which is a function of the electron energy ϵ.

The 2D current density response to an applied electric field E and temperature gradient ∇T takes the form

$$\mathbf{J} = \sigma \mathbf{E} + L \nabla T. \tag{1}$$

Here the conductivity σ and the coefficient L are both scalars. They are easily evaluated in the diffusive regime using either the electron Boltzmann equation or Kubo formulae. The low temperature results are particularly well known (Gallagher and Butcher, 1992; Butcher, 1992). They are: $\sigma = \sigma(\epsilon_F)$ and

$$L = \frac{\pi^2}{3e} k_B^2 T \, d\sigma(\epsilon_F)/d\epsilon \tag{2}$$

Here ϵ_F is the Fermi level and $\sigma(\epsilon) = n_e(\epsilon)e^2 \tau(\epsilon)/m^*$ is the conductivity at absolute zero when the Fermi level is at ϵ. ($n_e(\epsilon)$ denotes the corresponding 2D electron density). The Seebeck coefficient in the diffusion regime, S_d, is the ratio of $\mathbf{E}$ to ∇T when $\mathbf{J} = 0$. We see immediately from equation (1) that

$$S_d = -L/\sigma = -\left(\pi^2 k_B^2 T/3e\right) \left(d\ell n \sigma(\epsilon)/d\epsilon\right)_{\epsilon_F} \tag{3}$$

For most energies and most scattering mechanisms $\sigma(\epsilon)$ increases with ϵ. Equation (3) then shows that S_d has the negative value which is commonly observed for electrons. However, when ϵ moves into a new subband in a 2DEG there is a sudden increase in the density of final states available for scattering. Consequently $\tau(\epsilon)$ shows a sharp decrease and so does $\sigma(\epsilon)$. When broadening effects are ignored equation (3) therefore predicts that S_d will exhibit a positive δ-function like peak when ϵ_F enters a new subband. In reality the peaks are broadened both thermally and by electron scattering (Cantrell and Butcher, 1985). The effect has been observed by Ruf et al (1989). The knee expected in σ when the second subband begins to populate has also been observed recently by Fletcher et al (1991).

3. THE THERMOPOWER OF A QUANTUM POINT CONTACT

A quantum point contact can be regarded as an experimental realisation of a 1D electron waveguide (Beenakker and van Houten 1991). It has sub-micron dimensions and is fabricated in a 2DEG with a mean free path in the order of $10\mu m$ at 100mK. Consequently the electrons move ballistically through the structure in all the waveguide modes which propagate at energies below the Fermi level. A simple calculation (Beenakker and van Houten, 1991; Gallagher and Butcher, 1992; Butcher, 1992) shows that each propagating mode contributes $(2e^2/h)$ to the conductance when spin degeneracy is assumed. Consequently the quantum point contact has a conductance $G = N (2e^2/h)$ where N is the number of propagating modes. It is possible to vary N by altering the Fermi level or, more simply, by varying the gate voltage which controls the width of the quantum point contact (van Wees et al, 1988; Wharam et al, 1988).

The transport situation in a quantum point contact is closely analogous to that discussed in Section 2 for a 2DEG. Everytime a new mode begins to propagate there is a jump of $2e^2/h$ in the conductance. In a quantum point contact the thermopower (which we continue to denote by S_d) is defined as the ratio of the voltage difference to the temperature difference between the terminals. To calculate it we continue to use equation (3) with $\sigma(\epsilon)$ replaced by G when $\epsilon_F = \epsilon$. Since G goes up when a new mode begins to propagate we predict a series of negative peaks for S_d. The width of the peaks is in the order of $k_B T$ because the electrons are ballistic and the broadening is due to thermal processes. A formal analysis of semiconductor microstructures confirms the qualitative picture developed here. (Butcher, 1990; van Houten et al, 1992; Gallagher and Butcher, 1992; Butcher, 1992). The predicted behaviour of the thermopower has been confirmed in a series of experiments in which electrical conductivity, thermal conductivity and the Peltier coefficient are also measured (van Houten et al, 1992; Molenkamp et al, 1992).

4. THE PHONON DRAG THERMOPOWER OF A 2D ELECTRON GAS

4.1 Introduction

It is useful to begin by developing a macroscopic argument for estimating the phonon drag thermopower S_g^{3D} for a 3D system in which the electron density is n_{3D} (Guenault, 1971). In a steady state the electrical force density $-en_{3D}E$ and the rate $\alpha P/\tau_p$ at which phonon momentum density P is transferred to the electrons add up to zero. Here τ_p is the phonon lifetime and α is the fraction of the phonon momentum loss which is transferred to the electrons. Thus we have $-en_{3D}E + \alpha P/\tau_p = 0$ so that the emf $E = \alpha P/en_{3D}\tau_p$. To relate E to an applied temperature gradient ∇T we note that at low temperatures only acoustic phonons are involved. Considering unit volume and assuming for simplicity a single, constant sound velocity v_s we easily verify that

$$P = \sum_q N_q \hbar q = Q/v_s^2 \tag{4}$$

where

$$Q = \sum_q N_q \hbar \omega_q v_s (q/q) \tag{5}$$

is the phonon heat flux. In these equations q is the phonon wave vector, $\omega_q = v_s q$ is the phonon frequency and N_q is the phonon distribution function. Now, we can also write $Q = -\kappa \nabla T$ where κ is the thermal conductivity due to phonons which is given by the well known formula $\kappa = C_g v_s^2 \tau_p/3$ where C_g is the phonon specific heat (Kittel 1976). The end result of these simple manipulations is $E = S_g^{3D} \nabla T$ with

$$S_g^{3D} = -\alpha C_g/3n_{3D}e \tag{6}$$

In the systems of interest to us there is a planar sheet of electrons with a 2D number density n_e. The phonons, however, remain essentially 3D because the interfaces which confine the electrons are poor reflectors of sound waves. We may modify equation (6) so that it describes this hybrid situation by replacing n_{3D} by n_e/L_z and 3 by 2. Here L_z is the width of the phonon bath in the direction perpendicular to the electron gas. Thus we arrive at a simple formula for the phonon drag thermopower of a 2DEG:

$$S_g = -\alpha C_g L_z/2n_e e \tag{7}$$

The dominance of S_g when T increases significantly above 1K is due to the cubic dependence of C_g on T (Kittel, 1976) which ensures that S_g rapidly overtakes S_d which equation (3) shows to be linear in T.

4.2 The Semiclassical Formula for S_g

To calculate α in equation (7) we use coupled 2D electron and 3D phonon Boltzmann equations (Cantrell and Butcher, 1987; Smith and Butcher, 1989a; Butcher 1992). The final formula may be put in the transparent form of equation (7) with

$$\alpha = -\Omega^{-1} \sum_q \frac{dN_q^o}{dT} \hbar \omega_q \alpha(q)/C_g \qquad (8a)$$

where Ω is the 3D volume occupied by the phonons and C_g is the 3D phonon specific heat per unit volume:

$$C_g = \Omega^{-1} \sum_q (dN_q^o/dT) \hbar \omega_q \qquad (8b)$$

We see by inspection that α is an appropriate average of the quantity $\alpha(q)$ which we still have to define. It is the fraction of q-phonon emission processes due to electron-phonon scattering in thermal equilibrium. For weak electron-phonon scattering: $\alpha(q) = \tau_p W_{ep}^{em}(q)$ where $W_{en}^{em}(q)$ is the rate of q phonon emission. To complete the calculation of α and of S_g we write $\tau_p = \lambda_p/v_s$ where λ_p is the phonon mean free path obtained by measuring the thermal conductivity. To calculate W_{ep}^{em} we also require the screened matrix elements of the electron-phonon scattering processes. Full details are given in Smith and Butcher (1989), Qin et al (1993) and Fromhold et al (1993).

4.3 Comparison of the Theory with Experimental Data

Figure 1 is taken from Gallagher et al (1990). The curves show the temperature dependence of $-S_g/T^3$ calculated for a Si MOSFET. The points are experimental values of $-S_t/T^3$ where S_t is the total measured thermopower. Results are given for three electron densities. Available experimental values were used for all the parameters required to calculate the theoretical curves. Their excellent agreement with the data points

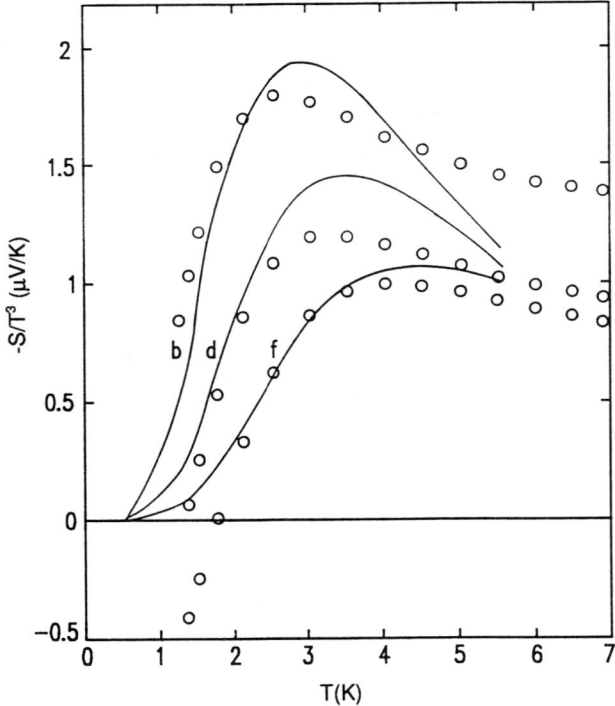

Figure 1. Plots of $-S/T^3$ against T for a Si MOSFET. Curves: calculated values. Points: experimental data. Sheet electron density $n_e = 5.47 \times 10^{15}$ m^{-2} (b), 8.35×10^{15} m^{-2}, (d) and 14.1×10^{15} m^{-2} (f).

with regard to both T and n_e dependence shows that S_g is dominant over most of the temperature range considered and that equations (7) and (8) describe the behaviour of S_g very well for Si MOSFETS. This is also true for GaAs/AlGaAs heterojunctions (Lyo, 1988; Smith and Butcher, 1989b).

According to the elementary discussion given in Section 4.1 we might expect $-S_g/T^3$ to be independent of T. In reality the q dependence of the summand in equation (8) involves two factors: the first is a phonon distribution factor which has a peak at a dominant value of q proportional to T. The second is a transition rate factor which has a peak when q is equal to twice the Fermi wave number. The peaks in the calculated curves occur at the temperature at which the peaks of both factors coincide. (Cantrell and Butcher 1987).

5. THE PHONON DRAG MAGNETOTHERMOPOWER OF A 2DEG

5.1 Introduction

When a magnetic field **B** is applied perpendicular to the plane of the 2DEG (Oxy) all the transport coefficients become 2D tensors with symmetry properties which are the same as those of the resistivity tensor: $\rho_{xx} = \rho_{yy}$ and $\rho_{xy} = -\rho_{yx}$. (Fletcher et al, 1986; Fromhold et al, 1993). The experimental data with which we are concerned is taken in quantising magnetic fields (B > 2T) at temperatures which are high enough to ensure that the magnetothermopower is dominated by the phonon drag contribution.

The calculation of the phonon drag magnetothermopower tensor presents a formidable problem. A different approach is required from that used in Section 4. This is because a quantising magnetic field localises the electrons in the Oxy plane so that we can no longer formulate an electron Boltzmann equation. To avoid this problem it is convenient to calculate the transport tensor M which determines the heat flux **Q** in an electric field **E** via the equation **Q** = M**E**. Moreover, the relationships between the transport coefficients (Fletcher et al, 1986) show that the elements of the magnetothermopower S are related to those of ρ and M through the equations

$$T S_{xx} = \rho_{xx} M_{xx} - \rho_{yx} M_{yx} \qquad (9a)$$

$$T S_{yx} = \rho_{yx} M_{xx} + \rho_{xx} M_{yx} \qquad (9b)$$

We make use of these equations in two different ways which are both useful. Experimental data is available for both S and ρ. We use it to obtain experimental data for M by solving equations (9). The results may then be compared directly with the outcome of our theoretical calculation of M. An alternative procedure is to substitute the calculated M into equations (9) so as to arrive at theoretical estimates of S which can be compared directly with the raw magnetothermopower data. In that case <u>experimental</u> values of ρ are used because there are no theoretical formulae for ρ which adequately account for its observed behaviour in the quantum Hall regime.

5.2 The formula for M

We suppose that the electric field **E** is turned on adiabatically. Then it produces a change of the one-electron energies which is easily evaluated (Lyo, 1988; Fromhold et al, 1993). The adiabatic assumption ensures that the occupation probabilities of the perturbed states remain frozen at their initial values (when **E** = 0). They are given by Fermi functions involving the electron energies when **E** = 0. Consequently, the electron-phonon scattering terms in the *phonon* Boltzmann equation are thrown out of balance because the energies in the Fermi functions are different from those in the energy conserving δ-functions. When the other phonon mechanisms are described by a phonon relaxation time τ_p we may immediately solve the phonon Boltzmann equation to obtain the perturbation N_q^1 of the phonon distribution function produced by the electric field. The phonon drag contribution to M may then be calculated by evaluating the 2D phonon heat flux $\mathbf{Q}_g$ to first order in $\mathbf{E}$:

$$\mathbf{Q}_g = \frac{L_z}{\Omega} \sum_q N_q^1 \hbar\omega_q (v_s q/q) = \mathbf{M}\,\mathbf{E} \qquad (10)$$

The final results are $M_{xx} = 0$ and

$$M_{yx} = -\,(\ell_B^2\, 2e/k_B TA) \sum_{qs} \lambda_{qs}\, \hbar\omega_{qs}\, (q_y^2/q)\, \Gamma\,(qs) \qquad (11)$$

where A is the area of the 2DEG, $\ell_B = (\hbar/eB)^{1/2}$ is the magnetic length, λ_{qs} is the mean free path of phonons with wave vector q in mode s and Γ(qs) is the rate of transfer of electrons produced by these phonon in thermal equilibrium as a result of phonon emission processes.

To evaluate M_{yx} we use the same parameters as in the case B=0, and set λ_{qs} equal to the measured phonon mean free path. We also need to specify the Landau level line shape which enters into Γ(qs). A Gaussian line shape is assumed with a line width parameter $\gamma = CB^{1/2}$ meV (Fromhold, et al 1993). The quantity C is the only adjustable parameter in the theory. We set C = 0.5 meV T$^{-1/2}$ so as to achieve the best fit to the experimental data for $|M_{yx}|$ discussed below. Some details of the method of calculation are given by Kubakaddi et al (1988) and by Lyo (1989). A full account is to be found in Fromhold et al (1993). These three papers all concentrate on GaAs/AlGaAs heterojunctions. Similar conclusions to those described below are reached in a similar study of a Si MOSFET by Qin et al (1993).

5.3   Discussion of the Results

Figs.2 to 4 give results for a GaAs/AlGaAs heterojunction with $n_e = 5.2 \times 10^{15}$ m^{-2} (Fromhold et al 1993). Figure 2 shows the behaviour of M_{xx} and M_{yx} when

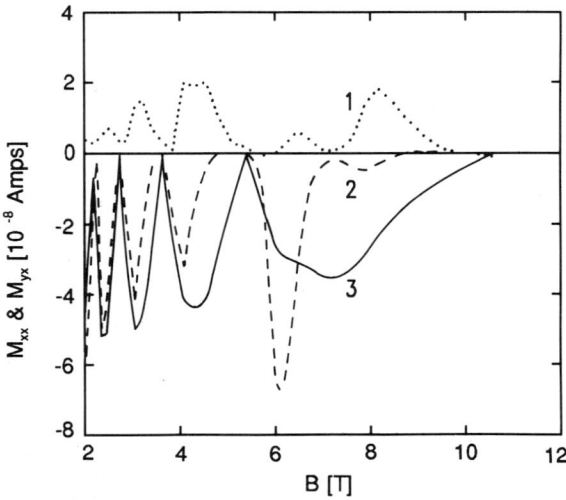

Figure 2. Comparison of experimental and theoretical values of M_{yx} and M_{xx} for a GaAs/AlGaAs heterojunction at 1.275 K. Curve 1: experimental values of M_{xx}. Curve 2: experimental values of M_{yx}. Curve 3: theoretical values of M_{yx}. For curve 3: $\lambda_p = 0.75$ mm.

T = 1.275 K. Curves 1 and 2 show experimental values of M_{xx} and M_{yx} respectively. Curve 3 shows the theoretical value of M_{yx}. The $|M_{yx}|$ values oscillate in phase with the magnetically modulated density of states at the Fermi energy. The theory predicts $M_{xx} = 0$ because it is only asymptotically correct at high fields when the phonon heat flux, like the Hall current, is perpendicular to **E**. This result is not born out very well by the experimental data. The expression (11) for M_{yx}, on the other hand, is in reasonable accord with data. Figure 3 shows the behaviour of S_{xx} when T = 2.937 K.

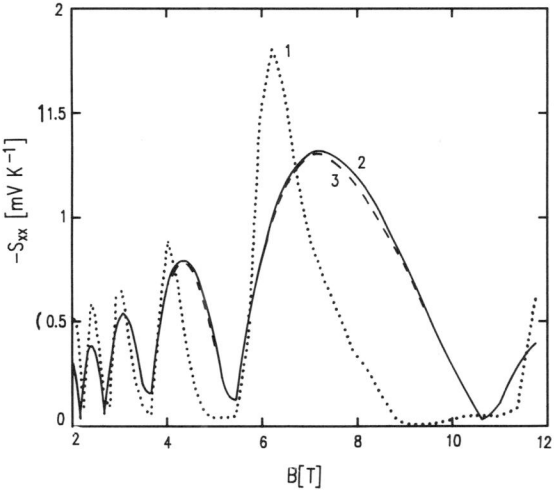

Figure 3. Comparison of experimental and theoretical values of $-S_{xx}$ for a GaAs/AlGaAs heterojunction at T = 2.937 K. Curve 1: Experimental values of $-S_{xx}$. Curve 2: theoretical values of $-S_{xx}(M_{xx} = 0)$. Curve 3: $-S_{xx}$ calculated using empirical M_{xx} data. For curves 2 and 3: $\lambda_p = 0.6$mm.

Curves 1 and 2 show experimental and theoretical values of $-S_{xx}$ respectively. Curve 3 is a "hybrid" value of $-S_{xx}$ calculated from equation (9a) when M_{xx} is given its *experimental* value. It is very close to the theoretical curve 2 for which $M_{xx} = 0$. We conclude that $-S_{xx}$ is insensitive to the precise value of M_{xx}, i.e. the second term dominates on the right hand side of equation (9a) and the failure of the theory to predict M_{xx} properly is of little consequence as for as $-S_{xx}$ is concerned. Figure 4 shows the behaviour of S_{yx} when T = 2.937K. Curves 1 and 2 show experimental and theoretical values respectively. Curve 3 is a "hybrid" value of S_{yx} calculated from equation (9b) when M_{xx} is given its *experimental* value. It is completely different from the theoretical curve 2 and very close to the experimental curve 1. We conclude that S_{yx} is very sensitive to the precise value of M_{xx}, i.e. the first term on the right hand side of equation (9b) is important and the failure of the theory to predict M_{xx} properly means that it cannot describe the behaviour of S_{yx} either.

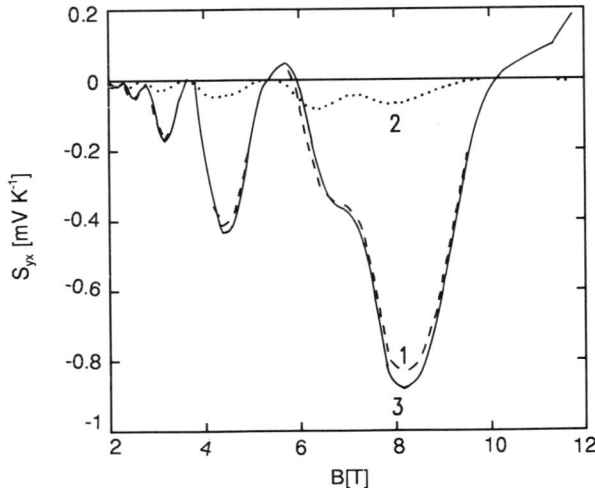

Figure 4. Comparison of experimental and theoretical values of S_{yx} for a GaAs/AlGaAs heterojunction at T = 2.937 K. Curve 1: Experimental values of S_{yx}. Curve 2: theoretical values of S_{yx} ($M_{xx} = 0$). Curve 3: S_{yx} calculated using empirical M_{xx} data. For curves 2 and 3: $\lambda_p = 0.6$ mm.

6. PHONON DRAG MAGNETOTHERMOPOWER OF 2D HOLE GASES

Magnetotunnelling experiments on GaAs/(AlGa)As resonant-tunnelling diodes containing a lightly p-doped emitter have confirmed that the lateral dispersion relations $\epsilon(\underline{k})$ for 2D holes gases (2DHGS) are strongly perturbed by the 1D confining potential (Hayden et al, 1991). The degeneracy of the light and heavy hole bands at $\mathbf{k} = 0$ is removed. The in-plane effective mass of the upper dispersion curve (the lowest energy hole states) $m^*_{\parallel} \simeq 0.3\, m^o$ is field dependent and in the order of five times larger than for a 2D electron gas. The density of states for the 2DHGS is therefore also approximately five times higher than for a 2DEG. As a consequence, hole gases will screen electrical potential fluctuations much more effectively.

We present here new measurements of the magnetothermopower S and the heat transport tensor M for a high mobility 2DHG in a GaAs/(AlGa)As heterojunction grown on the (311) crystal plane. This growth direction produces higher hole mobilities than the more usual (100) substrate orientation (Wang et al, 1986) and the present samples have values up to $50 m^2\, V^{-1}\, s^{-1}$. The quantising magnetic field **B** is applied perpendicular to the plane of the 2DHG ($\mathbf{B} \parallel [311]$). Details of the sample preparation and experimental techniques are given by Barraclough et al, (1993).

Curve 1 of figure 5 shows the observed off-diagonal component M_{yx} of the heat transport tensor for a 2DHG with sheet density $n_h = 1.63 \times 10^{15}\, m^{-2}$ when T = 1.722K. At this temperature the heat flux is dominated by phonon drag. The M_{yx} values are obtained from equation (9) using the empirical ρ and S data. Note that $M_{yx} < 0$ for both electrons and holes. The magnitudes of M_{yx} are much larger than for the 2D electron gases discussed in Section 5. This is because the carrier-phonon scattering rate is proportional to the density of states at the Fermi energy $D(E_F)$ and is consequently higher for the holes than for electrons, even though the phonon potentials are more effectively screened by the 2DHG.

To calculate M_{yx} for the hole gas we have adapted the model for the phonon drag magnetothermopower of 2DEGS presented in Section 5. The following modifications are made to account for the different mass, charge and crystal orientation of the 2DHG.

(a) The Landau levels for the 2DHG are $E_n = (n+½)\hbar Be/m^*_{\parallel}(B)$ where the lateral field-dependent effective mass $m^*_{\parallel}(B) = (0.22 + 0.0217 B) m^o$ is determined from cyclotron resonance data (Chamberlain, 1992).

(b) We assume that the system is isotropic under rotations in the (311) plane and therefore average the anisotropy factors A_ℓ and A_t for the carrier-phonon scattering (equation 55 of Price, 1981) so that they depend only on $q_\parallel$ and q_z.

(c) Because of the high in-plane mass the system is non-degenerate at T = 1.722 K. We therefore use temperature-dependent Thomas-Fermi screening calculated within the

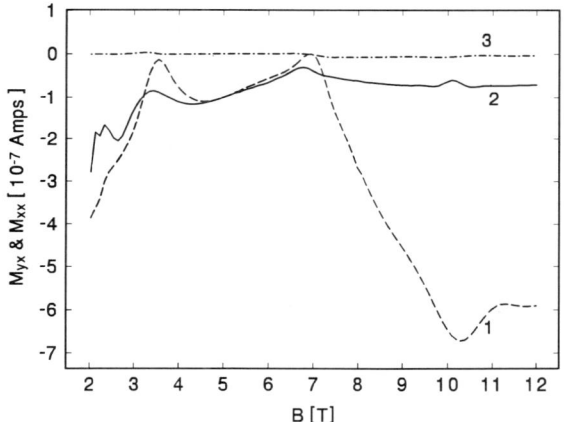

Figure 5. Experimental and theoretical values of M_{yx} together with empirical M_{xx} data for a 2DHG. Curve 1: experimental M_{yx} values. Curve 2: theoretical M_{yx} values. Curve 3: experimental M_{xx} values. For curve 2: $\lambda_p = 0.87$ mm.

Sommerfeld approximation (Ziman, 1989). The dielectric function for the 2DHG is then obtained from equation (8) of Qin et al, (1993) by replacing $D(E_F)$ by $D(E_F^o) + (\pi kT)^2 D^{(2)}(E_F^o)/6$ where E_F^o is the Fermi energy at absolute zero and $D^{(2)}$ denotes the second derivative of the density of states.

(d) The parameters used to calculate the magnetothermopower of the hole gases are identical to the electronic case (Fromhold et al 1993) except that $m_\perp^* = 0.4\ m^o$, $D = 5.64$ eV and $h_{14} = 0.73 \times 10^9$ V m^{-1}. Few estimates of D and h_{14} are available for hole gases. Our choice of D is slightly smaller than the value of 6.7 eV used by Walukiewicz (1985) to fit the measured mobility of a 2DHG in a GaAs/(AlGa)As heterojunction. In contrast to this work we find that piezoelectric scattering is not negligible and must be included in order to account for the magnitude and oscillatory amplitudes of the M_{yx} data.

We compare the theoretical and experimental results in Figures 5 to 7. In every case the sheet hole density $n_h = 1.63 \times 10^{15}$ m^{-2}, T = 1.722K and the Landau level width parameter is taken as $\gamma = 0.12$ B^{½} meV. This is comparable to the value $\gamma = 0.2$ B^{½} meV given by the self-consistent Born approximation for the sample mobility $\mu = 18.63$ m^2 V^{-1} s^{-1}. Theoretical values of M_{yx} are shown in curve 2 of figure 5. The $|M_{yx}|$ values oscillate in phase with the magnetically-modulated density of states at the Fermi energy. The deep minima in $|M_{yx}|$ are associated with integer filling factors and occur at B values for which $D(E_F)$ is also a minimum. In the quantum limit (B > 7T) the predicted M_{yx} values are lower than observed. This is probably because our Gaussian model line shape underestimates the density of states in the tails of the Landau levels in this high-field regime.

Curve 3 of figure 5 shows experimental values of the diagonal heat flux component M_{xx}. The ratio $|M_{xx}|/|M_{yx}|$ is close to zero and much smaller than for the 2DEGS discussed in Section 5. This is because there is so little disorder in the p-type heterostructures that scattering between adjacent localised Landau states, which produces the heat and charge fluxes along the E-direction, is very weak.

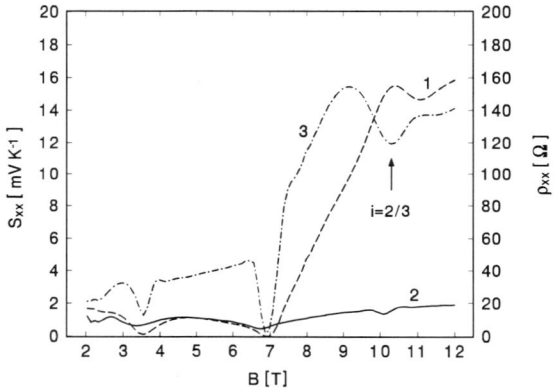

Figure 6. Experimental and theoretical values of S_{xx} together with empirical ρ_{xx} data for a 2DHG. Curve 1: experimental S_{xx} values. Curve 2: theoretical S_{xx} values. Curve 3: experimental ρ_{xx} values. For curve 2: and $\lambda_p = 0.87$ mm.

Experimental and theoretical values of the diagonal magnetothermopower $S_{xx}(B)$ are shown by curves 1 and 2 of figure 6. Since $|M_{xx}| \ll |M_{yx}|$ and $|\rho_{xx}| \ll |\rho_{yx}|$, S_{xx} is dominated by the second term on the right-hand side of equation (9a); just as for the 2DEGS. For B < 7 T the predicted S_{xx} curve accurately reproduces the experimental data. Discrepancies for B > 7 T reflect those in M_{yx} explained above. The deep minima in S_{xx} coincide with integer minima in both the ρ_{xx} data (curve 3 of figure 6) and $D(E_F)$. However the 2/3 fractional minimum in ρ_{xx}, marked by the arrow in figure 6, coincides with a clear *peak* in S_{xx}.

We suggest that this anti-phase behaviour is due the strongly correlated motion of holes in the fractional quantum Hall regime. The hole gas is 'stiffened' and responds less easily to the lattice polarisation produced by the phonons. The 2DHG therefore screens the phonon potentials less effectively close to the fractional minimum in ρ_{xx}. The hole-phonon scattering rate consequently increases here and produces a peak in S_{xx}. Away from the fractional regime, the carrier correlation is much weaker and the holes readily screen the phonon potentials.

Anti-phase oscillations in ρ_{xx} and $|S_{xx}|$ have been predicted for high mobility 2DEGS by Fromhold et al (1992) but the effect is much weaker for the electronic system and has not yet been observed. We stress that the phase shift predicted for the electron gases does not involve strong carrier correlation and originates from very high screening of the phonon potentials when $D(E_F)$ is maximal.

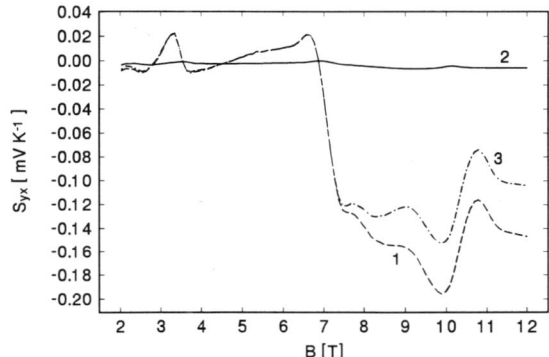

Figure 7. Comparison of experimental and theoretical S_{yx} values for a 2DHG. Curve 1: experimental S_{yx} values. Curve 2: theoretical S_{yx} values with $M_{xx} = 0$. Curve 3: S_{yx} calculated using empirical M_{xx} data. For curves 2 and 3: $\lambda_p = 0.87$ mm.

Experimental and theoretical S_{yx} curves are shown in figure 7. The $|S_{yx}|$ values are typically between one and two orders of magnitude smaller than the corresponding S_{xx} values. This contrasts with earlier measurements which show that $|S_{xx}| \sim |S_{yx}|$ for 2DEGS (Fletcher et al, 1986; Fromhold et al, 1992, 1993). Curve 2 of figure 7 is calculated from equation (9b) using experimental values for ρ_{xx} and ρ_{xy} with $M_{xx} = 0$. It is in very poor agreement with the experimental data: the predicted S_{yx} values are generally less than 5% of the observed values. If experimental M_{xx} data is substituted into equation (9b) the agreement increases dramatically as shown by curve 3 of figure 7. This clearly demonstrates that the first term dominates the right-hand side of equation (9b). We see that $|S_{yx}|$ is much smaller for 2DHGS than for 2DEGS simply because $|M_{xx}|$ is also smaller as shown in Figure 5.

7. CONCLUSION

The semiclassical formulae for the thermopower in the diffusive regime are well established and are valid for all dimensionalities. They give a good description of the quantum size effects observed in 2DEGS and the quantisation effects observed in quantum point contacts when B=0.

The semiclassical theory of the phonon drag contribution to the thermopower is also in good agreement with experimental data when B=0 in 2DEGS. When B≠0 the theoretical situation is less satisfactory. Attention has been focused on the behaviour predicted in quantising magnetic fields when an electric field **E** is turned on adiabatically to produce a phonon heat flux **Q** = M**E**. The theory predicts that $M_{xx} = 0$. This is a reasonable approximation in 2DHGS but is very wide of the mark in 2DEGS. In both cases the theory gives a poor account of S_{yx} data which is very sensitive to M_{xx}. On the other hand, the behaviour of S_{xx} is fairly well described by the theory in both 2DEGS and 2DHGS because S_{xx} is overwhelming dependent on M_{yx} for which the theoretical predictions are in reasonable accord with the data.

The most pressing problem in achieving a full understanding of phonon drag magnetothermopower in quantising magnetic fields is to calculate M_{xx}. What is required is a theory which goes beyond the assumption of adiabatic switching and yet remains simple enough to be useful in interpreting data for well characterised 2DEGS and 2DHGS.

REFERENCES

Barraclough, R.J., Fromhold, T.M., Rodgers, P.J., Gallagher, B.L., Jezierski, A., Henini, M., Butcher, P.N., and Hill, G., 1993, in preparation for *Phys. Rev. B*.

Beenakker, C.W.J. and van Houten, H., 1991, Quantum transport in semiconductor nanostructures in Solid State Physics Vol 44, Ehrenreich and Turnball, eds, Academic Press, New York.

Butcher, P.N., 1990, Thermal and electric transport formalism for electronic microstructures with many terminals, *J.Phys: Condens Matter* 2: 4869.

Butcher, P.N., 1992, Theory of electron transport in low dimensional structures, in Physics of Low -Dimensional Semiconductor Structures, P.N. Butcher, N.H. March and M.P. Tosi, eds, Plenum, New York.

Cantrell, D.G. and Butcher, P.N., 1985, Lifetime broadening of the subband structure in the thermopower of narrow channel systems, J. Phys. C: Solid State Phys. 18:6639.

Cantrell, D.G. and Butcher, P.N., 1987, A calculation of the phonon drag contribution to the thermopower of quasi-2D electrons coupled to 3D phonons *J. Phys. C: Solid State Phys.* 20: 1985, 1993.

Chamberlain, J.M., 1992, private communication.

Fletcher, R., Harris, J.J. and Foxon, C.T., 1991, The effect of second subband occupation on the thermopower of a high mobility GaAs-$Al_{0.33}$ $Ga_{0.33}$ As heterojunction, *Semicond. Sci. Technol.* 6: 54.

Fletcher, R., Maan, J.C., Ploog, K., and Weimann, G., 1986, Thermoelectric properties of GaAs-$Ga_{1-x}Al_x$As heterojunctions at high magnetic fields, *Phys. Rev. B* 33:7122.

Fromhold, T.M., Butcher, P.N., Qin, G., Mulimani, B.G., Oxley, J.P., and Gallagher, B.L., 1992, 180° phase shift of phonon drag magnetothermopower oscillations in high mobility 2DEGS, *Surf. Sci.* 263:183.

Fromhold, T.M., Butcher, P.N., Qin, G., Mulimani, B.G., Oxley, J.P., and Gallagher, B.L., 1993, Phonon drag magnetothermopower oscillations in GaAs/AlGaAs heterojunctions, submitted to *Phys. Rev. B*.

Gallagher, B.L. and Butcher, P.N., 1992, Classical transport and thermoelectric effects in low dimensional and mesoscopic semiconductor structures, in Handbook on Semiconductors Vol.1, P.T. Landsberg, ed., North Holland, Amsterdam

Gallagher, B.L., Oxley, J.P., Smith, M.J. and Butcher, P.N., 1990, The phonon drag and diffusion thermopower of a Si inversion layer, *J. Phys.: Condens. Matter* 2:775.

Guenault, A.M., 1971, A physical picture of phonon drag thermoelectric power, *J. Phys. F: Metal Phys.* 1:373.

Hayden, R.K., Maude, D.K., Eaves, L., Valadares, E.C., Henini, M., Sheard, F.W., Hughes, O.H., Portal, J.C., and Cury, L., 1991, Probing the hole dispersion curves of a quantum well using resonant magnetotunneling spectroscopy, *Phys. Rev. Lett.* 66:1749.

Kittel, C., 1976, Introduction to Solid State Physics, John Wiley, New York.

Kubakaddi, S.S., Butcher, P.N. and Mulimani, B.G., 1989, Phonon drag thermopower of a 2D electron gas in a quantising magnetic field, *Phys. Rev. B.* 40:1377.

Lyo, S.K. 1988, Low temperature phonon drag thermoelectric power in heterojunctions, *Phys. Rev. B* 38:6345

Lyo, S.K., 1989, Magnetothermopower oscillations of the phonon drag thermoelectric power in heterojunctions, *Phys. Rev. B* 40:6458

Molenkamp, C.W., Gravier, Th., van Houten, H., Buijk, O.J.A. and Mabesoone, M.A.A., 1992, Peltier coefficient and thermal conductance of a quantum point contact, *Phys. Rev. Lett.* 68:3765.

Price, P.J., 1981, Two-dimensional electron transport in semiconducting layers, 1, phonon scattering, *Ann. Phys. (N.Y.)* 133:217.

Qin, G., Fromhold, T.M., Butcher, P.N., Mulimani, B.G., Oxley, J.P., and Gallagher, B.L., 1993, Magnetothermopower in silicon MOSFETS, *J. Phys.: Condens. Matter* 5:1355.

Ruf. C., Brummell, M.A., Gmelin, E. and Ploog, K. 1989, The influence of subband structure on the thermopower of $GaAs-Al_xGa_{1-x}As$ heterostructures, *Superlat. and Microstruct.* 6:175.

Smith, M.J., and Butcher, P.N., 1989a, A calculation of the effect of screening on phonon-drag thermoelectric power in a Si MOSFET, *J. Phys.: Condens. Matter* 1:1261a

Smith, M.J. and Butcher, P.N., 1989b, Inelastic scattering and the temperature dependence of thermoelectric power in quasi-2D systems, *J. Phys. C: Condens. Matter* 1:4859.

van Houten, H., Molenkamp, L.W., Beenakker, C.W.J. and Foxon, C.T., 1992, Thermo-electric properties of quantum point contacts, *Semicond. Sci. Technol.* 7:B215.

van Wees, B.J., von Houten, H., Beenakker, C.W., Williamson, J.G., Kouwenhoven, L.P., van der Marel, D. and Foxon, C.T., 1988, Quantised conductance of point contacts in a two-dimensional electron gas, *Phys. Rev. Letts*, 60:848.

Walukiewicz, W., 1985, High mobility in modulation-doped heterostructures: GaAs-AlGaAs, *Phys. Rev. B* 31:5557.

Wang, W.I., Mendez, E.E., Iye, Y., Lee, B., Kim, M.H., and Stillman G.E., 1986, High mobility two-dimensional hole gas in an $Al_{0.26}Ga_{0.74}As$/GaAs heterojunction, *J. Appl. Phys.* 60:1834.

Wharam, D.A., Thornton, T.J., Newbury, R., Pepper, M., Ahmed, H., Frost, J.E.F., Hasko, D.G., Peacock, D.C., Ritchie, D.A. and Jones, G.A.C., 1988, One dimensional transport and the quantisation of the ballistic resistance, *J. Phys. C: Solid State Phys.* 21:L209.

Ziman, J.M., 1989, "Principles of the Theory of Solids", Cambridge University Press, Cambridge.

ENERGY RELAXATION VIA ACOUSTIC PHONONS IN 2D AND 1D ELECTRON SYSTEMS

A.Shik

Ioffe Physic-Technical Institute
194021 St-Petersburg, Russia

1. INTRODUCTION

The processes of momentum and energy relaxation play the central role in kinetic effects. The energy dependence of the momentum relaxation rate, usually described with the relaxation time τ_p, determines the temperature and concentration dependence of most linear kinetic coefficients. The energy relaxation processes (ESP) are of most importance in nonlinear effects such as hot electron phenomena. ERPs are usually described in terms the energy loss rate per one electron $Q(T)$ which is the average energy lost by an electron with effective temperature T_e exceeding the lattice temperature T.† The knowledge of $Q(T)$ allows us to determine most characteristics of hot electrons. For instance, in a system heated by stationary current, the energy balance equation

$$e\mu E = Q(T_e), \tag{1}$$

(μ is the electron mobility) determines the dependence of electron temperature T on the applied electric field E.

In the present paper we give a survey of the main regularities of the energy loss rate $Q(T_e)$ for electron systems of different dimensionality.

We restrict ourselves to the case of relatively low electron temperatures ($T \leq 30 - 50K$) where optical phonon emission is of little importance and ERPs are due to the interaction with acoustic phonons.

2. CONSERVATION LAWS

Let us discuss the qualitative character of the distribution of phonons emitted by an electron gas of different dimensionality with effective temperature T. It is governed by energy and momentum conservation.
First of all, the typical energy of emitted phonons cannot exceed the thermal

---

† The energy relaxation time τ_ε also widely used in many papers is the result of linearization of $Q(T)$ for low heating (i.e. for $T_e - T \leq T$) : $Q(T) \sim k(T_e - T)/\tau_\varepsilon$

energy of electrons:

$$\hbar s q \leq k_B T_e. \tag{2}$$

(s is the sound velocity). As to momentum conservation, the final conclusions depend on the dimensionality of the electron gas.

For a 3D electron system, the phonon wave vector **q** is unambiguously determined by the initial **p** and final **p'** electron momenta:

$$\mathbf{p} - \mathbf{p'} = \pm \hbar \mathbf{q}, \tag{3}$$

where the signs + and − correspond, respectively, to phonon emission and absorption. It results from (3), that q cannot noticeably exceed typical electron momenta (thermal or Fermi momenta):

$$\hbar \mathbf{q} \leq (m \quad \max(k_B T_e, E_F)), \tag{4}$$

where E_F is the Fermi energy.

For a 2D electron system lying in the xy-plane, strict momentum conservation (3) and, hence, the restriction (4) will take place only for x- and y-components. In z-direction there is no momentum conservation but only qualitative estimates based on the uncertainty principle. If a_z is the thickness of 2D electron layer, then

$$|q_z| \leq \pi/a_z. \tag{5}$$

For a 1D electron wire oriented along the y-axis, similar arguments give the following restrictions:

$$|q_x| \leq \pi/a_x; \quad \hbar q_y \leq (m\max\{kT_e, E_F\})^{1/2}; \quad |q_z| \leq \pi/a, \tag{6}$$

where a_x and a_z are the characteristic confinement widths along the corresponding axes.

3. PHONON EMISSION SPECTRA

Let us consider the above-mentioned restrictions for the momentum of emitted phonons in more detail and draw some conclusions about the phonon distribution function. Some different cases will be considered separately but T_e will be always assumed to exceed $ms^2/2k$ which for most semiconductors is less than 1 K.

3.1. Non-degenerate case

In this case the momentum-based restriction (4) is more severe than the energy-based one (2). Phonons emitted by a 3D electron gas have typical energy $\sim (mkT_e)^{1/2} s \gg k_B T$ which means that **electron-phonon scattering is quasi-elastic** [1].

For 2D electrons $|q_{x,y}| \sim (mk_B T_e)^{1/2}/\hbar$, as for the 3D case, but $|q_z|$ is much larger which means that **phonons are emitted almost normal to the plane of the 2D gas**. If $ms^2 \ll k_B T_e \ll \hbar s/a_z$ then $|q_z| \sim k_B T_e/\hbar s$ whereas for $k_B T_e \gg \hbar s/a_z$, $|q_z| \sim 1/a_z$. In the latter case the phonon energy $\hbar s q \ll k_B T_e$ and the scattering is again quasi-elastic. For $k_B T_e \ll \hbar s/a_z$ this is not the case [2].

The 1D case has much in common with the 2D one. In the direction of free electron motion (y-axis) the limiting value for **q** is determined by the electron momentum: $|q_y| \sim (mk_BT_e)^{1/2}/\hbar$ whereas $|q_{x,z}| \sim \min\{1/a_{x,z}; k_BT_e/\hbar s\} \gg |q_y|$. As a result, **phonons are emitted almost normally to the electron wire.**

Note that phonon energies for 2D and 1D electrons exceed the 3D value $(mk_BT_e)^{1/2}s$. Therefore, **low-dimensional systems are characterized by more effective energy relaxation.**

3.2. Degenerate case, low temperatures $[ms^2 \ll k_BT_e \ll (mE_F)^{1/2}s]$

Here the energy-based restriction (2) results in the of the phonon momentum being small compared to the electron Fermi momentum E_F. This means that **phonon emission is a small-angle scattering process.** The energy lost in any such event is $\Delta\varepsilon = (\hbar pq\cos\theta)/m$ where θ is the angle between electron and phonon momenta. Since $p \simeq (2mE_F)^{1/2} \equiv p_F$, $q \simeq k_BT_e/\hbar s$ and $\Delta\varepsilon \simeq k_BT_e$, we obtain

$$\cos\theta \simeq ms/p_F \ll 1. \tag{7}$$

So, **phonons are emitted almost normal to the direction of the electron momentum.** In a 1D system the phonon momentum may be in any direction in the plane xz normal to the wire axis [3]. In a 2D system the total emission is strongest normal to the plane xy containing the electrons since this is the only direction in which the total intensity is the sum of the emission intensities from individual electrons.

3.3. Degenerate case, high temperatures $[k_BT_e \gg (mE_F)^{1/2}s; \hbar s/a]$

In a 3D system the maximal **q** is determined by the Fermi momentum: $\hbar q \leq 2p_F$. In low-dimensional cases the q-components of the free motion direction are also restricted by $2p_F$ whereas in the other directions $|q_i| \sim 1/a_i$.† We see that for any dimensionality of the system $\Delta\varepsilon \sim \max\{p_Fs; \hbar s/a_i\} \ll k_BT_e$ and, hence, the scattering, as in the non-degenerate case, is **quasi-elastic.**

3.4. Degenerate case, intermediate situations

So far we have considered p_F and all the $\hbar/a_i$ to be of the same order of magnitude. In principle, other situations are also possible. For instance, in most quantum wire structures the confinement lengths a_x and a_z differ considerably and the condition $\hbar s/a_x \ll k_BT_e \ll \hbar s/a_z$ may also occur. The situation here recalls that of Sec.3.2 with the only difference that $|q_z|$ will be limited by $1/a_x$ rather than $k_BT_e/\hbar s$ [3].

For a low electron concentration the situation $p_Fs \ll k_BT_e \ll \hbar s/a_z$ may also occur. This intermediate case in the 2D gas is characterized again by **phonon emission almost normal to the plane of 2D gas** with $|q_x|, |q_y| \sim p_F/\hbar; |q_z| \sim k_BT_e/\hbar s$

---

† In the 1D case there are two types of electron-phonon scattering processes giving a comparable contribution to Q: weak scattering with $|q_y| \sim mk_BT_e/\hbar p_F \ll p_F/\hbar$ and back scattering with $q_y = -2p_F/\hbar$ [3].

[2]. In this case a single scattering event changes the electron momentum by $\sim p_F$ and the energy by $\sim k_B T_e$. As a result, **the momentum and energy relaxation times will be of the same order.** In the 1D case $|q_x| \sim \min(k_B T_e/\hbar s; 1/a_x)$; $|q_y| \leq 2p_F$ (see footnote †); $|q_z| \sim k_B T/\hbar s$.

4. QUANTITATIVE EXPRESSIONS FOR $Q(T_e)$

Quantitative formulae for $Q(T_e)$ can be obtained from the general

$$Q = -\frac{1}{N} \sum_{\mathbf{p},\mathbf{q},\mathbf{p}'} (\varepsilon_{\mathbf{p}'} - \varepsilon_{\mathbf{p}}) W_{\mathbf{p} \to \mathbf{p}'}(\mathbf{q}) f_{T_e}(1 - f_{T_e}) \frac{1 - \exp[\hbar s q(1/kT_e - 1/kT)]}{1 - \exp(-\hbar s q/k_B T)}. \quad (8)$$

where N is the total number of electrons, f_{T_e} is the Fermi function with the temperature T_e and T is the ambient lattice temperature. The transition probability

$$W_{\mathbf{p} \to \mathbf{p}'}(\mathbf{q}) = \frac{2\pi}{\hbar} |V_\mathbf{q}|^2 \delta(\varepsilon_p - \varepsilon_{p'} \mp \hbar s q) \delta_{\mathbf{p},\mathbf{p}' \pm \hbar \mathbf{q}}$$

is determined by the square of electron-phonon matrix element which can be written as [1]:

$$|V_\mathbf{q}|^2 = C q^\gamma. \quad (9)$$

For the deformation and piezoelectric potential γ is equal, to +1 and -1, respectively.

We shall be especially interested in the dependence of Q on the temperature T_e and Fermi energy E_F of the electrons. The results of

Sect. 3 allow us to predict qualitatively these dependencies. Consider, for example, the low-temperature 2D degenerate case (Sect. 3.2). Here the difference $(\varepsilon_{p'} - \varepsilon_p)$ is of order of the mean energy loss per collision $\Delta \varepsilon \sim k_B T_e$. The sum over $\mathbf{q}$ is proportional to the phase volume $\Delta q_x \Delta q_y \Delta q_z \sim (k_B T_e/\hbar s)^2 (m k_B T_e/\hbar p_F)$. The sums over $\mathbf{p}$ and $\mathbf{p}'$ are canceled by the δ-function and δ-symbol in the transition probability and the matrix element $|V_\mathbf{q}|^2$ gives an additional factor $(k_B T_e/\hbar s)^\gamma$. The factor N in the denominator is proportional to E_F and, eventually, $Q(T_e) \sim T^{\gamma+4}/E_F^{3/2}$. If we take into account phonon absorption as well as the emission, the dependence of Q on T_e and E_F can be expressed in the general form

$$Q(T_e) \sim (T_e^\alpha - T^\alpha) E_F^\beta, \quad (10)$$

where in our particular case $\alpha = \gamma + 4$; $\beta = -3/2$.

Here all restrictions for $\mathbf{q}$ are connected with temperature rather than with the confinement lengths. As a result, the formula (10) **is valid for any dimensionality of the system.**

Other cases considered in Sect. 3. can be treated in a similar way. The main qualitative results including the values of α and β in (10) are summarized in the table:

The exact quantitative formulae can be found in [1,4] (3D case), [2,5,6] (2D case) and [3] (1D case).

5. EFFECTS OF SCREENING

The expression (9) for the matrix element is valid for the unscreened piezoelectric potential. The screening suppresses processes with small q. Formally this is described by the additional factor $[q^2/(q^2+q_s^2)]^2$ for 3D and $[q_\parallel/(q_\parallel+q_s)]$ for the 2D case† where q_s is the screening length.

These effects are of importance at very low temperatures where a typical q appears to be considerably less than q_s. For a degenerate 3D gas $q_s^2 = 8m\sqrt{\varepsilon_B E_F}/\pi\hbar^2$ whereas in the 2D case $q_s = 2\sqrt{2m\varepsilon_B}/m$ (ε_B is the effective Bohr energy). Screening is essential at $k_B T_e \ll \hbar s q$ and in this temperature region the dependence $Q(T_e) \sim (T_e^3 - T^3)$ is to be replaced by $Q(T_e) \sim (T_e^7 - T^7)$ for the 3D case or $Q(T_e) \sim (T_e^5 - T^5)$ for the 2D case [5,7].

Table 1. ERP regularities for an electorn gas of different dimnensionality

	3 D	2 D	1 D
degenerate $k_B T_e \ll p_F s$; $k_B T_e \ll \hbar s/a$	$q \sim \frac{k_B T_e}{\hbar s}$; $\cos(\mathbf{p}\cdot\mathbf{q}) \sim ms/p_F \ll q$; $\Delta\varepsilon \sim k_B T_e$ $\alpha = \gamma + 4; \beta = -3/2$		
degenerate $p_F s \ll k_B T_e$; $k_B T_e \ll \hbar s/a$	$q \sim p_F/\hbar$ $\cos(\mathbf{p}\cdot\mathbf{q}) \sim ms/p_F \ll 1$ $\Delta\varepsilon \sim p_F s \ll k_B T_e$ $\alpha = 1; \beta = \gamma/2$	$q_{x,y} \sim p_F/\hbar \ll q_z$ $q_z \sim k_B T_e \hbar s$ $\Delta\varepsilon \sim k_B T_e$ $\alpha = \gamma + 3; \beta = -1$	$q_x \sim 1/a_x \ll q_z$ $q_y \sim p_F/\hbar \ll q_z$ $q_z \sim k_B T_e/\hbar s$ $\Delta\varepsilon \sim k_B T_e$ $\alpha = \gamma + 3; \beta = -3/2$
degenerate $p_F s \ll k_B T_e$; $\hbar s/a \ll k_b T_e$	$q \sim p_F/\hbar$ $\cos(\mathbf{p}\cdot\mathbf{q}) \sim ms/p_F \ll 1$ $\Delta\varepsilon \sim p_F s \ll k_B T_e$ $\alpha = 1; \beta = \gamma/2$	$q_{x,y} \sim p_F/\hbar \ll q_z$ $q_z \sim 1/a_z$ $\Delta\varepsilon \sim \hbar s/a_z \ll k_B T_e$ $\alpha = 1; \beta = -1$	$q_y \sim p_F \hbar \ll q_{x,z}$ $q_{x,z} \sim \hbar s/a_z$ $\Delta\varepsilon \sim \hbar s/a_z \ll k_B T_e$ $\alpha = 1; \beta = -3/2$
nondegenerate $\hbar s/a \ll k_B T$	$q \sim \sqrt{m k_B T_e}/\hbar$ $\Delta\varepsilon \sim \sqrt{m k_B T_e} \ll k_B T_e$ $\alpha = 1$	$q_{x,y} \sim \sqrt{m k_B T_e}/\hbar \ll q_z$ $q_z \sim \hbar s/a_z$ $\Delta\varepsilon \sim \hbar s/a_z \ll k_B T_e$ $\alpha = 1$	$q_y \sim \sqrt{m k_B T_e}/\hbar \ll q_{x,z}$ $q_{x,z} \sim \hbar s/a_{x,z}$ $\Delta\varepsilon \sim \hbar s/a_z \ll k_B T_e$ $\alpha = 1$

---

† In the 1D case, screening is very weak and does not change the scattering processes considerably.

6. INTERSUBBAND TRANSITIONS

Discussing ERP in low-dimensional systems, we have considered pure 1D or 2D cases where only one subband related to quantum size of the system is occupied. Very often, however, a so-called quasi-1D (or quasi-2D) situation with several occupied subbands is realized. We consider, as an example, a 2D electron gas with two occupied subbands. The net Q will consist of four terms: $Q = Q_{11} + Q_{22} + Q_{12} + Q_{21}$ where Q_{ij} describes ERP due to the electron transitions from $i - th$ into $j - th$ subband.

Intersubband transitions ($i \neq j$) are caused by phonons with relatively large wave vectors: $q \sim 1/a$. Therefore, **at low temperatures, $k_B T_e < \hbar s/a$, these processes are absent**. Nevertheless, even under these conditions $Q(T_e)$ can be different in samples with one or several occupied subbands [8]. This results from the fact that E_F, the distance between the Fermi energy and the subband edge, is different for different subbands. Therefore, processes in different subbands may correspond to different cases of Sec.3 and be described by different formulae for $Q(T_e)$.

Note that the higher subbands with lower electron concentration may even dominate the ERP since they may have less restrictions for the wave vector of emitted phonons (compare, e.g., Sec.3.2 and 3.4).

7. ENERGY RELAXATION IN STRONG MAGNETIC FIELD

Application of a strong magnetic field **H** normal to a 2D electron system transforms its energy spectrum into the series of discrete Landau levels: $\varepsilon = \hbar\omega_c(N + 1/2)$ (where $\omega_c = eH/mc$, $N = 0, 1, 2...$) causing the quantum Hall effect and other unusual electron properties. At first glance, this will suppress ERP dramatically since the probability of inter-Landau-level transitions becomes exponentially small $\sim exp(-\hbar\omega_c/k_B T)$.

However, this is not the case near the sample edges where current-carrying edge states exist [9]. These states can be considered as 1D channels and, hence, **the main regularities of ERP in 2D systems in strong magnetic field must be similar to those in 1D systems without field** [10].

If we place a 1D system in a magnetic field, the latter will increase the energy level separation and the effective mass along the wire axis, $m^\star$. At low temperature (Sec.3.2) this will increase ERP since in this case $Q \sim m^\star$ and make the emitted phonon distribution more isotropic. Eventually, however, $Q(H)$ reaches a maximum and then falls because the growth of $m^\star$ reduces the Fermi velocity below s which makes the phonon emission impossible [3]. The field-dependent subband separation also leads to oscillations in $Q(H)$.

8. EXPERIMENTAL METHODS OF MEASURING T_e

The most straightforward method of fitting our theoretical predictions to the experiment consists in comparing the electric field dependence of the electron temperature given by (1) with the corresponding experimental data. Most important here is the problem of the experimental measurement of the effective electron temperature T_e. Comparing the electric field dependence of this value at the lowest lattice temperature T with its T-dependence at low E (in the linear regime) one readily obtains the effective temperature T_e corresponding to any value of E.

The following T_e-sensitive effects are used to measure the $T_e(E)$ dependence:
a) the amplitude of the Shubnikov oscillations;
b) the width of the quantum Hall plateaus;
c) the intensity of infrared radiation from hot electrons;
d) the quantum interference effects: the logarithmic temperature dependence of the conductivity and negative magnetoresistance. To use the last effects, we must be sure that the phase relaxation time τ_ϕ playing the central role in the theory of weak localisation is connected with electron-electron rather than electron-phonon interactions since only in this case the conductivity depends on T_e but not on the lattice temperature T.

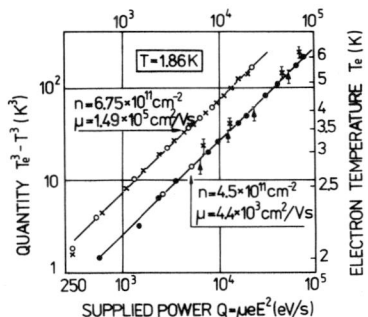

Fig. 1. Experimental dependence of Q on the electron temperature T_e for two GaAs heterostructures. The experimental points are obtained from Shubnikov-de Haas oscillations ($\times$), logarithmic temperature dependence of conductivity ($\bullet$), negative magnetoresistance ($\triangle$) and quantum Hall effect ($\circ$).

9. EXPERIMENTAL DATA FOR $Q(T_e)$

Fig.1 shows the curves $Q(T_e)$ obtained with the help of (1) from $T_e(E)$ dependencies measured by three different methods in a 2D electron gas in GaAs/AlGaAs heterostructures. One can see that $Q(T_e) \sim (T_e^3 - T^3)$. Additional measurements at different electron gas density n_s have shown that $Q \sim E_F^{-3/2}$ [8,11]. Comparing these facts with the Table 1, we conclude that **energy relaxation in this temperature range is caused by small-angle acoustic phonon emission due to piezoelectric interaction.**

At higher temperatures the contribution of deformation interaction becomes significant [4] and, finally, at $T_e > 50K$ optical phonon emission becomes the dominant ERP mechanism [12]. At very low temperatures, $T_e < 1K$, the dependence $Q \sim (T_e^5 - T^5)$ has been observed [13], in agreement with the predictions of Sect. 5.

Note that Fig. 1 **demonstrates the same $Q(T_e)$ dependence for both weak ($\triangle$) and strong ($\times$, o) magnetic fields as well as at H=0 (for ($\bullet$))**. This is in agreement with the conclusion of Sect. 7 on the similarity of $Q(T_e)$ in a strongly magnetized 2D gas and in a 1D gas with $H = 0$. The latter, in turn, at low T_e must be proportional to the same factor $(T_e^3 - T^3)$ as for the 2D gas with $H = 0$ (see Sect. 4).

The investigation of $Q(T_e)$ in 2D GaAs-structures with different electron density n_s show a large rise in Q near $n_s \simeq 7 \times 10^{11} cm^{-2}$ due to the second subband occupation [14]. At the same time the exponent α in (10) decreased abruptly from 3 to 2 [8,11]. This is explained by the onset of the Q_{22} term which, due to the small value of E_F in the second subband, corresponds to the case 3.3 and, according to Table 1, must have $\alpha = 1$, contrary to the $Q_{11} \sim (T_e^3 - T^3)$ in this temperature region. Hence, the sum $Q_{11} + Q_{22}$ (the temperature is low enough for Q_{12} and Q_{21} to be negligibly small) must have an intermediate effective lying between 1 and 3.

REFERENCES

[1] V.F. Gantmakher and Y.B. Levinson. Carrier Scattering in Metals and Semiconductors, North-Holland, 1987.
[2] V. Karpus, *Energy and momentum relaxation of 2D carriers by the interaction with deformation acoustic phonons*, Sov. Phys. Semicond. **20**, 6 (1986).
[3] A. Shik and L.J. Challis, Phonon emission from a quasi-one dimensional electron system in zero and quantizing magnetic fields, *Phys. Rev. B* (1993) (in press).
[4] Sh.M. Kogan, Theory of hot electrons in semiconductors, *Sov. Phys. Sol. State* **4**, 1813 (1962).
[5] V. Karpus, Energy relaxation of 2D electrons for piezoelectric scattering, *Sov. Phys. Semicond.* **22**, 268 (1988).
[6] Y.H. Xie et. al., Power loss by 2D holes in coherently strained GeSi/Si heterostrutures, *Appl. Phys. Lett.* **49**, 283 (1986).
[7] P.J. Price, Hot electron in a GaAs heterolayer at low temperature, *J. Appl. Phys.* **53**, 6863 (1982).
[8] A.M. Kreshchuk et. al., Role of higher subbands in relaxation of the energy of a two-dimensional electron gas, *Sov. Phys. Semicond.* **22**, 377 (1988).
[9] M. Büttiker, Absence of backscattering in the quantum Hall effect in multiprobe conductors, *Phys. Rev. B* **38**, 9375 (1988).
[10] A.Ya. Shik, Edge energy relaxation in the quantum Hall effect, *Sov. Phys. Semicond.* **26**, 481 (1992).
[11] A.M. Kreschuk et. al., Energy relaxation of 2D electron gas at an AlGaAs/GaAs heterojunction at helium temperatures, *Solid State Comm.* **65**, 1189 (1988).
[12] P. Hawker et. al.. Changeover from acoustic to optic phonon emission by a hot 2D electron gas in the GaAs/AlGaAs heterojunction, *Semicond. Sci. Technol.* **7**, B29 (1992).
[13] A.K.M. Wennberg et. al., Electron heating in a multiple-quantum- well structure below 1K, *Phys. Rev. B* **34**, 4409 (1986).

[14] Y. Ma *et. al.*, Energy-loss rates of 2D electrons at a GaAs/AlGaAs interface, *Phys. Rev. B* **43**, 9033 (1991).

A THEORY OF THE SUPPRESSION OF THE ELECTRON-PHONON INTERACTION

M.V. Entin, O.V. Kibis

Institute of Semiconductor Physics
Siberian Branch of Russian Academy of Sciences
13 Lavrent'eva, 630090 Novosibirsk, Russia
E-mail: entin@isph.nsk.su

ABSTRACT

The possibilities for decreasing the electron-phonon interaction are discussed. Among them there are the control of the lattice constants, mixing of electron states with different signs of interaction by means of fields, quantization etc., and the screening of the long-range electric fields responsible for the polar interaction.

INTRODUCTION

The purpose of the present study is to find ways to control such fundamental parameters of a material as the electron-phonon coupling constants. More precisely we want to suppress them.

It is well known that many parameters of solid-state electronic devices are limited by relaxation processes. Among them there are the operation rate and losses of the device. The operation rate is usually determined by the ratio of the carrier drift velocity to the length of the device and so it is directly connected with the mobility of carriers.

Some relaxation processes, like impurity scattering, can be reduced by purification of the crystal. Another way, which solves the contradiction between the necessity of having a finite number of carriers in the semiconductor and therefore implying the same number of impurities on the one hand and a vanishing number of scatterers on the other hand is a selective doping method. This method is now widely using in HEMT devices. This goal is achieved by confining the electrons through space quantization and separating them from impurities by means of spacers.

The problem is that devices are to operate at room temperature and in strong electric fields when the electrons are hot. In such conditions the mobility is determined by electron-phonon scattering rather than impurity scattering. Thus the suppression of the integral mobility is not too effective. One can get rid of electron-phonon scattering only by minimizing the number of phonons by lowering the temperature, the electron mass and the electron-phonon interaction. We shall study the last way.

Generally speaking, there exist two kinds of electron-phonon interaction, – short and long range. The first appears due to the shift of a band extremum by the crystal deformation u_{ij}. In the case of degenerate minima, band splitting also appears. This is determined by the shift of the initial energy level by the deformation. The second way is the action of an electric field resulting from a deformation or sublattice shift. It is typical for polar crystals.

The position of the bottom of any band is a constant functionally depending on the deformation and other parameters. Expanding it in powers of u_{ij} and retaining only the first terms we get

$$E_0 = E_{00} + \Lambda_{ij} u_{ij}. \tag{1}$$

Generally, the deformation constants Λ_{ij} are functions of the same parameters λ_k, e.g. the static deformation, fields, content ratio for compounds and space quantization. To suppress the electron-phonon interaction one must ensure $\Lambda_{ij} = 0$. This gives $d(d+1)/2$ conditions, where d is the dimension of the space. Broadly speaking, this can be satisfied by the same number of parameters. The parameters may or may not change the symmetry of the crystal. In the first case the number of conditions coincides with the number of independent components of Λ_{ij}, determined by the symmetry of the crystal. Otherwise a change of the parameters may decrease the symmetry and in general the number of necessary parameters is $d(d+1)/2$.

MOLECULAR MODEL

To argue the possibility of a vanishing deformation potential we shall start from the molecular model of a crystal. It is well known that the positions of band extrema at symmetric points can be described as molecular terms. Each atomic level yields some extremum of energy. The extrema are shifted by the crystalline potential and the overlapping of states.

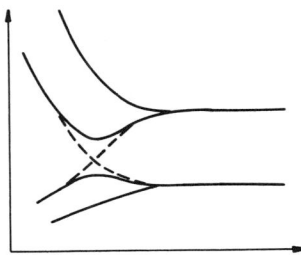

Fig. 1. Electronic terms of a system with initially degenerate levels. The splitting of the upper and lower states leads to quasicrossing of central states (dashed lines). The small splitting of crossing terms (solid lines) results in extrema.

Let us start from two atomic states of the same symmetry, which are not very far from each other. In a molecule composed of these two atoms both of these states will be split into two states higher and lower than the initial energy. If the shift is greater than the distance between the initial states, the terms should cross. This is shown in Fig. 1 by dashed lines. In fact, it is not a crossing which is forbidden by the Landau theorem, but a pseudocrossing (solid lines), the small interaction of levels will split them. This unfailingly results in the appearance of extrema on the curves $E_{2,3}(l)$. The band edges for the crystal should behave like these terms. This proves the possibility of suppression of electron-phonon interaction.

DEFORMATION POTENTIAL AT THE Γ_1 POINT

Consider the Γ_1 representation of a cubic semiconductor like $A_3 B_5$, for example GaAs. Its energy spectrum is connected with the formula of the tight binding model [4]:

$$E(\Gamma_{1c,v}) = \frac{(A+B)}{2} \pm \left[\left(\frac{A-B}{2}\right)^2 + P_5^2\right]^{1/2}, \tag{2}$$

where $A = P_1 + 3P_{18}$, $B = P_2 + 3P_{19}$,

$$P_1 = E_{ss}(000)0, \quad P_2 = E_{ss}(000)1, \quad P_5 = 4E_{ss}(1/2,1/2,1/2)$$
$$P_{18} = 4E_{ss}(110)0, \quad P_{19} = 4E_{ss}(110)1.$$

The subscript s means an atomic orbital, the numbers in brackets denote the lattice vector, the numbers 0 and 1 out of the brackets correspond to anions and cations respectively. The quantities $E_{ss}(000)$ are the sums of initial atomic state energies $E_\pm$ and their electrostatic shifts in the presence of an opposite ion $\pm\alpha e^*/d$, where α is the Madelung constant, e^* is the effective charge of the ion, $d = a\sqrt{3}/2$ and a is the lattice constant.

According to [5], we can consider $P_5 = 1/a^2 \eta_{ss\sigma}$ with $\eta_{ss\sigma}$=const.

The strong Coulomb intra-atomic repulsion leads to the difference of $E_{ss}(000)$ for states with an additional electron (the bottom of the conduction band) and a hole (lower lying minimum). The sign of the shift depends upon what state we want to study. The lowest state is a binding one. It corresponds to an electron being closer to an anion than to a cation. $E_{ss}(000)0$ should have a negative shift, and $E_{ss}(000)1$ a positive one. The upper state is an antibonding one. The electron is closer to a cation, so signs should be opposite. We shall omit P_{18} and P_{19} in (2). One can see the lower level has a minimum as a function of a at

$$a = 6\left(\frac{\eta_{ss\sigma}^2}{\Delta E e e^*}\right)^{1/3}, \tag{3}$$

Numerically we find $a = 2.645$ Å, which is 10% larger than the equilibrium value 2.45 Å.

Of course this is only a rough estimate. Nevertheless the inclusion of P_{18} and P_{19} results only in a small change of this value. We see that for a conduction band suppression is impossible, at least in this model of a cubic semiconductor.

Similar to eqn. (2) one can find the same expression for the valence band maximum:

$$E(\Gamma_{15c,v}) = \frac{(C+D)}{2} \pm \left[\left(\frac{C-D}{2}\right)^2 + P_8^2\right]^{1/2}, \tag{4}$$

where $C = P_3 + 2P_{14} + P_{10}$, $D = P_4 + 2P_{15} + P_{11}$,

$$P_3 = E_{xx}(000)0, \quad P_4 = E_{xx}(000)1, \quad P_5 = 4E_{xx}(1/2,1/2,1/2)$$
$$P_{14} = 4E_{xx}(110)0, \quad P_{15} = 4E_{xx}(110)1,$$
$$P_{10} = 4E_{xx}(011)0, \quad P_{11} = 4E_{xx}(011)1.$$

The same considerations show that both $E(\Gamma_{15c})$ and $E(\Gamma_{15v})$ have no extrema at least for the parameterization of P_5 chosen.

Our conclusions are true for the Γ point of zinc blend semiconductors only. We hope that for some other situations the top of the valence band and the bottom of the conduction band will have such extrema.

THE SUPPRESSION, CAUSED BY MIXING OF STATES

We can obtain another proof if we consider the deformation potential of a state which appears after the mixing of two edge states with the same symmetry by any mixing field.

Let us consider 2 states of the same symmetry, e.g. Γ_1, E_c and E_v and a deformation u_{ii} which shifts them by $\Lambda_c u_{ii}$ and $\Lambda_v u_{ii}$, respectively, We shall assume that these states are mixed by some mixing field Δ, as a first step leaving the type of field unspecified. The Hamiltonian is

$$\mathcal{H} = \begin{vmatrix} E^c + \Lambda_c u_{ii} & \Delta \\ \Delta & E_v + \Lambda_c u_{ii} \end{vmatrix}.$$

Expanding the bottom energy in powers of the small deformation one obtains:

$$\Lambda_{\text{eff}} = \alpha \Lambda_c + (1-\alpha)\Lambda_v,$$

where $\alpha = \frac{1}{2}\left[1 \pm (E_c - E_v)/\sqrt{(E_c - E_v)^2 + 4\Delta^2}\right]$, the signs correspond to the mixed states $\frac{1}{2}\left[(E_c + E_v) \pm \sqrt{(E_c - E_v)^2 + \Delta^2}\right]$, The values of the effective deformation constants Λ_{eff} are within the range $\Lambda_c \leq \Lambda_{\text{eff}} \leq \Lambda_v$ or $\Lambda_c \geq \Lambda_{\text{eff}} \geq \Lambda_v$.

If Λ_c and Λ_v have different signs, a zero value of Λ_{eff} appears to be possible. The field Δ should be high enough, generally speaking of the same order as the band gap $E_c - E_v$. It should not change the symmetry of the crystal. These conditions are fulfilled if the mixing is induced by the addition of some contents of atoms as in the continuously mixed crystal $Ga_x Al_{1-x} As$.

SUPPRESSION OF THE PIEZOPOTENTIAL.

To suppress the piezopotential we can screen the electric field. First it can be done by screening by mobile charges. The second way is to change the structure of the crystal in such way that the directions of electrical fields produced by long-wave phonons alternate even in the infinite wave length limit.

The piezopotential originates from the polarization of the medium by a deformation: $P_i = \chi_{ikl} u_{kl}$. Substituting it into the Laplace equation and the Hamiltonian we find the potential $\phi = 4\pi \chi_{ikl} k_i k_k u_l / \varepsilon k^2$ and $H = \{\Sigma_k c_k^+ a_{p-k}^+ a_p 4\pi \chi_{ikl} k_i k_k e_l / (\varepsilon k^2 2\rho \omega_k)$ + c.c., where e_l and ω_k are the polarization and the frequency of a phonon respectively and ε is the dielectric constant.

If we have an electric charge distributed with wave vector $\mathbf{k}$ the action of mobile carriers can be described by the factor $1/\varepsilon(\omega, \mathbf{k})$. One can distribute carriers in such way that they will be isolated from the conduction band but will be mobile enough to screen fields. For example we can make a 1D periodic structure with isolating layers, conducting layers, which are intercalated by metallic screens (Fig. 2). In the limit of small ω and $\mathbf{k}$, $\varepsilon(\omega, \mathbf{k})/\varepsilon = (k^2 + \kappa^2)/k^2$, where κ is the inverse Debye radius. In the case of thin layers with one level of transversal quantization κ is collecting from the screening of the individual layers [7]. We found $\kappa^2 = 4me^2/(\hbar^2 \varepsilon l)$, where m is electron mass, and l is the distance between layers. This reduces the piezopotential interaction by a factor $\varepsilon/\varepsilon(\omega, \mathbf{k})$.

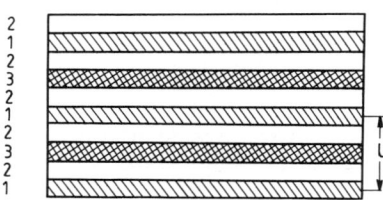

Fig. 2. The periodic structure for screening of the piezopotential interaction. 1) semiconductor active layers, 2) isolating layers, and 3) metallic screens.

Another problem is a 2D layer surrounded by two 2D layers of thickness h. We will also assume that only the central layer is conducting and the role of the side layers is to screen fields. A simple solution gives the corrections to the 2D Fourier transform of the potential:

$$\frac{\varepsilon}{\varepsilon(\omega, K)} = \left[1 - \frac{e^{-Kh}(u + 2K)}{(u + K)\cosh(Kh) + K\sinh(Kh)}\right]. \quad (5)$$

where $u = 4me^2/(\hbar\varepsilon)$, K is the plane Fourier component. Like the previous case, the factor $\varepsilon/\varepsilon(\omega, K)$ gives the change of the piezopotential. The only difference is that K is the plane component of **k**.

Both cases result in the removal of the divergence of the potential for small k. It is this divergence which gives the general part of polar scattering without screening. The static case we considered here is fulfilled if $\omega \ll kv_F$ with the Fermi velocity in a screening layer $v_F \cong 10^7 cm/s$.

RESIDUAL TWO-PHONON SCATTERING

If we can suppress the first order e-ph interaction, the other terms in the Hamiltonian will still remain. Continuing the expansion (1), we have

$$E_0 = E_{00} + \Lambda_{ij}u_{ij} + \Lambda^{(2)}_{ijkl}u_{ij}u_{kl} + \Lambda^{(3)}_{ijk}\partial_k u_{ij} + \Lambda^{(4)}_{ijkl}\partial_l\partial_k u_{ij} + \ldots \quad (6)$$

The second term gives the nonlinear e-ph interaction, the third and the fourth give nonlocal interactions. We shall estimate the scattering conditioned by the second term. There are 3 two-phonon processes: absorption, scattering and emission (Fig. 3).

The probabilities for them are given in first order perturbation theory using the second term in (6). We shall consider the case of a simple spherical central minimum with nondegenerate electrons. The temperature is assumed to be less than the Debye temperature. For the emission rate we find

$$\frac{1}{\tau_e} = 1.35 \times 10^{-6}\frac{\Lambda_2^2 p^{11}}{\pi^3 \rho^2 s^7 \hbar^7 m^4}, \quad (7)$$

where ρ is the density of the crystal, s is the phase velocity of longitudinal acoustic phonons and the energy is assumed to be much greater than temperature T. The same quantity we find for absorption and scattering rates while for the emission rate for energy $p^2/2m \sim T$, p^{11}/m^4 is replaced by $T^5 pm$.

Note that these quantities are much smaller than the first order scattering rate. This means that the suppression is sufficient although this is not an exact but a perturbative result.

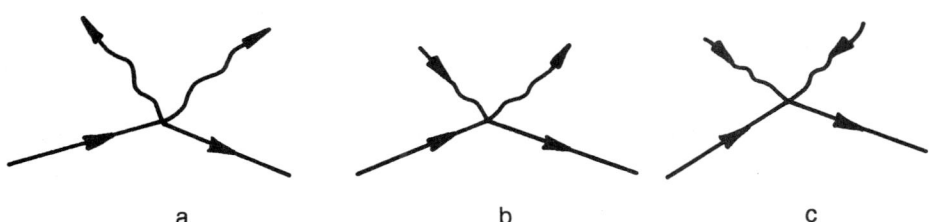

Fig. 3. Diagrams for the second order electron-phonon processes. a) two-phonon emission, b) scattering, and c) absorption. The wavy lines correspond to phonons.

SUPPRESSION, CAUSED BY MIXING OF BANDS IN MAGNETIC FIELD

The e-ph interaction constants for different bands differ in magnitude and sign. If one proceeds to the semiconductor by outer fields, this will mix the states of various bands. The resulting constants should be also some mixture of the unperturbed ones. This gives the hope of compensating them by mixing different bands using a static field. This idea was exploited by the authors in [3].

We shall consider a crystal in a magnetic field, subjected to a static deformation. In the general case this gives us 9 independent parameters. Our purpose is to suppress the 6 independent constants corresponding to the lowest Landau subband of the conductivity band. Of course we consider a narrow-gap semiconductor to do the action of deformation and field comparable with the band gap.

We used the Cohn-Luttinger 3-band Hamiltonian for cubic conduction and valence bands [8] with corrections caused by deformation (Table 1) for simplified case of magnetic field $\mathbf{H} \parallel (001)$.

Table 1

	1/2, 1/2	3/2, 3/2	3/2, -1/2	3/2, 1/2	3/2, -3/2	1/2, -1/2
1/2, 1/2	F	L	M	N	O	O
3/2, 3/2	L*	K	I	Q	0	0
3/2, -1/2	M*	I*	T	0	-Q	N
3/2, 1/2	N*	Q*	0	T	I	-M
3/2, -3/2	0	0	-Q*	I*	K	-L
1/2, -1/2	0	0	N*	-M*	-L*	F

All quantities are defined by the following formulae

$$\bar{F} = E_0 J + E_g + a_6(2u_\perp + u_\parallel) \pm E_0,$$
$$\bar{K} = E_0 J + a_8(2u_\perp + u_\parallel) + b(u_\perp - u_\parallel) \pm E_0,$$
$$\bar{T} = E_0 J + a_8(2u_\perp + u_\parallel) - b(u_\perp - u_\parallel) \mp E_0,$$
$$\bar{L} = P(1 - 2u_\perp)a/a_H \quad \bar{M} = P(1 - 2u_\perp)a^+/a_H\sqrt{3},$$
$$\bar{I} = \bar{N} = \bar{Q} = 0,$$
$$E_0 = \frac{1}{2}m_0 a_H^2, \quad \tilde{u}_{ii} = \tilde{u}_{xx} + \tilde{u}_{yy} + \tilde{u}_{zz},$$

$$\tilde{F} = a_6 \tilde{u}_{ii},$$
$$\tilde{K} = b/2(\tilde{u}_{xx} + \tilde{u}_{yy} - 2\tilde{u}_{zz}) + a_8 \tilde{u}_{ii},$$
$$\tilde{T} = -b/2(\tilde{u}_{xx} + \tilde{u}_{yy} - 2\tilde{u}_{zz}) + a_8 \tilde{u}_{ii},$$
$$\tilde{L} = -P/a_H[(\tilde{u}_{xx} + i\tilde{u}_{yx})(a + a^+) + (\tilde{u}_{yy} - i\tilde{u}_{xy})(a - a^+)],$$
$$\tilde{M} = -P/(a_H\sqrt{3})[\tilde{u}_{xx} + \tilde{u}_{yy})a^+ + (\tilde{u}_{xx} - \tilde{u}_{yy})a - 2i\tilde{u}_{xy}a)],$$
$$\tilde{N} = 2iP/(a_H\sqrt{3})[\tilde{u}_{zx} + \tilde{u}_{yy})(a^+ + a) - i\tilde{u}_{zy}(a - a^+)],$$
$$\tilde{I} = \sqrt{3}\frac{b}{2}(\tilde{u}_{xx} + \tilde{u}_{yy}) - id\tilde{u}_{xy},$$
$$\tilde{Q} = -d(\tilde{u}_{yz} + i\tilde{u}_{xz}),$$

$$\mathcal{H} = \bar{H} + \tilde{H}, \qquad u_{ij} = \bar{u}_{ij} + \tilde{u}_{ij}. \tag{8}$$

The deformation is separated into two parts, the predetermined one, considered as a control parameter, $\bar{u}_{ij}$, and an arbitrary deformation of the phonon field $\tilde{u}_{ij}$. The corresponding separation is done in the Hamiltonian. Its components are defined Table 1 and are denoted by a dash or a tilde, respectively. The $\pm$ signs in the matrix elements correspond to the diagonal elements of the upper and lower blocks of $\mathcal{H}$. We shall deal with the static deformation $u_{xx} = u_{yy} \equiv u_\perp, u_{zz} \equiv u_\parallel$. All other components are assumed to vanish. The quantity $a_H = (\hbar c/eH)^{1/2}$ is the magnetic length, the creation and destruction operators a and a^+ satisfy the equalities:

$$a^+\phi_n = \sqrt{n+1}\phi_{n+1}, \qquad a\phi_n = \sqrt{n}\phi_{n-1}, \tag{9}$$

$J = 2a^+a + 1$, ϕ_n is the wave function of the Landau subband with number n for a free electron. We shall neglect the quantities $\tilde{L}, \tilde{M}, \tilde{N}$ by use of the small parameter a_0/a_H, where a_0 is the lattice period.

The eigenfunctions of the Hamiltonian are

$$\psi_{1n} = \begin{vmatrix} 0 \\ 0 \\ 0 \\ \phi_{n+2} \\ \phi_n \\ \phi_{n+1} \end{vmatrix}, \qquad \psi_{2n} = \begin{vmatrix} A_{2n}\phi_{n+1} \\ B_{2n}\phi_{n+2} \\ C_{2n}\phi_{n+1} \\ 0 \\ 0 \\ 0 \end{vmatrix}, \tag{10}$$

where the constants satisfy the algebraic equations following from the effective mass equations:

$$\bar{\mathcal{H}}\psi_{1n} = \varepsilon_{1n}\psi_{1n}. \tag{11}$$

Considering $\tilde{\mathcal{H}}$ as a perturbation we find the shift of energy to the first order in the deformation,

$$\delta\varepsilon_{1n} = (\psi_{1n}\tilde{\mathcal{H}}\psi_{1n}) = \Lambda_{ij}^{1n}\tilde{u}_{ij}, \tag{12}$$

where $\Lambda_{xy}^{1n} = \Lambda_{xz}^{1n} = \Lambda_{zy}^{1n} = 0$.

We can suppress the interaction with phonons for one of the electron states at a time. The most interesting is the suppression of coupling constant for the lowest state which will be populated at low temperatures. This means that we should consider the zero momentum state. The tensor Λ_{ij}^{1n} determines the change of energy in the linear approximation for the deformation. It represents the required effective constants of the deformation potential for the electronic state ε_{1n}. Demanding $\Lambda_{ij}^{1n} = 0$, we obtain the system:

$$A_{1n}^2 a_6 + B_{1n}^2(a_8 + \frac{b}{2}) + C_{1n}^2(a_8 - \frac{b}{2}) = 0,$$
$$A_{1n}^2 a_6 + B_{1n}^2(a_8 - b) + C_{1n}^2(a_8 + b) = 0, \tag{13}$$

with the solutions

$$A_{1n} = (-1)^m \sqrt{2(\frac{-a_8}{a_1})} C_{1n}, \qquad B_{1n} = (-1)^l C_{1n}, \tag{14}$$

where $m, n = 1, 2$. This determines 4 wave functions ψ_{1n}, which correspond to zero Λ_{ij}^{1n}. The wave function should satisfy Eq. (11). This gives us 4 sets of static deformation with vanishing deformation potential and corresponding positions of Landau subband minima ε_{1n}

$$u_\| = -r - (-1)^m P \frac{[-a_8/a_6]^{1/2}}{3\sqrt{2}a_H} \left\{ \left[(-1)^l \sqrt{(n+1)} + \sqrt{\frac{(n+2)}{3}} \right] \frac{(a_6+a_8)}{a_8(a_6-a_8)} \right.$$
$$\left. + (-1)^l \left(\frac{2\sqrt{n+1}}{b} \right) - \frac{2\sqrt{n+2}}{b\sqrt{3}} \right\} + \mathcal{O}\left(\frac{\sqrt{(n+1)}a_{cr}}{a_H} \right), \quad (15)$$

$$u_\perp = -r - (-1)^m P \frac{[-a_8/a_6]^{1/2}}{3\sqrt{2}a_H} \left\{ \left[(-1)^l \sqrt{(n+1)} + \sqrt{\frac{(n+2)}{3}} \right] \frac{(a_6+a_8)}{a_8(a_6-a_8)} \right.$$
$$\left. - (-1)^l \left(\frac{2\sqrt{n+1}}{b} \right) + \frac{2\sqrt{n+2}}{b\sqrt{3}} \right\} + \mathcal{O}\left(\frac{\sqrt{(n+1)}a_{cr}}{a_H} \right), \quad (16)$$

$$\varepsilon_{1n} = -3ra_8 - (-1)^m \frac{P\sqrt{2}}{a_H(a_6-a_8)\sqrt{-a_8/a_6}} \left[(-1)^l a_6\sqrt{n+1} + a_1 \sqrt{\frac{n+2}{3}} \right]$$
$$+ \mathcal{O}\left(\frac{\sqrt{n+1}a_{cr}}{a_H} \right), \quad (17)$$

with $r = \frac{E_g}{3(a_6-a_8)}$.

An analogous result can be obtained for ψ_{2n}. The bottom of the lowest subband of the conduction band corresponds to $n = 0, l = 2, m = 1$, so these numbers should be substituted into (17).

The deformation (15), (16) eliminates the short range interaction with longitudinal acoustic phonons with momenta $q \ll 1/a_H$. The scattering of electrons on these phonons dominates in the ultra-quantum limit for small temperature $T \ll s/a_H$.

CONCLUSIONS

We see that a situation, in which the electron-phonon interaction constants are suppressed is not impossible. Any kind of interaction can be destroyed in some way. It is a pity that we could not find a way to do this in the most interesting case of high mobility conduction band electrons without magnetic field. The suppression we found is a relative reduction of the interaction only. Nevertheless the partial success inspires us to search for a way reduce in absolute magnitude the electron-phonon interaction.

Acknowledgement: This work was supported in part by the G. Soros Foundation.

References

1. O.V. Kibis, M.V. Entin: in "Reports of 2 International Symposium on surface waves in solids and layered structures", Varna, 1989.
2. O.V. Kibis, M.V. Entin: in 14 Soviet Union conference in acoustoelectronics and physical acoustic of solids, Kishinev, 1989.
3. O.V. Kibis, M.V. Entin, submitted to JETP, (1992).
4. D.N. Talwar, C.S. Ting. Phys. Rev. B **25**, 2660 (1982).
5. W.A. Harrison. Phys. Rev. B **31**, 2121 (1985).
6. Li Yang, P.J. Lin-Chung. J. Phys. Chem. Solids. **46**, 241.
7. A.V. Chaplik, M.V. Entin. Soviet Physics JETP, **61**, 2496 (1971).
8. G.L. Bir, G.E. Picus, *Symmetry and Strain-Induced Effects in Semiconductors*. Wiley, New York, 1974.

CARRIER CAPTURE BY QUANTUM WELLS VIA 3D→2D AND 2D→2D CHANNELS

V. Karpus
Semiconductor Physics Institute
Goštauto 11
2600 Vilnius
Lithuania

INTRODUCTION

Efficient charge carrier collection in quantum wells has been observed in a number of experiments (see, e.g. references in the review of Ridley 1991). In theoretical studies carrier capture has been usually investigated (Kozyrev and Shik, 1985; Brum and Bastard, 1986) as electron transitions from three-dimensional (3D) to two-dimensional (2D) states (see Fig. 1) due to optical phonon emission.

Electrons in the initial state, i.e. above the well ($E > 0$), can occupy both three-dimensional and two-dimensional states. In the latter case the kinetic energy of the 2D electrons exceeds their binding energy ε_b, and, therefore, the total energy of these 2D electrons is positive $E = \varepsilon_k - \varepsilon_b > 0$. Thus, apart from the 3D→2D transitions there exists a 2D→2D capture channel, the contribution of which should be also examined. Carrier collection by a QW from the 2D states has an advantage over the usual 3D→2D collection. Indeed, the rate of 3D→2D transitions is reduced by the reflection of 3D electrons from the QW and hence the 3D→2D capture is hindered. Reflection does not affect transitions between 2D states, and one can expect the 2D→2D capture channel to be more effective than the 3D→2D channel in spite of the small number of initial 2D states as compared with that of 3D states.

Another interesting problem arising here is the role of the redistribution of electrons between 3D and 2D states above the well. The redistribution is determined by elastic scattering, which is usually much faster than energy relaxation. However, the isoenergetic 3D↔2D transitions are also slowed down by the reflection from QWs and one can expect the situation, in which the 3D↔2D redistribution will proceed slower than the 2D→2D energy relaxation, but faster than the 3D→2D energy relaxation. Then, the bottleneck of the capture will be the supply of electrons to the 2D states above the well, and elastic scattering will directly determine the capture rate.

In what follows we shall examine the roles the of 3D→2D and 2D→2D channels as well as that of 3D↔2D redistribution played in the capture, promoted by electron-phonon, electron-electron and electron-impurity scatterings. To examine the relevant transitions, we shall choose the simplest models for both the electron and phonon subsystems. Namely, we shall study thin quantum wells and shall examine the transitions induced by bulk phonons or uniformly distributed impurities. Consequently, many interesting problems, first of all capture rate oscillations (Kozyrev and Shik, 1985; Brum and Bastard, 1986) and the influence of the QW on a phonon subsystem (see, e.g., Ridley, 1991 and Babiker, 1992), will be beyond the scope of our discussion.

SCHEME OF CAPTURE RATE CALCULATIONS

The rate of capture by isolated quantum wells is proportional to their number N_{QW}, and can be presented in the form

$$\frac{1}{\tau_c} = n_{QW} <v> \gamma, \qquad (1)$$

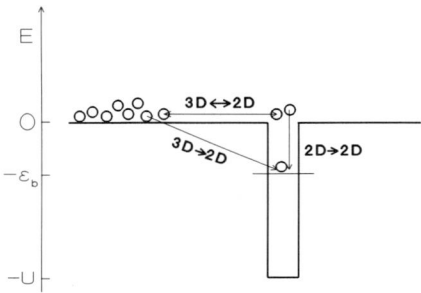

Figure 1. The scheme of 3D→2D and 2D→2D capture channels and 3D↔2D isoenergetic redistribution.

where τ_c is the capture time, $n_{QW} = N_{QW}/L$ is the concentration of the wells, L is the width of the structure, $<v> = (2k_B T_e/\pi m)^{1/2}$ is the thermal velocity of 3D electrons, and T_e is their temperature. Electrons encounter a QW with frequency $n_{QW} <v>$, and, therefore, the introduced capture coefficient γ represents the probability that an electron will drop into the well after having encountered it.

To find the time of carrier capture by a quantum well the collection flow J_c has to be calculated $\tau_c^{-1} = J_c/N$, where N is the number of electrons above the well. The schemes of collection flow calculations are different for the cases of inelastic and quasielastic scattering.

Inelastic Capture

When the electron interaction with scattering particles is inelastic and the characteristic energy loss of electrons exceeds their temperature, $\Delta\varepsilon \gg k_B T_e$, an electron will be captured by a quantum well after one elementary act of the interaction. Then the collection flow is simply the sum of transition probabilities over all final and initial states, weighted by the electron distribution function:

$$J_c = \sum_{\lambda_{3D}} f_{3D} \sum_{\lambda'_{2D}} W_{\lambda_{3D} \to \lambda'_{2D}} + \sum_{\lambda_{2D}} f_{2D} \sum_{\lambda'_{2D}} W_{\lambda_{2D} \to \lambda'_{2D}}. \qquad (2)$$

Here λ_{3D} and λ_{2D} denote the quantum states of 3D and 2D electrons, W are the transition probabilities and f_{3D} and f_{2D} are the electron distribution functions. The first and second terms in the expression correspond to electron capture via the 3D→2D and 2D→2D channels, respectively. The summation over λ_{2D} in the second term is carried out over the initial 2D states above the well ($E > 0$).

The transition probabilities W are determined by Fermi's golden rule. The wavefunctions and energy spectrum for a thin quantum well can be found from the usual quantum mechanical solutions of the rectangular potential well problem in the limit $2md^2 U/\pi^2\hbar^2 \ll 1$, where d and U are the width and depth of the well. The thin quantum well energy spectrum, corresponding to motion perpendicular to the QW, consists of the single bound level $-\varepsilon_b = -\hbar^2\alpha^2/2m$ and of the continuum of free states $\varepsilon_\kappa = \hbar^2\kappa^2/2m$. The corresponding wavefunctions are

$$\varphi_b = \sqrt{\alpha}e^{-\alpha|z|}, \qquad \varphi_\kappa = \frac{1}{\sqrt{L}}\left(e^{i\kappa z} + re^{i\kappa|z|}\right), \qquad (3)$$

where $r = -\alpha/(\alpha + i\kappa)$ is the reflection amplitude.

To calculate the collection flow (2), the electron distribution functions f_{3D} and f_{2D} should be defined. We shall assume that the 3D electrons are thermalized and their distribution function has the usual Maxwellian form

$$f_{3D} = \frac{n_{3D}}{2} \left(\frac{2\pi\hbar^2}{mk_B T_e}\right)^{3/2} \exp\left(-\frac{E}{k_B T_e}\right). \qquad (4)$$

The distribution function of 2D electrons f_{2D} above the well ($E > 0$) can be found from the kinetic equation including the term of isoenergetic 3D↔2D redistribution. In the case of inelastic scattering $\Delta\varepsilon \gg k_B T_e$, the equation has the form

$$\frac{\partial f_{2D}}{\partial t} = -\frac{f_{2D}}{\tau_{2D}(E)} + \frac{f_{3D} - f_{2D}}{\tau_0(E)}, \qquad \frac{1}{\tau_{2D}(E)} = \sum_{\lambda'_{2D}} W_{\lambda_{2D} \to \lambda'_{2D}}, \qquad (5)$$

where τ_{2D} corresponds to the time of 2D→2D energy relaxation and τ_0 is the time of 3D↔2D elastic scattering.

When the 3D↔2D redistribution is faster than the 2D→2D transitions, $\tau_0 \ll \tau_{2D}$, the distribution function of 2D electrons is equal to that of 3D electrons, $f_{2D} = f_{3D}$.

In the case of a slow 3D↔2D redistribution, when $\tau_0 \gg \tau_{2D}$, the steady-state solution of Eq.(5) is $f_{2D} = f_{3D}\tau_{2D}/\tau_0 \ll f_{3D}$. Introducing it into expression (2) we find the following formula for the capture coefficient

$$\gamma = \frac{\pi\hbar}{k_B T_e} \int_0^\infty \frac{dE}{k_B T_e} \frac{\exp(-E/k_B T_e)}{\tau_0(E)}. \qquad (6)$$

As seen from this expression the capture coefficient is directly determined by the elastic scattering time τ_0 under the considered conditions.

Cascade Capture

When electrons loose their energy in small portions as compared to their temperature, $\Delta\varepsilon \ll k_B T_e$, i.e. when the electron scattering is quasielastic, a number of elementary acts of interaction is needed for the capture.

Quasielastic scattering of an electron leads to its diffusion in energy space. Slightly changing its energy in an elementary act of interaction, an electron wanders on the staircase of energy states, and the capture is a diffusive descent to sufficiently deep energy levels.

The diffusive collection of 3D carriers is described by the theory of cascade capture (Lax, 1960; Pitayevsky, 1962; Abakumov et al, 1978). Here we shall employ it to describe the capture of 2D electrons. The distribution function of 2D electrons in the case of quasielastic scattering obeys the Fokker-Planck equation:

$$\frac{\partial}{\partial t} f_{2D} = -\frac{1}{R_{2D}} \frac{\partial}{\partial E} J, \qquad (7)$$

where R_{2D} is the density of 2D states and J is the flow of electrons in energy space. In the case of electron-phonon interaction, when the characteristic loss of electron energy is smaller than $k_B T$, the flow J can be presented in the form

$$J = -R_{2D} <\Delta E/\Delta t> \left(f_{2D} + k_B T \frac{\partial f_{2D}}{\partial E} \right), \qquad (8)$$

where $<\Delta E/\Delta t>$ is the rate of energy loss due to the spontaneous emission of phonons.

The steady-state solution of Eq. (8) has the following form in the $E < 0$ region:

$$f_{2D} = const \, \exp(-E/k_B T) \int_{-\infty}^{E} dE \, \frac{\exp(E/k_B T)}{k_B T R_{2D} <\Delta E/\Delta t>}. \qquad (9)$$

The integration constant can be found by matching the solutions of the $E > 0$ and $E < 0$ regions. In the case of fast 3D$\leftrightarrow$2D redistribution the distribution function of 2D electrons above the well is equal to that of 3D electrons.

Matching the functions (9) and (4) we determine the integration constant and, subsequently, find the collection flow $J_c = -J|_{E=0}$ and the cascade capture coefficient

$$\gamma = \frac{\pi \hbar k_B T}{(k_B T_e)^2} \left[\int_{-\infty}^{0} dE \, \frac{\exp(E/k_B T)}{<\Delta E/\Delta t>} \right]^{-1}. \qquad (10)$$

In the case of slow 3D$\leftrightarrow$2D redistribution the f_{2D} function in the $E > 0$ region must be found from the kinetic equation (7) including the redistribution term $(f_{3D} - f_{2D})/\tau_0$. It can be shown, that the treatment of slow 3D$\leftrightarrow$2D redistribution in the case of quasielastic scattering leads to the same result (6), obtained in the case of inelastic scattering.

Further we shall present and discuss the final results of capture time calculations, obtained employing formulae (2), (6) and (10).

OPTICAL PHONON-ELECTRON INTERACTION

The 3D→2D carrier capture promoted by polar (PO) and deformation (DO) optical phonon-electron interaction has been studied by Brum and Bastard (1986) and Kozyrev and Shik (1985).

In the range of lattice and electron temperatures $T, T_e \ll \varepsilon_b, \hbar\omega_o$, where ε_b is the binding energy of the QW and $\hbar\omega_o$ is the energy of the optical phonon, electrons will be captured by the QW after one elementary act of spontaneous emission of an optical phonon. Therefore, the formula (2) should be used to calculate the capture rate.

The following expressions for the 3D→2D capture coefficients due to Fröhlich PO interaction are obtained:

$$\gamma = \bar{\gamma}_{PO}\frac{\hbar\omega_o}{k_B T_e} \exp\left(-\frac{\hbar\omega_o - \varepsilon_b}{k_B T_e}\right) y_1, \qquad \varepsilon_b < \hbar\omega_o, \tag{11}$$

$$\gamma = \bar{\gamma}_{PO}\left(\frac{k_B T_e}{\hbar\omega_o}\right)^{1/2} y_2, \quad y_2 = \frac{\pi^{3/2}}{2b^2}\frac{2[1+(1-b^{-1})^{1/2}]^2 + b^{-1}}{2[1+(1-b^{-1})^{1/2}]^4}, \qquad \varepsilon_b > \hbar\omega_o, \tag{12}$$

where the nominal capture coefficient $\bar{\gamma}_{PO} = [2m/(\hbar\omega_o)]^{1/2} e^2/(\bar{\kappa}\hbar)$ is introduced, $\bar{\kappa}^{-1} = \kappa_\infty^{-1} - \kappa_0^{-1}$, κ_0 and κ_∞ are the static and high-frequency dielectric constants, and y_i are dimensionless functions of the binding energy $b = \varepsilon_b/\hbar\omega_o$

$$y_1 = \pi(1-b)^{3/2} \int_0^1 dx \frac{x^2}{b+(1-b)x^2} \frac{1 + 2b + 4[b(1-b)(1-x^2)]^{1/2}}{\{1 + 2[b(1-b)(1-x^2)]^{1/2}\}^2}. \tag{13}$$

The capture coefficient corresponding to the 2D→2D channel, when electrons are captured by the QW from 2D states in the case of slow 3D↔2D redistribution, has the form

$$\gamma = \bar{\gamma}_{PO}\frac{\hbar\omega_o}{k_B T_e} \exp\left(-\frac{\hbar\omega_o - \varepsilon_b}{k_B T_e}\right) y_3, \quad y_3 = \frac{\pi^2}{2}\sqrt{b}\frac{1+4\sqrt{b}}{(1+2\sqrt{b})^2}, \qquad \varepsilon_b < \hbar\omega_o, \tag{14}$$

$$\gamma = \bar{\gamma}_{PO}\frac{\hbar\omega_o}{k_B T_e} y_4, \quad y_4 = \frac{\pi}{2\sqrt{b}}\int_0^\infty \frac{dx}{(1+x^2)^2[(x^2-(4b)^{-1})^2 + x^2]^{1/2}}, \qquad \varepsilon_b > \hbar\omega_o. \tag{15}$$

The capture coefficients due to PO and DO optical phonon-electron interactions as functions of the QW binding energy are presented in Fig. 2 for the fixed electron temperature $k_B T_e = 0.1 \hbar\omega_o$. The temperature dependences of the PO capture times are schematically shown[1] in Fig. 3 for two GaAs QWs with binding energies larger, $\varepsilon_b = 50\,meV$ (Fig. 3a), and smaller, $\varepsilon_b = 10\,meV$ (Fig. 3b), than the optical phonon energy $\hbar\omega_o = 36.3\,meV$. (To estimate the capture times the $n_{QW} = 10^4\,cm^{-1}$ value of the QW concentration, corresponding to a single well in $1\,\mu m$, was assumed.)

The decrease of the capture rate with decreasing binding energy as well as with decreasing temperature for the QW with $\varepsilon_b < \hbar\omega_o$ (Figs. 2 and 3b) is determined by the exponential factors in (11) and (14), because only an exponentially small number of electrons from the high-energy tail of the distribution function can emit optical phonons under the conditions $k_B T_e \ll \varepsilon_b < \hbar\omega_o$. As a result, the optical phonon contribution to

---

[1]The following set of GaAs parameters is used: $m = 0.07\,m_e$, $\hbar\omega_o = 36.3\,meV$, $\bar{\gamma}_{PO} = 0.14$, $2ms^2 = 0.244\,K$, $\bar{\gamma}_{DA} = 4.2\,10^{-3}$, $s_T = 0.59\,s$, $\bar{\gamma}_{PA} = 5.4\,10^{-3}$, $\bar{\gamma}_B = 0.27$, and $n_0 = 5.8\,10^{17} cm^{-3}$.

the capture is exponentially reduced for both the 3D→2D and 2D→2D channels, and one can expect other scattering mechanisms to play an important role in the capture by QWs with binding energies $\varepsilon_b < \hbar\omega_o$.

When the binding energy exceeds $\hbar\omega_o$, all electrons above the well can participate in the capture. However, the 3D→2D transitions, presented in Figs. 2 and 3a by solid curves, now are hindered due to reflection from the QW, and the decrease of the 3D→2D capture coefficient with the increase of ε_b as well as the unusual, decreasing T_c temperature dependence is caused by the reflection. The PO 2D→2D capture channel presented in Figs. 2a and 3a by the dashed curves is essentially more effective than the PO 3D→2D channel for QWs with binding energies $\varepsilon_b > \hbar\omega_o$.

As seen from the comparison of Figs. 2a and 2b, the contribution of the DO interaction to the carrier capture exhibits the same features as the contribution of the PO interaction.

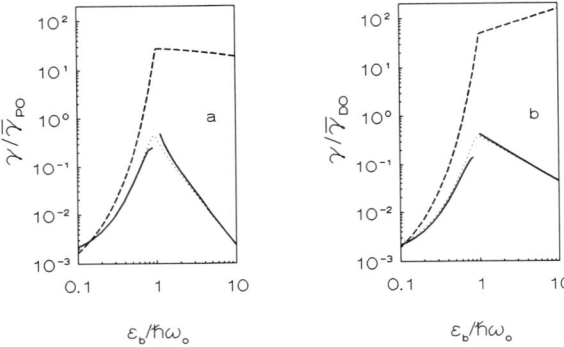

Figure 2. Normalized capture coefficients due to polar (a) and deformation (b) optical phonon-electron interactions as a functions of QW binding energy ε_b for the fixed electron temperature $T_e = 0.1\,\hbar\omega_o$. Solid and dashed curves correspond to 3D→2D and 2D→2D capture channels, respectively.

ACOUSTIC PHONON-ELECTRON INTERACTION

The characteristic longitudinal wave-vector $\vec{q}_\parallel$ of the acoustic phonons involved in the 3D→2D or 2D→2D transitions is determined by the law of momentum conservation, $\vec{q}_\parallel = \pm(\vec{k} - \vec{k}')$, and is of the order of the electron wave-vector $\vec{k}$ or $\vec{k}'$ of the motion longitudinal to the well, $q_\parallel \sim k' = (2m\varepsilon_{k'})^{1/2}/\hbar$. The characteristic transverse wave-vector of the phonons is determined by the binding energy of the QW, and is of the order of $q_\perp \sim \alpha = (2m\varepsilon_b)^{1/2}/\hbar$. Therefore, the characteristic energy of the acoustic phonons involved in the transitions is $\hbar sq \sim (2ms^2\varepsilon_b)^{1/2}$, where s is the longitudinal sound velocity. The characteristic energy $(2ms^2\varepsilon_b)^{1/2} = k_B T_0$ is small as compared with the binding energy ε_b; however, it can be either smaller or larger than the electron temperature. ($T_0 \simeq 5.3\,K \cdot (10\,meV/\varepsilon_b)^{1/2}$ for GaAs.) Thus, there exist two regimes of carrier capture due to acoustic phonons which occur, respectively,

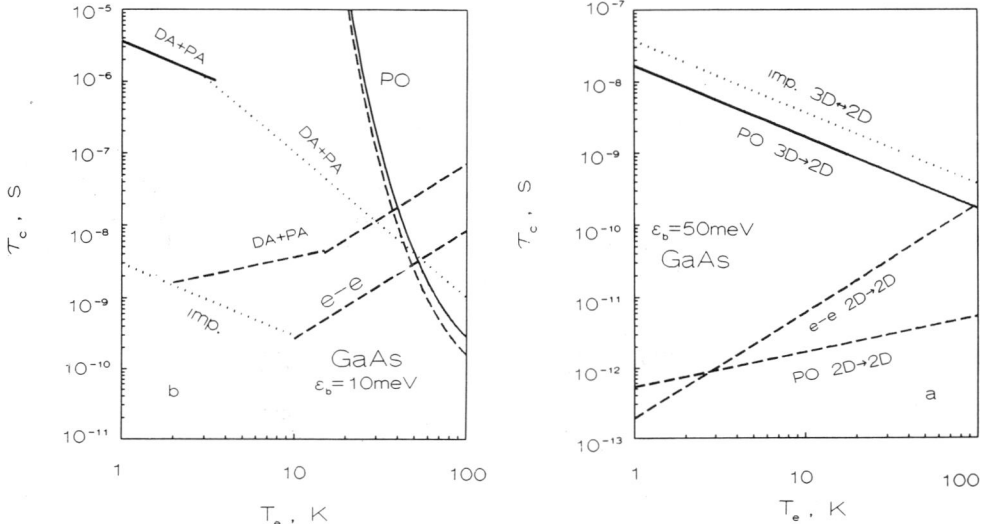

Figure 3. The temperature dependences of capture times for GaAs QWs. Solid, dashed and dotted curves represent 3D→2D, 2D→2D capture channels and 3D↔2D redistribution channel, respectively. $n_{QW} = 10^4\,cm^{-1}$. (a): $\varepsilon_b = 50\,meV$, $n_{imp} = 3 \times 10^{17}\,cm^{-3}$, $n_{3D} = 10^{15}\,cm^{-3}$, (b): $\varepsilon_b = 10\,meV$, $n_{imp} = 3 \times 10^{16}\,cm^{-3}$, $n_{3D} = 10^{13}\,cm^{-3}$.

in the two temperature regions $T_e \ll T_0$ and $T_e \gg T_0$. In the $T_e \ll T_0$ region, the characteristic energy loss of electrons exceeds their temperature, and the electrons are captured by the QW after one elementary act of acoustic phonon-electron interaction. In the $T_e \gg T_0$ region, a number of elementary acts is needed for the capture, and the regime of cascade capture sets up.

$T_e \ll T_0$. Inelastic Scattering

In this temperature region electrons emit acoustic phonons the energy of which exceeds the electron temperature, and, therefore, one elementary act of interaction is needed for the capture. The capture coefficients for the 3D→2D channel can be presented in the form

$$\gamma = \bar{\gamma}_{DA}\left(\frac{2ms^2}{\hbar\omega_o}\right)^{1/2}\left(\frac{k_BT_e}{\hbar\omega_o}\right)^{1/2}y_9, \quad y_9 = \frac{2\sqrt{\pi}}{3}\left(\frac{\hbar\omega_o}{\varepsilon_b}\right)^{1/2}, \quad (16)$$

$$\gamma = \bar{\gamma}_{PA}\left(\frac{2ms_T^2}{\hbar\omega_o}\right)^{1/2}\left(\frac{k_BT_e}{\hbar\omega_o}\right)^{1/2}y_{10}, \quad y_{10} = \frac{4\sqrt{\pi}}{15}\left(\frac{\hbar\omega_o}{\varepsilon_b}\right)^{3/2}, \quad (17)$$

for deformation (DA) and piezoelectric (PA) coupling, respectively. Here the nominal capture coefficients

$$\bar{\gamma}_{DA} = m^{3/2}E_1^2(2\hbar\omega_o)^{1/2}/(\pi\varrho s^2\hbar^3), \quad \bar{\gamma}_{PA} = m^{1/2}(eh_{14})^2/(\pi\varrho s_T^2\hbar(2\hbar\omega_o)^{1/2}) \quad (18)$$

are introduced, E_1 is the constant of the deformation potential, h_{14} is the component of piezoelectric tensor, ϱ is the density and s_T is the velocity of transverse sound. The optical phonon energy $\hbar\omega_o$ is formally introduced into expressions (16), (17) and (18) to make the comparison of optical and acoustic phonon contributions easier.

The capture coefficients corresponding to the 2D→2D channel, calculated under the assumption of a fast 3D↔2D redistribution, have the form

$$\gamma = \bar{\gamma}_{DA}\left(\frac{2ms^2}{\hbar\omega_o}\right)^{1/2}\frac{\hbar\omega_o}{k_BT_e}y_{11}, \quad y_{11} = \frac{\pi^2}{2}\frac{\varepsilon_b}{\hbar\omega_o}, \quad (19)$$

$$\gamma = \bar{\gamma}_{PA}\left(\frac{2ms_T^2}{\hbar\omega_o}\right)^{1/2}\frac{\hbar\omega_o}{k_BT_e}y_{12}, \quad y_{12} = \frac{3\pi^2}{16}. \quad (20)$$

Having compared the contributions of the 3D→2D (Eqs.(16) and (17)) and the 2D→2D (Eqs.(19) and (20)) channels, we find that the ratio of capture coefficients $\gamma_{3D\to 2D}/\gamma_{2D\to 2D}$ is of the order of $(k_BT_e/\varepsilon_b)^{3/2} \ll 1$ for both DA and PA interactions. As was the case for the optical phonon-electron interaction, the 2D→2D channel is more effective due to the reflection of 3D electrons from the QW.

$T_e \gg T_0$. Quasielastic Scattering

In this temperature region electrons loose their energy in small portions as compared with their temperature, and the cascade mechanism of capture takes place. Employing the scheme of calculations outlined for cascade collection, we obtain the following formulae for the capture coefficients

$$\gamma = \bar{\gamma}_{DA}\frac{2ms^2}{\hbar\omega_o}\left(\frac{\hbar\omega_o}{k_BT_e}\right)^2 y_{13}, \quad y_{13} = \frac{3\pi^2}{2}\left(\frac{\varepsilon_b}{\hbar\omega_o}\right)^{3/2}, \quad (21)$$

$$\gamma = \bar{\gamma}_{PA}\frac{2ms_T^2}{\hbar\omega_o}\left(\frac{\hbar\omega_o}{k_BT_e}\right)^2 y_{14}, \quad y_{14} = \frac{\pi^2}{4}\left(\frac{\varepsilon_b}{\hbar\omega_o}\right)^{1/2}. \quad (22)$$

The presented expressions correspond to capture via the 2D→2D channel in the case of a fast 3D↔2D redistribution of electrons above the well. In the considered temperature region the transitions from 3D to 2D states promoted by acoustic phonons contribute only to the 3D↔2D redistribution due to the quasielastic character of the interaction. Calculations of the capture coefficients in the case of slow redistribution performed according to formula (6) lead to the following expressions for the capture coefficients due to deformation and piezoelectric interactions

$$\gamma = \bar{\gamma}_{DA} \frac{k_B T \sqrt{k_B T_e}}{(\hbar\omega_o)^{3/2}} y_{15}, \quad y_{15} = \frac{\pi^{3/2}}{4} \frac{\hbar\omega_o}{\varepsilon_b}, \qquad (23)$$

$$\gamma = \bar{\gamma}_{PA} \frac{k_B T \sqrt{k_B T_e}}{(\hbar\omega_o)^{3/2}} y_{16}, \quad y_{16} = \frac{\pi^{3/2}}{8} \left(\frac{\hbar\omega_o}{\varepsilon_b}\right)^2. \qquad (24)$$

As seen from the comparison of the latter formulae with formulae (21) and (22), the 3D↔2D redistribution due to acoustic phonons is slower than the 2D→2D transitions, when the lattice and electron temperatures meet the requirement $(k_B T)^2 (k_B T_e)^5 < (2ms^2)^2 \varepsilon_b^5$.

The temperature dependences of the acoustic capture times are shown in Fig. 3b. The contributions of the DA and PA interactions are comparable for the considered QW, and their overall contribution (PA+DA) is presented in Fig. 3b. The acoustic 2D→2D channel, presented by the dashed lines, is more effective than the 3D→2D channel, presented by the solid line. With the increase of the temperature the acoustic phonon-electron interaction changes its character from inelastic to quasielastic, and the 2D→2D capture time changes its temperature dependence when entering the region of cascade capture. The acoustic 3D→2D capture channel actually ceases to exist in the quasielastic region, $T_e \gg T_0$, since the quasielastic transitions from 3D to 2D states contribute to isoenergetic redistribution of electrons only. The capture time due to the acoustic 3D↔2D redistribution is presented by the dotted DA+PA line in Fig. 3b. It corresponds to the formulae (23) and (24), where weak electron heating $T_e \simeq T$ was assumed.

Acoustic phonons can play an important role in the capture, when the QW binding energy is smaller than the optical phonon energy $\hbar\omega_o$. When the binding energy exceeds $\hbar\omega_o$, the contribution of acoustic phonons to the capture is lower than the PO contribution by several orders of magnitude and is not presented in Fig. 3a.

IMPURITY SCATTERING

As has been already pointed out, the isoenergetic transitions between the 3D and 2D states above the well can significantly affect the capture. Elastic carrier scattering in semiconductors is usually determined by shallow Coulomb impurities. Assuming a uniform distribution of the impurities in the barrier, one can calculate the elastic scattering time τ_0 and, employing the formula (6), find the capture coefficient due to the slow impurity induced 3D↔2D redistribution

$$\gamma = \bar{\gamma}_B \frac{n_{imp}}{n_0} \left(\frac{k_B T_e}{\hbar\omega_o}\right)^{1/2} y_{19}, \quad y_{19} = \frac{5\pi^{3/2}}{32} \left(\frac{\hbar\omega_o}{\varepsilon_b}\right)^3, \qquad (25)$$

where $\bar{\gamma}_B = 8me^4/(3\pi\kappa_0^2 \hbar^3 \omega_o)$ is the nominal capture coefficient, n_{imp} is the impurity concentration and $n_0 = (2m\hbar\omega_o)^{3/2}/(3\pi^2\hbar^3)$ is the characteristic concentration.

The temperature dependence of the capture time due to $n_{imp} = 10^{17}\, cm^{-3}$ impurity induced 3D↔2D redistribution above a QW with $\varepsilon_b = 50\, meV$ is presented in Fig. 3a. As seen from the figure, elastic electron-impurity scattering is not sufficient to supply electrons to the PO or e-e 2D→2D channels for the assumed n_{imp} value. Therefore, the capture times via the 2D→2D channel presented in the figure is just the theoretical limit for τ_c, provided that there are no other more effective elastic scatterers present in the structure.

The dotted "imp." line in Fig. 3b corresponds to the 3D↔2D redistribution due to shallow impurity scattering induced by $n_{imp} = 10^{16}\,cm^{-3}$ impurity concentration. As seen from the figure the impurities can directly control the capture in the region of low temperatures for the considered QW with $\varepsilon_b = 10\,meV$ and the assumed value of impurity concentration. In the remaining temperature region where 3D↔2D redistribution occurs more rapidly than the energy relaxation via 2D states, impurity scattering plays an important role by supplying electrons to the 2D states above the well, and thus enabling the fast 2D→2D channel to operate.

ELECTRON-ELECTRON SCATTERING

Electron collection by a QW can be promoted by the electron-electron interaction. The probe electron colliding with the electron "bath" can drop into the QW passing its energy to other electrons. To estimate the e-e capture rate let us consider the interaction of the probe electron with the bulk thermal "bath" of 3D electrons in a barrier. The features of the 3D→2D and 2D→2D transitions induced by this "bath" differ from those for the usual 3D-3D e-e interaction as well as from the features of the scattering of 2D electrons by a "bath" of 2D electrons (Esipov and Levinson, 1987). In our case the transitions of 3D electrons to 2D states induced by the 3D "bath" are essentially inelastic and the 2D→2D transitions induced by 3D "bath" are quasielastic. Having slightly modified the schemes of calculations described above for electron-phonon interaction to take into account the e-e interaction specifically, we obtain the following expressions for the capture coefficients

$$\gamma = \bar{\gamma}_B \frac{n_{3D}}{n_0}\left(\frac{k_B T_e}{\hbar\omega_o}\right)^{1/2} y_{21}, \quad y_{21} \simeq 1.46\left(\frac{\hbar\omega_o}{\varepsilon_b}\right)^3, \tag{26}$$

$$\gamma = \bar{\gamma}_B \frac{n_{3D}}{n_0}\Lambda\left(\frac{\hbar\omega_o}{k_B T_e}\right)^2 y_{22}, \quad y_{22} = \pi\left(\frac{\hbar\omega_o}{\varepsilon_b}\right)^{1/2}, \tag{27}$$

$$\tag{28}$$

which correspond to the 3D→2D and 2D→2D e-e capture channels, respectively. Here n_{3D} is the 3D electron concentration in a barrier and Λ is the Coulomb logarithm. It should be noted, that the presented results (26) and (27) are preliminary, since the expressions describe the transitions induced by bulk $3D$ electrons, which are not affected the by QW.

As seen from expressions (26) and (27), the e-e 2D→2D capture channel is essentially more effective than the 3D→2D channel. (The ratio of the capture coefficients $\gamma_{3D\to 2D}/\gamma_{2D\to 2D}$ is of the order of $(k_B T_e/\varepsilon_b)^{5/2} \ll 1$.) The temperature dependence of the e-e capture time via the 2D→2D channel is shown in Figs. 3a and 3b for a QW with $\varepsilon_b = 50\,meV$ and at the $n_{3D} = 10^{15}\,cm^{-3}$ value of 3D electron concentration, and for a QW with $\varepsilon_b = 10\,meV$ and $n_{3D} = 10^{13}\,cm^{-3}$, respectively. (The $\Lambda = 2$ value is assumed). As seen from the figures, the e-e 2D→2D channel can play important role in the capture at the considered values of the 3D electron concentrations, provided it will be sufficiently fast supplied with electrons as a result of 3D↔2D redistribution.

SUMMARY

We have considered the contributions of 3D→2D and 2D→2D transitions to the carrier collection by thin QWs, induced by electron-phonon, electron-impurity and electron-electron scattering. The performed analysis shows that the 2D→2D capture channel is more effective than the usual 3D→2D channel. When the binding energy of the QW exceeds the optical phonon energy a decisive role in the capture is played by optical phonons. When the binding energy is smaller than $\hbar\omega_o$, the acoustic phonon capture channels dominates over optical ones. The efficient 2D→2D capture channel can manifest itself when it is sufficiently quickly supplied by electrons due to $3D \leftrightarrow 2D$ electron redistribution above the well. The shallow Coulomb impurity scattering is too slow to supply electrons to 2D→2D channel even at the $n_{imp} \sim 10^{17}\, cm^{-3}$ impurity concentration for the QWs with binding energies $\varepsilon_b > \hbar\omega_o$. However, the shallow impurities are quite efficient at "opening" the 2D→2D capture channel in the QWs with $\varepsilon_b < \hbar\omega_o$.

REFERENCES

Abakumov, V.N., Perel', V.I., and Yassievich, I.N., 1978, Trapping of charge carriers by attracting centers in semiconductors, *Sov.Phys.-Semicond.* **12**, 1

Babiker, M., 1992, Coupling of polar optical phonons to electrons in superlattices and isolated quantum wells, *Semicond.Sci.Technol.* **7**, B52

Brum, J.A., and Bastard, G., 1986, Resonant carrier capture by semiconductor quantum wells, *Phys.Rev.B* **33** 1420.

Esipov, S.E, and Levinson, Y.B., 1987, The temperature and energy distribution of photoexcited hot electrons, *Adv.Phys.* **36** 331.

Kozyrev, S.V, and Shik, A.Ya., 1985, Trapping of carriers into quantum wells of heterostructures, *Sov.Phys.-Semicond.* **19** 1024.

Lax, M., 1960, Cascade capture of electrons in solids, *Phys.Rev.* **119** 1502.

Pitayevsky, L.P., 1962, Recombination of electrons in a monoatomic gas, *Sov.Phys.-JETP* **15** 919.

Ridley, B.K., 1991, Hot electrons in low-dimensional structures, *Rep.Prog.Phys.* **54** 169.

APPLICATION OF PHONON PHYSICS TO CRYOGENIC DETECTORS

Hans Kraus

Technische Universität München
Physik-Dept. E 15
James-Franck-Straße
D-8046 Garching bei München

1. INTRODUCTION

1.1 Motivation for Cryo Detectors

The name 'Cryogenic Detectors' is generic for a class of detectors which require low temperatures for their operation.[1-6] The operating principle may, for example, involve superconductivity, phonon focussing, or the reduction of the specific heat of materials at very low temperatures.

Many experiments in particle physics could benefit from new detection techniques. Sensitivity for non ionizing events combined with a large mass of the detectors and a low threshold is very important. Phonons play an important role in many of these detector concepts. Many cryogenic detectors are still at an early stage of their development, some of them already provide performance characteristics superior to presently available conventional detectors.[7,8] Experiments which motivate cryogenic detectors are briefly discussed in the next paragraphs.

Coherent Elastic Scattering of Neutrinos. The coherent elastic scattering of neutrinos off nuclei is a process predicted by the Standard Model of Weak Interaction.[9] If the wave length of a neutrino is long compared to the dimension of a nucleus the scattering amplitudes of neutrino–nucleon scattering may add coherently, resulting in a significant increase of the interaction cross section σ_{tot}.[10,11]

$$\sigma_{tot} \approx \frac{G_F^2}{4\pi} N^2 E_\nu^2 \qquad (1)$$

G_F is the Fermi constant, N the number of neutrons in the nucleus, and E_ν is the energy of the incident neutrino. The high cross section permits a reduction of the active detector mass by some orders of magnitude in typical experiments with low energy neutrinos (sun, fission reactor).[12,13] A drawback, however, is the requirement to detect the very low non ionizing recoil energies of the nuclei. In the case of a Silicon absorber and neutrinos of energy 3 MeV, the recoil energy is 220 eV (non ionizing). Phonon mediated detectors could be well suited for this application.

Dark Matter. Also benefitting from the ability to register non ionizing events with low threshold is the search for WIMPS (Weakly Interacting Massive Particles).[14] There exists the problem of 'Dark Matter'.[15] The motion of the galaxies and celestial objects suggest that the invisible (dark) matter exceeds the visible matter by a factor of 10 to 100.[16] Some dark matter candidates are under discussion and experiments have to test hypotheses and have to exclude regions of a parameter space for those dark matter candidates. At our position in the galaxy, a dark matter density of $\rho_{DM} \approx 300$ MeV c^{-2}cm^{-3} is passing through the universe with an average relative velocity of $v_{rms} \approx 260 .... 470$ km/s. Measuring the small nuclear recoil energies, nuclei receive after an interaction with a dark matter candidate, permits the restriction of characteristics for those Dark Matter candidates.

Heavy Ion Physics. Detectors employed in Heavy Ion Physics need to be radiation hard.[17] This requirement is incompatible with the highly structured composition of a standard semiconductor detector. Depositing high energies in such a detector destroys the sensitive diffusion layers essential for detector operation. Consequently, the performance of those detectors is severely degraded within a very short time. Using cryogenic detectors with large uniform absorbers could reduce the impact of defects and thus prolong the detector life time. If the sensitive element of a detector is already a highly disordered metal film, a few defects more should not matter at all.[18]

X-Ray Astronomy. X-rays are an important tool in the diagnosis of extra terrestrial X-ray sources. With plasma diagnosis information can be revealed on elemental abundance, mass motion, mass distribution, electron temperature, and many more features. A detector is necessary which is capable of resolving emission line features across an X-ray continuum.[19] The difficulty in this field is that many X-ray sources are so faint that a crystal spectrometer can no longer be used for efficiency reasons. On the contrary, the energy resolution of semiconductor diodes is too poor, especially for the spectroscopy of low Z elements. An instrument is needed with an energy resolution close to that of a crystal spectrometer but combined with the high quantum efficiency of a semiconductor diode. Superconducting tunnel junctions for X-ray detection could be used for this application.[20]

$\beta\beta$-Decay. The process of $\beta\beta$-decay involves a triplet of isobars (A, Z), (A, Z+1), and (A, Z+2) where the single β-decay to (A, Z+1) is forbidden or at least strongly suppressed.[21,22] Such a situation is shown in figure 1.

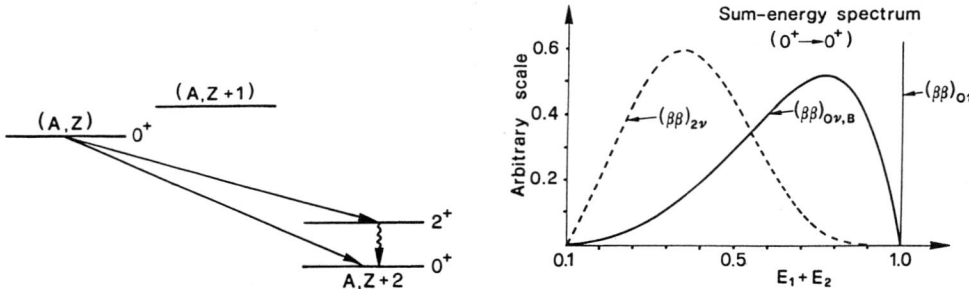

Figure 1. Double beta decay may occur if the final energy level of the nearest neighbor daughter nucleus is higher than that of the decaying nucleus. The resulting energy spectrum for the sum energy of the two emitted electrons is shown for the various decay channels.

The nucleus (A, Z) may decay into (A, Z+2) via one of the following channels:

$$(A, Z) \longrightarrow (A, Z+2) + 2e + 2\bar{\nu}_e$$
$$(A, Z) \longrightarrow (A, Z+2) + 2e$$
$$(A, Z) \longrightarrow (A, Z+2) + 2e + X$$

The first decay is allowed by all present laws of elementary particle physics. The energy released is shared between four particles, leading to a distribution for the sum energy of the two electrons. The second decay channel is the so called 'neutrinoless' $\beta\beta$-decay and violates lepton number conservation. Because the energy is almost completely (nuclear recoil) shared only between the two electrons, this decay should manifest itself in a single line in the sum energy spectrum. The third channel is also neutrinoless and a neutral Goldstone boson, called Majoron is emitted. A continuum results.

With the presently available methods it was possible to study only a very limited number of $\beta\beta$-decay candidates. This situation may change with the availability of low temperature calorimeters which allows a wide choice of absorber materials.[23]

1.2 Overview on Types of Cryogenic Detectors

Superconducting Tunnel Junctions for X-ray detection. Semiconductor detectors have been developed over a long period of time. As a consequence, these detectors have indeed reached the level of energy resolution given by the statistics of charge carriers.[24] The contribution to the line width by electronic noise has been made negligibly small. Therefore, to improve further, the number of charge carriers per deposited energy needs to be increased. This may be accomplished by the use of superconducting materials with an energy gap smaller by more than three orders of magnitudes.[25,26] The excitation caused by energy deposition in a superconductor can be measured with superconducting tunnel junctions.

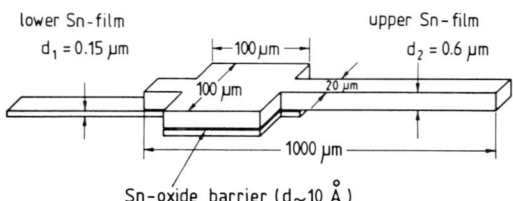

Figure 2. A superconducting tunnel junction consists of two superconducting metal films separated by a thin oxide barrier through which tunneling of quasiparticles is possible. The number of excess quasiparticles is proportional to the energy deposited in the tunnel junction.

A superconducting tunnel junction consists of two superconducting metal layers separated by a thin tunnel barrier. Energy absorption breaks Cooper pairs, thereby creating an excess density of quasiparticles. These quasiparticles can tunnel causing an excess tunnel current which is registered with a suitable preamplifier. Detectors exploiting the small energy gap for direct absorption of particles are suited very well for application to X-ray astronomy. Usually such detectors consist of a thin film absorber for the absorption of photons or particles with tunnel junctions attached to the absorber to measure the number of excited quasiparticles.

Superconducting Tunnel Junctions for Phonon Detection. For the detection of weakly interacting particles large absorber masses are required. In addition, the interaction of such particles manifests itself as a nuclear recoil which exhibits an ionisation efficiency, strongly reduced compared to a charged particle of the same energy.

Crystals of dielectric materials with volumes of some cm^3 are covered with several superconducting tunnel junctions.[27-29] The tunnel junctions register the non equilibrium phonons created by a recoiling nucleus as long as the energy of these phonons is larger than the energy gap of the sensitive superconducting film. An aim in constructing such detectors is a covering of the crystal surface as complete as possible. Several possibilities are under discussion:

– Single tunnel junctions with large area
– Series connected tunnel junctions
– Superconducting thin film absorbers with small tunnel junctions

A detector is pursued with an energy threshold well below 1 keV and with a volume of at least several cm^3.

Massive calorimeters. In earlier experiments monolithic calorimeters were used. Absorber and thermometer are identical in such a device. Monolithic calorimeters have the disadvantage that the choice of detector material is limited to very few elements such as Si and Ge. Composite calorimeters avoid this restriction by a separation of absorber and thermometer.[30-32]

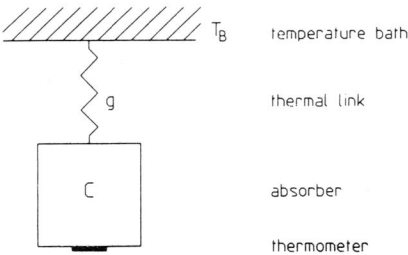

Figure 3. A composite calorimeter consists of an absorber in which particles are absorbed, a separate thermometer to measure the temperature change of the absorber, and a thermal link to a constant temperature bath to return the calorimeter to equilibrium.

Calorimeters exploit the reduction of specific heat of materials at low temperatures. In a very simple model, the temperature increase ΔT of a dielectric absorber for the absorption of an energy E is $\Delta T = E/C_A$. The specific heat of a dielectric absorber C_A is proportional to T^3 (ideally). Lowering the operating temperature permits the construction of large mass calorimeters. As absorber, materials like Silicon, Sapphire, Diamond, Tellurium Oxide, or Germanium are being investigated, just to name a few of them. For the thermometers NTD (Neutron Transmutation Doped) Germanium or the phase transition between the superconducting and normalconducting state of metals are the most popular ones.[30,33] To reduce the read out noise arising from the Johnson noise in the thermometer, also capacitive[34] and inductive[35] methods are under investigation. But to date, none of the two approaches was able to demonstrate superiority over resistive thermometers.

Micro Calorimeters for X-ray detection. Calorimetric detectors can be used not only to provide large absorbers for the detection of weakly interacting particles. They may also be employed for the detection of low energy X-rays in very small calorimeters with very high energy resolution. The energy resolution of a calorimeter is given by $\Delta E_{rms} = \xi \cdot \sqrt{kT_0 \cdot CT_0}$, with k, C, and T_0 the Boltzmann constant, the specific heat of the system, and the operating temperature, respectively. ξ is a parameter with a value between one and ten depending on the actual calorimeter. A calorimeter with a HgTe absorber and a Silicon thermometer with sensitive area of about $0.25 \, mm^2$ may exhibit a ΔE_{rms} in the 1 eV range.[36]

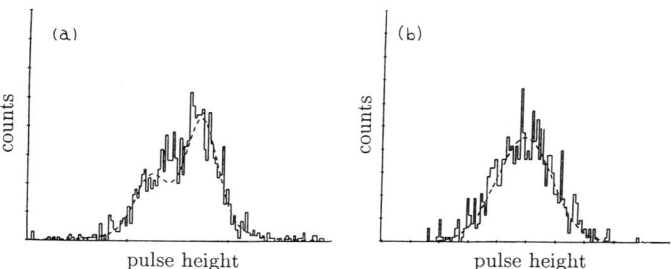

Figure 4. Energy spectrum obtained with a micro calorimeter consisting of a HgTe absorber and a doped silicon thermometer operating at T = 80 mK. Irradiation was carried out with the X-rays of a ^{55}Fe-source, emitting two X-ray lines at energies 5.89 keV and 6.49 keV, respectively. (a) shows the 5.89 keV line fitted with the $K_{\alpha 1}$ and $K_{\alpha 2}$ energies of the ^{55}Fe source. The energy resolution is $\Delta E = 7.35$ eV (FWHM) for the $K_{\alpha 1}$ line. (b) shows the base line fluctuation with $\Delta E = 4.6$ eV (FWHM).

Such calorimeters provide an energy resolution somewhat better than detectors with superconducting tunnel junctions. However, due to the calorimetric mode they do not offer position resolution within a single pixel. These detectors are most suited for the observation of integral radiation, such as the diffuse X-ray background.

Superheated – Supercooled Granules. This type of detector was highly favoured in the early days of low temperature detector developments.[37–40] It was the simplicity of the idea which was so inspiring. Small spheres of type I superconductor with radius $\sim 10 \, \mu m$ are suspended in a colloid. In type I superconductors, there is no single critical magnetic field, but there are regions in which superheating and supercooling exists. Operating grains in the superheating region while the grain is initially superconducting leads to a phase transition to the normalconducting state either by a sufficiently high temperature increase (thermal nucleation) or by an increase of the magnetic field (magnetic nucleation). The phase transition is registered by a change in the susceptibility due to the Meissner effect. The idea is elegant and simple, but there are some technical difficulties in the fabrication of grains of the required quality.

2. DETECTORS WITH PHASE TRANSITION THERMOMETERS

2.1 General Experiments and Geometry

A calorimeter consists of an absorber, a thermometer, and a thermal link to a constant temperature bath. In the situation discussed here, the absorber is Silicon, the thermometer either pure Iridium or a Gold Iridium proximity sandwich. The thermal link is accomplished via Gold bond wires. The operating temperature of the detector is determined by the phase transition thermometer which is best operated at its steepest section of the resistance–temperature characteristics.

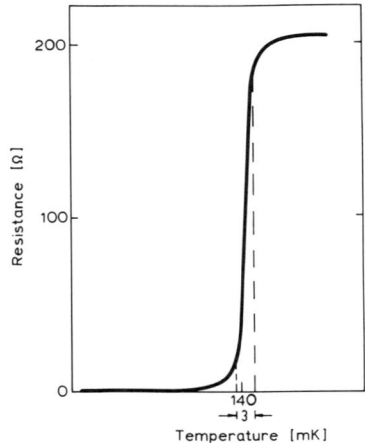

Figure 5. The transition curve between the normalconducting and the superconducting state of an Iridium stripe used as a thermometer. The thermometer is operated at its steepest section of the R–T curve and the change in resistance R as a consequence of a change in temperature T is measured.

In a very simple model, it is assumed that the deposition of energy ΔE in the absorber (heat capacity C) causes an increase in temperature ΔT according to $\Delta E = C \cdot \Delta T$. This formulation assumes total thermalisation of the phonons in the absorber immediately after energy deposition. It will be shown later that this is not true in reality. The equation describing the calorimeter is:

$$P(t) = C \cdot \dot{T} + g(T - T_b), \qquad (2)$$

with g the thermal coupling between calorimeter and temperature bath of temperature T_b. With a δ-like (in time) power input $P(t) = \Delta E \cdot \delta(t)$ the time dependent temperature of the calorimeter is:

$$T(t) = \frac{\Delta E}{C} \cdot e^{-t/\tau} + T_b \qquad \text{with} \qquad \tau = \frac{C}{g}. \qquad (3)$$

A single exponentially decaying pulse is expected. This simple model being only a first approximation does not include non equilibrium phonons, phonon propagation, nor details of the transmission of phonons into the thermometer.

The temperature change is in addition assumed to be small so that the specific heat C is constant and the change in temperature is directly proportional to the change in resistance of the thermometer.

2.2 Results from Irradiation

Experiments were performed with a calorimeter consisting of a 19 g Si absorber and an Ir thermometer operating at T = 220 mK.[41] The irradiation of the calorimeter with 5.5 MeV-α-particles gave a result different from that predicted by the simple model. Instead of just one decay time constant of the pulse, there were two decays, a fast one

($\tau_{\text{fast}} = 0.4\,\text{ms}$) and a slow one ($\tau_{\text{slow}} = 290\,\text{ms}$). This is not an unexpected result since the formulation of the problem (cf. eq. 2) assumed thermalized phonons which can not be true in a Si crystal at the time scales relevant here.

The relatively broad transition of the Ir thermometer gave the opportunity to investigate the temperature dependence of the calorimeter signal in a temperature range of 210 mK to 230 mK. It has been shown that the amplitude of the slow component had a temperature dependence consistent with the assumption of a Debye heat capacity to within 10% error. Thus, the slow component may be identified as the thermal component of the signal.[42]

A different calorimeter with an identical absorber but with a Au-Ir thermometer was fabricated to investigate the temperature range between 20 mK and 120 mK. The transition was inadvertently very broad. Lowering the operating temperature should increase the amplitude of the slow component as the specific heat of the dielectric absorber decreases $\propto T^3$ (ideally). This behavior was observed to some extent at an operating temperature of 120 mK (compared to 220 mK with an Ir thermometer). But

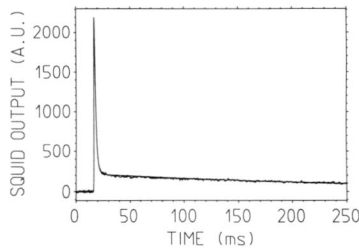

Figure 6. A typical pulse resulting from the absorption of a 5.5 MeV-α-particle in a calorimeter with a Si absorber and equipped with an Ir transition edge thermometer. The pulse exhibits two decay time constants. The longer decay is due to the thermal signal.

instead of a further increase of the amplitude of the slow component a reduction while lowering the operating temperature was observed. At a temperature of 17 mK the amplitude was lower by almost a factor of 100 compared to the expectation resulting from the Debye heat capacity. The relaxation time of the fast component was unaffected by the temperature variations.

2.3 Model of the Calorimeter

To understand the operation of the calorimeter, a model including non equilibrium processes had to be developed.

General Model and Ideas. The absorption of a particle with energy much higher than typical phonon energies leads to the creation of mainly optical phonons. These optical phonons decay within $\sim 100\,\text{ps}$ to acoustical phonons. A Monte Carlo simulation was performed including the anharmonic decay,[43] isotope scattering, and surface scattering. The starting frequency was 7.5 THz. Rapid down conversion to frequencies below 700 GHz occurs within some 10 μs slowly approaching $\sim 300\,\text{GHz}$ persistent for ms. Phonons are being homogenized across the absorber volume but thermalisation cannot be reached within a reasonable time scale.

Rapid thermalisation, however, can occur in the metal film of the thermometer. In metals there exists a strong electron phonon coupling and the mean free path of a phonon is inversely proportional to its frequency. Thus, high frequency phonons are easily absorbed and down converted to thermal energies in a metal film.

In a thermal model of the calorimeter it is assumed that the high frequency phonons act as a power input to the electronic subsystem of the thermometer. The electron system in the thermometer works as a perfect thermalizer radiating thermal phonons back into the absorber. The phonon system of the absorber is coupled to the phonons in the thermometer via the Kapitza conductance G_K. The electron system of the thermometer is connected to the phonons in the thermometer via G_{e-p}, the electron phonon conductance, and to the thermal bath via the coupling G_{e-b}. These conductances are temperature dependent as follows:

$$G_K \propto T^3 \qquad G_{e-p} \propto T^4 \qquad G_{e-b} \propto T \qquad (4)$$

The reduction of the thermal signal's amplitude of the calorimeter may have several reasons:

— Phonons thermalized in the thermometer may escape from the thermometer into the temperature bath. Thus they do not heat the absorber and the temperature increase is less than expected.

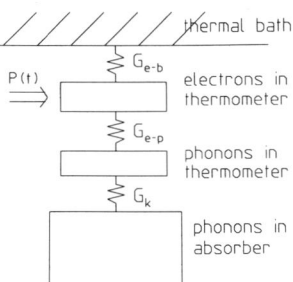

Figure 7. Thermal model of the calorimeter. Phonons in the thermometer couple to phonons in the absorber via the Kapitza conductance G_K and to the electron system of the thermometer G_{e-p}. High frequency phonons from the absorber are viewed as a power input to the electrons in the absorber.

— At very low temperatures the coupling between electrons in the thermometer and the temperature bath may be much better than the coupling between electrons and phonons in the thermometer ($G_{e-b} \gg G_{e-p}$). Therefore, the thermometer does not see the full temperature rise of the absorber.

A Si crystal (absorber) is mounted in a copper frame (temperature bath). A metal stripe (thermometer) is evaporated onto the surface of the absorber. Aluminium bond wires (electrical contacts) and one Gold wire (thermal link) are attached to the thermometer.

In the following paragraphs the various processes and contributions important for the understanding of the chain of events from the initial energy absorption to the thermometer signal are discussed.

Electron Phonon Conductance in the Thermometer. To calculate the conductance between the electron and the phonon system $G_{e\text{-}p}$ we consider the heat flow from phonons to electrons:

$$\dot{Q}_{p \to e} = \sum_{q,m} \frac{\hbar \omega_m v_m}{\ell_m} \cdot F(\omega_m) \implies G_{e\text{-}p} \cdot \Delta T \tag{5}$$

v_m is the velocity of mode m, ℓ_m is the mean free path for scattering off electrons, and F_m is the occupation of the phonon state m. The sum runs over all wave vectors q and phonon modes m.

The energy transfer to the thermometer is dominantly carried out by high frequency longitudinal phonons. The mean free path of L-phonons ℓ_L with energy ω is:

$$\ell_L = (\eta \omega)^{-1} \tag{6}$$

Considering only longitudinal phonons the heat transfer between electrons and phonons in the thermometer is given by:[44,45]

$$\dot{Q} = \frac{V\eta k_B}{2\pi^2 v_L^2} \cdot \left(\frac{k_B}{\hbar}\right)^4 \cdot (T_p^5 - T_e^5) \cdot \int_0^{x_D} \frac{x^4}{e^x - 1} dx \tag{7}$$

The substitution $x = \hbar \omega / kT$ has been used, V is the volume of the metal film, and v_L is the sound velocity of longitudinal phonons. At low temperatures, the upper limit of the integral may be replaced by ∞. The temperature difference $T_p - T_e$ is ΔT.

$$\dot{Q} = \frac{V\eta k_B}{2\pi^2 v_L^2} \cdot \left(\frac{k_B}{\hbar}\right)^4 \cdot 5T^4 \cdot 4! \cdot 1.037 \cdot \Delta T = \underbrace{\beta T^4}_{G_{e\text{-}p}} \cdot \Delta T \tag{8}$$

with

$$\beta = \frac{5 \cdot 4! \cdot 1.037 \, V\hbar}{2\pi^2 v_L^2} \cdot \left(\frac{k_B}{\hbar}\right)^4 \cdot \eta \tag{9}$$

The absorption of a high frequency phonon transfers energy to a single electron. A very strong Coulomb interaction among the electrons leads to a rapid distribution of energy among many electrons, thereby thermalizing the electrons. Thus, the efficient absorption of high frequency phonons may be considered as a heat input P(t) to the electrons in the thermometer.

Transfer of Phonons across an Interface. The heat transfer per area and time across an interface from medium 1 to medium 2 can be written as:

$$\dot{Q} = \frac{C_1}{2} \langle v_{g\perp} \alpha \rangle \Delta T = G_K \Delta T \tag{10}$$

C_1 is the heat capacity of the phonon system per volume in medium 1, $v_{g\perp}$ the component of the group velocity perpendicular to the interface, α the probability for the transmission of phonons from medium 1 to medium 2, ΔT is the temperature difference between the two media, and G_K is the Kapitza conductance. <> indicates the averaging over a distribution of $\vec{q}$-vector directions and phonon modes in medium 1 according to the density of states for phonons. As $\langle v_{g\perp} \alpha \rangle$ is independent of temperature G_K exhibits the temperature dependence of C_1, i.e. $G_K \propto T^3$. In earlier publications[46] the two media were considered as isotropic continua. Here, the anisotropic elastic theory was used. This also includes effects such as phonon focussing. As our evaporated metal films are poly crystalline, an averaging over all crystal orientations was performed for the metal films. As a consequence of the phonon focussing effect, energy transport across the interface can occur even for $\vec{q}$-vectors pointing away from the interface.

Mode conversion can occur at the interface. An incident wave of a distinct phonon mode splits into three transmitted and three reflected waves of the three modes. The parallel momenta of the phonons are conserved and mode conversion occurs according to the phase spaces available for the phonon modes. In a real experiment, the situation may be more difficult because of a possible oxide layer on the crystal.

Calculation of the Thermometer Response. The thermometer is treated in a one dimensional model. It is assumed that the change in resistance ΔR which can be measured is proportional to the change in mean temperature $\overline{\Delta T_e}$ of the electrons.

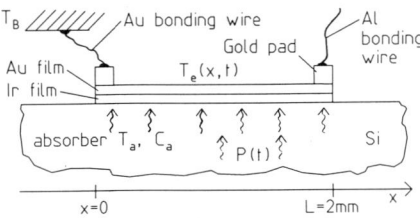

Figure 8. Geometry of an Ir-Au proximity effect thermometer on a Silicon absorber. A thermal link to a constant temperature bath is established via a Au bond wire attached to the left side of the thermometer. Electrical contact to the thermometer is provided by superconducting Al wires attached to both sides of the thermometer.

Two equations are formulated. One equation describes the heat diffusion along the metal stripe, the radiation of heat from the thermometer to the absorber crystal, and the power input by high energy phonons.

$$C_e \frac{\partial T_e}{\partial t} - \chi V_e \frac{\partial^2 T_e}{\partial x^2} + G_{e-a}(T_e - T_a) = P(t) \tag{11}$$

The second equation describes the heating of the absorber by the thermometer.

$$C_a \frac{dT_a}{dt} = G_{e-a} \cdot \left(\overline{T_e} - T_a\right) \tag{12}$$

The effective thermal coupling between the phonons in the absorber and the electrons in the thermometer G_{e-a} is given by:

$$\frac{1}{G_{e-a}} = \frac{1}{G_K} + \frac{1}{G_{e-p}} \tag{13}$$

C_e, χ, and V_e are the specific heat of the electrons in the thermometer, the heat conductivity, and the thermometer volume, respectively. With the parameters of the calorimeter discussed here, it is justified to discard the first term in equation 11 ($C_e \dot{T}_e$).

The thermometer of length L is thermally coupled to the temperature bath of temperature T_b only at one end ($x = 0$) via a Gold bond wire of heat conductance G_{Au}. Thus, the boundary conditions are:

$$G_{Au} \cdot (T_e(x=0) - T_b) = -wd\chi \frac{dT_e}{dx}\bigg|_{x=0} \quad \text{and} \quad \frac{dT_e}{dx}\bigg|_{x=L} = 0 \quad (14)$$

w and d are the width and thickness of the thermometer stripe. With the assumption of a power input such as

$$P(t) = P_0 \cdot \exp(-t/\tau) \cdot \theta(t) \quad (15)$$

the solution for the average temperature increase of the electron system in the thermometer is:

$$\overline{\Delta T_e(t)} = \frac{1-\gamma}{G_{e-a}} \cdot P(t) + (1-\gamma)\Delta T_a(t). \quad (16)$$

The temperature increase of the absorber ΔT_a is:

$$\Delta T_a(t) = \frac{E}{C_a} \cdot \frac{1-\gamma}{1 - \frac{\tau}{\tau_{slow}}} \cdot \left(e^{-t/\tau_{slow}} - e^{-t/\tau}\right) \cdot \theta(t) \quad (17)$$

with E the energy deposited in the absorber. The constants γ and τ_{slow} are given by:

$$\gamma = \left(\frac{G_{e-a}}{G_{Au}} + \frac{\lambda L}{\tgh(\lambda L)}\right)^{-1} \quad \lambda = \sqrt{\frac{G_{e-a}}{\chi V_e}} \quad \tau_{slow} = \frac{1}{\gamma} \cdot \frac{C_a}{G_{e-a}} \quad (18)$$

The result is a signal which has indeed two decay time constants. The fast component stems from the initial thermalisation of high energy phonons entering the thermometer. The slow component arises from heating the absorber with thermalized phonons radiated back from the thermometer into the absorber.

Calculation of the Power Input. It is assumed that the high frequency phonons in the absorber are homogeneously distributed in the absorber within a time scale of several µs. It is further assumed that thermalisation takes place only in the thermometer and that a fraction ξ ($0 < \xi < 1$) of the phonons entering the thermometer is absorbed. With the time dependent power input of:

$$P(t) = P_0 \cdot e^{-t/\tau} \cdot \theta(t) \quad (19)$$

the initial power P_0 and the time constant τ are given by:

$$P_0 = \frac{1}{2} \cdot \frac{E}{V_a} \cdot \langle v_{g\perp}\alpha\rangle \cdot A \cdot \xi \quad (20)$$

and

$$\tau = \frac{E}{P_0} = \frac{2V_a}{A \cdot \langle v_{g\perp}\alpha\rangle \xi}. \quad (21)$$

E is the energy deposited in the absorber, V_a is the absorber volume, and A is the area of the interface between absorber and thermometer.

2.4 Comparison with the Experiment

To compare the model with the experiment, data concerning the coupling constants are required. The thermal conductance along the thermometer is given by:

$$\chi V_e = 2.4 \cdot 10^{-13} \left(\frac{T}{1\,K}\right) \frac{Wm^2}{K}. \quad (22)$$

273

The geometry of the thermometer enters and the heat conductance is obtained from the electrical resistance via the Wiedemann-Franz-law. Due to the broad transition of the phase transition thermometer the heat conductance may be overestimated by a factor of four at the lowest temperature. The thermal conductance along the gold wire is determined with the electrical resistance and the Wiedemann-Franz-law.

$$G_{Au} = 9.2 \cdot 10^{-7} \left(\frac{T}{1\,K}\right) \frac{W}{K} \tag{23}$$

The Kapitza conductance is calculated using phonon transmission across the interfaces Si → Au and Si → Ir:

$$G_K = 1.44 \cdot 10^{-3} \left(\frac{T}{1\,K}\right)^3 \frac{W}{K} \tag{24}$$

The only free parameter of the model β appears in the electron phonon conductance:

$$G_{e-p} = \beta \left(\frac{T}{1\,K}\right)^4 \tag{25}$$

Amplitude of the Slow Component. The temperature increase of the absorber which is indeed 'seen' by the thermometer is the cause for the slow component of the signal. Its amplitude is given by:

$$A_{slow} = \frac{E}{C_a} \cdot \frac{(1-\gamma)^2}{1 - \frac{\tau}{\tau_{eff}}}. \tag{26}$$

The model was fitted to the experimental amplitude A_{slow} as a function of temperature and the parameter β was adjusted.

$$\beta = 2.4 \cdot 10^{-3} \left(\frac{T}{1\,K}\right)^4 \frac{W}{K} \tag{27}$$

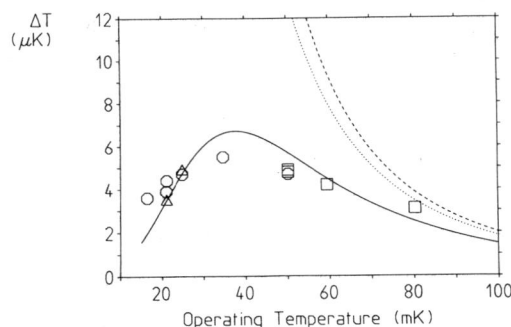

Figure 9. The temperature change 'seen' by the thermometer as a function of operating temperature. The solid line is a fit with the parameter $\beta = 2.4 \cdot 10^{-3}\,(T/1\,K)^4\,(W/K)$. The dotted curve represents rigid coupling between electrons and phonons, the dashed curve is the temperature increase expected from the Debye model heat capacity.

gave the optimal fit. Inspecting figure 9 shows that the amplitude A_{slow} determined for different models (present model, rigid coupling between electrons and phonons, and Debye heat capacity) gives almost the same model independent result at higher temperature. A difference is observed only at the lower temperatures.

Using a thermometer volume of $V_e = 2 \cdot 10^{-7}\,\text{cm}^3$ for the calculation of the normalized electron phonon resistance $R_{e-p}V_e T^4$ gave a value of $0.83 \cdot 10^{-4}\,\text{K}^5\text{cm}^3\text{W}^{-1}$ for the Au-Ir thermometer. A comparison with other materials, such as Rh:Fe ($1.5 \cdot 10^{-4}\,\text{K}^5\text{cm}^3\text{W}^{-1}$)[47] or Cu ($2.0 \cdot 10^{-4}\,\text{K}^5\text{cm}^3\text{W}^{-1}$) supports the fit result.

The experimental data show that the fraction of energy transmitted to the thermometer must be clearly smaller than unity ($\xi < 1$, cf. eq. 20). The anharmonic decay of phonons causes a phonon energy distribution with a maximum at a frequency of 300 GHz after some ms. If an identical scattering rate for all three phonon modes is assumed the mean free path for phonons of 300 GHz in the thermometer is 700 nm. With a film thickness of just $d \approx 100$ nm the metal film is too thin to absorb all phonons. It may be assumed that only longitudinal phonons are absorbed in the thermometer. The mean free path for longitudinal phonons of frequency 300 GHz is $\ell \approx 70$ nm. The thermometer is thick enough to absorb all longitudinal phonons and as 22% of the phonons transmitted into the thermometer are finally of the longitudinal mode, an absorption probability of $\xi = 0.22$ is assumed.

Relaxation Time of the Slow Component. The shape of the experimental curve is in good agreement with the model prediction (fig. 10) but the absolute value of the slow time constant is overestimated by a factor of two.

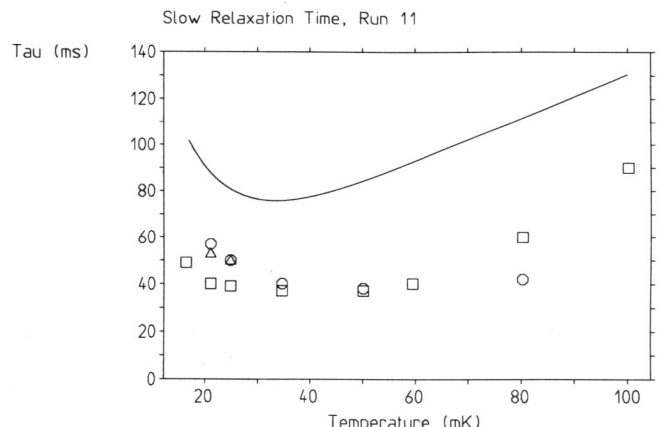

Figure 10. Relaxation time of the slow component as a function of temperature. Squares and circles correspond to read out currents of $100\,\mu\text{A}$ and $50\,\mu\text{A}$, respectively. The solid curve is the model prediction.

A possible reason might be the electron phonon interaction in the contact pad to which the gold bond wire is attached. This contact pad has about 1/2 of the volume of the thermometer itself. The present model does not include such additional insensitive films.

Amplitude of the Fast Component. The fast component results from the power input of the initially high energetic phonons. Its amplitude is given by:

$$A_{\text{fast}} = (1-\gamma) \cdot \frac{P_0}{G_{\text{e-a}}}. \tag{28}$$

Again, $\beta = 2.4 \cdot 10^{-3}\,\text{WK}^{-5}$ was used yielding a value of $\eta = 1.2 \cdot 10^{-5}\,\text{m}^{-1}$ (cf. eq. 9) and with $\ell_L = (\eta\omega)^{-1}$ a mean free path of $\ell_L \approx 72$ nm results. The probability that a phonon transmitted into the thermometer is of the longitudinal mode is calculated as the average of transmission across Si → Au (24.6%) and Si → Ir (17.4%) including mode conversion. Thus $\xi \approx 0.22$ is obtained, explaining well the data in figure 11. But for a given energy density in the absorber the decay time of the fast component should increase by a factor of $(0.22)^{-1}$. Such a long decay time, however, is in clear disagreement with the experiment. It is inevitable that phonon down conversion at surfaces of the absorber is included.

Relaxation Time of the Fast Component. τ_{fast} is determined by the relaxation time of the power input P(t). A 60 keV γ photon absorbed in a Si absorber of volume 8 cm³ creates a power input to the thermometer of $P_0 = 2.8$ pW ($\xi = 1$, cf. eq. 20). To calculate the relaxation time according to equation 21 the following numbers are used:

| Si → Ir | $\langle v_g \perp \alpha \rangle = 1289 \, \frac{m}{s}$ | $A = 2 \, \text{mm}^2$ |
| Si → Au | $\langle v_g \perp \alpha \rangle = 2069 \, \frac{m}{s}$ | $A = 1 \, \text{mm}^2$ |

With $\xi = 0.22$ a relaxation time of 15.4 ms results. The experimentally obtained time constant is $\tau = 1.5$ ms. Thus, the experimental time constant is a factor of ten shorter

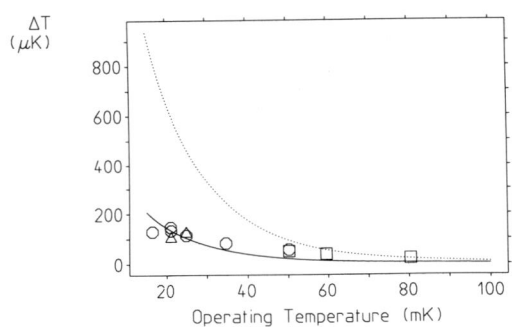

Figure 11. Amplitude of the fast component as a function of temperature. The dotted curve corresponds to a situation where all phonons entering the thermometer are indeed absorbed. The experimental data are much better explained by assuming only 22% of the phonons being absorbed.

than expected and one has to conclude that only $\sim 10\%$ of the high frequency phonons are thermalized in the thermometer. The remaining 90% are thermalized by surface scattering and surface adsorbate in the absorber.

Rise Time of the Fast Component. As a consequence of the omission of the term $C_e \dot{T}$ in equation 11 the present model does not predict the rise time of the fast component. It may be assumed, however, that in the early phase total coupling of electrons exists and that $\Delta T_a(t=0) \approx 0$ holds. Thus, the rise time of the fast component can be obtained.

$$\tau_{\text{rise}} = (1 - \gamma) \cdot \frac{C_e}{G_{\text{e-a}}}. \tag{29}$$

A comparison with the experiment gave good agreement.

Heating of Film and Substrate by the Read Out Current. The presence of a read out current I causes an additional power input to the calorimeter. The dissipated power is not constant as the resistance of the thermometer changes during a power input induced by the absorption of energy in the absorber. Electro–thermal feedback is introduced by modifying the power input:

$$P(t) \longrightarrow P(t) + I^2 R(T). \tag{30}$$

After this modification and with the parameters χ, G_K, and $G_{e\text{-}p}$ being temperature dependent, it is impossible to give an analytical solution for the thermometer response. Thus, a numerical solution has to be found. The presence of a read out current leads in any case to a reduction of thermometer sensitivity at otherwise identical external parameters.

2.5 Summary

There exists a good understanding of detectors with dielectric absorbers and superconducting phase transition thermometers. The model describes well the presence of the two components in the signal arising from the non equilibrium phonons absorbed in the thermometer (fast component) and from the heating of the absorber (slow component). The relative weight of the two components may be controlled by thermometer design. To explain the experimental data with the present model, the inclusion of surface thermalisation is necessary. An energy resolution of $\sim 1\,\text{keV}$ (FWHM) was obtained at an energy of 60 keV in a 19 g Si absorber. This very good result was obtained with a broad transition. Further progress in thermometer fabrication will lead to an improvement in energy resolution and threshold.

3. DETECTORS WITH SUPERCONDUCTING TUNNEL JUNCTIONS

3.1 General Experimental Setup and Geometry

The detector discussed in this section consists of a dielectric absorber of volumes presently in the cm^3 range with superconducting tunnel junctions on its surface. Interaction of a particle or radiation in the absorber causes the creation of phonons. The frequency distribution is not well known. Phonons propagate in the absorber partly diffusively and partly ballistically. Phonons with energies $\hbar\omega$ above the detection threshold of a tunnel junction 2Δ may be absorbed therein and contribute to the signal current.

The creation, propagation, and down conversion at surfaces in the absorber is not different from the situation in a detector with a phase transition thermometer. However, the tunnel junction does have a threshold and is not measuring the thermal signal as the phase transition thermometer does in a calorimeter. The signal expected for a detector with tunnel junctions as sensitive elements should be similar to the fast component of the calorimeter signal.

It has proven important that as much as possible of the absorber surface should be covered with sensitive elements. By coupling the absorber strongly or weakly to a temperature bath, or by varying the ratio of sensitive to insensitive metal films on the absorber surface, the ratio of ballistic phonons and homogenized phonons contributing to the signal can be adjusted.[48]

3.2 Phonon Propagation in Crystals

Elastic Theory. The initially created phonons are optical phonons which decay rapidly ($\sim 100\,\text{ps}$) to acoustical phonons and after further decays, the phonon wave length is long enough to consider the absorber crystal as an elastic continuum in the non

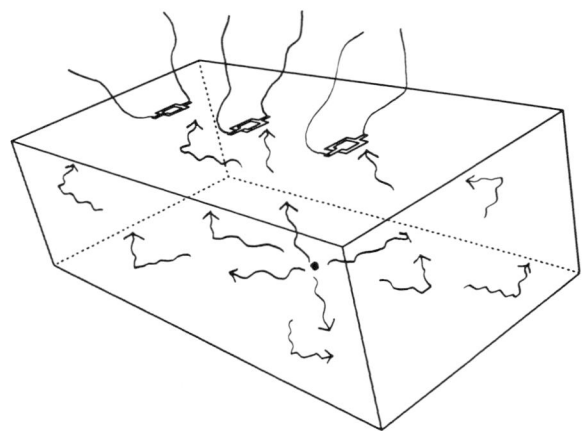

Figure 12. Radiation emitted by a radioactive source is absorbed in an absorber where non equilibrium phonons are created. These phonons propagate through the absorber while undergoing scattering and down conversion in energy. Eventually a fraction of them is registered in superconducting tunnel junctions on the surface of the absorber crystal.

dispersive regime ($\omega = ck$).[49,50] The problem can be formulated by using a general displacement tensor ϵ_{ij} and a stress tensor σ_{ij}:

$$\epsilon_{ij} = \frac{1}{2}\left(\frac{\partial u_i}{\partial x_j} + \frac{\partial u_j}{\partial x_i}\right) \quad \text{and} \quad \sigma_{ij} = \frac{\partial W}{\partial \epsilon_{ij}}. \tag{31}$$

W is the phonon energy density. The stress and displacement tensors are connected by a generalized Hook law and an equation of motion can be formulated with ρ the density and c_{ijlm} the elastic constants of the absorber material.

$$\rho \frac{\partial^2 u_i}{\partial t^2} = \sum_j \frac{\partial \sigma_{ij}}{\partial x_j} = \sum_{jlm} c_{ijlm} \frac{\partial^2 u_l}{\partial x_j \partial x_m} \tag{32}$$

A solution with the ansatz of plane waves yields the dispersion relations for the three acoustical phonon modes. The anisotropy of a crystal is reflected in the shape of the slowness surfaces, surfaces in $\vec{k}$-space with $\omega = $ constant.

Phonon Focussing. The energy transport by phonons is in direction of the group velocity $\vec{v}_{gr} = \vec{\nabla}_k \omega$. The unit vector of $\vec{v}_{gr}$ is normal on the slowness surfaces. A shape of the slowness surfaces deviating from isotropy causes the energy to be directed into preferential directions, which is referred to as phonon focussing.[51]

Phonon focussing can be exploited to provide position sensitivity of the detectors.[52] Experiments have shown, however, that other processes such as isotope scattering and the decay of phonons greatly smear out the focussing patterns.[48]

Decay and Scattering of Phonons. Anharmonicity of a crystal permits 3-phonon processes, i.e. decay and coalescence of phonons. Considering isotropic media and only colinear processes restricts the allowed decay modes of phonons to:

$$\text{LA} \longrightarrow \text{LA} + \text{TA}$$
$$\text{LA} \longrightarrow \text{TA} + \text{TA}$$

LA and TA are the longitudinal acoustical and transverse phonon modes. Optical phonons decay within a very short time scale to acoustical phonons. With above assumptions only LA phonons decay, TA phonons are stable in this model. Typical decay

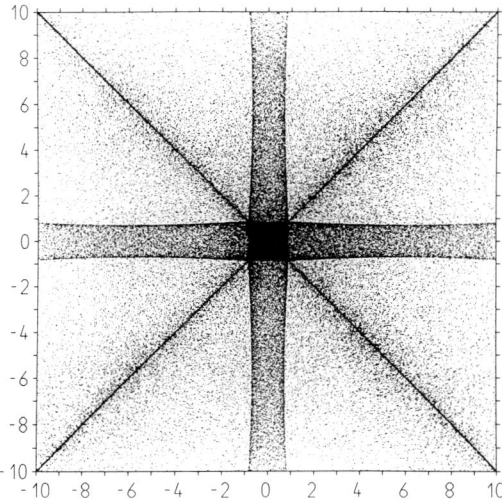

Figure 13. Monte Carlo simulation of a phonon focussing pattern on the surface of a Si crystal. The phonon source is located in the center of the area and 10 mm below the surface shown. Dimensions are given in mm.

constants for LA phonons with frequency ν in Si are:[43]

$$\Gamma_{\text{dec}}(\nu) = \frac{1}{\tau_{\text{dec}}} = \frac{1}{8\,\mu s} \cdot \left(\frac{\nu}{\text{THz}}\right)^5 \tag{33}$$

There exists isotope scattering with a scattering rate of:[43,53,54]

$$\Gamma_{\text{iso}}(\nu) = \frac{1}{\tau_{\text{iso}}} = \frac{1}{0.4\,\mu s} \cdot \left(\frac{\nu}{\text{THz}}\right)^4. \tag{34}$$

Natural Si has an isotope composition of 92.2% ^{28}Si, 4.7% ^{29}Si, and 3.1% ^{30}Si. In isotope scattering or surface scattering mode conversion occurs. This mode conversion maintains a population of the phonon modes according to their phase space, leading to an abundance of 9.4% of longitudinal (L), 37.5% fast transverse (FT), and 53.1% slow transverse (ST) phonons. Thus, also TA phonons decay with a decay time $\tau_{\text{TA}} \approx 10 \cdot \tau_{\text{LA}}$.

Quasidiffusive Propagation. In the present case, where the life time of a phonon generation is much longer than the scattering time, a frequency dependent diffusion constant is established. If it is assumed further that a phonon decays into two phonons each with half the energy of the original phonon, the mean free path for each phonon generation is about 22 times longer than the mean free path of the parent generation.[55] Even in an optimal case, the phonon source of the last generation before reaching the tunnel junctions must have a spatial extension of at least 1/22 of a linear detector dimension. Thus, the phonon focussing pattern must be washed out.

To investigate the processes of phonon propagation in the absorber, a Monte Carlo simulation has been performed. Fig. 14 shows the results for pure ballistic propagation of phonons with the inclusion of phonon decay and scattering, and with surface scattering.

The average energy of phonons arriving at the surface along a line on which the tunnel junctions are situated is also calculated. It is evident, that the detection threshold of tunnel junctions, which is typically some 100 GHz, does not pose a problem.

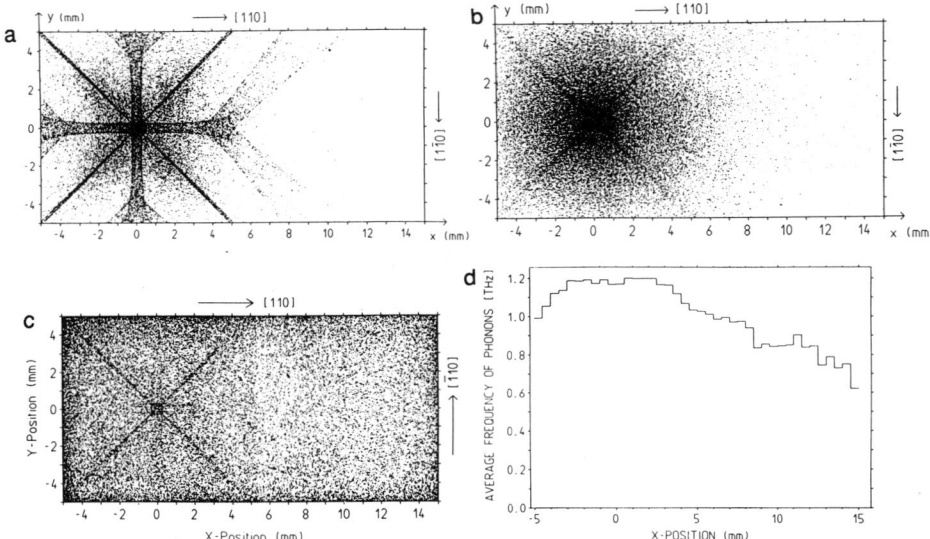

Figure 14. (a) Monte Carlo simulation of the number density of phonons arriving at the top surface of a Si crystal with dimensions $20 \times 10 \times 3\,\text{mm}^3$ for pure ballistic propagation of phonons. The coordinates are given in mm, the phonon source is located at $(0,0)$ 3 mm below the surface shown. The phonon frequency is low enough ($\nu \ll 1\,\text{THz}$) for ballistic phonon propagation. Reflections at the crystal surface are not taken into account. Three tunnel junctions are located at positions $(0,0)$, $(5,0)$, and $(10,0)$. (b) Monte Carlo simulation of the phonon number density including scattering and decay of phonons during propagation. The starting frequency for phonons was $\nu = 7.5\,\text{THz}$. The geometry is equivalent to (a). Reflections of phonons at surfaces are still neglected. The focussing pattern is largely smeared out due to isotope scattering. (c) Monte Carlo simulation of the phonon pattern for pure ballistic propagation including elastic surface reflections but neglecting inelastic surface processes and phonon escape. The geometry is equivalent to (a) and (b). Mode conversion in specular surface reflections is treated according to the anisotropic continuum elasticity theory. In diffuse reflections k vector and mode of the outgoing phonon are chosen according to the density of states. A ratio of specular to diffuse reflection of 0.3:1 was assumed for the polished top and bottom surfaces. Pure diffuse reflection was assumed for the unpolished side walls. A point source located as in (a) and (b) emits a pulse of ballistic (low frequency) phonons at $t = 0$. All phonons which hit the surface within the following 15 μs are shown. On average a phonon suffers 21 surface scattering within these 15 μs. The chosen ratio of specular to diffuse surface reflection has very little effect of the result. (d) Average frequency of phonons as in (b) along the line $y = 0$.

3.3 Superconducting Tunnel Junctions as Phonon Sensors

Operating Principle. A tunnel junction consists of two superconducting metal layers separated by a thin tunnel barrier.[56] The signal from a tunnel junction is a change in tunnel current I_T.

$$I_T = e\gamma_T \cdot (n_{Th}V + N_{excess}) \tag{35}$$

The thermally excited quasiparticle density n_{Th} is a strong function of temperature (cf. eq. 38), thus it is advantageous to operate at low temperatures.[57] There, the thermally excited quasiparticles are negligible against the quasiparticles induced by energy absorption N_{excess}. e is the electronic charge, γ_T is the tunnel rate (cf. eq. 40), and V is the volume of the tunnel junction. There exists a threshold for the detection of phonons in a tunnel junction. Typical values are 90 GHz for Aluminium and 300 GHz for Tin.

Energy Calibration. A silicon substrate of thickness 0.38 mm was used with Tin tunnel junctions evaporated onto the top surface. An α-source illuminated the bottom surface of the crystal to create phonons. An X-ray source (6 keV) irradiated the tunnel junctions directly. Thus, it was possible to relate the energy detected via phonons propagating through the crystal to the energy of the 6 keV X-rays. It was shown that the energy detected via the mediation by phonons even for assumed isotropic propagation was less by more than a factor of 2 compared to the expectation. Many reasons can account for this result, such as an oxide layer on the Silicon, or a reduced transmission of phonons across the interface between absorber and tunnel junction.

Signal Response and Measurement System. A tunnel junction is represented by a current source delivering an excess current, shunted by a dynamical resistance R_D at the bias point and a capacitance C_D (see fig. 15). A charge sensitive preamplifier which is commonly used to read out tunnel junctions in detector applications can be represented by a voltage amplifying block with open loop gain G and a feedback resistor R_F divided by G in parallel with a feedback capacitor C_F multiplied by G ($R_F/G \parallel C_F \cdot G$).

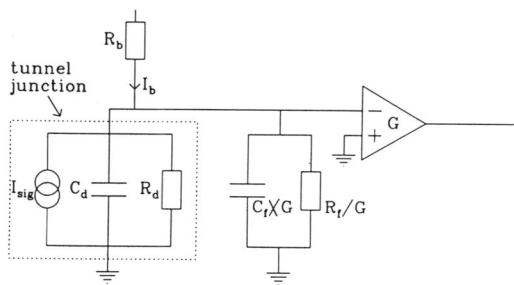

Figure 15. Equivalent circuit of the tunnel junction with a charge sensitive preamplifier and the bias network attached. Energy absorption causes a pulse of the current source $I_{Sig.}(t)$.

The preamplifier output voltage U_A is the convolution of the excess tunnel current $I_{Sig.}$ with the response function of the preamplifier loaded with the detector. The excess tunnel current is caused by excitations in either of the two layers of the tunnel junction (a_l, a_r) convolved with the response function of the tunnel junction.

$$I_{Sig.}(t) = e\,\gamma_T \cdot \int_{-\infty}^{t} e^{-\gamma_V(t-\tau)} [a_l(\tau) + a_r(\tau)]\,d\tau \tag{36}$$

$$U_A(t) = \frac{1}{C_F} \cdot \int_{-\infty}^{t} e^{-\gamma_E(t-\tau)} \cdot I_{Sig.}(\tau)\,d\tau \tag{37}$$

γ_V is the confinement time of an excess density of quasiparticles in the tunnel junction and $\gamma_E = 1/(RC)$ is the electrical time constant, where $R = (R_F/G \parallel R_D)$ and $C \approx C_F \cdot G$. A high value of R_D is essential to maintain a long integration time and to keep the electronic noise low. For high signals, the ratio of γ_T/γ_V needs to be high. Thus, the tunnel barrier of the tunnel junction has to be thin, but also leakage free. A small value of γ_V is achieved by low operating temperatures, clean, pure, and homogeneous films, and shielding against magnetic flux.

3.4 Experiments with Strong Coupling

Comparison of Ballistic and Diffusive Signals. This series of experiments was carried out to investigate the difference of phonon propagation in focussing direction and out of a main focussing direction. To provide α-particle absorption at two specific spots, a copper sheet with two holes (see fig. 16) was used as an aperture. The small hole (diameter: 0.2 mm) is situated such that tunnel junction 1 is in focussing direction [001], while junction 2 is out of a focussing direction. Junction 2 is in a main focussing direction [111] with respect to the large hole (diameter: 0.3 mm). No focussing is expected for junction 1 for absorption in the large hole.

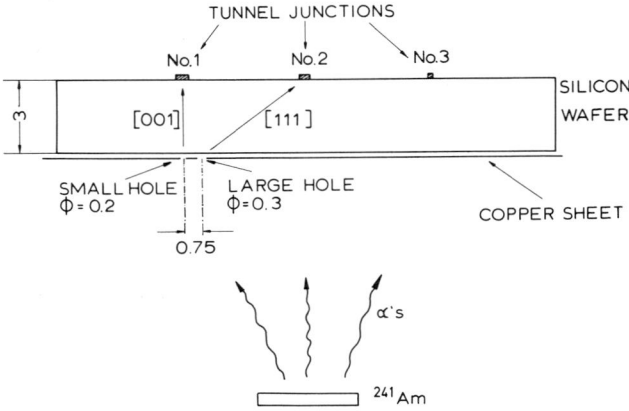

Figure 16. Schematic cross section of the experiment to investigate the effect of focussing properties in certain crystal directions. A Si single crystal of dimensions $20 \times 10 \times 3$ mm^3 with three Al tunnel junctions evaporated onto its top surface has been used to absorb α-particles at two distinct spots, separated by 0.75 mm, on its bottom surface. The operating temperature was 370 mK.

The absorber crystal was coupled strongly to the temperature bath and large contact pads for the tunnel junctions were evaporated on the crystal surface. The result is a pronunciation of the ballistic phonon component because the time phonons are confined to the absorber is too short to favor homogenisation. The dynamical resistance R_D of the tunnel junctions at a temperature of 370 mK was ≈ 10 kΩ leading to an integration time of $\tau_{int} \approx 80\,\mu$s. Compared to the short quasiparticle life time of $\tau_{QP} \approx 30\,\mu$s the integration time is still sufficiently long.

Figure 17 shows the result obtained with the experiment depicted in figure 16. For junction 1 the clear separation of events resulting from energy deposition in the large and small holes is apparent. The pulse height ratio from the experiment is 2.3:1. The ratio expected from pure phonon focussing is 5.5:1, assuming quasi diffusive propagation it is 1.5:1, and with just an isotropic flux 1.1:1 results.

This is a clear indication that phonon focussing must be taken into account to explain the results. Ignoring surface reflections, the pulse height for the large signals requires that at least 70% of the phonons causing the signal must be ballistic. Thus, 12% of the energy is carried by ballistic phonons after creation. Including the reflection of phonons the signals of junction 1 are also consistent with being totally caused by ballistic phonons. The signals of junction 3 can only be explained if surface reflections

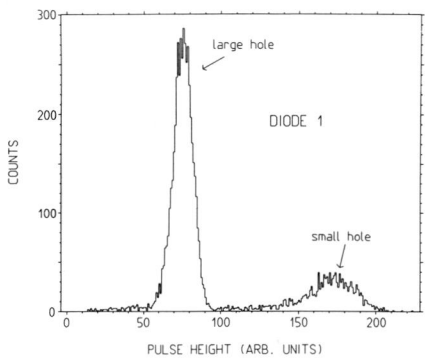
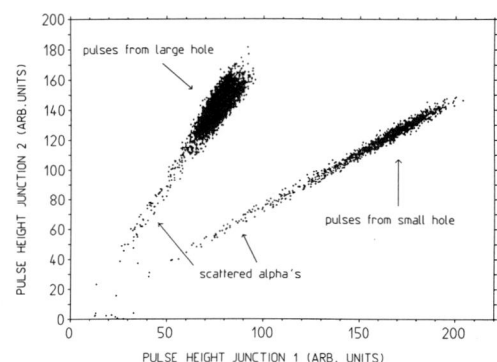

Figure 17. Left: Pulse height spectrum of junction 1. The clear separation of events stemming from α-absorption in the small or in the large hole clearly demonstrates the presence of a large fraction of ballistic phonons in the signals. Right: Pulse height of junction 2 versus pulse height of junction 1. The position information is clearly much better than 0.5 mm the shortest distance between points in the large and the small hole). This high position resolution is inconsistent with pure quasidiffusive propagation.

are included which increase the amount of energy directed towards the distant junction 3.

Residual Fits and Time Differences. To determine the time difference between the signal onsets, the pulses were fitted with an exponential onset. For the signals stemming from absorption in the large hole, the time difference was 1.6 μs, the time difference resulting from absorption in the small hole was 2.5 μs. The expected time difference for even the slowest phonon mode (ST) is 0.4 μs and 0.6 μs. The results may be explained by quasi diffusive propagation of phonons. By comparing the deviation of the registered pulses with the fit function a small fraction of quasi ballistic phonons for junction 2 can be found.

Scattering of phonons off point defects and decay of phonons lead to a situation where the focussing patterns are smeared out and the propagation of phonons is quasi diffusive.

Signal Outside Focusing Directions. The aim of a second set of experiments was to investigate phonon propagation outside any main focussing direction. Irradiation of a Si crystal was permitted at five distinct spots.

The crystal is still strongly coupled to the temperature bath, thus phonons do not stay long in the crystal. As the phonons may not homogenize, a pronounced dependence of the signals on the actual absorption position occurs. Plotting the signals of junction 3 versus junction 1, absorptions stemming from the five holes can be clearly distinguished and a position resolution better than 0.6 mm is deduced.

From the spectrum of time differences, a time resolution of 1.3 μs is obtained and with the time differences plotted against the path length difference for the phonons, an effective velocity of $v \approx 500$ m/s results. The velocity for quasi diffusive propagation is a factor of ≈ 10 smaller than the velocity of sound.

The threshold for energy absorption anywhere in the crystal is 250 keV, in a focussing direction the threshold is as low as 20 keV.

3.5 Experiments with Weak Coupling

Limitation of the Detector Sensitivity. The sensitivity of a phonon mediated

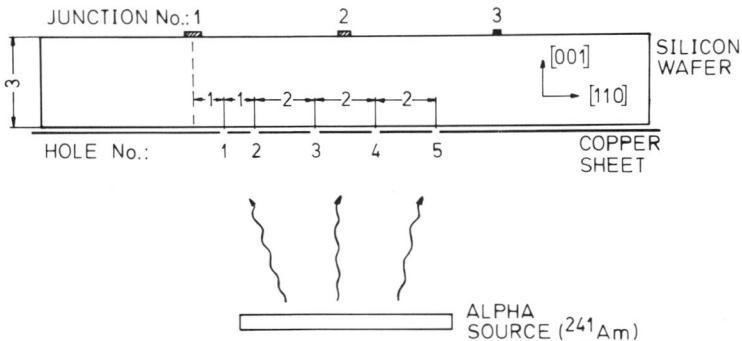

Figure 18. Schematic cross-sectional view of the experimental setup to investigate signals from energy absorption at points outside of any main focussing direction. The diameters of the absorption spots are 0.2 mm.

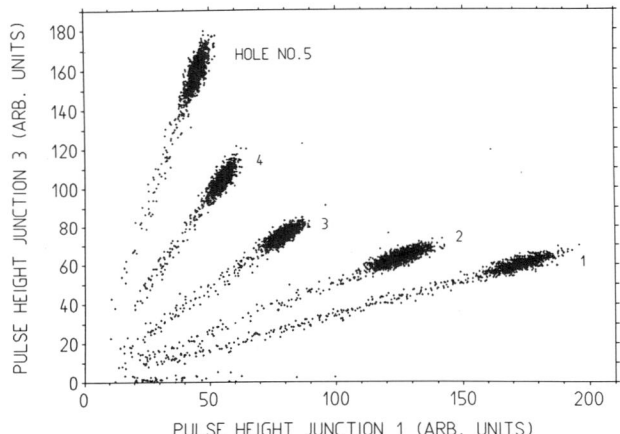

Figure 19. Pulse height of junction 3 vs. pulse height of junction 1. The position resolution is better than about 0.6 mm.

detector could be increased significantly if the phonons created in the absorber cannot leave without being registered by a tunnel junction. Thus, insensitive additional metal films on the absorber surface have to be avoided and coupling to the temperature bath must be weak. Lowering the operating temperature of the tunnel junctions also leads to an improved performance.

With the small tunnel junctions and the absorbers discussed in the previous sections, the probability for the registration of a phonon hitting a surface is $\sim 2 \cdot 10^{-4}$. With a transit time of a phonon across the absorber thickness of $\sim 1\,\mu s$ it takes of order 10 ms to collect a substantial amount of phonons. These 10 ms have to be compared to other time scales:

1.) Preamplifier integration time: $\tau_{int} = R_D G C_F \approx 80\,\mu s$. By reducing the operating temperature of the tunnel junction and by a higher quality, the dynamical resistance of $\sim 10\,k\Omega$ is increased and thus the integration time will be prolonged.

2.) Lowering the operating temperature should also increase the life time of quasiparticles in a tunnel junction from the value of $\tau_{QP} \leq 30\,\mu s$ at a temperature of 370 mK.

The yield of charge presented by the tunnel junction after energy absorption is thereby increased.

3.) Large contact pads used for Indium press contacts cause an escape of phonons with a time scale of $\tau_{esc} \leq 200\,\mu s$. By using wire bonding, the In is avoided and by small contact pads the phonon loss is greatly reduced.

4.) Phonons may decay below the threshold of the tunnel junctions. Very little is known about τ_{inel}, but it may be assumed to be of order $\tau_{inel} \approx 1000\,\mu s$.[58]

Experimental Setup. In contrast to the experimental setup of the experiments discussed in the previous section, the idea of 'weak' coupling of the absorber to the temperature bath is realized in these experiments. The phonons shall stay as long as possible in the absorber and if they are absorbed in a metal film on the absorber surface, a signal is generated.

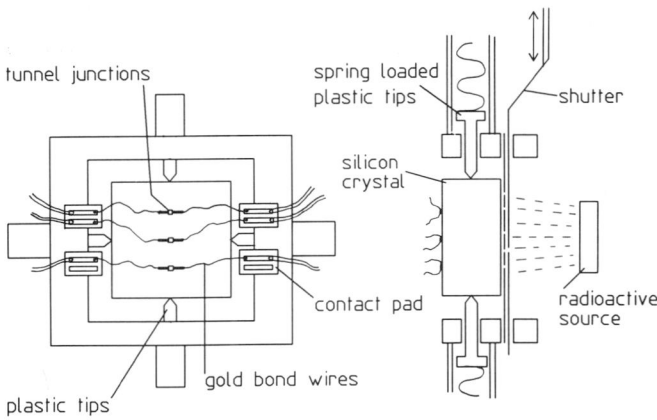

Figure 20. A Si single crystal is supported by plastic tips to minimize heat conduction. In addition, Au bond wires have been used instead of Indium press contacts. The thermal time constant of this setup was of order 1 ms.

The contact pads are reduced in size from $9\,mm^2$ to $0.1\,mm^2$, while the tunnel junction size was even slightly enlarged compared to experiments with 'strong' coupling. A copper bar supporting the crystal was replaced by plastic tips and the In press contacts were replaced by Au bond wires. A thermal time constant of $\tau \approx 1\,ms$ results. The operating temperature was lowered from 370 mK to 60 mK.

Results. The absorber crystal was illuminated at seven positions. As a consequence of the weak coupling the duration of the pulses of the tunnel junctions increased from $\sim 40\,\mu s$ to $\sim 500\,\mu s$. Homogenisation of the phonons took place. Plotting the pulse heights of junction 1 versus junction 2, the pulses were identical for any of the absorption positions except for those where one of the tunnel junctions laid in a focussing direction.

In the experiments with strong coupling, 11% of the signals was caused by reflected phonons. Introducing weak coupling, no change for the ballistic phonons should occur, whence the signals obtained for focussing directions may be referenced to each other. The ratio of homogenized to ballistic phonons increased from 0.12 in the strong coupling case to 5.2 in the weak coupling case. Although homogenizing phonons in the absorber decreases the position resolution, while emphasizing good energy measurement, position information is still contained in the onset of the pulses. The onset of the pulse is still caused by the ballistic phonons and it is just a question of the elapsed time after the pulse onset at which the pulse is analyzed if position or energy measurement is favored.

An improvement in threshold was not yet demonstrated due to some experimental difficulties with the tunnel junctions. Later experiments, however, gave a threshold already in the keV range.

3.6 Summary

Detectors with dielectric absorber crystals with volumes of some cm^3 and superconducting tunnel junctions on their surfaces are detectors well suited for the registration of non ionizing events. Applications may be found in experiments such as the detection of dark matter candidates or as with low energy neutrinos and the exploitation of the coherent elastic scattering of neutrinos. A good understanding of the detector exists. By design, mainly energy information or position information may be emphasized. It is important to increase the sensitive area of the absorber covered with tunnel junctions. This can be accomplished by large single junctions, series connected junctions, or quasiparticle trapping elements. An energy threshold below 1 keV per several cm^3 appears conceivable.

4. SUPERCONDUCTING TUNNEL JUNCTIONS IN GENERAL

4.1 General Detector Principle

A superconducting tunnel junction may be considered as a device measuring energy. The electrical signal, which is the voltage output of a preamplifier, is proportional to the energy deposited in the tunnel junction.

A superconducting tunnel junction may eventually replace surface barrier semiconductor detectors in some applications. In this detector concept the existence of the small energy gap is exploited to increase the number of charge carriers created per energy deposition. In the case of a semiconductor detector research and development over decades has led to a situation where the energy resolution of such detectors is indeed limited by the statistics of charge carriers.[59] The contribution by electronics has been made negligibly small. Thus, the only way to improve further is to use a material with a smaller energy gap. Superconducting tunnel junctions are well suited for this application, but due to the thin tunnel barrier, the electrical capacitance of such devices is very high and thus the electronic noise limits the energy resolution (cf. eq. 49).

This situation could be eased to some extent by fabricating detectors utilizing quasiparticle trapping.[60] This design permits smaller tunnel junctions without loosing sensitive area. Quasiparticles created in a separate absorber diffuse therein and get trapped by attached tunnel junctions.[61]

In a different detector application superconducting tunnel junctions are fabricated on the surface of a dielectric single crystal to register the non equilibrium phonons created by a recoiling nucleus in the absorber. Such a detector is sensitive also to non ionizing events of low energy (see section 3 of this paper).

Relaxation phonons may also be detected as long as their energy is above the threshold of the tunnel junction. Such phonons are emitted while a quasiparticle relaxes down in energy.[62]

4.2 Superconducting Tunnel Junctions

A superconducting tunnel junction consists in our experiments of two superconducting metal films separated by a thin oxide layer acting as tunnel barrier. The tunnel barrier is as thin as 10 Å to 20 Å to allow quantum mechanical tunneling of quasiparticles. Quasiparticles are created by the breaking of Cooper pairs, requiring the energy of 2Δ, the energy gap in the superconductor. Energy to the superconducting tunnel junction may be transferred by the absorption of a particle directly in a tunnel junction, the absorption of phonons from the crystal onto which the tunnel junction is fabricated, or by injection of quasiparticles created in a separate absorber. In addition to the induced quasiparticles there exists always a density of thermally excited quasiparticles with a temperature dependence as:

$$n_{Th} \propto \sqrt{T} \cdot \exp(-\Delta/kT) \tag{38}$$

The thermally excited quasiparticles have two detrimental effects on the performance of a tunnel junction as particle detector.

— The tunnel current at the bias point is proportional to the thermal density of quasiparticles in the tunnel junction. With a high quasiparticle density a high bias current is required, leading to a high shot noise contribution (cf. eq. 48).

— Excess quasiparticles may recombine to form Cooper pairs with thermally excited quasiparticles, thereby reducing the life time of the excess quasiparticles.

The impact of both processes can be greatly reduced by lowering the operating temperature of the tunnel junction to values below 1/10 of the critical temperature T_C of the junction material.

4.3 Quality of Superconducting Tunnel Junctions

A tunnel junction with the two metal films made of the same material should have a pronounced increase of the tunnel current at a voltage of $2\Delta/e$. Below this voltage there should exist a thermal tunnel current with a temperature dependence given by $I_{Th} \propto n_{Th}$ (cf. eq. 38). In reality, however, there exist additional features.[20]

At very low temperature an appearant increase of the tunnel current at a voltage Δ/e is visible and also some ohmic behavior may be seen. This is an indication that besides the superconductor-superconductor tunnel junction there is also a fraction of the tunnel junction which is represented by a superconductor-normalconductor characteristics. To an even lower fraction there is also some normalconductor-normalconductor tunnel junction with a temperature independent tunnel current. The deviation from the ideal behavior of a tunnel junction is caused by material impurities and inhomogenities.[63]

4.4 Tunnel Process and Initial Experiments

Two tunnel processes exist for transferring quasiparticles from one metal film to the other one and vice versa. Although, quasiparticles may travel in either direction one electric charge is always transferred in the direction of applied voltage.

By the two tunnel processes a and b (cf. fig. 22) quasiparticles may tunnel in either direction, whence tunneling itself is in general not a loss process. The excess tunnel current resulting from the excitation of N_L or N_R quasiparticles in the left or right layer is:

$$I_{Sig.} = e \cdot (\gamma_{T_{LR}} \cdot N_L + \gamma_{T_{RL}} \cdot N_R). \tag{39}$$

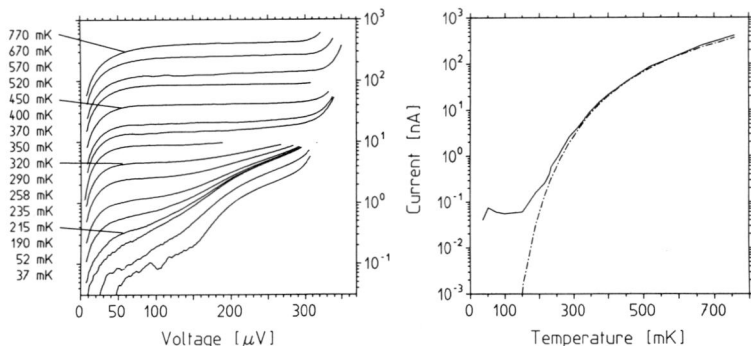

Figure 21. In the left part of the picture, several branches of current voltage traces are shown for various operating temperatures. The vertical axis is logarithmic. In the right diagram, the current level at fixed voltage from the experiment (solid line) together with the calculation according to the BCS-theory dot dashed line) is plotted versus temperature. The deviation at low temperatures may be attributed to the presence of non idealities in the tunnel junction.

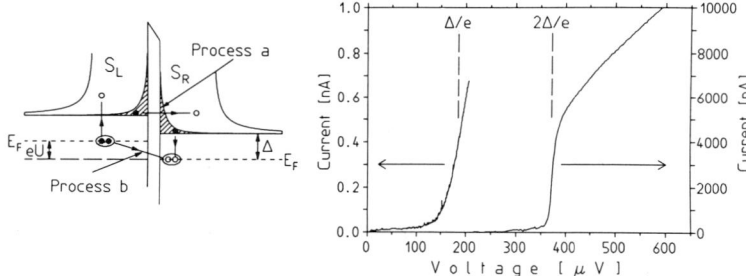

Figure 22. On the left, the two dominant tunneling processes in a superconducting tunnel junction are shown. By process a, a quasiparticle tunnels from left to right. Process b transfers a quasiparticle from right to left via the exchange of a virtual phonon. In both cases, one electrical charge flows from left to right. On the right, a current voltage characteristics of one of our superconducting Aluminium tunnel junctions is shown. The operating temperature is 60 mK. Two branches of the same device are shown with different current scales. The increase in current below the energy gap of the low current (left) branch may be due to non ideal properties of the superconducting metal films.

The tunnel rate constants are:

$$\gamma_{T_{xy}} = \frac{1}{4\,e^2\,n(E_F)} \cdot \frac{1}{R_{NN}\,A\,d_x} \cdot \overbrace{F(\Delta, U)}^{\approx 1.5\ \text{(typ.)}} \qquad (40)$$

$n(E_F)$, R_{NN}, A, and F are the single spin density at the Fermi level, the normal conducting resistance of the tunnel junction, the tunnel area, and a function reflecting the density of states in the superconductor, respectively. For a given amount of energy deposited in a tunnel junction the pulse height is inversely proportional to the thickness d_x of the film in which the energy absorption took place. The possibility to distinguish between absorption in the two layers is appreciable only for test purposes of tunnel junctions. The ambiguity in signal response should be avoided in real detector applications. Thus, the separation of absorber and tunnel junctions is advantageous.

4.5 Quasiparticle Trapping

Potential wells for quasiparticles are created if superconductors of different energy gaps are metallically connected. Quasiparticles relax to the locally lowest energy state with a rate proportional to E^3, with E the excitation energy above the lowest state.[64]

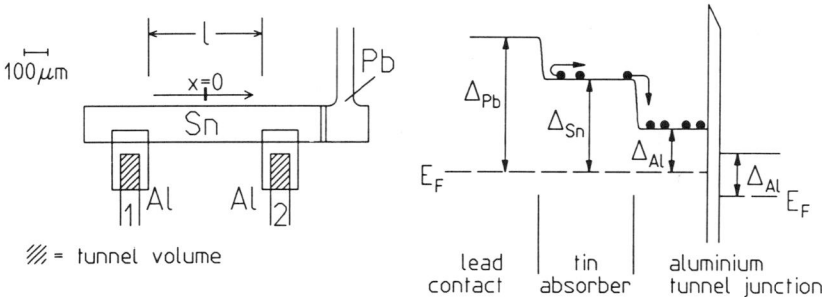

Figure 23. On the left side, the arrangement of the detector is shown: The tunnel junctions consist of two Aluminium films separated by a thin tunnel barrier. The right side shows the energy level diagram for quasiparticles: Quasiparticles created in the Tin absorber are trapped by one of the Aluminium tunnel junctions where they have to tunnel before having a chance to escape into the contact pad of the particular tunnel junction.

Tin with an intermediate energy gap is used as absorber for X-ray photons. Aluminium tunnel junctions with a small energy gap are attached to the absorber to register the quasiparticles created in the absorber. A Lead contact with high energy gap is attached to the Tin absorber to provide a common electrical contact.

The quasiparticles created by X-ray absorption in the absorber propagate by diffusion while being submitted to a loss of single quasiparticles. The equation for the quasiparticle density $n(x,t)$ is:

$$\dot{n}(x,t) = D \cdot \Delta n(x,t) - \gamma_D \cdot n(x,t) \tag{41}$$

Assuming a δ-like initial quasiparticle distribution and complete trapping into one film of the superconducting tunnel junctions a relation between the deposited energy E_0 and the absorption position x_0 with the measured pulse heights Q_+ and Q_- can be found:

$$E_0 = \frac{\omega}{K} \cdot \sqrt{Q_+^2 + Q_-^2 + 2Q_+Q_-\cosh\alpha} \tag{42}$$

ω is the effective energy required to create one quasiparticle and K is the relative yield of quasiparticles ($0 < K \leq 1$).

$$x_0 = \frac{\ell}{2\alpha} \cdot \ln\left(\frac{Q_+ \cdot e^{\alpha/2} + Q_- \cdot e^{-\alpha/2}}{Q_+ \cdot e^{-\alpha/2} + Q_- \cdot e^{\alpha/2}}\right) \tag{43}$$

The parameter $\alpha = \ell/\sqrt{D\tau_D}$ contains the detector parameters ℓ absorber length, D diffusion constant, and τ_D the life time of quasiparticles in the absorber.

An energy resolution of 56 eV (FWHM) at an energy of 5.89 keV was obtained and a relative position resolution of better than 1% was demonstrated.

4.6 Electronic Noise

The capacitance of semiconductor detectors is small and the charge created by absorbed energy is released in a very short time. Thus, it was possible to render the electronic contribution to the total line width negligible. However, the situation is different with superconducting tunnel junctions. The electrical capacitance C_D is much higher, the

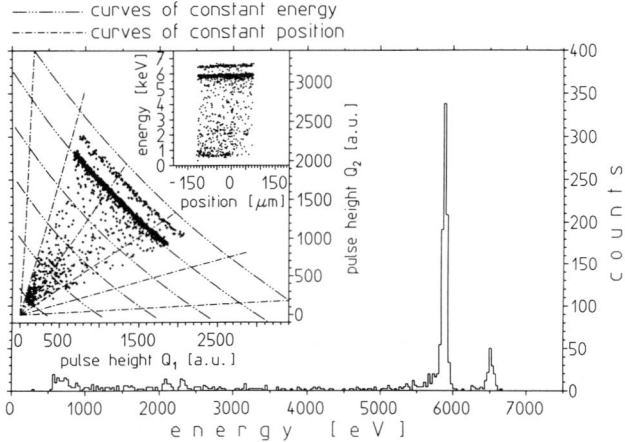

Figure 24. The energy spectrum resulting from the irradiation of part of the absorber with X-ray photons emitted by an ^{55}Fe-source. The energy resolution is better than 60 eV (FWHM) for the K_α line. In the inset, the original data are shown in the Q_+ vs. Q_- plane superimposed with lines of constant energy and constant position and the transformation into the energy vs. position plane.

dynamical resistance at the bias point R_D is much lower, and the time required for signal collection τ_p is much longer compared to semiconductor detectors.

To specify the electronic line width the signal to noise ratio $\mathcal{R}_{SN}$ with optimal filtering is calculated.

$$\mathcal{R}_{SN}^2 = \frac{1}{2\pi} \cdot \int \frac{U_{sig.}^2}{\mathcal{N}(\omega)} d\omega \qquad (44)$$

Signal voltage $U_{sig.}$, spectral power density of noise $\mathcal{N}(\omega)$, series noise **a**, and parallel noise **b** are given by:

$$U_{sig.} = I_{sig.} \cdot Z_F \qquad (45)$$

$$\mathcal{N}(\omega) = \Big(\underbrace{\mathbf{b} + \frac{\mathbf{a}}{R_D^2}}_{\mathbf{b_a}} + \mathbf{a}\omega^2 C_D^2\Big) \cdot Z_F^2 \tag{46}$$

$$\mathbf{a} = 4kT \cdot \frac{0.65}{g_m} \propto \frac{1}{C_D} \tag{47}$$

$$\mathbf{b} = 4kT \cdot \Big(\frac{1}{R_B} + \frac{1}{R_F}\Big) + 2e \cdot (I_G + I_B) \tag{48}$$

T, Z_F, g_m, R_B, R_F, I_G, and I_B are temperature, feedback impedance of the preamplifier, forward transconductance of the amplifier's input FET, bias resistor for the detector, feedback resistance, gate leakage current, and detector bias current, respectively. The optimal signal to noise ratio for a signal charge Q can be written as:

$$\mathcal{R}_{SN} = \frac{Q}{\sqrt{\sqrt{\mathbf{a}\,\mathbf{b_a}}\,C_D + \mathbf{b_a}\,\tau_p}} \tag{49}$$

It is evident from eq. 49 that the high electrical capacitance C_D of a superconducting tunnel junction and the long pulse length τ_p are detrimental for the signal to noise ratio. The best energy resolution obtained so far with superconducting tunnel junctions is limited by electronic noise as a consequence of the achieved parameters for such detectors.

To improve further, the life time of quasiparticles in the tunnel junction τ_V and the quality factor $\mathcal{Q} = R_D/R_{NN}$ have to be maximized, while the tunnel time τ_T and the tunnel area A_t need to be minimized.

4.7 Comparison with Phonon Experiments in Solid State Physics

The aim of experiments in solid state physics is different from that in particle physics. A brief comparison between the two fields is given here:

In solid state physics, one aim of research is to study the dynamics of phonons, while phonons are just a transport medium for energy in particle detectors. To study phonons, one usually chooses an excitation energy suitable for the signal to noise ratio of the experiment (typ. nJ) and one has a start signal available from triggering a laser, heater, or other energy transducer. It is possible to average over many signals to improve the signal to noise ratio and thereby to detect very weak effects. In the field of cryogenic detectors, one's interest is to detect single particles of very low energy (typ. fJ to pJ, or even less) stemming from a particle interaction or a radioactive decay. Thus, the time an event occurs is in general not under control of the experimentalist and signal averaging is impossible. There is always a trade off between time resolution of a detector and its sensitivity. Solid state physicists want to study the time structure of some feature, and thus operate tunnel junctions at higher temperature where the quality of a tunnel junction is of less importance. With cryogenic detectors, highest sensitivity is required and the quality of the tunnel junctions is very important.

4.8 Summary

Tunnel junctions with very high quality suitable for the detection of soft X-ray photons are available. The quality factor $\mathcal{Q} = R_D/R_{NN}$ is higher than $\sim 10^5$. Quasiparticle trapping by attaching superconductors of different energy gaps works. An X-ray detector with a Tin absorber and two Aluminium tunnel junctions attached to it exhibits an energy resolution of 56 eV (FWHM) at an energy of 5.89 keV. In addition, relative position resolution better than 1% was demonstrated. These results are limited by electronic noise. Tin as absorber material is cumbersome, thus, Vanadium absorbers have been grown epitaxially. The mean free path of electrons in our Vanadium films against scattering off impurities and defects is already longer than 0.6 μm.

5. REVIEW ON OTHER CRYOGENIC DETECTOR CONCEPTS

5.1 Calorimetric Detectors

Among the calorimetric detectors there is a very wide span in absorber mass. From the μg range with extremely high energy resolution to masses of several 100 g with still a low threshold. There is a variety of absorber materials as well as thermometers. The overview, which is by far incomplete, is based on information contained in ref. [4].

A resolution of 10.5 keV (FWHM) for 5.8 MeV α-particles has been demonstrated by N. Coron et al. with a 3 mg composite calorimeter. The absorber was sapphire and the thermometer NTD Germanium. The baseline noise is 600 eV. The same group has fabricated a 24 g sapphire calorimeter operating at a temperature of 55 mK. They achieve an energy resolution of 3.7 keV (FWHM) for 59.6 keV γ-quanta from an ^{241}Am source.

The group of E. Fiorini uses calorimeters with TeO_2 absorbers of masses from some g up to 73 g. The thermometer again is NTD Germanium. A relative energy resolution of better than 1% is obtained for energies above 2 MeV. These detectors already operate in the Gran Sasso National Laboratory in Italy in $\beta\beta$-decay experiments.

B. Sadoulet and his group at the Center for Astroparticle Physics investigate 60 g Germanium crystals where they simultaneously measure signals from phonons (with an NTD Ge sensor) and from ionisation in the Germanium. They report an energy resolution of 0.8 keV (FWHM) at 60 keV obtained with the phonon signal.

At the Technische Universität München and the Max-Planck Institut the thermometers used are transition edge thermometers, consisting either of pure Iridium or of a Gold Iridium sandwich. Absorber materials are Silicon, Vanadium, or Molybdenum. For a 35 g Molybdenum calorimeter, the relative energy resolution is $\sim$ 10% and for a 15 g Vanadium calorimeter it is 1.2%, in both cases for α-particles of energy 5.8 MeV. With a 19 g Silicon calorimeter operating at a temperature of 220 mK an energy resolution of 1.2 keV (FWHM) was obtained for 60 keV γ-quanta.

D. McCammon from the Goddard-Wisconsin collaboration presents a calorimeter with doped Silicon as thermometer and HgTe as absorber with an active area of 0.25 mm^2. An energy resolution of 7.35 eV (FWHM) for X-ray photons of energy 5.89 keV is demonstrated. At lower energies, the baseline fluctuations of 4.6 eV limit the energy resolution.

5.2 Magnetic Bolometers

In a magnetic bolometer, the very small elementary excitation stemming from the Zeeman splitting between the two states of a Kramers doublet is utilized. The energy required for such an excitation is of order 10^{-7} eV. E. Umlauf from the Walther Meissner Institute in Garching has operated a compound detector of 120 g mass. A baseline noise of 8 eV/$\sqrt{Hz}$ (rms) is published but due to experimental problems an energy spectrum has not yet been presented.[65]

5.3 Rotons in Superfluid Helium

Neutrinos may also be detected in large mass of superfluid Helium. Recoiling nuclei create rotons which have a long mean free path in superfluid Helium. On the surface, rotons evaporate Helium atoms which are subsequently registered by calorimeters above the liquid. Initial tests have been performed and the efficiency of the process is discussed.[66]

6. CONCLUDING REMARKS

The initial motivation for the development of cryogenic detectors was initiated by particle and nuclear physics. Many new detector concepts have been proposed since the early days of low temperature detectors and also many of these ideas have been discarded or proven infeasible. Nevertheless, the number of people developing cryogenic detectors is growing.

There are detectors with superconducting tunnel junctions, calorimetric detectors, superheated supercooled granules, and other miscellaneous detector concepts. The detectors become more and more elaborated and sophisticated, whence the help of physicists working in the traditional fields of solid state physics is highly appreciated. The elements used in cryogenic detectors are well known for a long time in solid state physics. But in the detector development discussed here, the sensitivity of the devices has been pushed to their limits, expressed by the small energies being registered (some orders of magnitude smaller than in solid state physics).

This interdisciplinary field has introduced new viewpoints on detector development. Some of the detectors exhibit features never attained before and initial experiments are already in progress. Still, there are many opportunities for further improvement.

7. ACKNOWLEDGEMENTS

Part of the work reviewed in this article is supported by the German Bundesministerium für Forschung und Technologie. I want to thank Dr. F. Pröbst from the Max-Planck Institute in München for providing the material used in section 2 and for valuable discussions. The support of J. Jochum, B. Kemmather, and M. Gutsche is gratefully acknowledged.

8. REFERENCES

[1] Proceedings of Low Temperature Detectors for Neutrinos and Dark Matter, edited by K. Pretzl, N. Schmitz, and L. Stodolsky, Springer-Verlag Berlin Heidelberg (1987)

[2] Low Temperature Detectors for Neutrinos and Dark Matter II, edited by L. Gonzalez-Mestres and D. Perret-Gallix, Editions Frontières (1988)

[3] Low Temperature Detectors for Neutrinos and Dark Matter III, edited by L. Brogiato, D.V. Camin, and E. Fiorini, Editions Frontières (1990)

[4] Low Temperature Detectors for Neutrinos and Dark Matter IV, edited by N.E. Booth and G.L. Salmon, Editions Frontières (1992)

[5]   R.L. Mössbauer, J. Phys. G: Nucl. Part. Phys. **17** (1991) S1
[6]   E. Fiorini, Physica **B 169** (1991) 388
[7]   H. Kraus et al., Nucl. Instr. Meth. **A 326** (1993) 172
[8]   M. Juda et al., in: EUV, X-Ray, and Gamma-Ray Instrumentation for Astronomy III, Oswald H. W. Siegmund, Editor, Proc. SPIE 1743 (1992) 398
[9]   L. Stodolsky, Comments Nucl. Part. Phys. **18** (1988) 157
[10] D.Z. Freedman et al., Ann. Rev. Nucl. Sci. **27** (1977) 167
[11] D.Z. Freedman, Phys. Rev. **D 9** (1974) 1389
[12] G. Zacek et al., Phys. Rev. **D 34** (1986) 2621
[13] P. Anselmann et al., Physics Lett. **B 258** (1992) 376
[14] B.V. Pritychenko, Nucl. Instr. Meth. **A 314** (1992) 390
[15] L.M. Krauss, "Dark Matter in the Universe", Scientific American, Dec 1986, p.50
[16] V. Trimble, Ann. Rev. Astron. Astrophys. **25** (1987) 425
[17] A. v. Kienlin et al., in ref. [4] p. 377
[18] A. Gabutti et al., Nucl. Instr. Meth. **A 289** (1990) 425
[19] S.S. Holt, Astro. Lett. and Commun. **26** (1987) 61
[20] H. Kraus et al., in: EUV, X-Ray, and Gamma-Ray Instrumentation for Astronomy III, Oswald H. W. Siegmund, Editor, Proc. SPIE 1743 (1992) 36
[21] W.C. Haxton and G.J. Stephenson Jr., Progr. in Part. and Nucl. Phys. **12** (1984) 409
[22] G.F. Dell'Antonio and E. Fiorini, Suppl. Nuovo Cim. **17** (1960) 132
[23] A. Alessandrello et al., Nucl. Instr. Meth. **A 314** (1992) 595
[24] W. Schreiber, Diploma Thesis, "Grenzen der Energieauflösung röntgenempfindlicher pn-CCDs", MPE Report 224, Jan. 1991
[25] H. Kraus et al., Europhys. Lett. **1** (1986) 161
[26] D. Twerenbold, Europhys. Lett. **1** (1986) 209
[27] Th. Peterreins et al., Phys. Lett. **B 202** (1988) 161
[28] M. Kurakado, Nucl. Instr. Meth. **A 314** (1992) 252
[29] N.E. Booth et al., Nucl. Inst. Meth. **A 315** (1992) 201
[30] W. Seidel et al., Phys. Lett. **B 236** (1990) 483
[31] A. Alessandrello et al., Nucl. Instr. Meth. **A 295** (1990) 405
[32] N. Coursol et al., Nucl. Instr. Meth. **A 312** (1992) 24
[33] E.E. Haller et al., in: "Neutron Transmutation Doping of Semiconductor Materials" edited by R.D. Larrabee, Plenum Press New York (1984) 21
[34] E.H. Silver et al., Nucl. Inst. Meth. **A 277** (1989) 657
[35] D.G. McDonald, Appl. Phys. Lett. **50** (1987) 775
[36] D. McCammon et al., Nucl. Inst. Meth. **A 326** (1993) 157
[37] A.K. Drukier and L. Stodolsky, Phys. Rev. **30** (1984) 2295
[38] B.G. Turrell et al., Nucl. Inst. Meth. **A 289** (1990) 512

[39] D. Hueber et al., Nucl. Inst. Meth. **167** (1979) 201

[40] W. Seidel et al., Rev. Sci. Instrum. **58** (1987) 1471

[41] F. Pröbst et al., in ref. [4] p. 193

[42] F. Pröbst, "Phonon Mediated Detection of Particles", to be published in the "Proceedings of the 7th International Conference on Phonon Scattering in Condensed Matter", Ithaca, Aug 3–7, 1992

[43] S. Tamura, Phys. Rev. **B 31** (1985) 2574

[44] J.P. Harrison, J. Low Temp. Phys. **37** (1979) 467

[45] G. Bergmann et al., Phys. Rev. **B 41** (1990) 7386

[46] S.B. Kaplan, J. Low Temp. Phys. **37** (1979) 343

[47] E.T. Swartz and R.O. Pohl in A.C. Anderson and J.P. Wolfe (eds.): Proceedings of "Phonon Scattering in Condensed Matter V", Springer Series in Solid State Science (1986) 68

[48] Th. Peterreins et al., J. Appl. Phys. **69(4)** (1991) 1791

[49] R. Baumgartner et al., Phys. Lett. **A 94** (1983) 55

[50] R.S. Markiewicz, Phys. Rev. **B 21** (1980) 4674

[51] B. Taylor et al., Phys. Rev. **B 3** (1971) 1462

[52] B. Cabrera et al., Nucl. Instr. Meth. **A 275** (1989) 97

[53] D.V. Kazakovtesev and Y.B. Levinson, phys. stat. sol. **b 136** 1986) 425

[54] S. Tamura, Phys. Rev. **B 27** (1983) 858

[55] W.E. Bron et al., Phys. Rev. Lett. **49** (1982) 209

[56] I. Giaever and K. Megerle, Phys. Rev. **122** (1961) 1101

[57] W. Eisenmenger et al., Appl. Phys. **11** (1976) 307

[58] W. Knaak et al., in A.C. Anderson and J.P. Wolfe (eds.): Proceedings of "Phonon Scattering in Condensed Matter V", Springer Series in Solid State Science (1986) 174

[59] J. Kemmer et al., Nucl. Instr. Meth. **A 288** (1990) 92

[60] N.E. Booth, Appl.Phys.Lett. **50** (1987) 293

[61] H. Kraus et al., Phys. Lett. **B 231** (1989) 195

[62] D.J. Goldie et al., Physica **B 169** (1991) 443

[63] K.E. Gray, J.Physics F (Metal Physics) **1** (1971) 290

[64] S.B. Kaplan et al., Phys.Rev. B **14** (1976) 4854

[65] E. Umlauf and M. Bühler, in: EUV, X-Ray, and Gamma-Ray Instrumentation for Astronomy III, Oswald H. W. Siegmund, Editor, Proc. SPIE 1743 (1992) 391

[66] S.R. Bandler et al., Phys. Rev. Lett. **68** (1992) 2429

CRYOGENIC PARTICLE DETECTORS: PHONON PHYSICS IN NIOBIUM

Richard Gaitskell

Department of Physics
University of Oxford
Nuclear Physics Laboratory
Keble Road
Oxford OX1 3RH
UK

INTRODUCTION

In this paper we will detail some of the work being conducted in the Cryogenic Detector Group, at the Nuclear and Particle Physics Laboratory in Oxford, UK, in order to realise a new generation of particle detectors.

In particle physics it is becoming increasingly difficult to probe further into the structure of matter. Accelerator experiments are used to create particles at high energies. However, a number of fundamental problems in cosmology, neutrino and weak interaction physics, that do not require high energy accelerators, have not been resolved for lack of sensitive particle detectors. Our work on the propagation of non–equilibrium phonon signals in single crystal absorbers and their detection with superconducting tunnel junctions has been catalysed by such problems.

Dark Matter Detector based on Single Crystal Niobium

The search for weakly interacting massive particles (WIMPs) [1] by observing their scattering in underground detectors will require a range of target materials. Nb is a good candidate given it is mono–isotopic with a large neutron content (N=52) and non–zero nuclear spin (9/2). This makes it sensitive both to the coherent scattering of WIMPs through vector interactions and to axial couplings.

We have been studying the propagation of phonons and quasiparticles produced in single crystal superconducting Nb by laser and particle interactions. Superconducting Tunnel Junctions (STJs) and Series Arrays of Al Superconducting Tunnel Junctions (Al SASTJs) fabricated on the surface of Nb crystals were used to detect these signals.

The demands of a low temperature Nb based detector for dark matter are exacting. If such a device is to be realised a detailed understanding of the behaviour of phonons and quasiparticles in bulk Nb single crystals has to be achieved. In particular, the detector

---

[1] These particles have been proposed as a solution to the "missing" mass or dark matter problem of the Universe [1, 2].

would be required to differentiate between nuclear and electron recoils. This may be possible by identifying differences in the yield of phonons and quasiparticles or the energy spectrum of phonons from the initial interaction [3]. Considering this point has raised a number of interesting questions about the way in which interactions couple to the crystal lattice and superconducting state and how this will influence the subsequent spread of the non–equilibrium state throughout the crystal [4, 5].

Cryogenic Detector as a ß-Decay Spectrometer

We are also developing a ß–decay spectrometer based on low temperature detectors that will address some of the problems present in the existing detector technologies. It is our objective to carry out a definitive experiment using an entirely new technique. In this experiment the ß–particles are measured via the phonons they produce in the absorber. The phonons are detected by SASTJs [6]. Such a detector can be constructed to avoid back–scattering effects associated with separated source–detector geometries by constructing it as a sandwich surrounding the thin source layer hermetically [7]. Work at this laboratory by Goldie et al on 2 g InSb absorbers using SASTJs has shown resolution FWHM of 300 eV on 22 keV X–rays at 70 mK [8].

Global Development of Low Temperature Detectors

Our work should be viewed in the context of the wider development of particle detectors based on low temperature (less than 4 K) techniques. These new detectors can be divided into two broad categories which are based either on the detection of thermal phonon distributions (calorimeters) or more complex non–equilibrium signals. Our work falls into the latter category. The dividing line is, however, becoming increasingly blurred. The field is reviewed elsewhere in these proceedings [9].

INVESTIGATION OF TRANSPORT PHENOMENA IN NIOBIUM

Introduction

Relatively little is known about the non–equilibrium dynamics of phonon and quasiparticle propagation in bulk single crystal Nb. Existing studies have concentrated on the measurement of the specific heat [10-16] and thermal conductivity [17-21] of bulk Nb: properties characterising the behaviour of near–equilibrium distributions of the phonons and electrons. Measurements of the propagation of phonons using ultrasonic techniques have also been made [22-27].

Nb is now in common usage in the manufacture of thin film superconducting devices. This work, however, is still in its infancy as far as the detailed understanding of quasiparticle dynamics in Nb. In principle, the use of bulk single crystal Nb should provide a simple system in which to study such behaviour. However, our investigations in this area have been plagued by the complexity of the surface physics of Nb (see the comprehensive summary by Halbritter [28]). Similar problems have hampered the development of the Nb thin film devices and have also been the major barrier to previous investigations of superconducting tunnel junction fabrication directly on Nb single crystals (see [29] and references therein).

Phonon propagation in bulk superconducting Nb is limited by an upper energy, determined by $2\Delta_{Nb} \sim 3$ meV. Phonons above this threshold will rapidly break Cooper

Pairs in the Nb itself creating quasiparticles (above–gap phonon m.f.p. ~ 100 Å). At temperatures significantly below the T_c of Nb, phonons at energies below the superconducting energy gap will propagate freely, unhindered by scattering from electrons. Given that Nb has the largest energy gap of the elemental superconductors it's phonon propagation properties at low temperatures are more akin to those of dielectric crystals.

Ballistic Phonon Propagation

The leading edge of non-equilibrium signals traversing high purity Nb crystals [2] generated by a laser excitation at the surface and detected by an Al STJ or SASTJ on an opposite face, are characterised by the ballistic propagation of phonons. Along the [110] axis the phonons will propagate in three non–degenerate modes L (longitudinal), FT (fast transverse) and ST (slow transverse).

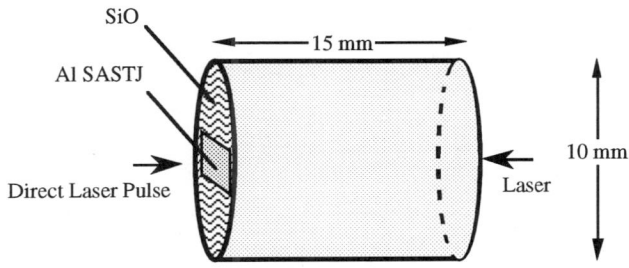

Figure 1 Diagram of Nb single crystal (15 mm x ø10 mm) with position of laser fibres and SASTJ. The cylinder axis lies in the [110] direction.

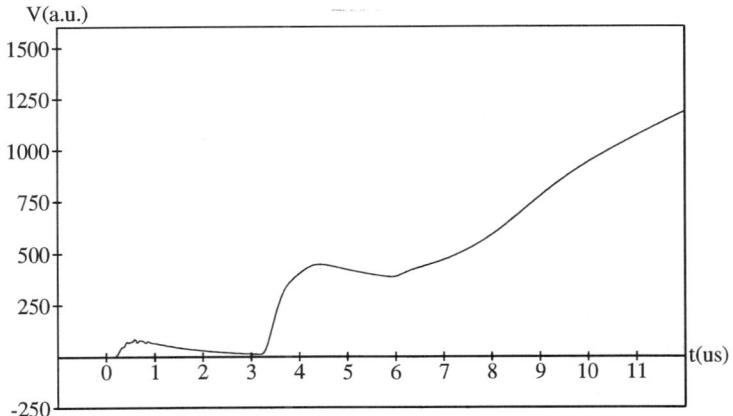

Figure 2 Voltage signal from a current biased Al SASTJ (shown in Fig. 1) following a laser pulse on the far face of the Nb crystal.

Figure 2 shows the first 12 µs of the voltage signal from an Al SASTJ in the geometry shown in Fig. 1 at 500 mK [30]. The voltage signal shape does not directly represent the phonon flux (energy $\Omega \geq 2\Delta_{Al}$) entering the thin films of the junctions (and breaking quasiparticles) because of the integrating effect of the finite quasiparticle

---

[2]Residual Resistance Ratio ~ 10^4.

recombination time in the Al (in this case, $\tau_r = 1.30$ μs), and the RC effect arising from the electronic characteristics of the SASTJ, leads and head amplifier ($\tau_{RC} = 0.20$ μs). We will discuss the response of the STJ further before returning to the Nb phonon properties.

Superconducting Tunnel Junction Characteristics

The electrical response of a constant current biased (I_B) STJ to a pulsed injection of quasiparticles is summarised in Fig. 3. The increase in the quasiparticle density within the STJ films effectively shifts the IV characteristic to one corresponding to a slightly higher temperature. The increase in the thermal tunneling is described by the short–circuit signal

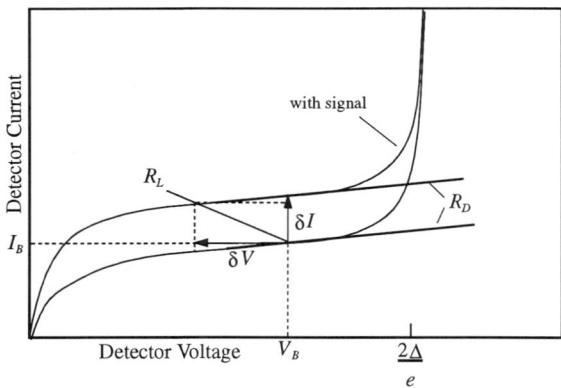

Figure 3 Load line consideration for STJ detector signal in a current bias circuit. Symbol definitions are in the text.

current, δI. The effect of the constant current bias circuitry is, however, to make this current appear as a shift in the apparent voltage bias (V_B) by an amount, δV. In the limit that the load resistor, R_L, of the biasing circuitry is chosen to be very much greater than the dynamic impedance, R_D, of the STJ the signal voltage is then given by the relation

$$\delta V \cong -R_D \delta I. \qquad [1]$$

At large injections of quasiparticles, $\delta N_{qp} \sim N_{qp}$ (the thermally generated number) the non–linearities of the I–V characteristic forbid this relationship. The electronic detector time constant is given by

$$\tau_{RC} \cong C R_D, \qquad [2]$$

where C is the junction capacitance (in parallel with the amplifier input and lead capacitances). Typical values, for single STJs, in this work were, a dynamic resistance of 30 Ω and a capacitance 3 nF [3] giving a time constant less than 100 ns.

Relationship Between Voltage Signal from STJ and Non-Equilibrium Phonon Flux Entering Films

If non–equilibrium phonons, from an excitation in the substrate, pass from the substrate into the thin films (energy gap, Δ) of the STJ, those phonons with energies $\Omega \geq 2\Delta$ will break Cooper Pairs creating additional quasiparticles. The quasiparticles will

---

[3] Oxide barriers on Al have a typical capacitance of 6 μFcm^{-2}.

tunnel through the junction barrier leading to a change in the voltage bias of the STJ as described in the previous section. The excess quasiparticles will ultimately be lost, due to recombination with the thermal quasiparticle population [4].

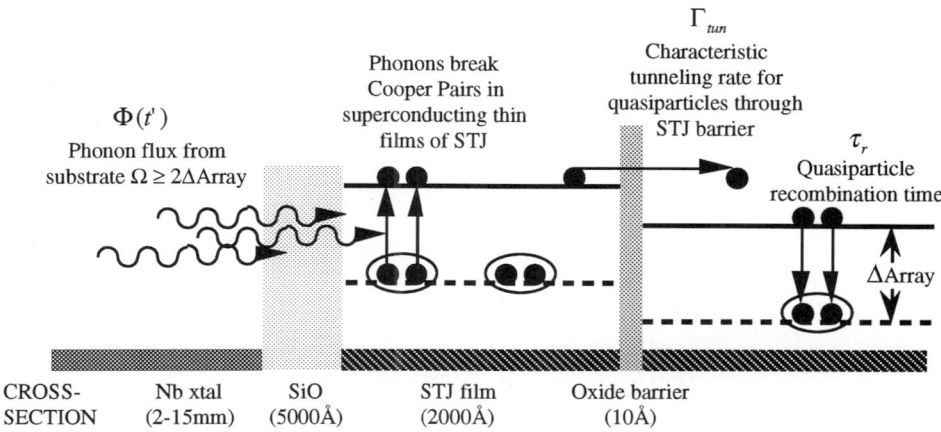

Figure 4 Schematic of the detection of phonons (energy $\Omega \geq 2\Delta$) in an STJ.

If the phonon flux creating excess quasiparticles in the superconducting films of the junction is $\Phi(t)$ then the time development of signal voltage is given by

$$\delta V(t,T) = \{e\Gamma_{tun} R_D(T)\} \int_{-\infty}^{t} dt' \, \Phi(t') \alpha(T, t-t') \, 2\beta(T,\Omega) \qquad [3]$$

where T is the equilibrium temperature of the crystal and STJ, $R_D(T)$ is the dynamic resistance of the junction at the bias point, Γ_{tun} is the tunnelling rate of the quasiparticles across the junction, ß is the efficiency with which phonons (energy Ω) break Cooper Pairs creating two quasiparticles and $\alpha(T,t)$ is the response function of the STJ.

The coefficient ß contains a number of different effects. Clearly, for $\Omega < 2\Delta$, it will be zero due to the cut off in the Pair breaking energy. Above this threshold the number of quasiparticles created will show a non–linear dependence with Ω. This is illustrated in Fig. 5 and discussed in detail by Eisenmenger [31].

The coefficient ß also contains a factor related to the probability of an above gap phonon pair breaking before it either decays in the film or leaves the film back into the substrate. The pair breaking rate itself also has a small temperature dependence, however for $T < 0.5T_c$ (of the film), this can be assumed to be constant. In the experiments we conducted the film thicknesses were typically in excess of the m.f.p. for pair breaking for the materials used.

The transmission probability of phonons from the substrate in to the film will also be dependent on the phonon energies and modes. An adjustment could be made to ß using the acoustic mismatch model [32, 33], however, the results will not be significantly altered.

---

[4]Quasiparticle loss will also take place due to their out diffusion down the STJ leads. In addition, their effective lifetime will be enhanced by pair breaking of the recombination phonons. We will assume all these effects are taken into account in the quoted lifetime, τ_r.

For the purposes of evaluating Eq. 3 we will assume ß is independent of temperature and ~1 bearing in mind that the STJ becomes a phonon counter rather than measuring the total energy in the phonon flux. In the following calculations we will simply use a factor of 2Δ to convert the phonon number back into an energy flux.

α(T,t-t') is a function describing the response of the detector arising from the combined effects of the effective recombination time of the quasiparticles in the STJ films (τ_r) and the electronic RC time constant (τ_{RC}). It has the form

$$\alpha(t) = \frac{\tau_r}{(\tau_{RC} - \tau_r)}\left[e^{-\frac{t}{\tau_{RC}}} - e^{-\frac{t}{\tau_r}}\right] \qquad [4]$$

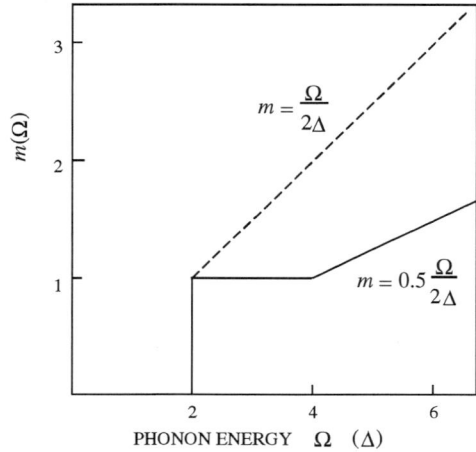

Figure 5 Detector phonon counting efficiency m as a function of phonon energy Ω. The case where m=1 corresponds to the excitation of two quasiparticles by Cooper Pair breaking in the energy range 2Δ < Ω < 4Δ. For higher energies the counting rate is expected to increase as m = 0.5 Ω/2Δ by additional 2Δ–phonons produced by quasiparticle relaxation in the detector. The dashed line corresponds to a counting rate increasing in proportion to energy with m = 1 at Ω = 2Δ. Figure is from Eisenmenger [31] after Forkel.

Both τ_r and τ_{RC} are function of temperature. At a given temperature $R_D(T)$ and $\tau_r(T)$ (see Fig. 6) can be measured directly and Γ_{tun} determined from the normal state resistance of the STJ.

In order to extract variations in the phonon flux which are on time–scales comparable to τ_r, it is necessary to return to the form of Eq. 3. Within numerical factors this is a convolution of the functions α(t) and Φ(t). Using Fourier Transforms (FT) it is possible to show that the phonon flux is given by

$$\Phi(t) \sim FT^{-1}\left[\frac{FT[\delta V(t')]}{FT[\alpha(t'')]}\right]. \qquad [5]$$

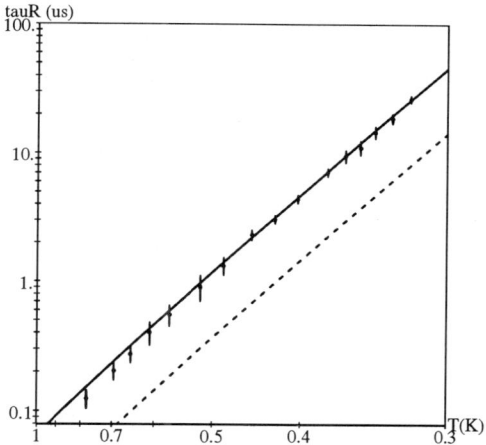

Figure 6 Experimental recombination times of quasiparticles in Al film of Al SASTJ. Dotted line is (half) recombination time calculated using theory from Kaplan et al. [34] with $\tau_0^{qp} = 110$ ns and $\Delta = 220$ μeV. Experimental recombination times were extracted from direct illumination pulses and fit using a convolution algorithm to take account of electronic RC effects.

This analysis can also be applied to a SASTJ with a few modifications. The details of the electronic equivalent circuit represented by an SASTJ will not be discussed in detail here [35]. In summary, the SASTJ behaves electronically with a dynamic resistance, R_D, and capacitance, C, such that [5]

$$R_D^{(Array)} = nR_D^{(Junction)}, \quad [6]$$

$$C^{(Array)} = \frac{C^{(Junction)}}{n}, \quad [7]$$

where n is the number of series connected junctions and R_D and C (on the right–side of the equations) are the dynamic resistance and capacitance of each junction. Thus the effective electronic RC time constant of the array is identical to that of a single junction. This is important since we are observing signals on time–scales ~ 100 ns. The SASTJ covers a much larger area of the Nb crystal than a single STJ leading to a proportionately larger voltage signal from a uniform above–gap phonon flux (assuming the junction characteristics are the same).

We were able to check the enhanced sensitivity of the SASTJ by comparing results with a single Al–Ox–Al STJ in a similar geometry [30]. The phonon flux entering a

---

[5] The full form of the equations are $R_D^{(Array)} = \frac{\left(nR_D^{(Junction)}\right)R_L}{R_L + \left(nR_D^{(Junction)}\right)}$, and $C^{(Array)} = \frac{C^{(Junction)}}{n} + C_i$ where R_L is the AC load–line given by the parallel combination of the bias resistance and amplifier input impedance, and C_i is the sum of the amplifier input capacitance and stray capacitance. The reduced form of the equations is valid in the limit that $R_L \gg R_D^{(Array)}$ and $C^{(Array)} \ll C_i$.

single junction of an SASTJ will be related to the observed voltage signal, δV, across the entire SASTJ where R_D becomes the dynamic resistance of the SASTJ [6].

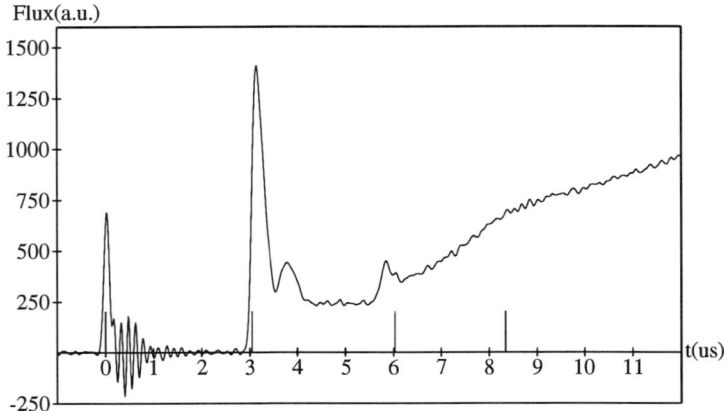

Figure 7 Signal representing the phonon flux entering a SASTJ, obtained from trace in Fig. 2 using deconvolution technique outlined in the text. Vertical marks give calculated phonon arrival times, based on elastic constants given in Fig. 9, for the L (5.05 kms^{-1}), FT (2.56 kms^{-1}) and ST (1.85 kms^{-1}) modes. The spike at t = 0 arises due to scattered light from the fibre on the far face falling directly onto the SASTJ.

The result of this deconvolution given by Eq. 3 is shown in Fig. 7. Figure 8 shows the detector response function used to obtain the transformation of the STJ voltage signals pulse to the phonon flux pulse.[7]

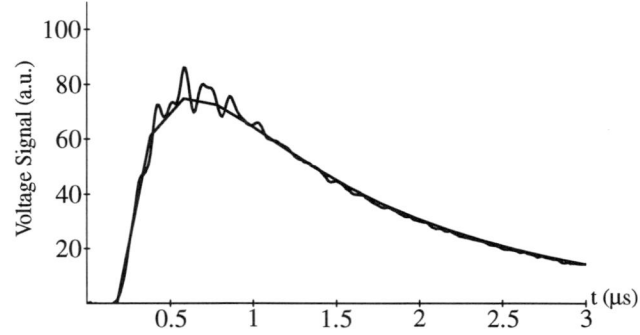

Figure 8 The detector response function for the STJ used in the ballistic phonon experiment of Fig. 1. The pulse was obtained by direct illumination of the STJ by the laser. The line fit is $\alpha(t)$ with $\tau_r = 1.30$ µs, $\tau_{RC} = 0.20$ µs (and an offset of 0.18 µs).

Ballistic Phonon Velocities in Niobium and Phonon Focusing

In Fig. 7 a small signal at t = 0 arises due to scattered light from the fibre, aligned with the far face, falling directly on the SASTJ. It's presence as a narrow spike

---

[6] Alternatively, the voltage signal across each junction of the array can be thought of as $\delta V/n$ across a resistance $R_D^{(Array)}/n$.

[7] This response can be measured by direct illumination of the SASTJ using a second laser fibre positioned directly above it.

(FWHM ~ 100 ns) indicates that the deconvolution has worked. The following oscillation is electrical pick up from the laser drive mechanism.[8]

The L phonon peak is observed with a leading edge arrival time of 3.06 ± 0.01 μs for the propagation distance of 15.3 ± 0.1 mm. This gives a velocity of 5.00 ± 0.05 kms^{-1} which is close to that determined by Weber [27] from ultrasonic measurements on Nb.

The second peak is caused by the reflection of L phonons from the surface of Nb crystal. Unfortunately, we cannot obtain a precise estimate of the surface reflection coefficient implied by the signal size because of the complexities presented by the cylindrical profile of the crystal. A theoretical calculation of the expected reflection coefficient will have to take into account potential mode conversion which would also reduce the L phonon flux [36].

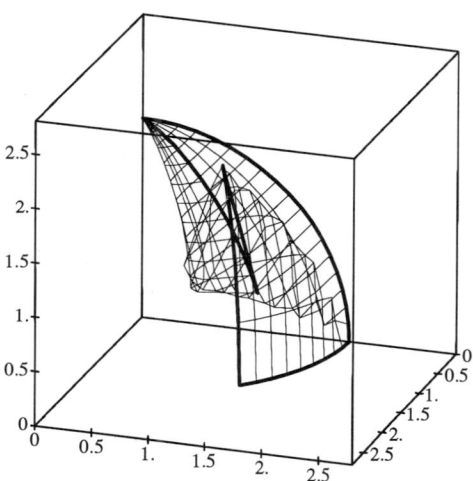

Figure 9 Slow transverse phonon group velocity surface (kms^{-1}) in Cartesian co-ordinate system. Only half of one octant mapped from k-space is plotted for simplicity. The high degree of symmetry means that the full surface can be created by the reflection of this surface into the other sectors. The lines are spaced in 5 degree intervals of θ and φ in k-space. Areas were they are more closely spaced compared with the normal isotropic pattern (or where they double back) indicate directions of increased phonon focusing.[9]

The third peak in Fig. 7 arises from the [110] FT phonons and corresponds to a velocity of 2.65 ± 0.03 kms^{-1}. The forth peak arising from ST phonons appears considerably smoothed out at around 8.5μs (1.8 kms^{-1} ± 0.1). The rising background on which it appears is due to reflected FT phonons which arrive before the expected ST peak. ST phonons preferentially propagate in the [111] direction due to phonon focusing effects leading to a reduction in the number observed along the [110] direction.[10] The

---

[8]It is interesting to note how the deconvolution (Fig. 7) emphasises the high frequency pick up when compared to the previous Fig. 6
[9]The techniques to obtain this surface are well documented and will not be discussed further here [37-39]. The second order elastic constants c11, c12 and c44 used are 246, 134 and 29.4 10^{10} dynes cm^{-2}, respectiveley, taken from Weber [27].
[10]Private Communication: Professor B. Cabrera, Stanford, USA.

group velocity surface for this phonon mode in Nb is shown in Fig. 9. The importance of these focusing patterns will have a number of ramifications for the detection of the position of particle interactions within the bulk of Nb crystals.

Total Ballistic Phonon Energy

Figure 10 shows the variation in the total L and T ballistic phonon energy (which is the flux integrated over the time 0–6 µs) detected by an Al STJ in the temperature range 0.3–1.0 K following laser pulse illumination of the far face of the Nb crystal (length

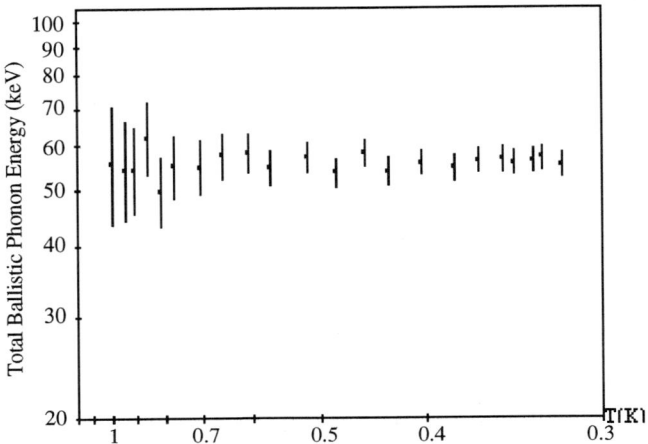

Figure 10 Ballistic Phonon energy (0–6 µs) versus temperature. The temperature dependence of the Al STJ has been adjusted for. The data are consistent with no temperature dependent attenuation of ballistic phonons below 1 K.

10.8 mm). The phonon energy was calculated from the voltage signal of the STJ using Eq. 3 assuming ß = 1 and that each detected phonon has energy $\Omega = 2\Delta$.

In the temperature range 0.3–1.0 K there is no variation in the attenuation of L or T ballistic phonons. This is in agreement with a calculation of the expected scattering length of phonons from thermally generated quasiparticles in the Nb single crystal **[40]** which would become comparable with the propagation distance (10 mm) at 1.5 K for $\Omega = 400$ µeV.

The STJ (300 x 160 µm) subtends an angle from the point of interaction of 5×10^{-4} sterads. The total flux of order 50 keV would imply a source strength of ~1 GeV into 2π sterads. isotropically. The estimated energy of the source (calculated from the temperature rise of the crystal) is 18 GeV. The ballistic flux deficit arises in part due to an overestimation of the STJ phonon detection efficiency but could also arise due to mechanisms within the Nb itself such as focusing, details of the initial phonon spectrum and phonon/quasiparticle yield immediately after the interaction. We will discuss this again later.

Phonon Reverberation Signal and Rapid Thermalisation

Figures 11 and 12 show the variation in the pulse shape (0-300 µs), for temperatures 0.3–1.0 K, in the same geometry as the previous section, after the deconvolution of the detector response function. The constancy of the ballistic feature at the leading edge can be seen. At 0.317 K the signal reaches a maximum at around 25 μs, shortly after the ballistic peak. It then decays with a time constant of 145 μs. At 1.00 K the signal rises immediately after the ballistic signal with a time constant of 36 μs. Figure 12 shows the pulses on a semi–logarithmic plot in the range 0–1.6 ms. At long time–scales all the traces converge to a constant decay time of 10 ± 1 ms.

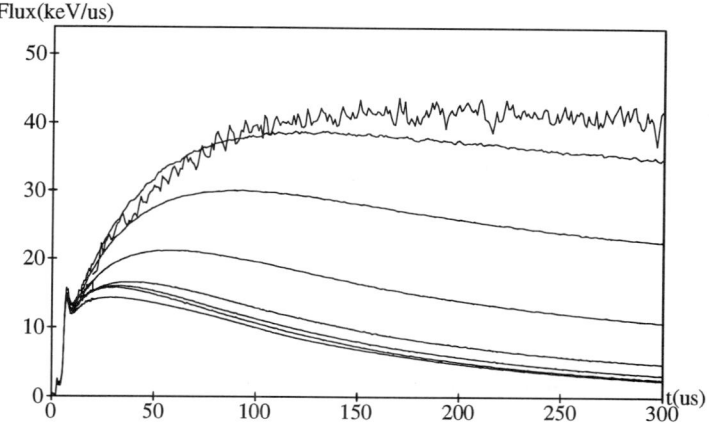

Figure 11 Phonon flux (in units of keVμs^{-1}) entering STJ, versus the time from the laser excitation, for temperatures, 0.317, 0.343, 0.380, 0.453, 0.571, 0.705, 0.858 and 1.00 K (curves run from bottom to top).

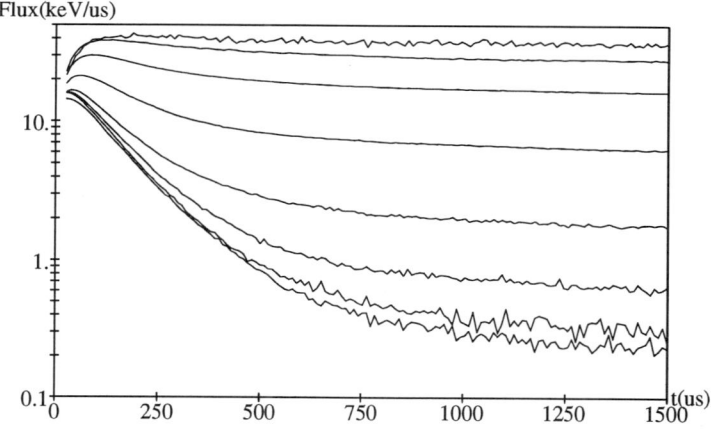

Figure 12 Log. plot of the phonon fluxes shown in Fig. 11.

After time–scales of around 100 μs–1 ms the "non-equilibrium" STJ signals [11] become consistent with those expected for a quasi–thermal shift in the STJ IV characteristic. The voltage change in an ideal STJ (at constant current bias) is given by

$$\delta V = \frac{(2\pi)^{\frac{1}{2}}}{e} k_B \delta T \, y^{\frac{3}{2}} e^{-y} \left(\frac{R_D}{R_{nn}} \right), \qquad y = \frac{\Delta}{k_B T}, \qquad [8]$$

---

[11] The plateau region evident in each trace on the right hand side of Fig. 12.

where δT is the change in temperature of the STJ and substrate (and R_{nn} is the normal state resistance of the junction). The temperature rise of the crystal is given by the Debye heat capacity of Nb ($c_v \sim T^3$) and an estimate of the total energy absorbed by the crystal from the laser pulse.

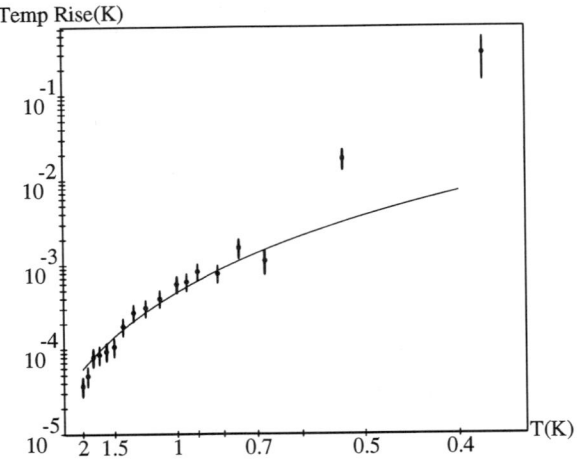

Figure 13 Temperature rise in Sn-Ox-Sn junction calculated using Eq. 8 plotted against the Nb crystal temperature. Line is expected temperature rise due to 10 GeV pulse energy deposition in the 3.4 g Nb crystal assuming a Debye lattice heat capacity.

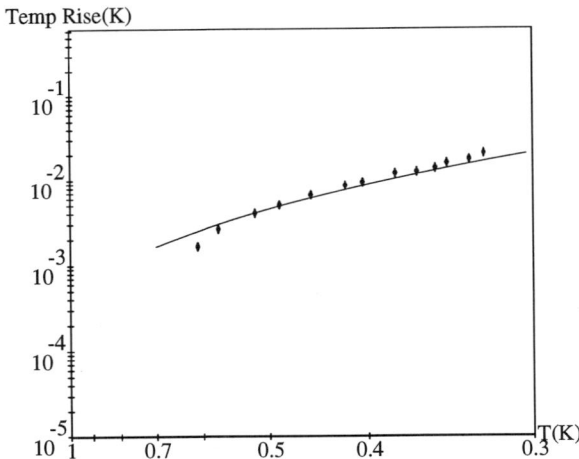

Figure 14 Temperature rise calculated as Fig. 13 for an Al–Ox–Al SASTJ. Line is expected temperature rise due to 30 GeV pulse energy deposition in the 10.1 g Nb crystal assuming a Debye lattice heat capacity.

Figure 13 shows a comparison of the temperature rise calculated from a Sn–Ox–Sn STJ voltage signal (at a time 200 μs after the laser pulse) against the temperature rise based on the Debye heat capacity (solid line). The only free parameter is the absolute value of the energy deposited by the laser. Figure 14 shows a similar comparison for a Al SASTJ on a Nb crystal (at a time of 1.5 ms after the laser pulse). In both cases the agreement is good.

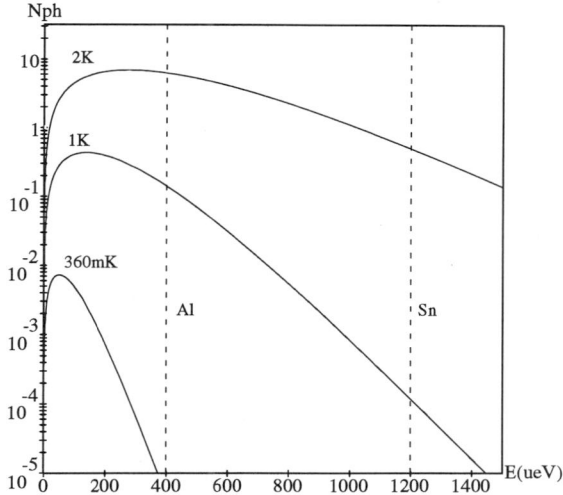

Figure 15 Phonon population against energy at 0.36, 1 and 2 K. The dashed lines show the threshold for detection with Al ($2\Delta = 400$ μeV) and Sn ($2\Delta = 1200$ μeV) STJs.

This approach represents a modification of previous models of the behaviour of STJs on bulk crystals [6]. As has already been mentioned [12] the tunneling current of a STJ is proportional to the number of phonons in the film above $2\Delta_{film}$. For an STJ on a ideal large single crystal it is assumed that phonons from an interaction will reverberate around the crystal. The above–gap phonons will be absorbed in the STJ films, on the crystal surface creating excess quasiparticles (above the thermal population). The quasiparticles will ultimately recombine in the films generating phonons at the gap energy. Many of these phonons will be retransmitted back into the bulk crystal and will be, subsequently, reabsorbed by the STJ film after traversing the crystal a number of times. The rate of creation of excess quasiparticles within the STJ films will therefore, to first order, be independent of temperature and so the signal tunneling current would become proportional to the recombination time of the quasiparticles in the thin films. The long decay time of the STJ signal would be determined by the anharmonic decay rate or loss rate of the above gap phonons from the bulk crystal. Observations of the STJ signal behaviour in Nb are at odds with this interpretation. The ballistic signal flux does appear constant, however, the amplitude of the tail flux is not independent of temperature. The phonons, above the threshold of the STJ, are being thermalised on time scales much shorter than our estimates from the anharmonic processes in a single crystal.

---

[12] Eisenmenger [31] pp110 discusses the sensitivity of STJs to phonons, $\Omega < 2\Delta$. He concludes that the ratio of the sensitivities for sub- to above-gap phonons is $< 10^{-4}$. This was verified experimentally by the failure to detect ballistic phonons, $\Omega \sim 400$ μeV, using Sn-Ox-Sn junctions [31, 41]. At higher temperatures, $T > 0.5 T_c$, increased phonon absorption can result in finite low energy phonon signal contributions as was found with Al detectors [31, 41]. The reduced sensitivity to sub–gap phonons at low temperatures can be seen to arise from two mechanisms. (i) The strong temperature dependence of the ratio of normal to superconducting absorption coefficients in the thin films and (ii) the rapid decay of quasiparticle excited states back to the gap edge which is the only possible way in which the quasiparticles in a superconductor can couple to sub-gap phonons. To first order the sub–gap phonons do not give rise to any additional tunnelling current since the number of quasiparticles has not changed. However, they may cause variations in the tunnelling current due to the change value of the density of states, in the opposite STJ film, available to the excited quasiparticles.

The number of phonons above 2Δ in a blackbody spectrum (see Fig. 15) is given by (assuming simple dispersion relation for phonons, energy Ω, of three modes with the same velocity, c, and Debye frequency, Ω_D)

$$N^{(phonon)}_{\Omega \geq 2\Delta} \cong \frac{3}{2\pi^2} \frac{1}{(\hbar c)^3} \int_{2\Delta}^{\Omega_D} \frac{d\Omega \Omega^2}{e^{\frac{\Omega}{k_B T}} - 1} \quad \left(\sim T e^{-\frac{2\Delta}{k_B T}}, \quad T \ll 2\Delta \ll \Omega_D \right). \qquad [9]$$

As the temperature falls the decrease in the phonon flux signal in the thermalised tails (shown in Fig. 12) is a consequence of the rapid reduction in the above gap phonon number in the crystal (and consequently any absolute change in the phonon number caused by a temperature rise δT). The analysis based on phonon flux (using Eqs. 3 and 5) is identical to that expressed in Eq. 8 when one considers the statistics governing the quasiparticle population in the superconducting thin films of the STJ and the phonon population in the substrate (Eq. 9). **[30]**

In summary, it can be seen that as the temperature of the STJ is lowered the sensitivity of the STJ falls for a fixed temperature rise ($\delta I/\delta T$), whereas, it increases for a fixed phonon flux ($\delta I/\delta \Phi$). This is shown in Fig. 16.

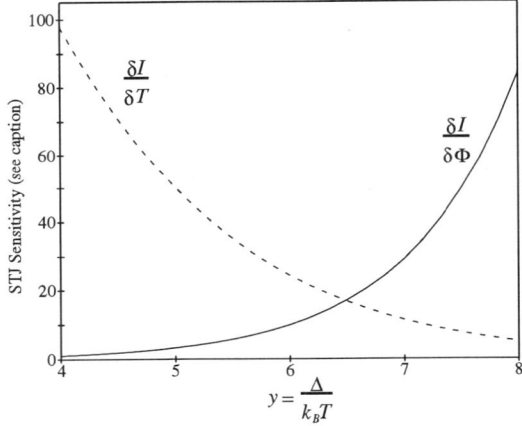

Figure 16 The relative sensitivity of the signal δI (=$R_D^{-1} \delta V$) from an STJ to a fixed above–gap phonon flux (line), $\delta \Phi$, and a temperature rise (dashed line),δT, plotted against reciprocal temperature.

The pulse shapes for the Al STJ, shown in Figs. 11 and 12, in the intermediate time region between the ballistic edge and the plateau (once thermalisation has occurred) will have a number of contributing factors. The leading edge will contain an additional component of higher energy phonons ($\Omega > 2\Delta_{Al}$) which diffuse from the interaction point due to their shorter m.f.p.'s. This will contribute to the structure of the first ~40 μs of the signal. There may also be a contribution from the quasiparticle signal in the Nb due to phonons emitted in their recombination. [13]

For the pulse recorded at 0.317 K the subsequent decay of the STJ signal must arise from the net loss of phonons of energies above the STJ threshold. The decay time is 145 μs. This is over an order of magnitude shorter than that estimated from a calculation of the anharmonic decay time for L phonons in Nb. It seems likely that the down–

---

[13] We will not discuss the contribution from the Nb quasipartilces further here. See [30]

conversion arises either in the Al films of the STJ or at the surfaces of the Nb crystal. If the effect arises from anharmonic decay and mode conversion of the phonons in the region of lattice distortion at the Nb surface, a 145 µs decay time in the 10 mm × ø10 mm Nb crystal would imply a probability of down–conversion at the Nb surface of 0.03 (per scattering) assuming the signal was entirely due to ballistic phonons (c = 2 kms^{-1}).

The pulse shape in the Al device at 1 K shows that the rate of quasiparticle creation in the Al films actually increases after the initial ballistic burst. The absence of this increase in the low temperature pulse shape indicates that the increase must arise from a change in the excitation phonon frequency distribution due to the presence of the thermal phonon distribution. Phonons that are initially above gap will still tend to decay. However, three phonon processes (at the crystal surface) will also tend to work in the other direction combining sub–gap phonons, from the non–equilibrium distribution, with sub–gap phonons from the thermal population, to produce new above gap phonons. Clearly the probability of this process occurring increases significantly as the mean energy of the population of thermal phonons increases.[14]

The source of these non–equilibrium phonons, that only make a contribution to the signal after ~10 µs, could arise from two sources. The initial excitation may be generating a significant number of below $2\Delta_{Al}$ phonons that are subsequently up–converted as previously described. Alternatively the initial excitation may create a significant number of high energy phonons. Those in the window $2\Delta_{Al} \leq \Omega < 4\Delta_{Al}$ will only break one Cooper Pair in the STJ [15]. As a consequence the full energy of this phonon distribution will only be manifest in the STJ when the phonons near 4Δ have redistributed their energy. At present we have insufficient data to distinguish between these possible mechanisms.

Inelastic Scattering at Crystal Boundaries

We believe that the thermalisation of the non–equilibrium distribution of phonons generated by the laser is taking place at the surface of the Nb crystals. Experimental results from Wigmore [42] using GaAs bolometers on sapphire suggest that phonon boundary scattering is not only diffuse, but also involves at least partial absorption and thermalisation of the incident phonon distribution. At each refection the re–emitted phonon distribution corresponds to that for a slightly lower temperature. He also finds that the thermalisation process appears to be independent of phonon frequency (in the range 1.9-5.0 K). Clearly, boundary scattering is a U-processes where crystal momentum is not conserved. The phonon population will tend to an equilibrium Plank distribution [43].

Attempts have been made at the quantitative analysis of the boundary scattering of thermal distributions of phonons, however, the application of these techniques to non–equilibrium distributions presents a number of difficulties and will not be proceeded further here.

Theoretical Phonon Decay Rates in Niobium

Theoretical anharmonic phonon decay rates have been studied by Klemens [44] and also in more detail by Tamura [45]. For high energy phonons $\Omega \gg k_B T$ the process of

---

[14] The average energy of the phonon populations is given by 2.70 kT.
[15] This will give rise to two quasiparticles, one at an energy Δ and the other at Ω-Δ. The latter will relax quickly to gap edge emitting a sub–gap phonon.

interest is the decay $\Omega_1 \to \Omega_2 + \Omega_3$ where the conservation rules for energy and momentum imply $\phi_L \leftrightarrow \phi_T + \phi_T$ or $\phi_L \leftrightarrow \phi_L + \phi_T$. For Nb a calculation based on the Klemens formula [32] leads to a LA lifetime of 16.3 μs $(\Omega/1 \text{ meV})^{-5}$. Tamura's analysis for the spontaneous decay of LA phonons, in terms of the second and third–order elastic constants shown in Table 1, gives decay times of 248 μs $(\Omega/1 \text{ meV})^{-5}$ and 27 ms $(\Omega/1 \text{ meV})^{-5}$ via the processes $\phi_L \leftrightarrow \phi_T + \phi_T$ or $\phi_L \leftrightarrow \phi_L + \phi_T$ respectively.

Table 1 Third order elastic constants in single crystal Niobium from Chang [46, 47]. Also shown are the phonon velocities, and density for Nb used in the calculation of the phonon decay rates. Second-order elastic constants used in calculation are from Weber [27].

c111	c112	c123	c144	c155	c456	c_L	c_T	ρ
		10^{10} dyne cm^{-2}				(kms^{-1})	(kms^{-1})	(gcm^{-3})
-2560	-1140	-467	-343	-168	+137	5.118	2.135	8.57

It is apparent that the decay of LA phonons is dominated by that to two TA phonons and that Tamura's calculation gives a lifetime over an order of magnitude longer than that of Klemens.

The Klemens and Tamura calculations were made assuming zero temperature. The effect of a thermal distribution of phonons at temperature T would be to enhance the decay rate by a factor

$$(1+n(\Omega_2))(1+n(\Omega_3))+U(\Omega,T),\qquad [10]$$

where Ω_2 and Ω_3 are the energies of the decay product phonons (which will most probably have energies $\Omega_2 = \Omega_3 = \Omega/2$). $U(\Omega,T)$ represents the contribution from $\phi + \phi \leftrightarrow \phi(\Omega)$ processes. For $\Omega \sim 2\Delta_{Nb}$ Eq. 10 will be ~1 at all temperatures of interest. At $\Omega = 2\Delta_{Al}$ at a temperature of 1 K the first terms of Eq. 10 have a value ~1.3.

The phonon decay times calculated in this section are clearly too long to be consistent with the thermalisation times discussed in the previous section.

THERMAL HEAT SINKING OF NIOBIUM CRYSTALS

As has already been suggested the long lived tails in the signals from STJs and SASTJs arises due to the sensitivity of STJs to changes in the above threshold component of thermal phonon distributions in the Nb (but with decreasing sensitivity as the temperature is lowered). The decay times on a number of Nb crystals were shown to be characteristic of the thermal heat sinking times,

$$\tau_{thermal} = \frac{C}{K}.\qquad [11]$$

For temperatures, $T \leq 2$ K, the heat capacity, C, of superconducting Nb is dominated by the phonon contribution. The thermal conductivity, K, of the heat sinking of our Nb crystals is more difficult to estimate. In general the crystals were clamped into OFHC Cu blocks machined to approximately fit their cylindrical profile (see Fig. 19). GE varnish and a layer of cigarette paper was used to provide electrical isolation of the crystal from the sample holder. The crystal was firmly clamped into the Cu block, however, given the difficulty of matching the deviations of the crystal from the profile of the clamp it is difficult to estimate the proportion of the crystal directly in contact with the GE/paper sandwich. Additional GE varnish was used to take up some of the slop.

An estimate of the heat sink thermal conductivity was made by introducing a constant heater power into the Nb crystal and measuring the temperature rise. The SASTJs on the surface of the crystals were used for both purposes.

The thermal heat sinking of the Nb crystal might be expected to be determined by the Kapitza resistance of those areas of the crystal touching the GE–Paper–GE–Cu crystal holder sandwich. Calculating the theoretical conductivity of such a composite would be invalidated by the uncertainties in the thermal characteristics of the materials and interfaces. We will, however, compare our results with those quoted for experiments on Nb-epoxy interfaces. The Kapitza thermal boundary resistance, R_K, of this interface is given by,

$$R_K = \frac{A \Delta T}{\dot{Q}}, \qquad [12]$$

where A is the effective area of the interface $\dot{Q}$ is the power generated by the SASTJ, and ΔT is the temperature drop across the Nb crystal–holder interface. In the limit of small temperature differences the Kapitza resistance between two solids (temperatures T_1, T_2) is expected to have a T^{-3} temperature dependence [48]. For larger temperature differences this becomes

$$R_K = \frac{1}{4\alpha}\left(\frac{T_1^4 - T_2^4}{4(T_1 - T_2)}\right)^{-1} = \frac{1}{4\alpha}(T^*)^{-3}. \qquad [13]$$

where α is a constant determined by summation (over all phonon energies and modes) of the transmission of phonons at the interface and T^* is defined by this equation. Figure 17 shows $(R_K (T^*)^3)$ plotted versus ΔT for the Nb crystal with a sample stage temperature of 0.360 K. $R_K (T^*)^3$ is a constant (= 0.0015 K⁴µW⁻¹) for temperature differences as large as 150 mK.

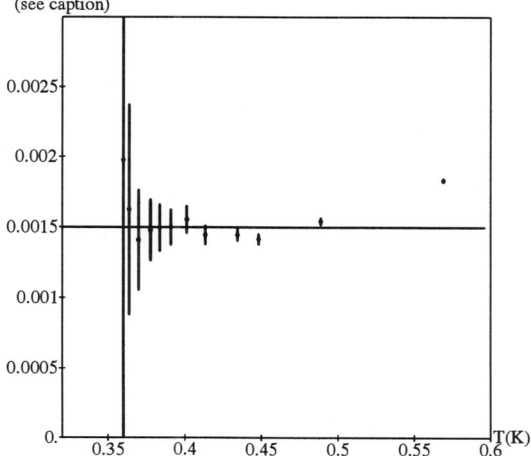

Figure 17 Plot of Kapitza Resistance function, $R_K(T^*)^3$ (K⁴µW⁻¹) against the temperature of the Nb crystal. (See text for details of T^*).

If we use Schmidt's value of R_K at 1 K (~ 10 cm^{2}KW^{-1}) for Nb-epoxy interfaces [49] we would have to assume that < 0.1 % of the surface area of our Nb crystal was contributing to the heat sinking. We estimate that the actual area in contact is around 10% and so conclude that the Ge–paper sandwich has a thermal resistance some two orders of magnitude worse than the systems studied by Schmidt.

At this stage of the development of the Nb single crystal detectors we have not been concerned with the quality of the thermal heat sinking. At low temperatures the response of STJs to thermal signals is vanishing small. The heat sinking is, however, important to the extent that poor sinking limits the bias power one can put into the SASTJs on the surface of the crystal when operating them to detect particles. Optimal voltage bias for the SASTJs appears to be a around 10% of the 2Δ sum and since the bias currents decrease exponentially as the temperature is lowered it is possible to operate arrays of individual junctions with mΩ normal state resistances without causing self heating. This is the case even with the poor heat sinking calculated above. In cases where the Nb crystals were only loosely clamped the SASTJ performance was adversely affected by self–heating.

Using the value for the thermal conductivity of the heat sinking calculated for the case detailed above we derive a thermal time constant for the crystal of mass 10.1 g of 15 ms. This is in close agreement to that experimentally observed using DC amplification of the STJ signal arising from the laser (14.8 ± 0.5 ms).

VARIATION OF NON-EQUILIBRIUM SIGNAL WITH TYPE OF INTERACTION

Introduction

In principal, it is possible to vary the method of excitation of a superconductor in order to generate different phonon and quasiparticle frequency distributions. For example, in thin films, quasiparticles can be injected directly into the superconductor using a tunnel junction [50, 51]. More suitable for investigation of transport phenomena in bulk crystals are microwave irradiation of the surface of the superconductor (photons of comparable energy to superconducting gap); the generation of phonons by STJs on the surface of the crystal [52, 53] ; high energy photo-excitation of the crystal surface or the injection of phonons using thin film heaters. We were able to investigate the signals in bulk Nb from the last two sources (which will be discussed here), as well as investigating the signals arising from α–particle and γ-ray irradiation. [3]

Phonons were produced in a thin film heater, on the insulated surface of the Nb crystal. A resistive film can be heated using an applied current pulse, however, this method introduces considerable electrical pick–up into the cryostat which mask the leading edge of any non–equilibrium signal detected in a separate STJ. Alternatively, the thin film can be heated using laser excitation. The spectrum of phonons emitted from the heater will probably be characteristic of the black–body spectrum at the heater temperature, T_H, with a peak in phonon number at an energy of, $\Omega = 1.6\, k_B T$ [54]. A common problem in quantitative calculations for the heat pulse technique is the uncertainty in T_H. Comparison of our work with that of Pannetier [55] indicated that the likely yield of phonons from our Cu film at the 2Δ_{Nb} threshold (35 K) will be negligible.

We have also studied extensively the non–equilibrium signals arising in bulk Nb from irradiation by a GaAs laser (λ ~ 900 nm). The photon energy is comparable to the

Fermi energy of the electrons in Nb and three orders of magnitude larger than the superconducting energy gap. The mechanisms of down–conversion of the energy initially deposited in a single free electron in a superconductor have been well described by Chang and Scalapino [56]. Initially the electron will partition its energy between a number of other electrons through Coulomb interactions. After a number of collisions, the electron energy is low enough ($E < (E_F\Omega_D)^{1/2}$) for collisions involving phonons to dominate. The relaxation phonons will themselves break more Cooper Pairs (if $\Omega > 2\Delta$) creating quasiparticles. Those phonons below the superconducting energy gap will be able to propagate without breaking Cooper Pairs. This cascade takes place extremely rapidly, on sub–ns time–scales. The yield of sub–gap phonons to quasiparticles from a photo–excitation is difficult to estimate theoretically, however, it is of critical interest to the design and use of superconducting absorbers for particle detection. [4, 5]

Experimental results

Figure 18 shows the time evolution of two signals, produced by (a) thin film heater and (b) photo–excitation, propagating 15 mm in the [110] direction of a Nb crystal at 0.7 K. The signals were detected using an Al SASTJ which is sensitive to phonons with energy greater than the Al energy gap ($\Omega \geq 2\Delta_{Al} = 340$ μeV).

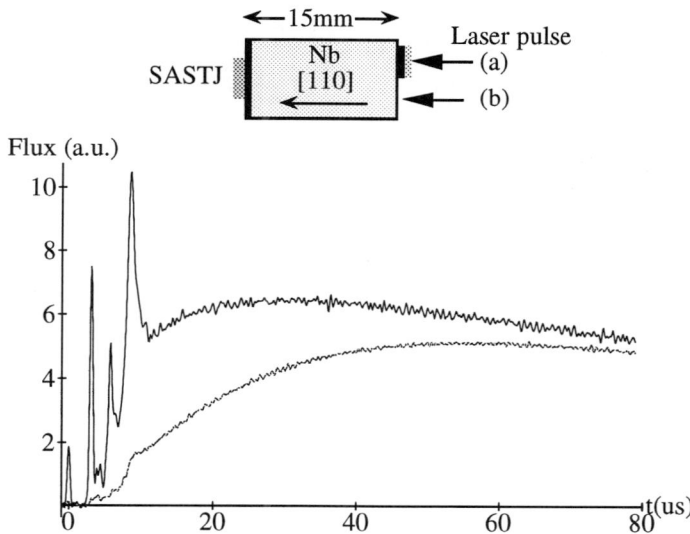

Figure 18 Signals due to laser excitation after propagating 15 mm in the [110] direction of a Nb crystal detected by Al SASTJ (Temperature 0.7 K). The signals have been deconvoluted from the SASTJ detector response as detailed earlier. (a) is the top (black line) trace due to laser incident on a Cu/SiO layer on the surface of the crystal. (b) is the bottom (grey line) trace due to laser incident directly on the surface of the crystal.

The upper pulse in Fig. 18 was generated by a laser pulse incident on a 350 Å Cu film insulated from the crystal surface by 5000 Å of SiO [57]. As discussed previously the Cu film thermalises the excitation due to the laser pulse, thereby producing low energy phonons ($\Omega \ll 2\Delta Nb = 3030$ μeV). The yield of quasiparticles in the Nb is expected to be extremely small. The sub–gap phonons will propagate ballistically through the Nb crystal, limited by impurity scattering. The structure in the signal in the first 15 μs arises from the differing speeds of propagation of the L, FT and ST phonon modes. The signal decays beyond 30 µs with a time characteristic of the thermalisation of the phonon population.

The lower pulse (b) in Fig. 18 was generated by the direct photo–excitation of the crystal surface by the laser. The yield of ballistic phonon is seen to be considerably smaller than that for pulse (a). The change in shape of the leading edge of the pulse arises in part because the average energy of the phonons produced in the initial interaction is much higher. The average impurity scattering m.f.p. of phonons is strongly energy

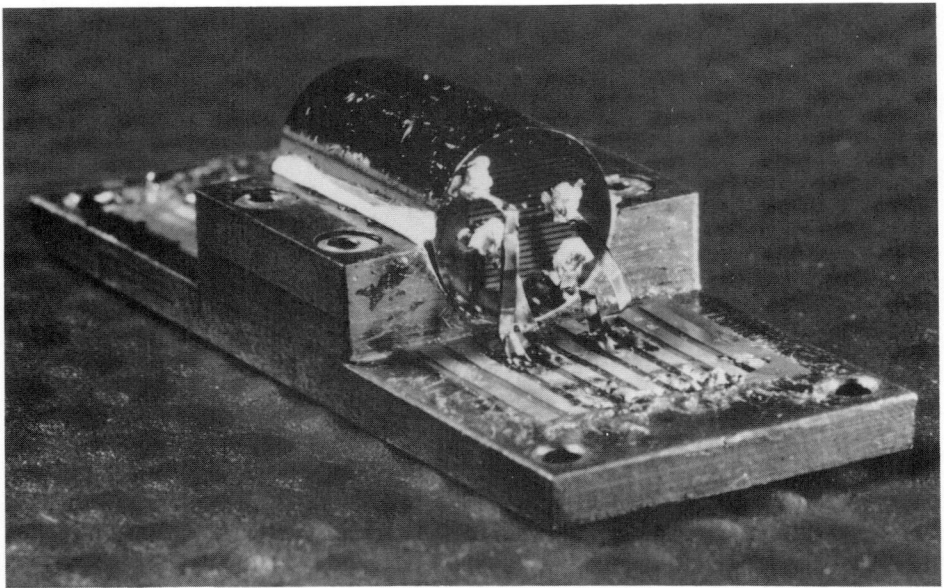

Figure 19 12 g Nb crystal (16 mm x 10 mm diameter) in Cu mounting block. Two series arrays of superconducting tunnel junctions (SASTJs) can be seen on the front face of the detector. A similar pattern is present on the far face.

dependent ($\sim \Omega^{-4}$) [58] leading to a more diffusion-like propagation for the phonon energy.

The question of the contribution of quasiparticles to the signal also arises. The unambiguous identification of the quasiparticle signal from the phonon signal in this trace is difficult because the characteristic propagation times[16] are comparable. We are continuing to investigate techniques for the simultaneous measurement of the separated quasiparticle and phonon components.

---

[16]Quasiparticle diffusion coefficient in high purity Nb is 0.7 m^{2}s^{-1}

CONCLUSION

The development of a dark matter detector based on Nb single crystals has been the catalyst for our detailed investigation of the properties of the Nb non–equilibrium superconducting state. Using SASTJs we are now close to the required level of sensitivity to detect the phonons from dark matter recoils (which are of the order of a few keV) in bulk Nb crystals. In order to improve on existing recoil–based dark matter detectors it is necessary for new detectors to be able to distinguish between nuclear recoil events (associated with dark matter interactions) and electron recoil events (arising from the majority of background sources). We already have some indications that the nature of the interactions does indeed effect the composition of the non–equilibrium signals in Nb. We intend to perform particle detection experiments with neutron sources, in order to mimic dark matter events, in the near future.

Acknowledgements

I would like to thank David Goldie and Andreas Hahn who have worked closely with me on a number of the experiments detailed in this paper. I would also like to thank N E Booth and G L Salmon for their advice and support. I thank M Broad, C Goodwin, B M Hawes and A Wire for their technical support. I acknowledge the financial support of the SERC through a research fellowship during part of this work.

BIBLIOGRAPHY

1. P. F. Smith and J. D. Lewin, *Phys. Reports* **187**, 203 (1990).
2. J. R. Primack, D. Seckel and B. Sadoulet, *Ann. Rev. Nucl. Part. Sci.* **38**, 751 (1988).
3. R. J. Gaitskell, N. E. Booth and G. L. Salmon, Ed. , N. E. Booth, G. L. Salmons, Low Temperature Detectors for Neutrinos and Dark Matter IV (Editions Frontières, Oxford, UK, 1991), pp. 435.
4. N. E. Booth, R. J. Gaitskell, D. J. Goldie, C. Patel and G. L. Salmon, Ed. , A. Barone, R. Cristiano, S. Paganos, X-Ray Detection by Superconducting Tunnel Junctions (World Scientific, Naples, 1991), pp. 125.
5. D. J. Goldie, N. E. Booth, R. J. Gaitskell and G. L. Salmon, Ed. , H. Koch, H. Lübbigs, Superconducting Devices and their Applications (Springer, Berlin, 1991), pp. 474.
6. D. J. Goldie, Ed. , A.Barone, R.Cristiano, S.Paganos, X-ray Detection by Superconducting Tunel Junctions (World Scientific, 1991), pp. 98.
7. N. E. Booth, *et al.*, Ed. , N. E. Booth, G. L. Salmons, Low Temperature Detectors for Neutrinos and Dark Matter IV (Editions Frontières, Oxford, UK, 1991), pp. 407.
8. D. J. Goldie, Private Communication (1993).
9. H. Kraus, (In these proceedings).
10. G. J. Sellers, A. C. Anderson and H. K. Birnbaum, *Phys. Lett.* **44A**, 173 (1973).
11. G. J. Sellers, A. C. Anderson and H. K. Birnbaum, *Phys. Rev.* **B 10**, 2771 (1974).

12. A. C. Anderson, R. E. Peterson and J. E. Robinchaux, *Rev. Sci. Instrum.* **41**, 528 (1970).
13. B. J. C. van der Hoeven Jr. and P. H. Keesom, *Phys. Rev.* **134**, A1320 (1964).
14. J. Bevk, *Philos. Mag.* **28**, 1379 (1973).
15. Y. Hiki, T. Maruyama and Y. Kogure, *J. Phys. Soc. Jap.* **34**, 725 (1973).
16. C. G. B. Baker, E. M. Forgan and C. E. Gough, *Physica* **108B**, 927 (1981).
17. S. G. O'Hara, G. J. Sellers and A. C. Anderson, *Phys. Rev.* **B 10**, 2777 (1974).
18. A. C. Anderson, C. B. Satterthwaite and S. C. Smith, *Phys. Rev.* **B 3**, 3763 (1971).
19. P. H. Kes, J. P. M. Van der Veeken and D. de Klerk, *J. Low Temp. Phys.* **18**, 355 (1975).
20. P. M. Rowell, *Proc. R. Soc. Lond.* **A254**, 542 (1960).
21. P. H. Kes, J. G. A. Rolfes and D. de Klerk, *J. Low Temp. Phys.* **17**, 341 (1974).
22. D. P. Almond, M. J. Lea and E. R. Dobbs, *Phys. Rev. Lett.* **29**, 764 (1972).
23. D. I. Bolef, *J. App. Phys.* **32**, 100 (1961).
24. F. Carsey and M. Levy, Ed. , K. D. Timmerhaus, W. J. O'Sullivan, E. F. Hammels, 13th International Conference on Low Temperature Physics (Plenum, New York, Boulder, Colorado, 1972), pp. 116.
25. E. M. Forgan and C. E. Gough, *J. Phys.* **F 3**, 1596 (1973).
26. W. F. Vinen, E. M. Forgan, C. E. Gough and M. J. Hood, *Physica* **55**, 94 (1971).
27. R. Weber, *Phys. Rev.* **133**, A1487 (1964).
28. J. Halbritter, *Appl. Phys.* **A 43**, 1 (1987).
29. M. H. Frommer, J. Bostock, K. Agyeman, R. M. Rose and M. L. A. MacVicar, *Solid State Commun.* **13**, 1357 (1973).
30. R. J. Gaitskell, D.Phil. Thesis, Oxford University (1993).
31. W. Eisenmenger, in *Physical Accoustics* W. P. Mason, R. N. Thurston, Eds. (Academic Press, London, 1976), vol. XII, pp. 79.
32. D. N. Langenberg, Ed. , M. Krusius, M. Vuorios, Proceedings of the Fourteenth International Conference on Low Temperature Physics (North Holland, Amsterdam, 1975),
33. S. B. Kaplan, *J. Low Temp. Phys.* **37**, 343 (1979).
34. S. B. Kaplan, et al., *Phys. Rev.* **B 14**, 4854 (1976).
35. N. E. Booth and D. J. Goldie, *Superconductor Science and Technology* , (to be published).
36. A. F. G. Wyatt, N. A. Lockerbie and R. K. A. Ziebeck, Ed. , L. J. Challis, M. C. Phillipss, Proceedings of the Second International Conference on Phonon Scattering in Solids (Plenum, Nottingham, 1975), pp. 40.
37. G. A. Northrop and J. P. Wolfe, *Phys. Rev.* **B22**, 6196 (1980).
38. G. P. Srivastava, *The Physics of Phonons* (Adam Hilger, New York, 1990).

39. B. Cabrera, J. Martoff and B. Neuhauser, *Nucl. Instrumen. and Methods* **A275**, 97 (1989).

40. R. J. Gaitskell, D. J. Goldie, N. E. Booth and G. L. Salmon, *Physica B* **167**, 445 (1991).

41. M. Welte, K. Lasmann and W. Eisenmenger, *Verh. Deut. Phys. Ges.* **6**, 699 (1972).

42. J. K. Wigmore, *Phys. Lett.* **37A**, 293 (1971).

43. J. Callaway, *Phys. Rev.* **113**, 1046 (1959).

44. P. G. Klemens, *J. App. Phys.* **38**, 4573 (1967).

45. S. Tamura, *Phys. Rev.* **B 31**, 2574 (1985).

46. R. F. S. Hearmon, in *Elastic, Piezoelectric, Pyroelectric, Piezooptic, Electrooptic Constants, and Nonlinear Dielectric Susceptibilities of Crystals* M. M. Choy, et al., Eds. (Springer-Verlag, New York, 1979), vol. 11,.

47. R. Chang, *Appl. Phys. Lett.* **11**, 305 (1967).

48. A. C. Anderson, in *Non-equilibrium Superconductivity, Phonons and Kapitza Boundaries* K. E. Gray, Eds. (Plenum, New York, 1980) pp. 1.

49. C. Schmidt, *Phys. Rev.* **B 15**, (1977).

50. K. E. Gray, A. R. Long and C. J. Adkins, *Philos. Mag.* **20**, 273 (1969).

51. K. E. Gray, *J. Phys. F:Metal Phys.* **1**, 290 (1971).

52. W. Eisenmenger, in *Tunneling Phenomena in Solids* E. Burnstein, S. Lundquist, Eds. (Plenum, New York, 1969) pp. 371.

53. H. J. Trumpp, K. Lassmann and W. Eisenmenger, *Phys. Lett.* **A 41**, 431 (1972).

54. O. Weis, *J. de Physique* **33**, C 4-48 (1972).

55. B. Pannetier, Ph.D. Thesis, L'Universite Pierre et Marie Curie, Paris 6 (1980).

56. J.-J. Chang and D. J. Scalapino, *Phys. Rev.* **B 15**, 2651 (1977).

57. J. P. Maneval and B. Pannetier, Ed., III International Workshop on Low Temperature Detectors for Neutrinos and Dark Matter (Editions Frontière, Gran Sasso, L'Aquilla, Italy, 1989),

58. W. E. Bron, in *Non-equilibrium Phonon Dynamics* W. E. Bron, Eds. (Plenum, New York, 1985), vol. 124, pp. 1.

PHONON SCATTERING AND HEAT TRANSFER IN SIMPLE MOLECULAR CRYSTALS

V.G. Manzhelii and V.A. Konstantinov

Institute for Low Temperature Physics
Academy of Sciences of Ukraine
Kharkov 310164, Ukraine

This report is dedicated to the study of high-temperature isochoric thermal conductivity and phonon scattering in solidified inert gases and simple molecular crystals in which the molecules rotate as a whole. The experimental data accumulated over the past 2-3 decades has cast doubt on the correctness of some ideas about the thermal conductivity of solids which had seemed to be established and, in particular, the description of the thermal conductivity of perfect crystals at temperatures above or of the order of the Debye temperature $(T \geq \Theta)$. Quite recently, it was not doubted that high-temperature thermal conductivity is proportional to the inverse temperature law, $\lambda \propto 1/T$. It was based on both the experimental data and assumptions being evident at first sight from which this dependence followed. In a simple kinetic model, the phonon thermal conductivity may be represented in the form

$$\lambda = \frac{1}{3}V \sum_{\mathbf{q},j} C(\mathbf{q},j) v(\mathbf{q},j) \bar{l}(\mathbf{q},j), \qquad (1)$$

where j is the branch of the vibrational spectrum; $\mathbf{q}$ is the wave vector; $C(\mathbf{q},j)$ is the contribution of the mode $(\mathbf{q},j)$ to the heat capacity; $v(\mathbf{q},j)$ and $\bar{l}(\mathbf{q},j)$ are the velocity and phonon free path length, respectively; V is the specimen volume.

At high temperatures, $C(\mathbf{q},j)$ and $v(\mathbf{q},j)$ may be considered to be constant and λ is determined by the temperature dependence of $l(\mathbf{q},j)$. If we take into account three-phonon processes only, then the mean free path $\bar{l}$ of phonons will be inversely proportional to the phonon density n. For $T \geq \Theta$, in the first approximation, $n \propto T$, so that $\bar{l}$ and λ are inversely proportional to the temperature.

In time, data on the deviation from $1/T$ dependence has accumulated, and in a number of cases some ideas explaining qualitatively the observed behaviour of thermal conductivity[1-10] have been proposed. The problem has been, however, that it is difficult to obtain the necessary information on phonon thermal conductivity in an explicit form from the available experimental data. In many cases, the phonons were not the only excitations determining the heat transfer and scattering processes. The most important reason making the interpretation of the results more complicated was that the studies were carried out at constant pressure. In this case, thermal expansion, being usually rather essential at high temperatures, leads as a rule to the increase of the temperature dependence of λ. It should be recalled that when deriving the dependence $\lambda \propto 1/T$, the crystal volume and, hence, its energy spectrum were assumed to be independent of temperature. This means that to test this dependence it is necessary to study isochoric thermal conductivity λ_v.

The most reasonable way to solve unambiguously the problem of the behaviour of high-temperature phonon thermal conductivity is to study the heavy solidified inert gases (Ar,Kr,Xe), as (i) the phonons are the only low-frequency collective excitations in these crystals; (ii) for the time being, the thermal conductivity measurements at constant volume have been made with high accuracy only in the case of solidified gases; (iii) heavy inert gas solids also melt at temperatures above or of the order of their Debye temperature.

The relative simplicity of the heavy inert gas solids plays a role in constructing a theory of thermal conductivity. Isochoric thermal conductivity measurements of solid Ar were made by Clayton and Batchelder[11]. They observed good agreement with the law $\lambda \propto 1/T$ over a wide temperature range but in the premelting region their results lay above the curve $1/T$. The authors did not pay enough attention to this effect and did not attempt to explain it. A detailed study of "the high-temperature region" was carried out in[10,12-14] by studying λ of solid Ar, Kr and Xe. In all cases, the temperature dependence of λ_v was weaker than $1/T$ and could be described by the expression

$$\lambda = A/T + B \qquad (2)$$

where A and B are constants not depending on the temperature.

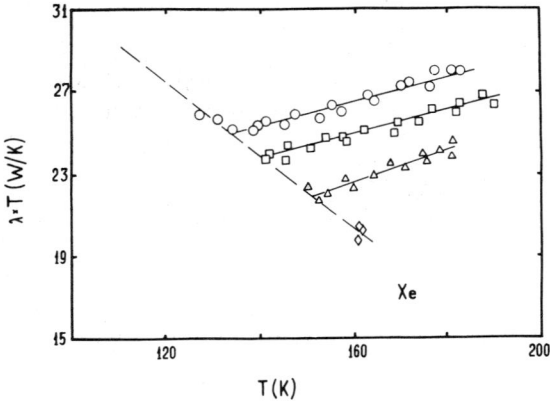

Fig. 1. The temperature dependence of λ_v of solid xenon at molar volumes V, $cm^3/mole$: 37.4 (o); 37.65 (□); 38.1 (△). The dashed line and the diamonds (◇) correspond to a free sample.

Fig. 1 gives the isochoric thermal conductivity λ_v for three molar volumes together with the thermal conductivity at equilibrium vapour pressure λ_s. The thermal conductivity λ_s displays a considerably stronger temperature dependence than $1/T$. It is connected with the change of the energy spectrum of the crystal due to thermal expansion. The deviation of λ_v from the $1/T$ law is also distinctly visible. The allowance of many-phonon interactions leads to a decrease of the thermal conductivity, and, therefore, does not explain this deviation. This becomes clear when we estimate the free path lengths of phonons of heavy inert gas solids in the premelting region. Let us use the experimental data on the isochoric thermal conductivity λ_v, sound velocity $\bar{v}$, isochoric heat capacity C_v and the simplified expression for thermal conductivity

$$\lambda = \frac{1}{3} C_v \bar{v} \bar{l} \qquad (3)$$

to show this. It turns out that in Ar, Kr, and Xe, near their melting temperatures, the magnitude of $\bar{l}$ exceeds merely twice the interatomic distance a and is less than the phonon wave lengths. Strictly speaking, the concepts of a phonon and the interaction of phonons which form the basis of the theory lose their sense in such a situation. Note that the observed temperature dependence of λ_v can be also explained formally in the framework of the phonon theory. With a temperature increase, the phonon mean free path length $\bar{l}$ decreases, and in the limiting case it can reach the minimum possible value being equal to the interatomic distance a. According to formula (3), from this moment taking into account the constancy of the heat capacity C_v and the weak temperature dependence of the sound velocity, the isochoric thermal conductivity reaches its minimum value and ceases to depend on temperature. Such a saturation must be preceded by a weakening of the temperature dependence of λ_v; this is observed experimentally.

The idea of a minimum free path length for phonons and the consequences following from it were discussed by a number of authors[1,3,4,6−8,15,16]. The following alternative assumptions form the basis of different expressions for this minimum thermal conductivity. The first is that the minimum length of excitation transfer is equal to the interatomic distance a. The second is that the minimum free path length of a phonon is equal to its wave length. Roufosse and Klemens[4] focused attention on the fact that there is no reason to prefer the second assumption. It is suitable only in the case when one can use wave representations. When it is impossible, one is to use the representation of localized excitations jumping from one lattice site to another. For such excitations, the limiting value l_{min} is constant.

The deviation from the $1/T$ dependance may arise when additional mechanisms of phonon scattering appear. It may be phonon scattering by other crystal excitations, different kinds of disturbances of the translational or orientational order in a lattice and impurities. The existence of long-range orientational order of the molecules leads to the fact that in molecular crystals along with phonons, librons are also elementary excitations of the lattice. The libron contribution to heat transfer is usually negligible because of the small width of libron zones whereas their role in phonon scattering processes may be rather considerable.[17] The disturbance of the orientational order may be another source of phonon scattering in molecular crystals.

The choice of simple molecular crystals as the objects of study makes easier the interpretation of the results. Isochoric thermal conductivity of solid CO_2 and N_2O was studied by Konstantinov et al.[8]. Table 1 gives the comparison of the values of Debye temperatures of CO_2 and N_2O in the high and low temperature limits, the molecular masses, the temperatures of triple points, and the parameters of the Lennard-Jones potential. Because of the strong anisotropic interaction, the orientational order of the molecules in these crystals remains up to their triple points.

Table 1. Some parameters of solid Xe, CO and $N\,O$.

substance	Θ_0	Θ_∞	T_{tr}	ε	σ	
	K				Å	M
CO_2	152	128	216.6	219	4.00	44
N_2O	141	120	182.4	236	3.80	44
Xe	64	54	161.4	230	3.92	131.2

Although CO_2 and N_2O crystals have many similar properties (see Table 1) and, in particular, they have a similar phonon spectrum, there is an essential difference between them. The molecules of N_2O have no inversion center and possess a small dipole moment in contrast to the molecules of CO_2. The orientational order in N_2O is determined by quadrupole forces. As concerns dipole moments, they are disordered.

Isochoric thermal conductivities of solid CO_2 and N_2O were compared with each other and with λ_v of the inert gas solids.

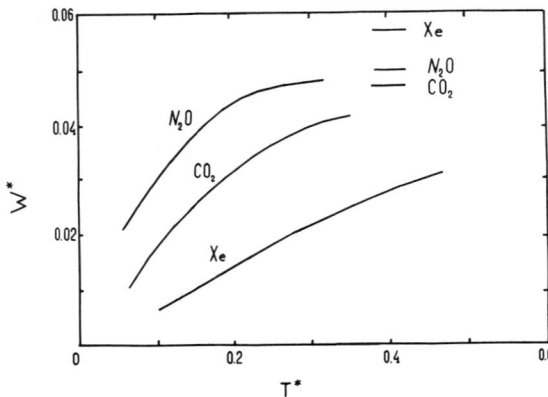

Fig. 2. Temperature dependence of isochoric thermal resistance of N_2O, CO_2 and Xe in the reduced variable $W^* = W/W_m$; $T^* = T/T_m$.

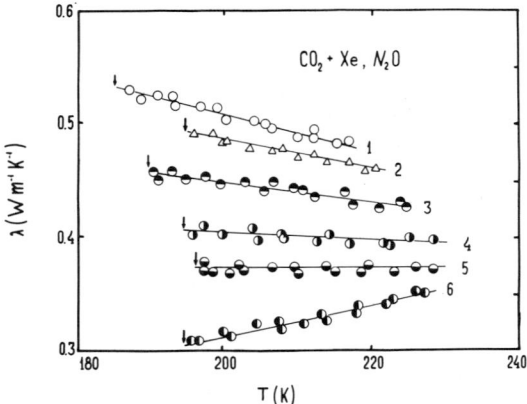

Fig. 3. Temperature dependence of the thermal conductivity of solid CO_2 samples: No. 1 (○, without impurity); No. 2 (△ 5.1 % N_2O); No.3 (○, 0.25 % Xe; No. 4 (○, 1.1 % Xe); No. 5 (○, 2.05 % Xe); No.6 (○, 9.1 % Xe). The arrows indicate the temperature T_0 at which the condition $V = $ const begins to be satisfied.

Such a comparison is of interest for the following reasons: the thermal resistance $W = 1/\lambda$ of an ideal crystal of an inert gas is due solely to phonon-phonon scattering. In CO_2, additional phonon scattering appears due to the interaction with a libron subsystem. In the case of N_2O the scattering resulting from dipole disordering is added to these two phonon scattering mechanisms. It is convenient to make a comparison between the crystals mentioned because all of them have a fcc lattice. In Fig. 2 we represent the isochoric thermal resistance in the reduced variables $W^* = W/W_m$; $T^* = T/T_m$ according to de-Boer[19]. This allows us to estimate, at least qualitatively, the different components of thermal conductivity. The reduction parameters are $T_m = \varepsilon/k$ and $W_m = \sigma^2/k\sqrt{m/\varepsilon}$ where m is the molecular mass; ε and σ are the parameters of the Lennard-Jones potential. As we would expect (see Fig. 2), the appearance of additional scattering mechanisms leads to a deviation from the $\lambda \propto 1/T$ (or $W \propto T$) dependence at lower temperatures and increases in magnitude with growing T.

Fig. 2 also gives the thermal resistance corresponding to the minimum of the high-temperature thermal conductivity calculated using the expression $l_{min} = \hbar v/k\Theta^1_\infty$. In CO_2 and N_2O crystals, the isochoric thermal resistance is rather close to the maximum possible already at the temperature of triple points. However, one must not attach too much importance to the good quantitative agreement keeping in mind the approximate character of the assumptions forming the basis of these theoretical evaluations.

The decrease of the mean free path length of phonons and, thus, the appearance of additional thermal resistance can arise from phonon scattering by impurities. Isochoric thermal conductivity of solid CO_2 with N_2O and Xe impurities was studied in.[20] The results are given in Fig. 3. At concentrations up to 5%, the impurity of N_2O does not affect significantly the thermal conductivity of solid CO_2. The molecules of both substances have equal masses and the parameters ε and σ of the Lennard-Jones potential[21] which differ from one another by not more than 10/impurity is introduced into CO_2. The atomic weight of the impurity is practically three times greater than the molecular weight of the matrix. The volume per Xe atom exceeds by 1.4 times the volume occupied by a CO_2 molecule so that the impurity strongly deforms the lattice in its neighbourhood. As is seen in Fig. 3, a 2% concentration of Xe impurities lowers the absolute value of the thermal conductivity by approximately 25 %. In this case, the thermal conductivity practically ceases to depend on temperature in accordance with the concept of a minimum thermal conductivity. With a 9.1% concentration of Xe impuritie, we observe a rise of the thermal conductivity with increasing temperature. We shall return to this unusual effect later.

In CO_2 and N_2O crystals, the rotational motion of the molecules up to the triple point is represented by small angle librations. As the next step, it seems natural to study λ_v for crystals with a reorientational motion of molecules, the limiting case of which is the free rotation of molecules. Ross et al.[22] studied isobaric thermal conductivity of adamantane $(CH)_4(CH_2)_6$ at different pressures. In the high-temperature region, the isochoric thermal conductivity of adamantane (obtained by calculation from the isobaric data does not decrease but increases with increasing temperature. This seemed surprising. The considerations given earlier and, in particular, the concept of a minimum thermal conductivity do not provide reasons to assume the increase of the thermal conductivity with temperature. At high temperatures, the rotational motion of the adamantane molecules can be described to a good approximation as jumps between various directions differing by 90°. At the melting temperature, the jump frequency is approximately one half the Debye frequency ν_d. This permits one to assume that the molecular rotation in crystalline adamantane is weakly hindered. To establish reliably whether the observed growth of the isochoric thermal conductivity is connected with a change in the character of rotational motion of molecules in the crystal, it is necessary to fulfill two conditions, namely, (i) to study a number of substances for which the rotational motion of the molecules is of a different character; (ii) to make direct studies of the isochoric thermal conductivity as its calculation from the data on isobaric thermal conductivity is inevitably accompanied by a decrease in accuracy. A series of studies of the isochoric thermal conductivity of the molecular crystals presented below have been made by Konstantiniv, Manzhelii, and Smirnov.

The first object of study was solid carbon tetrachloride CCl_4. It consists of a high-symmetric globular molecule, the noncentral interaction between of which is not large in comparison with the central one. This allows us to assume in principle that at high temperatures, the rapid reorientations of the CCl_4 molecules in a lattice will take place. In equilibrium with vapour, CCl_4 forms a monoclinic lattice with long-range orientational order of the molecules at temperatures up to 225 K (phase II). A high-temperature rhombohedral phase of carbon tetrachloride is orientationally disordered and exists above 225 K up to the melting temperature 250 K (phase Ib). The characteristic Debye and Einstein temperatures of the low-temperature phase of CCl_4 are equal to 96 K and 70 K, respectively[23]. Thus, it is possible to study the temperature range which is high with respect to both the phonon and libron spectra. We may state that in the solid phase near the melting temperature, the molecules reorient rapidly. In any case, the studies of the spin-lattice relaxation time of CCl_4[24] and the broadening of Raman lines[25] indicate that the character of molecular rotational motion changes slightly upon melting.

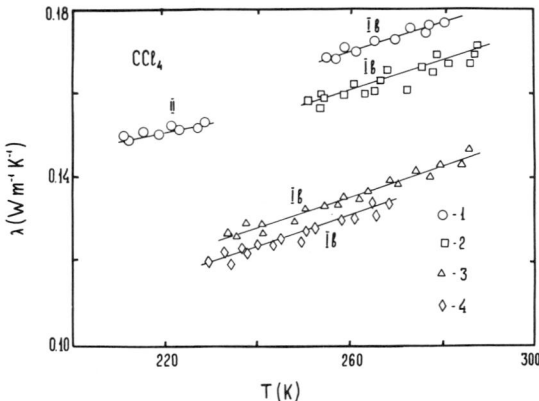

Fig. 4. Temperature dependence of the isochoric thermal conductivity of solid CCl_4 at molar volumes V, cm^3/mole: 81.5 (○); 82.6 (□); 85.2 (◇); 86.8 (△). Only single-phase regions are presented.

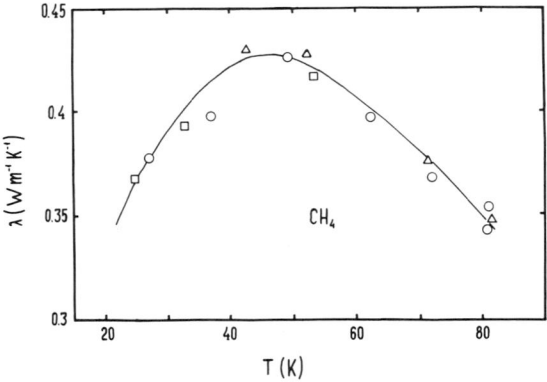

Fig. 5. Temperature dependence of the thermal conductivity of methane in equilibrium with vapour: monocrystalline (△) and polycrystalline (○, □).

The results of the studies of the isochoric thermal conductivity of CCl_4 with four molar volumes are given in Fig. 4.[26] The studies were carried out in the high-temperature phase. For the minimum molar volume, the measurements were also made in the low-temperature phase II near the phase transition temperature. Direct measurements show that the isochoric thermal conductivity of carbon tetrachloride increases with temperature both in phases II and Ib. From previous work, it follows that the additional contributions of radiation,[27] vacancies[10] or librons [17] to the heat transfer are negligible and can not account for the observed behaviour of λ.

In the framework of a simple kinetic model, the growth of the isochoric thermal conductivity with temperature can be explained by the increase of the phonon free path length because of a weakening of the influence of some phonon scattering mechanisms. It is natural to assume that the interaction of phonons with excitations of the rotational motion of molecules provides such a mechanism. These studies of adamantane and carbon tetrachloride were made in the temperature range where a gradual transition from the librational motion of molecules to their hindered rotation takes place. In contrast to the librations, free molecular rotation does not lead to phonon scattering. As the angular velocity of freely rotating molecules does not depend on their position, phonons can not interact with freely rotating molecules. From the above it follows that there is a temperature region where phonon scattering by excitations of the molecular rotational motion weakens with increasing temperature. This leads to a decrease of the component of the thermal resistance due to the phonon-rotational interaction with the increase of temperature. If this decrease is larger in absolute value than the increase of the thermal resistance resulting from the phonon-phonon interaction, then the total thermal conductivity of the crystal will rise with increasing temperature. Such a view was expressed by Manzhelii and Krupskii in their works on the study of isobaric thermal conductivity of high-temperature phase of solid methane.[28]

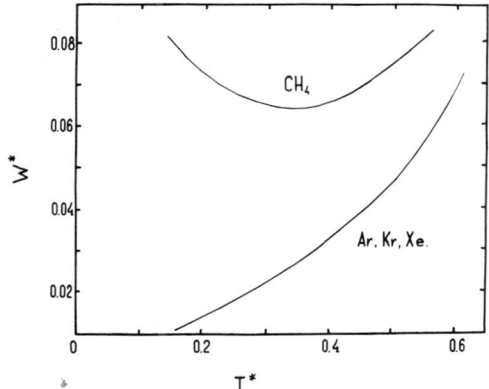

Fig. 6. Temperature dependence of thermal resistance of CH_4 and Ar, Kr, Xe in reduced variables $W^* = W^*(T^*)$.

There are numerous proofs that in the high-temperature phase of solid methane existing in the temperature range from 20 to 90 K, the molecule rotation varies from jump-like molecular reorientations to almost free rotation[29–32] with increasing temperature. Fig. 5 shows the temperature dependence of the thermal conductivity of methane in equilibrium with vapour. Note, that the well-known maximum of the thermal conductivity of dielectrics resulting from the competition between phonon-phonon scattering and phonon scattering at the sample boundaries is observed at considerably lower temperatures. It is assumed that the increase of thermal conductivity of solid methane in the temperature range from 20 to 45 K is the result of a weakening of phonon scattering by excitations of rotational motion of molecules. With a further temperature increase, the phonon rotational interaction weakens so that the behaviour of the thermal conductivity is determined to ever-increasing degree by the phonon-phonon interaction. This is distinctly visible in Fig. 6 when we compare the temperature dependence of the thermal resistance of methane with that of the inert gas solids in reduced variables. The data on Ar, Kr and Xe are described in these variables by a universal curve. At high temperatures, the role of the phonon-phonon interaction increases and the thermal resistance of methane approaches the thermal resistance of the inert gas solids. The results given in Figs. 5 and 6 refer to the isobaric thermal conductivity and isobaric thermal resistance. With the conversion to isochoric magnitudes, one expects that the steepness of the curves may change but that the main peculiarities such as the change of the sign of the temperature dependence of the thermal resistance and thermal conductivity of solid methane and the fact that the reduced thermal resistance of methane and the inert gas solids approaches closer to one another, will remain unchanged.

The situation at which the phonon-rotational interaction weakens so that thermal conductivity begins to decrease again with the increasing temperature, does not appear in solid CCl_4 up to melting. This is explained by the relatively important role played by noncentral forces of the intermolecular interaction in the case of CCl_4.

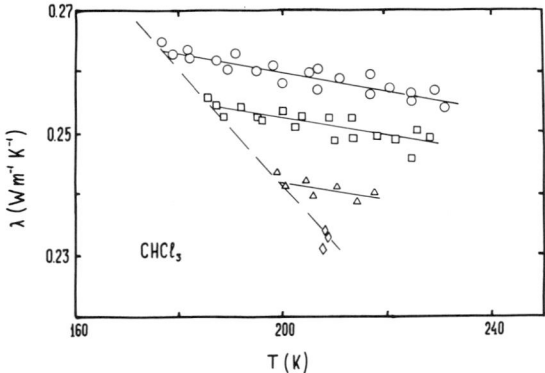

Fig. 7. Temperature dependence of the isochoric thermal conductivity of solid $CHCl_3$ at molar volumes V, cm^3/mole: 62.5 (o); 63.0 ($\square$); 63.5 ($\triangle$). The dashed line and the diamonds ($\diamond$) correspond to a free sample.

Previously we discussed the results of the thermal conductivity studies of two substances with a strong noncentral interaction, namely, solid CO_2 and N_2O in which the long-range orientational order is conserved and their λ_vs decrease up to the melting temperature. To reinforce the conclusions, it seemed interesting to us to study λ_v in solids with similar properties and structure to carbon tetrachloride but differing by a relatively stronger noncentral interaction. The isochoric thermal conductivities of solid chloroform ($CHCl_3$) and solid methylene chloride (CH_2Cl_2) were studied at premelting temperatures.[33] Molecules of these substances are more asymmetric than a molecule of CCl_4, and the forces acting between them are stronger. Both crystals have an orthorhombic structure[34,35] and the long-range orientational order is conserved in them up to melting. The rotational motion of molecules of these substances involves librations, and reorientation frequencies do not exceed $10^4 c^{-1}$.[36] This is lower than the characteristic frequencies of librations by many orders of magnitude. Figs. 7 and 8 show the temperature dependence of the isochoric thermal conductivities of solid chloroform and methylene chloride at three molar volumes. In contrast to carbon tetrachloride, λ_v for these systems does not increase with increasing temperature but decreases slowly, far weaker than a $1/T$ law. This agrees completely with the qualitative discussion above.

It is of interest to study the isochoric thermal conductivity of crystals in which orientational disorder and molecular rotation appear merely with respect to given directions. One of these substances is crystalline benzene C_6H_6. At pressures below 1.2 Gpa, benzene exists in one crystalline modification only (orthorhombic structure).[37,38] In the temperature range 90-120 K, the second moment of NMR in benzene undergoes a considerable fall[39]. This is explained by the formation of molecular reorientations in the plane of a benzene ring, that is around the axis of six-fold symmetry.[39]

The activation energy of the reorientational motion estimated from the spin-lattice relaxation time is equal to 3.7 kcal/mole. Near the triple point ($T_{tr} = 278$ K), the reorientation frequency reaches a magnitude[39] on the order of $10^{11}s^{-1}$ whereas the oscillation frequency of a benzene molecule around its axis of six-fold symmetry is equal to $1.05 \times 10^{12}s^{-1}$ at 273 K.[38]. Previously Eucken and Schröder[40] at equilibrium vapour pressure and Ross, Andersson, and Backstrom[41] at a pressure of 100 MPa observed good agreement of the high-temperature isobaric thermal conductivity

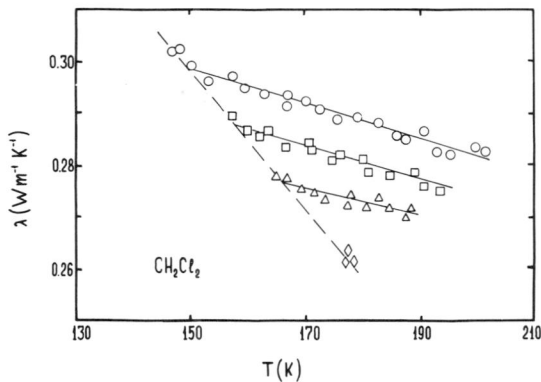

Fig. 8. Temperature dependence of the isochoric thermal conductivity of solid CH_2Cl_2 at molar volumes V, cm^3/mole: 49.1 (○); 49.4 (□); 49.8 (△). The dashed line and the diamonds (◇) correspond to the a sample.

of benzene with the $1/T$ dependence. The results of the last work, calculated using the constant pressure data at 100 MPa and 200K, and the thermal expansion data[39] do not depend on temperature. The results of the study of the isochoric thermal conductivity[42] with three molar volumes are presented in Fig. 9. The isochoric thermal conductivity of all three samples rises weakly with the increase of temperature ($d\lambda/dT = 1.110^{-4}$W m^{-1}K^{-1}). This speaks in favour of the considerations given. The effect appears as only marginal as the rotation takes place only in one plane.

The character of rotational motion of molecules and the degree of orientational ordering in a crystal are connected with one another in orientationally-ordered phases, molecules make librations in combination with jumps between different orientations. The amplitude of librations and the frequency of reorientations rise with temperature. In orientationally-disordered phases, the rotational motion of molecules turns from frequent reorientations to hindered rotation with the increasing temperature and sometimes, as in the case of solid methane, it turns to almost free rotation. Till now we have spoken mainly about the growth of the isochoric thermal conductivity with the temperature increase in the case of orientationally-disordered phases. However, the isochoric thermal conductivity of the orientationally-ordered phase in a narrow temperature region just below the temperature of orientational melting[26] was studied for one of the molar volumes of solid carbon tetrachloride. In this case, the orientational thermal conductivity rises with increasing temperature increase as well.

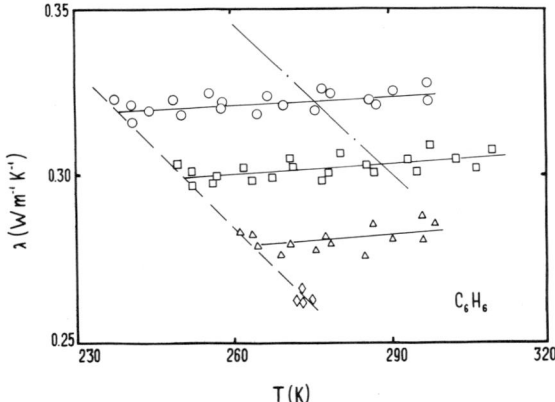

Fig. 9. Temperature dependence of the isochoric thermal conductivity of solid benzene at molar volumes V, cm^3/mole: 75.05 (O); 75.65 (□); 76.40 (△). The dashed line and the diamonds correspond to a free sample, the dash-dotted line correspond to the pressure of 100 MPa.

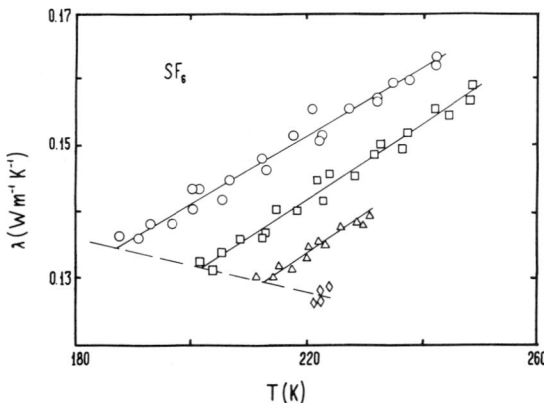

Fig. 10. Temperature dependence of the isochoric thermal conductivity of solid SF_6 at molar volumes V, cm^3/mole; 62.2 (o); 62.9 (□); 63.4 (△). The dashed line and the diamonds (◇) correspond to the free sample.

Recently, the isochoric conductivity of an orientationally-ordered phase of sulphur hexafluoride (SF_6) was studied over a relatively wide temperature range.[43] A molecule of SF_6 has octahedral symmetry. The high-temperature solid phase of SF_6 exists at vapour pressure over a very wide temperature range from $94.6K$ to $222.4K$

and has a body-centered cubic lattice. It is orientationally-ordered.[44] Fig. 10 shows the isochoric thermal conductivity for three molar volumes together with λ at equilibrium vapour pressure. The isochoric thermal conductivity rises rather rapidly with increasing temperature. Hence, it follows that phonon scattering by excitations of the rotational motion of molecules weakens markedly even in the orientationally-ordered phase. According to X-ray studies,[44] at temperatures above 140 K, the intensity of a series of reflections decreases which is indicative of an increase of the orientational disordering (while conserving the long-range orientational order). These processes are not the consequences of an increase of the libration amplitude only but they are also connected with a rapid reorientational motion of molecules. From neutron and Brillouin scattering experiments,[45,46] it is also known that there are no well-determined collective excitations connected with the rotational motion of molecules in SF_6 at premelting temperatures. This makes it possible to conclude that phonon scattering weakens with the dynamical destruction of collective orientational oscillations.

Previously we left without explanation the increase of the isochoric thermal conductivity of the orientationally-ordered crystal CO_2 with 9 % of Xe. Taking into account the above, the effect may be explained by assuming that the presence of the spherically symmetric atoms of Xe enhances the destruction of long-range orientational order with increasing temperature and hence decreases the phonon scattering by collective excitations of the rotational molecular motion.

Thus, the growth of the isochoric thermal conductivity with increasing temperature has a rather common character and it is evidently the consequence of the weakening of the phonon interaction with excitations (including collective ones) of the rotational motion of the molecules.

In conclusion it should be noted that the considerations of the presented work are of a qualitative character and are not supported by any microscopic theory.

REFERENCES

1. G.A. Slack, *Solid State Phys.* **34**, 1 (1979).
2. R. Berman, *Thermal Conduction in Solids*, Clarendon Press, Oxford (1976).
3. D.G. Cahill and R.O. Pohl, *Ann. Rev. Phys.* **39**, 93 (1988).
4. M.C. Roufosse and P.G. Klemens, *J. Geophys. Res.* **79**, 703 (1974).
5. P.G. Klemens, *High Temp.-High Pressures* **15**, 249 (1983).
6. A. Auerbach and P.B. Allen, *Phys. Rev. B.* **29**, 2884 (1984).
7. G.A. Slack and P. Andersson, *Phys. Rev. B.* **26**, 1883 (1982).
8. S. Petterson, *J. Phys.: Condens. Matter* **1**, 361 (1989).
9. V.E. Zinoviev and S.I. Masharov, *Thermal Properties of Solids*, Nauka, Moscow (1973).
10. V.A. Konstantinov, V.G. Manzhelii, M.A. Strzhemechny and S.A. Smirnov, *Sov. J. Low Temp. Phys.* **14**, 48 (1988).
11. F. Clayton and D.N. Batchelder, *J. Phys. C.: Solid State Phys.* **7**,1213 (1973).
12. A.I. Bondarenko, V.G. Manzhelii, V.A. Popov, M.A. Strchemechny, and V.G. Gavrilko, *Sov. J. Low Temp. Phys.* **8**, 617 (1982).
13. A.I. Bondarenko, V.G. Manzhelii, V.A. Popov, M.A. Strzhemechny, and V.G. Gavrilko, *Sol. St. Comm.* **45** N.4:387 (1983).
14. I.N. Krupskii and V.G. Manzhelii, *Sov. J. Exp. Theor. Phys.* **28**, 1097 (1969).
15. J.M. Ziman, *Electrons and Phonons*, Clarendon Press, Oxford (1960).
16. D.F. Spitzer, *J. Phys. Chem.* **31**, 19 (1970).
17. V.G. Manzhelii, V.B. Kokshenev, L.A. Koloskova, and I.N. Krupskii, *Sov. J. Low Temp. Phys.* **1**, 624 (1975).

18. V.A. Konstantinov, V.G. Manzhelii, S.A. Smirnov, and A.M. Tolkachev, *Sov. J. Low Temp. Phys.* **14**, 104 (1988).
19. J. de Boer, *Physica* **14**, 139 (1948).
20. V.A. Konstantinov, V.G. Manzhelii, and S.A. Smirnov, *Sov. J. Low Temp. Phys.* **14**, 412 (1988).
21. *Cryocrystals*, Ed. by B.I. Verkin and A.F. Prichotko, Naukova Dumka, Kiev (1983).
22. R.G. Ross, P.A. Andersson, B. Sundqvist, G. Backstrom, *Rep. Progr. Phys.* **47**, 1347 (1984).
23. M.I. Bagatskii and V.G. Manzhelii, *Ukr. Phys. J.* **15**, 1088 (1971).
24. D.F. O'Reilly, E.M. Peterson, and C.E. Schlie, *J. Chem. Phys.* **60**, 1603 (1974).
25. M. Djabonrov, C. Levy-Mannhein, J. Leblond, and P. Papon, *J. Chem. Phys.* **66**, 5748 (1977).
26. V.A. Konstantinov, V.G. Manzhelii, and S.A. Smirnov, *Phys. Stat. Sol. (b)* **163**, 369 (1991).
27. B.M. Mogilevskii and V.T. Surin, *Sov. J. Solid St. Phys.* **13**, 293 (1971).
28. V.G. Manzhelii and I.N. Krupskii, *Sov. J. Solid St. Phys.* **10**, 284 (1968).
29. J.H. Colwell, E.K. Gill, and J.A. Morrison, *J. Chem. Phys.*, **42**, 3144 (1965).
30. A. Anderson and R. Savoie, *J. Chem. Phys.*, **43**, 3468 (1965).
31. V.G. Manzhelii, A.M. Tolkachev, and E.G. Voitovich, *Phys. Stat. Sol.* **13**, 351 (1966).
32. V.G. Manzhelii, A.M. Tolkachev, and V.G. Gavrilko, *J. Phys. Chem. Solids*, **30**, 2759 (1969).
33. V.A. Konstantinov, V.G. Manzhelii, and S.A. Smirnov, *Sov. J. Low Temp. Phys.* **17**, 462 (1991).
34. R. Fourme and M. Renaud, *C.R. Acad. Sci. Paris* **263**, 69 (1966).
35. T. Kawaguchi, K. Tanaka, T. Takenchi, and T. Watanabe, *Bull. Chem. Soc. Japan*, **46**, 62 (1973).
36. H.S. Gutovsky and D.N. McCall, *J. Chem. Phys.*, **32**, 548 (1966).
37. E.G. Cox, D.W.S. Cruickshank, Y.A..S. Smith, *Proc. Roy. Soc.* **247**, N.1248:1 (1958).
38. G.E. Bacon, N.A. Curry, S.A. Wilson, *Proc. Roy. Soc.* **279**, N.1376:98 (1964).
39. E.R. Andrew, R.G. Eades, *Proc. Roy. Soc.* **218A**, 537 (1953).
40. A. Eucken, E. Schroder, *Anal. Phys.* **36**, 609 (1939).
41. R.G. Ross, P.A. Andersson, G. Backstrom, *Molec. Phys.* **38**, N.2:377 (1979).
42. R.G. Ross, P.A. Andersson, B. Sundquist, G. Backstrom, *Rep. Progr. Phys.* **47**, 1347 (1984).
43. V.A. Konstantinov, V.G. Manzhelii, S.A. Smirnov, *Ukr. Phys. J.* **37**, 737 (1992).
44. A.P. Isakina and A.I. Prokhvatilov, *Sov. J. Low. Temp. Phys.* **19**, N2 (1993).
45. M.T. Dove, G.S. Pavley, G. Dolling and B.M. Powell, *Mol. Phys.* **57** N4:865 (1986).
46. H. Kniefte, R. Peney, and M.J. Clouter, *J. Chem. Phys.* **88**, N9:5846 (1988).

FULLERENES

Kosmas Prassides

School of Chemistry and Molecular Sciences
University of Sussex
Falmer, Brighton BN1 9QJ
United Kingdom

ABSTRACT

Recent structural and dynamical studies of C_{60} and C_{70} fullerenes and their derivatives in the solid state are reviewed. Orientational ordering of the fullerene molecules is accompanied by drastic changes in the rotational dynamics and structural phase transitions. Solid C_{60} undergoes a first-order phase transition at 260 K from a face-centred to a simple cubic structure. The isotropic molecular reorientations change abruptly to quasi-random jump motion between nearly-degenerate orientations differing in energy by 11.4(3) meV. A transition to an orientational glass state occurs at 85 K. Static disorder dominates the crystal chemistry of C_{70}. The orientational ordering transitions from the high temperature face-centred cubic structure to the low temperature rhombohedral and monoclinic phases are accompanied by severe hysteresis effects. The rotational dynamic behaviour changes progressively on cooling from isotropic to anisotropic to uniaxial reorientations about the unique molecular axis. High pressure diffraction experiments reveal that C_{70} has a smaller compressibility than C_{60}. Disorder effects are also present well in the superconducting state for the potassium fulleride K_3C_{60} The C_{60}^{3-} units perform small-amplitude librational motions up to 650 K. Measurements of the phonon spectra of K_3C_{60} reveal substantial broadening of five-fold degenerate H_g intramolecular vibrational modes both in the low-energy radial and the high-energy tangential part of the spectrum. This provides evidence for a traditional phonon-mediated mechanism of superconductivity in the fullerides but with an electron-phonon coupling strength distributed over a wide range of energies (33-195 meV) as a result of the finite curvature of the fullerene spherical cage.

INTRODUCTION

Nearly twenty years after C_{60} was first conceived and five years since it was discovered[1], Buckminsterfullerene has been isolated[2] and its structure confirmed. As a result, a new field of carbon science has opened up overnight and catalysed immense research activity worldwide. The field was born out of a quest[3] for an understanding of the origins of carbon in space and in red giant stars. The fullerenes, a set of hollow, closed-cage molecules consisting purely of carbon, were serendipitously discovered and are currently modifying significantly our views on the chemistry, physics and materials science of carbon[4]. Notwithstanding the stability which led to their original detection, fullerenes are found to participate readily in a plethora of chemical reactions. Reactions can lead among others to the formation of: molecular derivatives $C_{60}X_n$ (for X= Br, bromides[5] with stoichiometries $C_{60}Br_6$, $C_{60}Br_8$ and $C_{60}Br_{24}$ and for X= H, a hydride[6] with stoichiometry $C_{60}H_2$ have been isolated and characterised), charge-transfer salts (reaction with ferrocene affords[7] the solid $[Fe(C_5H_5)_2]_2C_{60}$) ,and essentially ionic salts (reaction with alkali metals[8-10] leads to the formation of insulating solids with stoichiometries A_2C_{60}, A_4C_{60}, and A_6C_{60}, and superconducting solids with T_c's as high as 33 K and stoichiometry A_3C_{60}).

In this paper, we briefly review the evolution with temperature of the structural and dynamical properties of pristine C_{60} and C_{70} fullerenes in the solid state, drawing principally from the results of neutron scattering and μSR experimental techniques. The structural phase transitions accompanying orientational ordering of the molecules are sensitively related to deviations from isotropy arising through the presence of more than one type of carbon-carbon bonds which give rise to an anisotropic π electronic distribution on the fullerene surface. C_{70} shows a much richer and more complicated structural and dynamical behaviour than C_{60}, as expected from its more anisotropic molecular structure. Disorder effects are found to dominate the crystal chemistry both of pristine and intercalated fullerenes. Knowledge of the experimental vibrational properties of A_nC_{60} (n= 0, 3, 6), as obtained by neutron scattering measurements at non-zero momentum-transfer |**Q**| is very important for both the theoretical description of the electronic structure of fullerenes and the implications towards the possible mechanism for superconductivity in the fullerides, providing evidence for or against the participation of phonons in the pairing interaction. Selected radial and tangential vibrations of H_g symmetry are found to broaden substantially in the superconductors and are strongly involved in the pairing interaction. On the other hand, no evidence is found for a significant contribution to the total electron-phonon coupling strength from low-energy librational modes.

EXPERIMENTAL METHODS

Neutron Scattering

When a neutron impinges on a nucleus, several interactions lead to different processes[11]. The neutron may be transmitted, scattered, or absorbed. In a scattering experiment, two basic quantities are measured:
(i) the energy transfer, $\hbar\omega$, between the initial, E_i, and final, E_f, energies of the neutrons:

$$\hbar\omega = E_f - E_i = (\hbar^2/2m)(k_f^2 - k_i^2) \qquad (1)$$

where k_f and k_i are the wavevectors of the scattered and incident neutrons,

(ii) the scattering vector, Q, which is related to the momentum transfer

$$Q = k_f - k_i \qquad (2)$$

The scattering process may be either elastic ($\hbar\omega = 0$) or inelastic ($\hbar\omega \neq 0$), i.e. it may involve transfer of energy to (neutron energy gain, $\hbar\omega > 0$) or from the scattered neutron (neutron energy loss, $\hbar\omega < 0$). This is illustrated in Figure 1. The static

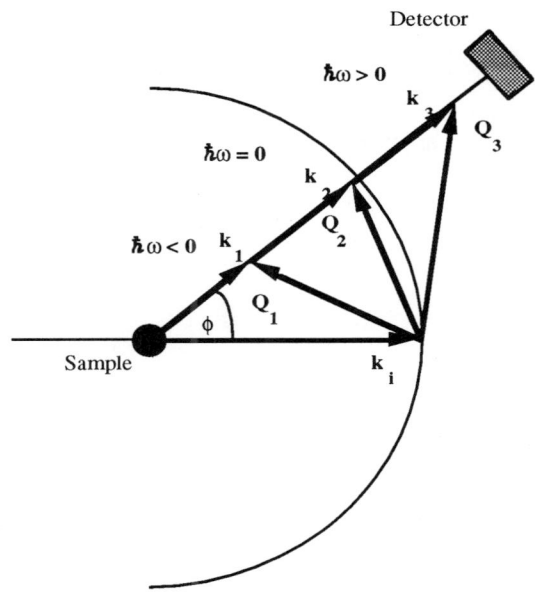

Figure 1. Schematic diagram illustrating neutron elastic and inelastic scattering events.

(infinite time) structure of a solid gives rise to elastic scattring; the wavelength of a thermal neutron, $\lambda = h/mv$ is of the same order of magnitude as the interatomic distances in solids and neutron diffraction may be employed in the study of crystal structures. The dynamical behaviour of solids gives rise to inelastic neutron scattering as energy is transferred from or to molecular or crystal vibrations. The use of neutrons in the study of molecular and crystal vibrations stems from the fact that their energies are of the same order of magnitude as those of the phonon modes in solids. The scattering may also be centred at $\hbar\omega = 0$ and possess a non-vanishing broadening, as a result of time-dependent differences in the atom-atom position correlation function; in such a case, it is termed quasielastic scattering and can give information about the characteristic times of the motion. Furthermore, the scattering processes can be either coherent, arising by scattering in a regular way from the lattice, or incoherent, arising from non-periodic fluctuations in scattering power.

Muon Spin Relaxation/ Rotation/ Resonance

Positive muons[12] (μ^+) are spin-$\frac{1}{2}$ elementary particles with a mass equal to $\frac{1}{9}$ the mass of a proton and a lifetime of 2.2 µs. They are produced as 100%-spin polarised in the course of pion decay. They may bind an electron to form a muonium atom (Mu ≡ μ⁺e⁻), the light isotope of hydrogen. Both the ionisation potential and the Bohr radius of Mu are within 0.5% of the H atom values. Mu is a two-spin-$\frac{1}{2}$ system with the hyperfine coupling giving rise to a singlet |F=0, m$_F$=0> and a triplet |F=1, m$_F$=0,±1> state in zero magnetic field. Vacuum Mu is characterised by the Hamiltonian (in units of h);

$$\mathcal{H}_{iso} = v_e S_z - v_\mu I_z + A_\mu \mathbf{S} \cdot \mathbf{I} \tag{3}$$

where $v_\mu = \gamma_\mu H = H \times 13.55$ kHz/G and $v_e = \gamma_e H = H \times 2.8025$ MHz/G are the muon and the electron Larmor frequencies, respectively, γ_μ and γ_e are the muon and electron gyromagnetic ratios, respectively, and $A_\mu = 4463$ MHz is the isotropic hyperfine interaction coupling the muon spin **I** and the electron spin **S**. The triplet degeneracy is lifted in a magnetic field (Figure 2); in low transverse fields, four spectroscopic transitions are magnetic dipole allowed, while in high field the number reduces to only two (purely nuclear) transitions.

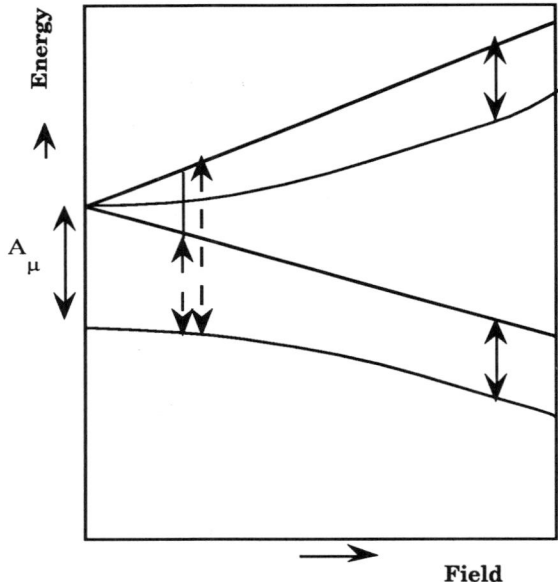

Figure 2. Schematic Breit-Rabi diagram for muonium in an applied magnetic field.

Mu mimics the behaviour of atomic hydrogen and reacts with unsaturated organic molecules through addition to multiple carbon-carbon bonds to give muonated radicals[12]. The muon spin label can then serve as a probe of the structure, kinetics and molecular dynamics of the radical. An important difference between vacuum Mu and chemically-bound Mu is that the electron spin density at the muon site is greatly reduced in the latter case, resulting in substantially smaller hyperfine coupling constants. In addition, for a static radical the hyperfine tensor is invariably anisotropic and only in the presence of fast tumbling (liquid-like) motion, isotropic coupling results. The evolution of the hyperfine anisotropy of muonated radicals with changing temperature can then be used to monitor their reorientational behaviour. Experimentally, the radicals may be identified by their characteristic frequencies in their muon spin rotation spectra (magnetic fields are applied perpendicular to the implanted muon spin direction, and the precession of the muon polarisation is monitored). Alternatively, muon spin relaxation experiments in longitudinal magnetic fields, applied parallel to the muon spin direction, yield information on the spin-lattice relaxation and can monitor the radical reorientational dynamics through the modulation of the hyperfine anisotropy. Finally, avoided crossings of magnetic energy levels at high fields can allow for the inversion of the muon spin, which leads to a resonance in the observed signal[13]. Of particular interest when studying the reorientational motion of radicals is the monitoring of the muon spin flip ($\alpha_\mu \rightarrow \beta_\mu$) $|\Delta M| = 1$ transition which is made allowed by the anisotropic (dipolar) part of the muon-electron hyperfine interaction. Such a resonance line disappears when *isotropic* reorientational motion sets in on a time scale comparable to the inverse anisotropy of the radical.

PRISTINE FULLERENES

Fullerene C_{60}

The icosahedral symmetry of the C_{60} molecules and its incompatibility with periodic translational symmetry make the determination of the structure of solid C_{60} of fundamental importance. Fleming et al.[14] have shown that the C_{60} I_h molecular symmetry is reduced to at most $m\bar{3}$ by a cubic crystalline field. The two standard

Figure 3. The two standard orientations of a C_{60} molecule placed at the origin of a cubic lattice.

orientations of the C_{60} molecule placed at the origin of a cubic lattice are shown in Figure 3. They are related by a 90° rotation about [100] axes which are now coincident with rotational 2-fold symmetry axes. However, the crystal symmetry may be further raised through disorder; superposition and equal fractional occupancy of the two standard orientations results in $m\bar{3}m$ symmetry with the [100] axes now possessing 4-fold rotational symmetry.

Early X-ray diffraction work[2] revealed that at room temperature the C_{60} crystalline powder consisted of spheroidal molecules of diameter 7.1 Å, forming a random mixture of hexagonal close-packed (hcp) and face-centred cubic (fcc) arrays. However, elimination of solvent molecules trapped in interstitial cavities leads only to a fcc crystal structure in which each C_{60} molecule is orientationally disordered. Powder neutron[15] and X-ray[16] diffraction data are consistent with the space group $Fm\bar{3}m$, with all four quasi-spherical C_{60} units being symmetry equivalent and a cubic lattice constant at 290 K of 14.1569(5) Å. A variety of experimental techniques revealed that in this plastic 'rotator' phase, the C_{60} molecules rotate effectively freely and randomly. Deviations from a uniform charge spherical distribution are however still present with excess electron density present along the <111> and a deficiency along the <110> directions[17]. Rotational correlation times τ_R of 9-12 ps at room temperature are only 3-4 times longer than expected for unhindered gas phase rotation and faster than any other known solid state rotor[18]. The high temperature description of pristine C_{60} that has emerged is of a prototypical plastic crystal (cf. adamantane, norbornane) with a well-defined translational order and a smooth rotational potential with many shallow minima ($E_a=35\pm15$ meV[19]), leading to continuous small-angle diffusive motions.

As the temperature is lowered, the rotational motion abruptly changes. The C_{60} units orientationally order and are no longer symmetry-equivalent; the crystal structure is now simple-cubic (space group $Pa\bar{3}$)[15,16,20]. The low-temperature structure of C_{60} can be traced to deviations from a fully isotropic electronic

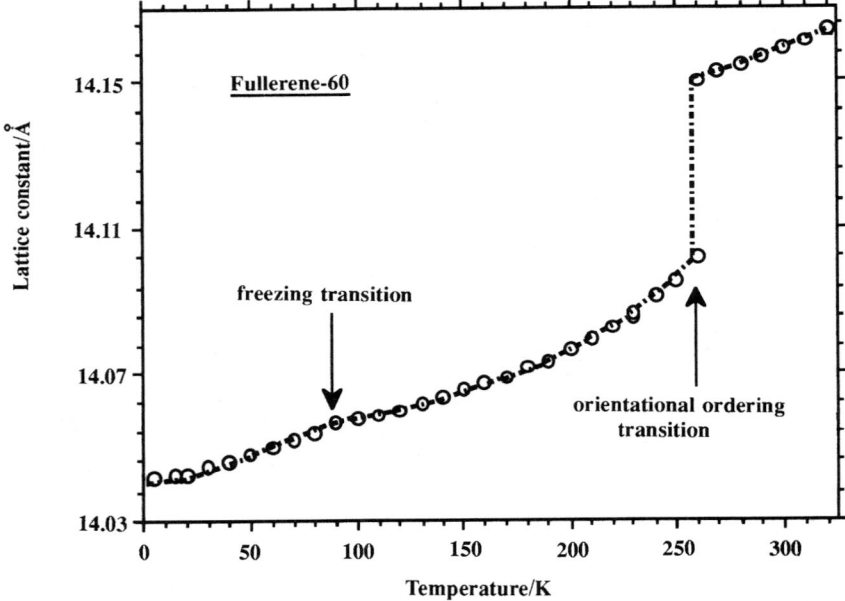

Figure 4. Temperature evolution of the cubic lattice constant of solid C_{60}. The first-order transition is clearly evident at 260 K and is accompanied by a decrease in the lattice constant of 0.0486(15) Å. Both fcc (~45%) and sc (~55%) phases coexist at this temperature. The signature of the glass transition appears as a cusp at ~90 K.

distribution which arises from the presence of two types of carbon-carbon bonds: long (fusions of five- and six-membered rings, 1.448(10) Å[21]) and short (fusions of two six-membered rings, 1.404(10) Å[21]). In addition to the predominant van der Waals interball interactions, intermolecular bonding is optimised by contributions arising through alignment of the "electron-rich" regions (short carbon-carbon bonds) of one molecule over the "electron-deficient" regions (pentagonal faces) of its near neighbours. The orientational ordering transition is of first-order[22] and occurs abruptly at 260 K in highly pure C_{60} (Figure 4). It is highly sensitive to the presence of impurities (e.g. solvent molecules) which may be retained tenaciously in the interball spacing*. It is also accompanied by a change in the nature of the rotational dynamics to jump reorientational motion, as the molecules "shuffle" between two *nearly*-degenerate orientations. An estimate[21] of the energy difference, $\Delta E = 11.4(3)$ meV between the two orientations (short C-C bonds facing pentagonal or hexagonal faces) may be obtained by considering the temperature evolution of the fraction, p of the molecules in the majority phase (short C-C bonds facing pentagonal faces) which is given at thermal equilibrium by:

$$p = \frac{1}{1 + \exp(-\Delta E/kT)} \quad (4)$$

The magnitude of the jump angle[23] may be also derived from the magnitude of the rotational barrier (~250-300 meV[18]) and of the librational energy (~2.5 meV[19]) as ~36-42° if we assume a simple sinusoidal hindrance potential (Figure 5)[23]. These

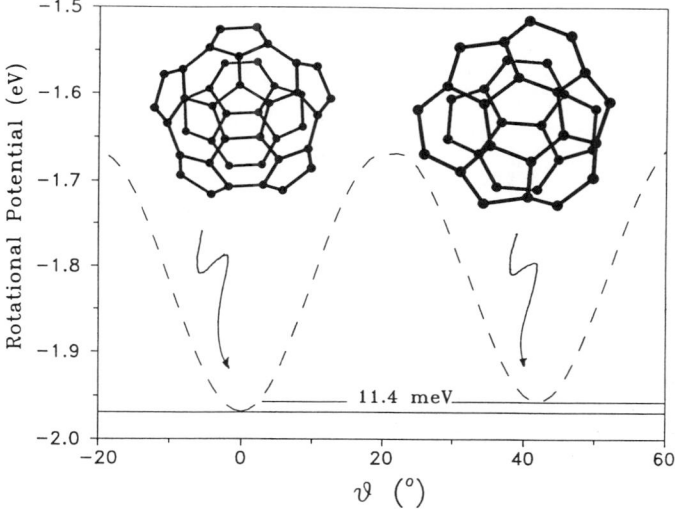

Figure 5. Schematic energy diagram for rotational motion about the [1$\bar{1}$0] direction. The two minimum energy near-neighbour orientations are also shown.

structural requirements are nicely satisfied if the molecules perform ~42° hops about that set of <1$\bar{1}$0> axes which straddle short C-C bonds (Figure 6). Moreover, such type of jump motion occurs in a *quasi-random* fashion and even though, it is still difficult to confirm unambiguously, it is noteworthy that both avoided-level-crossing muon spin resonance[24] and 2D ^{13}C NMR[25] studies are consistent with the implicit *quasi-isotropic* nature of the motion required by the hops about the <1$\bar{1}$0> axes of Figure 6.

---

* A variety of small molecules can be trapped in the interfullerene spacing in solid fullerenes, making the isolation of highly pure samples not a routine procedure. Solvent molecules are retained in the interfullerene spacing even after sublimation and long annealing of the solid samples. In our experience, we find that the hydrogen content of both sublimed and annealed C_{60} and C_{70} solids is no less than 1 hydrogen atom per ~2.5-3 fullerene molecules.

Finally, on further cooling[15] in the vicinity of 85-90 K, the molecular jump motion essentially freezes and a transition to an orientational glass phase occurs. This has now been confirmed by a variety of experimental techniques, including thermal conductivity[26], thermal expansion[27] and heat capacity[28] measurements.

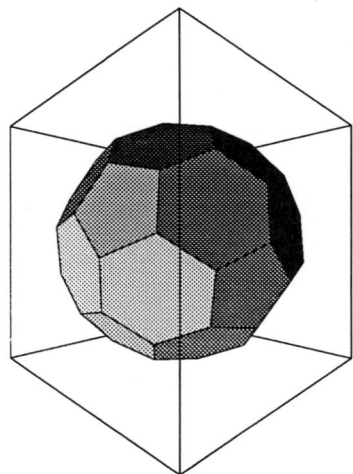

Figure 6. View along a <$1\bar{1}0$> axis straddling hexagon-hexagon fusions: the C_{60} molecule is at the origin of the $Pa\bar{3}$ unit cell in the majority phase.

Fullerene C_{70}

Understanding of the structural and dynamical properties of the less symmetric fullerene C_{70} whose molecular structure is that of an ellipsoid (point group D_{5h}) with five different types of carbon atoms in the ratio 1:1:2:2:1 is also developing at present. Large quantities of crystalline C_{70} samples have been more scarce while there is a greater complexity associated with the lower symmetry of the molecule. Early X-ray diffraction and electron microscopy work[29] had revealed the presence of more defects than in C_{60} and coexistence of cubic (*ccp*) and hexagonal close-packed (*hcp*) crystalline phases. Moreover, the minority *hcp* phase proved harder to eliminate and most reported studies were performed on mixed phase materials. At present, there exist only two diffraction studies[30,31] which have probed the temperature dependence of the most stable structural modification of solid C_{70}, namely the *ccp* one. At high temperatures, both neutron[30] and X-ray[31] diffraction reveal that C_{70} is *fcc* and adopts the space group $Fm\bar{3}m$. The cubic lattice constant (a= 14.976(7) Å at 430 K) scales with the long molecular axis, as the molecules tumble fast enough to average out the anisotropy. On cooling, the first evidence of extra reflections, not allowed in the *fcc* space group, appears at ~300- 280 K when a small peak starts growing on the low *d*-spacing side of the (220)$_c$ reflection. On further cooling, new peaks grow at the expense of the fcc reflections. The diffraction profile develops gradually down to 200 K. Little change occurs below this temperature down to 5 K. The observed diffraction profiles in the 280-200 K temperature range can be rationalised by considering the co-existence of a *fcc* and a rhombohedral (a_{rhomb}~ a_{cub}, α~ 85.6°, space group $R\bar{3}m$) phase; in the latter, the long molecular axis aligns towards the unit cell diagonal (Figure 7). What appears remarkable is that a fraction of the high symmetry disordered phase is still present even at very low temperatures. Below 180 K, the neutron diffraction profile is consistent with the existence of an orientationally ordered structure with monoclinic

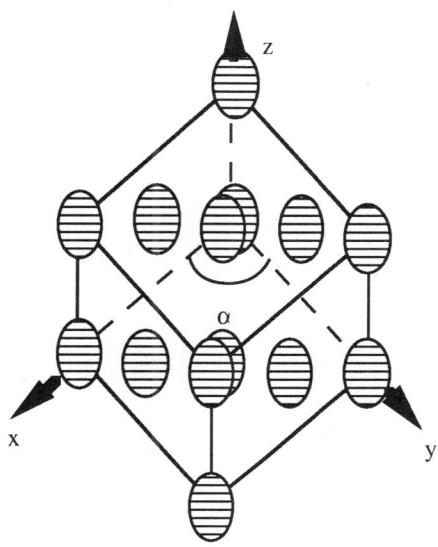

Figure 7. Schematic diagram of the rhombohedral unit cell of solid C_{70} ($\alpha \sim 85.6°$).

symmetry. These results are in agreement with the high resolution X-ray[31] powder diffraction measurements; these clearly identified the low temperature structure as monoclinic (space groups $C2$, Cm, $P2_1$, or Pm). Where there are differences between the two studies, are the temperature regimes over which each phase exists. The X-ray study finds that the fcc → rhombohedral phase transition occurs at 345 K and the rhombohedral → monoclinic one at 295 K. In contrast, no evidence was apparent in the neutron diffraction profiles of the presence of a non-cubic phase down to 280 K. It is noteworthy, however, that on heating, even though the rhombohedral → fcc transition again occurs at 280 K, the low-temperature structure does not disappear until the temperature is raised to 340- 350 K. The structural properties of C_{70} are thus sensitive to the high concentration of defects, associated with static and dynamic disorder, that lead to extensive "undercooling" for the fcc phase and "overheating" behaviour for the rhombohedral phase. Perhaps the hindrance potential barrier associated with the microstrains and defects is of the order of 350 K. Diffuse scattering arising from both static and dynamic disorder is present in the neutron diffraction profiles down to 10 K. Its intensity and Q-dependence changes little with temperature in sharp contrast with C_{60} in which the diffuse scattering present in the disordered phase is reduced substantially upon orientational ordering[19]. Figure 8 shows the temperature evolution of the unit cell diagonal in solid C_{70} which shows a drastic expansion of ~1.92(5) Å upon orientational ordering, consistent with the aspect ratio and the dimensions of individual C_{70} molecules ordering along the cell diagonal. This is also accompanied by an abrupt contraction along the close packing <110> direction (~0.37(3) Å). These changes in lattice dimensions may be contrasted with the more subtle changes accompanying orientational ordering of solid C_{60} (Figure 4).

Modest pressure at ambient temperature is also enough to cause an irreversible ordering transition at ~0.35 GPa[30], the resulting rhombohedral structure ($\alpha \sim$ 85.4(1)°) being slightly less compressible ($\kappa = 4.0 \pm 1.4 \times 10^{-2}$ GPa^{-1}) than the simple

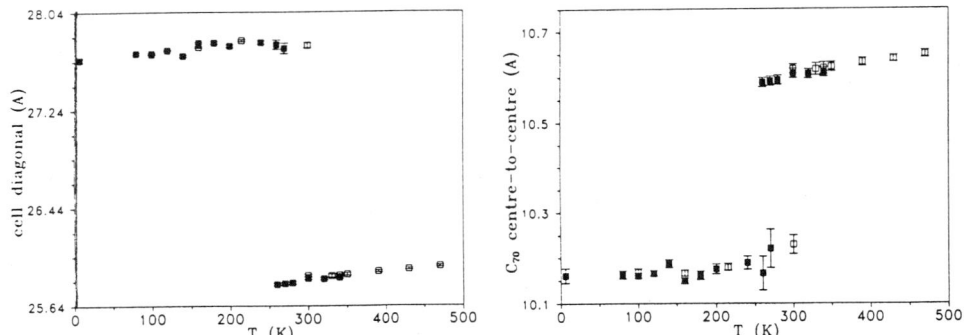

Figure 8. Temperature dependence of the unit cell diagonal and the centre-to-centre distance on the close-packing plane in solid C70. A pseudo-rhombohedral unit cell was used for the low-temperature phases.

cubic structure of C$_{60}$ ($\kappa = 5.5 \pm 0.6 \times 10^{-2}$ GPa^{-1}) (Figure 9). This may be related to the anisotropy of the C$_{70}$ molecules compared to the quasi-spherical C$_{60}$ units which can pack much more efficiently. No fcc fraction is present above 1 GPa. When pressures higher than 6.5 GPa were applied, a progressive reduction in the intensities of the 311, $31\bar{1}$, $11\bar{3}$, and $22\bar{2}$ reflections was observed and the solid becomes more compressible. This may be associated with a distortion of the rhombohedral unit cell, possibly driven by a reduced shear strength arising from the

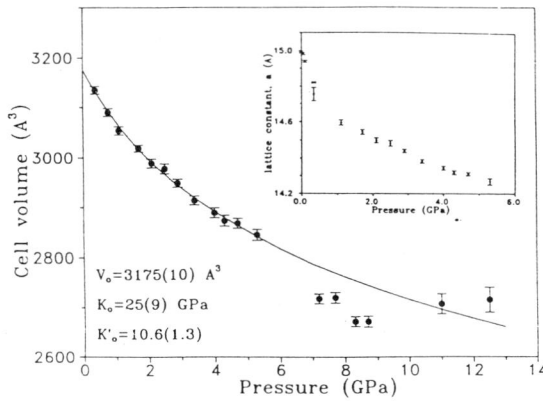

Figure 9. Pressure-volume plot for rhombohedral C$_{70}$. The solid line is a fit to the Murnaghan EOS with K$_O$= 25(9) GPa and K$_O$'= 10.6(1.3). Inset. - Pressure dependence of the lattice constant.

anisotropic structure of the molecules. Further increase of pressure to 11 GPa results in the appearance of an amorphous carbon phase; complete collapse of the Bragg intensity occurs by 18 GPa with amorphous carbon present to 25 GPa.

Experimental information on the rotational dynamics in both the ordered and disordered phases of C_{70} has come thus far from μSR[32,33] and inelastic neutron scattering[34,35] studies. Solid fullerenes have proven very effective media for the formation of muonium centres[36]. In contrast, no muonium signal has been observed in graphite, while both "normal" Mu and "anomalous" Mu* are encountered in diamond. In the μSR experiments, the muons act as non-perturbing spin labels, sensitive to the motion of the fullerene molecules. Two distinct sets of paramagnetic centres are observed: one corresponding to a Mu atom trapped inside the fullerene cage (Mu@C_n) and one to exohedral addition of Mu across the C-C short bonds to form muonated fullerene radicals (MuC_n) (Figure 10). The question whether the rotational dynamics of the MuC_{70} moities which are monitored in the mSR experiments are the same as those of C_{70} deserves some comment. The change in

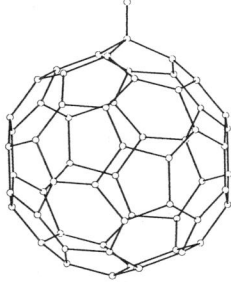

Figure 10. Optimised structure of the MuC_{60} radical.

moment of inertia upon Mu addition is insignificant, and in the absence of interaction with the lattice, the dynamics should be nearly identical. However, additional friction may arise from the Mu atom sticking out of the surface. The experience so far is that the dynamic information obtained from the Mu adducts to organic molecules, including fullerenes agrees well with that obtained by NMR on the unsubstituted molecules.

Zero-field experiments[32] on the endohedral muonium adduct Mu@C_{70} (forming at a yield of 27(4)%) reveal the appearance of a single frequency oscillation (ν=0.70(2) MHz) at the orientational ordering transition (~270 K), signaling the presence of an axially symmetric hyperfine interaction in the orientationally ordered phase and consistent with reorientational correlation times of the order of τ~ 30 ns at 200 K (Figure 11). The presence of anisotropy in the hyperfine interaction through dipolar magnetic interactions necessitates modification of the Hamiltonian (3). For a traceless anisotropic dipolar part **B** of the hyperfine tensor which is axially symmetric, (3) becomes:

$$\mathcal{H} = \mathcal{H}_{iso} + 2(3B)S_zI_z \qquad (5)$$

where $3B$ is the hyperfine anisotropy and z is the unique symmetry axis. Figure 2 thus should be modified to include the removal of the degeneracy of the triplet state even at zero applied field. The observed precession frequency provides an accurate measure of the anisotropic dipolar interactions. The endohedral Mu atom performs very fast anisotropic large-amplitude oscillatory motion inside the fullerene cage, preferably along the long molecular axis; since no anisotropic contribution is expected from the 1s state of the muonium atom, the value of the measured triplet precession reflects the $2p_z$ character of the wavefunction. The calculated dipolar interaction for a pure $2p_z$ state is 27.89 MHz, leading to a 0.092(2) % $2p_z$ admixture in the ground state wavefunction of Mu@C_{70}: $\psi_{Mu@C_{70}} = 0.9817\ \psi_{1s} + 0.0959\ \psi_{2p_z}$. In the vicinity of 160 K, there is evidence that the hyperfine tensor becomes fully anisotropic, signaling a further change in the molecular motion.

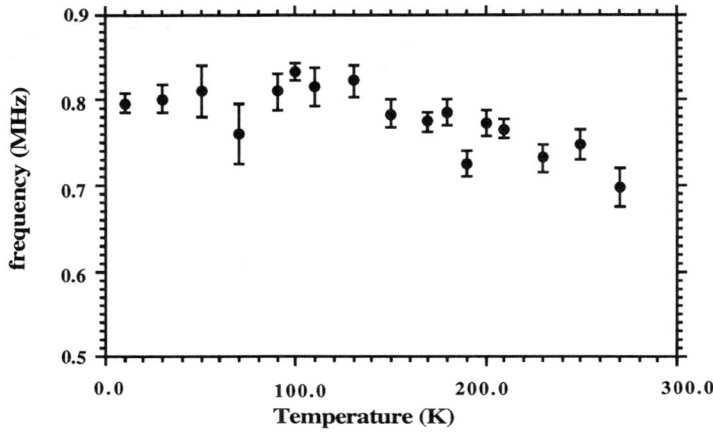

Figure 11. Temperature dependence of zero field precession frequency of Mu@C_{70}.

Avoided level crossing μSR[33] has been employed to search at high magnetic fields for the presence of the muon spin-flip transition in the temperature range 73-310 K; as described earlier, this transition is only allowed by the anisotropic (dipolar) part of the muon-electron hyperfine interaction and its observation provides a highly sensitive confirmation of the presence of anisotropy. The resonance disappears when *isotropic* reorientational motion sets in on a time scale $<<(2\pi\ D_{//})^{-1}$ ($D_{//}$ is the anisotropy of axial symmetry in frequency units). Our experimental data reveal that the expected anisotropy D_{zz} for the most abundant fully frozen MuC_{70} radical (Figure 12) is not attained down to 73 K. As a consequence, C_{70} may still be undergoing uniaxial rotation below 160 K unless D_{zz} is unusually low. As the temperature increases from 160 K to 270 K, $D_{//}$ (the anisotropy of axial symmetry) decreases smoothly, implying that there is an increasing degree of rotational disorder of the unique molecular axis which is progressively excited; on the other hand, the data cannot be explained using a reorientational model incorporating diffusive motions of the C_{70} molecules. The ALC resonance line is, however, still observed for T>270 K (Figure 12), signifying persisting anisotropy; this is, strictly speaking, incompatible with cubic symmetry. However, since diffraction experiments probe the average structure within the unit cell, the apparent observation of fcc symmetry does not preclude lower symmetry on a very short scale.

The evolution of the dynamics was followed to even higher temperatures (650 K) using inelastic neutron scattering[34] measurements. At these temperatures, the neutron experiments reveal the presence of broad quasielastic peaks and the C_{70} molecules are found to tumble fast enough, virtually independently of each other to average out any anisotropy present. The rotational potential resembles closely the

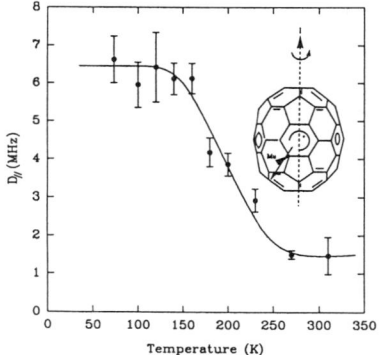

Figure 12. Temperature dependence of the axial hyperfine anisotropy of a MuC_{70} muonated radical shown in the inset.

one found in the high temperature phase of C_{60} and can be described well using a quasi-isotropic diffusive molecular motion between many shallow minima with an activation energy E_a= 32.0(6.5) meV. The coherent neutron scattering function written as:

$$S(Q,\omega) = \sum_{\ell=1}^{\infty} S_\ell(Q) L_\ell(\omega) \qquad (6)$$

is found to describe very well the Q-dependence of both the intensity and the full width of the quasielastic peak for an isotropic diffusion constant D_r= 1.5×10^{10} s^{-1} at 525 K. In eq. (6), $L_\ell(\omega)$ is a normalised Lorentzian with full width $\Gamma_\ell = \ell(\ell+1) D_r$ and $S_\ell(Q)$ is given by:

$$S_\ell(Q) = (2\ell+1) \sum_{v=1}^{70} \sum_{v'=1}^{70} j_\ell(QR_v) j_\ell(QR_{v'}) P_\ell(\cos\theta_{vv'}) \qquad (7)$$

where j_ℓ is a spherical Bessel function, P_ℓ is a Legendre polynomial, R_v is the distance of atom v from the molecular centre, and $\theta_{vv'}$ is the angle subtended by atoms v and v' at the molecular centre.

Below the order-disorder transition temperature[34,35] (~270 K), the quasielastic peak disappears and librational peaks appear at non-zero energy transfer. Thus in the ordered phase, the fullerene units undergo small-amplitude librations about their equilibrium positions, giving rise to librational peaks near 1.5 meV. The librational energy is smaller than in C_{60} and the peaks much broader, reflecting a weaker and more anisotropic interaction. The librations harden and their amplitudes decrease as the temperature decreases.

ALKALI FULLERIDES

The appearance of superconductivity in alkali fullerides with stoichiometry[8,9] A_3C_{60} led to considerable efforts in attempting to understand the electronic, structural, and vibrational properties of these materials and to elucidate the origin of their high T_c's and the mechanism responsible for the pair formation. A series of reviews, covering various aspects of alkali fulleride chemistry and physics already exists[21,37-42]. Even though many subtleties about their crystallographic structures exist, the broad premise, within which their structural properties can be discussed, consists of a face-centred-cubic array of essentially ionic C_{60}^{3-} units; the alkali metals then fully occupy all available tetrahedral and octahedral interstices. Each alkali ion either faces four hexagons or six hexagon-hexagon fusions of neighbouring fullerenes. Moreover, the fulleride ions are found to be merohedrally disordered[43] with equal occupancy of the two standard orientations of Figure 1. The resulting space group best describing their structure is thus $Fm\bar{3}m$.

Theoretical treatments of superconductivity in A_3C_{60} have considered purely electronic[44,45] as well as phonon-induced pairing, involving either low-energy intermolecular[46] or high-energy intramolecular modes[47-52]. The electronic mechanisms for pair-binding derive from the resonating valence bond (RVB) picture of Pauling and Anderson and focus on the strong intrafullerene electron correlations. Phonon-induced mechanisms follow the classic BCS route by ascribing the effective attractive interaction between electrons to their strong scattering near the Fermi surface by the vibrations of the ions.

Experimental evidence derived from the relation between lattice dimensions and superconducting transition temperatures in A_3C_{60} both at ambient[53] and at high pressures[54] is consistent with T_c being modulated by the density-of-states at the Fermi level, $N(\varepsilon_F)$. Moreover, no alkali-specific effect is observed. Using a BCS-type relation:

$$T_c \propto (\hbar\omega_{ph}) \exp[-\frac{1}{VN(\varepsilon_F)}] = (\hbar\omega_{ph}) \exp(-1/\lambda) \tag{8}$$

the observed high T_c's may be understood in terms of a high average phonon frequency $\hbar\omega_{ph}$, resulting from the light carbon mass and the large force constants associated with intramolecular modes, a high DOS at ε_F, resulting from the weak intermolecular interactions and strongly scattering intramolecular modes. Additional support for a phonon-mediated pair-binding interaction also comes from ^{13}C isotope-effect measurements which find exponents α ($T_c \propto M^{-\alpha}$) equal to 0.37(5) for Rb_3C_{60}[55] and 0.30(6) for K_3C_{60}[56]. The ideal BCS value of α is 0.5, the observed reduction probably associated with significant Coulomb interaction effects. No evidence for alkali metal isotope effect is found[57]. The value of the superconducting energy gap (2Δ) is also found to be close to what is expected from BCS theory in the weak limit which predicts $2\Delta/kT_c = 3.52$. Reflectivity measurements[58] give values of 3.6 and 2.98 for K_3C_{60} and Rb_3C_{60}, respectively, ^{13}C NMR measurements[59] result in 3.1 and 4.0, respectively, and μSR measurements[60] in 3.6 for Rb_3C_{60}. All these are close to the weak coupling limit; only STM point contact tunneling[61] has given a value of 5.2, close to strong coupling.

For a phonon-mediated superconductor, the superconducting properties may be derived from a knowledge of the Eliashberg spectral function, $\alpha^2 F(\omega)$, where $F(\omega)$ is the phonon density-of-states and α^2 is an effective electron-phonon coupling strength. This spectral function can be expressed in terms of the phonon linewidths γ_{vQ}, experimentally available through neutron scattering measurements at low temperatures:

$$\alpha^2 F(\omega) \propto N(\varepsilon_F) \omega \sum_{vQ} \gamma_{vQ} \delta(\omega - \omega_{vQ}) \quad (9)$$

The molecular nature of the superconducting fullerides results in fairly dispersionless phonon branches, and consequently, the electron-phonon Hamiltonian[47,48] has a particularly simple form. The total electron-phonon coupling constant λ may be simply expressed as a sum of partial contributions λ_v, associated with the each mode v mediating the pairing interaction, $\lambda = \Sigma_v \lambda_v$. Focusing on the occupied t_{1u} electronic states, to a first approximation the only relevant intramolecular modes in icosahedral symmetry that can couple to the t_{1u} conduction electrons are of A_g (for $Q \neq 0$) and H_g symmetry, Electron-phonon coupling should lead to broadening and softening of the affected modes in the superconducting fullerides, when compared to pristine C_{60}. Quantitatively such effects on the position and width may be estimated using the expressions:

$$\Delta \omega_v \approx -(\lambda_v/g_v) \omega_v \quad ; \quad \Delta \gamma_v \approx (\pi/g_v) N(\varepsilon_F) \lambda_v \omega_v^2 \quad (10)$$

where $\Delta \omega_v$ is the change in frequency upon reduction to C_{60}^{3-}, $\Delta \gamma_v$ is the increase in full width and g_v the degeneracy of the vth phonon.

The vibrational spectra of superconducting K_3C_{60}[62] and Rb_3C_{60}[63,64], and insulating C_{60}[65-67] and Rb_6C_{60}[68] have all been measured at low temperatures by inelastic neutron scattering at $Q \neq 0$. INS measurements are not restricted by the usual optical selection rules and consequently, information on all 46 intramolecular normal modes is obtained. Moreover, these are not expected to show significant dispersion, permitting direct comparisons between the INS and the zone-centre optical data to be made. For anisotropic oscillators in powder samples, the scattering law $S(Q,\omega)$ is given by :

$$S(Q,\omega)_v \propto (1/3) (Q^2 \, \mathrm{Tr} \mathbf{B}_v) \exp(-Q^2 \alpha_v) \quad (11)$$

$$\alpha_v \approx (1/5)\{\mathrm{Tr}\mathbf{A} + 2(\mathbf{B}_v:\mathbf{A} / \mathrm{Tr}\mathbf{B}_v)\}$$

where the total mean square displacement is given by $\mathbf{A} = \Sigma_v \mathbf{B}_v$, $\mathbf{B}_v$ is the mean square displacement tensor of the scattering atom in internal mode v and $Q=|\mathbf{Q}|$ is the momentum transferred during scattering. For a simple harmonic isotropic oscillator, eq. (11) is simplified to: $S(Q,\omega) = (Q^2 B) \exp(-Q^2 B)$, with the mean square displacement of the oscillator is given by: $B = (h/2\mu\omega)$ and μ the reduced mass.

Phonon modes are well separated in the low-energy (30-90 meV) range and changes in position and width can be followed as a function of the reduction level of the fullerene cage. The INS data provide evidence of strong electron-phonon coupling to the H_g modes. Indeed the most remarkable feature when comparing the K_3C_{60} and C_{60} INS spectra is the virtual disapperance of the $H_g^{(2)}$ mode (Figure 13) at 53.2 meV, consistent with strong electron-phonon coupling (Table 1); it reappears with virtually unchanged full width in insulating Rb_6C_{60}. Using eq 10, we

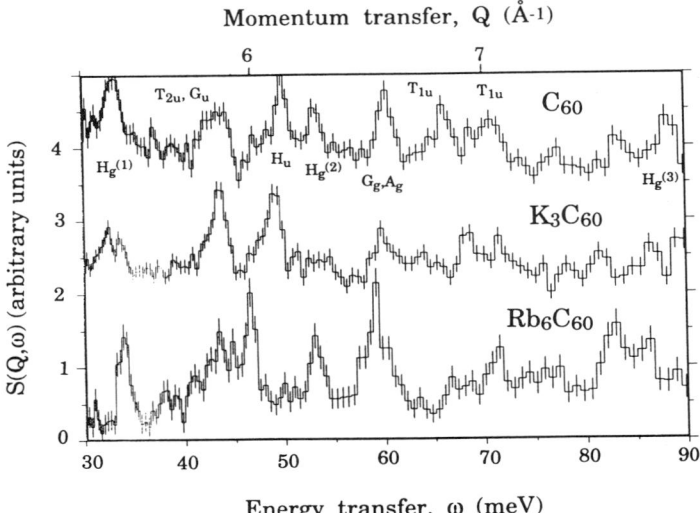

Figure 13. INS spectra of the intramolecular modes of C_{60}, K_3C_{60} and Rb_6C_{60} in the energy region 30-90 meV at low temperatures.

find an electron-phonon coupling constant for this mode of $\lambda(2) \approx 0.17$. By coincidence, the $H_g(2)$ mode occurs in a region of the spectrum where there are no neighbouring or overlapping peaks, making our conclusions on the effects of doping on its width and position unambiguous. Strong coupling is also evident for the $H_g(1)$ mode with an estimated $\lambda(1) \sim 0.10$. $H_g(4)$ is more weakly coupled ($\lambda(4) \sim 0.04$). The experimental data and derived electron-phonon coupling constants are collected in Table 1. Interpretation of the INS spectra in the tangential mode regime is, however,

Table 1. Observed positions, ω_v of the four radial H_g modes in C_{60} and their broadenings, $\Delta\gamma_v$ on reduction to $C_{60}{}^{3-}$ together with calculated electron-phonon coupling strengths, V_v in meV for a density-of-states $N(0)=14$ states/eV/spin/C_{60}.

Vibrational mode	$H_g(1)$	$H_g(2)$	$H_g(3)$	$H_g(4)$
ω_v / meV	33.1	53.2	88.1	96.2
$\Delta\gamma_v$ / cm^{-1}	8	35	1	25
V_v / meV	7.4	12.4	0.1	2.7

problematic because of worse resolution and intensity attenuation due to large Debye-Waller factors. Recent high resolution data[64] on Rb_3C_{60} reveal that the $H_g(6)$, $H_g(7)$ and $H_g(8)$ modes also show similar broadening effects, when compared to C_{60}, while $H_g(5)$ does not.

The experimental results thus reveal that electron-phonon coupling strength are distributed between both buckling and tangential modes. Using the calculated λ_v

values of Schlüter et al.[48] for the tangential modes, we find $\lambda=0.64$. This may be combined with estimates of the density-of-states $N(0)\approx 14$ states/C_{60}/eV/spin and the Coulomb pseudopotential $\mu^*\approx 0.15$ to give, using McMillan's formula,

$$T_c = \frac{\omega_{\log}}{1.2} \exp\left[-\frac{1.04\,(1+\lambda)}{\lambda-\mu^*(1+0.62\lambda)}\right] \quad (12)$$

an estimate of the order of magnitude of $T_c\approx 17$ K (for $\omega_{\log}= 1072$ K), in good agreement with experiment. Furthermore, we can estimate using the above set of values the isotope-effect exponent within McMillan's formalism; α is found to be 0.35 in good agreement again with experiment. A crucial point revealed by our results is the contrast with alkali-intercalated graphite, where the corresponding buckling modes do not couple to the π electrons because of symmetry restrictions. It appears that coupling to the radial modes, because of the finite curvature of the fullerene cage, accounts for roughly half the total electron-phonon coupling strength in A_3C_{60} and leads to a substantially increased T_c, compared to the lamellar graphite intercalates. Raman measurements performed on alkali fullerides give results in broad agreement with the neutron scattering ones[69-72].

We have also measured the low-energy INS spectra of superconducting K_3C_{60} as a function of temperature[73]. Well-defined peaks are observed in the vicinity of ~4 meV (Figure 14); the dependence of their intensities on the scattering vector shows that they are due to small-amplitude librational motion, in a similar fashion to the ordered phase of pristine C_{60}. However, the energy barrier for reorientation is estimated to be ~500 meV, i.e. twice as large as in C_{60}. Of particular interest is the behaviour of the librational modes in K_3C_{60} as the sample is cooled below T_c, since for phonons with energies less than $2\Delta(T)$, the contribution to their lifetime due to the electron-phonon interaction disappears in the superconducting state, and sharpening of the order of $\Delta\gamma_{ep}$ is expected. We find that within the experimental resolution, neither the energy nor the width of the librational excitation is affected. The changes in energy and width are $\Delta\omega(25\text{ K}-12\text{ K})= 0.06(5)$ meV and $\Delta\gamma(25\text{ K}-12\text{ K})= 0.02(13)$ meV, respectively. These are only consistent with small values of

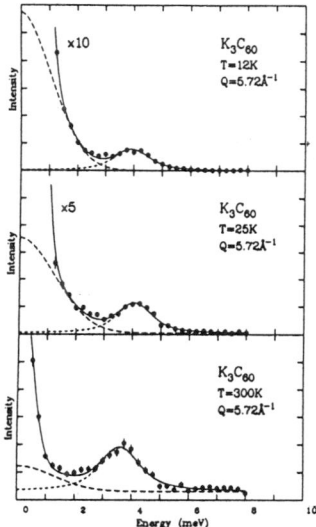

Figure 14. INS spectra of K_3C_{60} at constant Q= 5.72 Å^{-1}.

electron-phonon coupling for the librational modes of $\lambda_{lib} \leq 0.08$. This is in sharp contrast to the predictions of some theories of phonon-mediated pairing interactions in the fullerides[74], which necessitate a value for $\lambda_{lib} \sim 1.9-2.7$ and predict much larger changes in the librational linewidths. The experimental results clearly indicate that the coupling of electrons to librational modes does not substantially contribute to the formation of Cooper pairs.

POSTSCRIPT

Perhaps the most important and unexpected aspect of the fullerene discovery is the fact that the molecules form spontaneously. This has important implications for our understanding of the way in which extended carbon materials form and in particular the mechanism of graphite growth as well as the synthesis of large polycyclic aromatic molecules. However, a further surprise comes when it is realised how easily fullerenes react with a variety of reagents, leading to unexpected chemistry and novel materials. We have learned a lot in the last threeyears about the solid state properties of these fascinating systems: the way they crystallise, their dynamical behaviour, some fascinating properties of their derivatives (superconductivity, ferromagnetism). Their novelty has excited the imagination of diverse classes of scientists: from theorists and condensed matter scientists, to synthetic chemists and astrophysicists. We can safely predict that the road ahead is still full of surprises as the range of available materials expands and as our understanding of this novel round form of carbon deepens further.

REFERENCES

1. H.W. Kroto, J.R. Heath, S.C. O'Brien, R.F. Curl, and R.E. Smalley, *Nature* 318:162 (1985).
2. W. Krätschmer, L.D. Lamb, K. Fostiropoulos, and D.R. Huffman, *Nature* 347:354 (1990).
3. H.W. Kroto, *Angew. Chem.* 31:111 (1992).
4. K. Prassides and H.W. Kroto, *Physics World* 5:44 (1992).
5. P.R. Birkett, P.B. Hitchcock, H.W. Kroto, R. Taylor, and D.R.M. Walton, *Nature* 357:479 (1992).
6. C.C. Henderson and P.A. Cahill, *Science* 259:1885 (1993).
7. J.D. Crane, P.B. Hitchcock, H.W. Kroto, and D.R.M. Walton, *J. Chem. Soc. Chem. Commun.* 1764 (1992).
8. A.F. Hebard, M.J. Rosseinsky, R.C. Haddon, D.W. Murphy, S.H. Glarum, T.T.M. Palstra, A.P. Ramirez, and A.R. Kortan, *Nature* 318:600 (1991).
9. K. Holczer, O. Klein, S.-M. Huang, R.B. Kaner, K.-J. Fu, R.L. Whetten, and F. Diederich, *Science* 252:1154 (1991).
10. M.J. Rosseinsky, D.W. Murphy, R.M. Fleming, R. Tycko, A.P. Ramirez, T. Siegrist, G. Dabbagh, and S.E. Barrett, *Nature* 356:6368 (1992).
11. G.L. Squires. "Introduction to the Theory of Thermal Neutron Scattering," Cambridge University Press, Cambridge (1978).
12. E. Roduner. "The Positive Muon as a Probe in Free Radical Chemistry," Lecture Notes in Chemistry, Vol. 49, Springer, Heidelberg (1988).
13. E. Roduner, *Hyperfine Interactions* 65:857 (1990).
14. R.M. Fleming, et al., in: "Fullerenes: Synthesis, Properties, and Chemistry of Large Carbon Clusters", G.S. Hammond and V.J. Kuck, eds., ACS Sympos. Series 481:25 (1991).
15. W.I.F. David, R.M. Ibberson, T.J.S. Dennis, J.P. Hare, and K. Prassides, *Europhys. Lett.* 18:219 (1992); 735 (1992).

16. P.A. Heiney, J.E. Fischer, A.R. McGhie, W.J. Romanow, A.M. Denenstein, J.P. McCauley, A.B. Smith, and D.E. Cox, *Phys. Rev. Lett.* 66:2911 (1991).
17. P.C. Chow, X. Jiang, G. Reiter, P. Wochner, S.C. Moss, J.D. Axe, J.C. Hanson, R.K. McMullan, R.L Meng, and C.W. Chu, *Phys. Rev. Lett.* 69:2943 (1992).
18. R.D. Johnson, C.S. Yannoni, H.C. Dorn, J.R. Salem, and D.S. Bethune, *Science* 255:1235 (1992).
19. J.R.D. Copley, D.A. Neumann, R.L. Cappelletti, and W.A. Kamitakahara, *J. Phys. Chem. Solids* 53:1353 (1992).
20. W.I.F. David, R.M. Ibberson, J.C. Matthewman, K. Prassides, T.J.S. Dennis, J.P. Hare, H.W. Kroto, R. Taylor, and D.R.M. Walton, *Nature* 353:147 (1991).
21. K. Prassides, H.W. Kroto, R. Taylor, D.R.M. Walton, W.I.F. David, J. Tomkinson, M.J. Rosseinsky, D.W. Murphy, and R.C. Haddon, *Carbon* 30:1277 (1992).
22. K. Rapcewicz and J. Przystawa, contribution in this volume.
23. K. Prassides, *Int. J. Mod. Phys. B* 6:4007 (1992).
24. K. Prassides et al., unpublished results.
25. R. Blinc, J. Seliger, J. Dolinsek, and D. Arcon, *Europhys. Lett.* in press.
26. R.C. Yu, N. Tea, M.B. Salamon, D. Lorents, and R. Malhotra, *Phys. Rev. Lett.* 68:2050 (1992).
27. F. Gugenberger, R. Heid, C. Meingast, P. Adelmann, M. Braun, H. Wühl, M. Haluska, and H. Kuzmany, *Phys. Rev. Lett.* 69:3774 (1992).
28. T. Matsuo, H. Suga, W.I.F. David, R.M. Ibberson, P. Bernier, A. Zahab, C. Fabre, A. Rassat, and A. Dworkin, *Solid State Commun.* 83:711 (1992).
29. G.B.M. Vaughan, P.A. Heiney, J.E. Fischer, D.E. Luzzi, D.A. Ricketts-Foot, A.R. McGhie, Y.-W. Hui, A.L. Smith, D.E. Cox, W.J. Romanow, B.H. Allen, N. Coustel, J.P. McCauley, and A.B. Smith, *Science* 254:1350 (1991).
30. C. Christides, I.M. Thomas, T.J.S. Dennis, and K. Prassides, *Europhys. Lett.* in press.
31. G.B.M. Vaughan, P.A. Heiney, D.E. Cox, J.E. Fischer, A.R. McGhie, A.L. Smith, R.M. Strongin, M.A. Cichy, and A.B. Smith, to be published.
32. K. Prassides, T.J.S. Dennis, C. Christides, E. Roduner, H.W. Kroto, R. Taylor, and D.R.M. Walton, *J. Phys. Chem.* 26:10600 (1992).
33. T.J.S. Dennis, K. Prassides, E. Roduner, L. Cristofolini, and R. DeRenzi, *J. Phys. Chem.* to be published.
34. C. Christides, T.J.S. Dennis, K. Prassides, R.L. Cappelletti, D.A. Neumann, and J.R.D. Copley, to be published.
35. B. Renker, F. Gompf, R. Heid, P. Adelmann, A. Heiming, W. Reichardt, G. Roth, H. Schober, and H. Rietschel, *Z. Phys. B* 90:325 (1993).
36. E.J. Ansaldo, C. Niedermayer, and C.E. Stronach, *Nature* 353:129 (1991).
37. D.W. Murphy et al., *J. Phys. Chem. Solids* 53:1321 (1992).
38. K. Holczer and R.L. Whetten, *Carbon* 30:1261 (1992).
39. J.H. Weaver, *J. Phys. Chem. Solids* 53:1433 (1992).
40. O. Zhou and D.E. Cox, *J. Phys. Chem. Solids* 53:1373 (1992).
41. J.E. Fischer, P.A. Heiney, and A.B. Smith, *Acc. Chem. Res.* 25:112 (1992).
42. M. Schlüter, M. Lannoo, M. Needels, G.A. Baraff, and D. Tománek, *J. Phys. Chem. Solids* 53:1473 (1992).
43. P.W. Stephens, L. Mihaly, P.L. Lee, R.L. Whetten, S.-M. Huang, R. Kaner, F. Diederich, and K. Holczer, *Nature* 351:632 (1991).
44. S. Chakravarty and S. Kivelson, *Europhys. Lett.* 16:751 (1991); S. Chakravarty, M.P. Gelfand, and S. Kivelson, *Science* 254:970 (1991).
45. G. Baskaran and E. Tossati, *Curr. Sci.* 61:33 (1991).
46. F.C. Zhang, M. Ogata, and T.M. Rice, *Phys. Rev. Lett.* 67:3452 (1991).
47. C.M. Varma, J. Zaanen, and K. Raghavachari, *Science* 254:989 (1991).
48. M. Schlüter, M. Lannoo, M. Needels, G.A. Baraff, and D. Tománek, *Phys. Rev. Lett.* 68:526 (1992).

49. R.A. Jishi and M.S. Dresselhaus, *Phys. Rev. B* 45:2597 (1992).
50. L. Pietronero, *Europhys. Lett.* 17:365 (1992).
51. I.I. Mazin et al., *Phys. Rev. B* 45:5114 (1992).
52. Y. Asai and Y. Kawaguchi, *Phys. Rev. B* 46:1265 (1992).
53. R.M. Fleming, A.P. Ramirez, M.J. Rosseinsky, D.W. Murphy, R.C. Haddon, S.M. Zahurak, and A.V. Makhija, *Nature* 352:787 (1991).
54. O. Zhou et al.,*Science* 255:833 (1992).
55. C.-C. Chen and C.M. Lieber, *J. Am. Chem. Soc.* 114:3141 (1992).
56. A.P. Ramirez et al., *Phys. Rev. Lett.* 68:1058 (1992).
57. T.W. Ebbesen, J.S. Tsai, K. Tanigaki, H. Hiura, Y. Shimakawa, Y. Kubo, I. Hirosawa, and J. Mizuki, *Physica C* 203:163 (1992).
58. L. DeGiorgi, et al., *Phys. Rev. Lett.* 69:2987 (1992).
59. R. Tycko et al., *Phys. Rev. Lett.* 68:1912 (1992).
60. R.F. Kiefl et al., to be published.
61. Z. Zhang, C.-C. Chen, S.P. Kelty, H. Dai, and C.M. Lieber, *Nature* 353:353 (1991).
62. K. Prassides, J. Tomkinson, C. Christides, M.J. Rosseinsky, D.W. Murphy, and R.C. Haddon, *Nature* 354:462 (1991).
63. J.W. White, G. Lindsell, L. Pang, A. Palmisano, D.S. Sivia, and J. Tomkinson, *Chem. Phys. Lett.* 191:92 (1992).
64. C. Christides, K. Prassides, M.J. Rosseinsky, D.W. Murphy, and R.C. Haddon, to be published.
65. R.L. Cappelletti, J.R.D. Copley, W.A. Kamitakahara, F. Li, J.S. Lannin, and D. Ramage, *Phys. Rev. Lett.* 66:3261 (1991).
66. K. Prassides, T.J.S. Dennis, J.P. Hare, J. Tomkinson, H.W. Kroto, R. Taylor, and D.R.M. Walton, *Chem. Phys. Lett.* 187:455 (1991).
67. C. Coulombeau, H. Jobic, P. Bernier, C. Fabre, D. Schütz, and A. Rassat, *J. Phys. Chem.* 96:22 (1992).
68. K. Prassides, C. Christides, M.J. Rosseinsky, J. Tomkinson, D.W. Murphy, and R.C. Haddon, *Europhys. Lett.* 19:629 (1992).
69. M.G. Mitch, S.J. Chase, and J.S. Lannin, *Phys. Rev. Lett.* 68:883 (1992).
70. P. Zhou, K.A. Wang, A.M. Rao, P.C. Eklund, G. Dresselhaus, and M.S. Dresselhaus, *Phys. Rev. B* 46:2595 (1992).
71. R. Danieli et al., *Solid State Commun.* 81:257 (1992).
72. T. Pichler, M. Matus, J. Kürti, and H. Kuzmany, Phys. Rev. B 13841 (1992).
73. C. Christides, D.A. Neumann, K. Prassides, J.R.D. Copley, J.J. Rush, M.J. Rosseinsky, D.W. Murphy, and R.C. Haddon, *Phys. Rev. B* 46:12088 (1992).
74. I.I. Mazin, O.V. Dolgov, A. Golubov, and S.V. Shulga, *Phys. Rev. B* 47:538 (1993).

THEORETICAL INVESTIGATIONS OF THE ORIENTATIONAL ORDERING TRANSITION IN SOLID C_{60}

Krzysztof Rapcewicz

Instytut Fizyki Teoretycznej
Uniwersytet Wrocławski
Plac Maksa Borna 9
PL-50-204 Wrocław
Poland

INTRODUCTION

Solid Buckminsterfullerene C_{60} undergoes an orientational ordering transition[1-5] below T $\approx$ 260 K. The transition is from a high temperature structure with space group Fm3m to a low temperature ordered structure whose space group is Pa3. The transition is first order and is accompanied by a lattice contraction[6] of approximately 0.04 Å. The change in the transition temperature with pressure is $dT_c/dP = 10.4 \pm 0.2$ K/kbar.[7] Experimental aspects of the transition are discussed in more detail in an article in this volume by Kosmas Prassides.[5]

The following reviews a selection of the theoretical investigations of the orientational ordering transition in crystalline C_{60}. It begins by reviewing the Monte Carlo studies of Sprik, Chen, and Klein[8] and the total energy analysis of Li, Lu, and Martin.[9] The four Landau-type theories that have been presented in the literature are then discussed in varying levels of detail. The approach of Harris and Sachidanandam[10] which is microscopic is reviewed first; the theory of Michel, Copley, and Neuman,[11] being a systematic microscopic approach, is next discussed. Following this, the phenomenological approach used by Rapcewicz and Przystawa[12] is reviewed and, finally, the theory of Rasolt[13] is very briefly recalled.

MOLECULAR DYNAMICS SIMULATIONS

A series of molecular dynamics studies of the system using the constant-pressure Parrinello-Rahman molecular equations of motion were carried out by Chen and Klein[14,15] and Sprik, Chen, and Klein.[8] Chen and Klein began by investigating the properties of the high temperature disordered phase. The C_{60} intermolecular potential is assumed to be solely due to the interactions of the carbon atoms composing the

C_{60} molecules and the carbon-carbon interaction potential is taken to be a Lennard-Jones 6-12 (LJ) potential. This means that the C_{60}-C_{60} intermolecular potential is of the form

$$V_{12} = \sum_{\mu,\nu=1}^{60} 4\epsilon \left[\left(\frac{\sigma}{|\mathbf{r}_{1\mu} - \mathbf{r}_{2\nu}|} \right)^{12} - \left(\frac{\sigma}{|\mathbf{r}_{1\mu} - \mathbf{r}_{2\nu}|} \right)^{6} \right] \quad (1)$$

where $\mathbf{r}_{1\nu}$ ($\mathbf{r}_{2\nu}$) are the positions of carbon atoms in the C_{60} molecule labelled 1 (2); hence, the resulting interaction depends upon the relative orientations of the respective C_{60} molecules. Based upon the data from graphite, they chose $\epsilon = 2.4$ meV and $\sigma = 3.4$ Å. Carbon atom-atom potentials from graphite are used because the minimum carbon-carbon distance between two carbons atoms on two different C_{60} molecules (3 Å) is close to that in graphite (3.5 Å).[9]

With this potential, Chen and Klein[14,15] found that the minimum energy of the orientationally-dependent intermolecular potential occurs between 9.7 and 10.3 Å and that the energy of this minimum ranged from -.189 to -.250 eV. Further, with this potential, they found, considering the rotation of a single molecule about a fixed-centre for a separation of 10 Å that the potential barrier is between 8 and 25 meV.

Chen and Klein[14,15] carried out their molecular dynamics simulations with $2 \times 2 \times 2$ fcc unit cells meaning that their simulation contained 32 C_{60} molecules (or 1920 carbon atoms). For technical reasons (to make the calculation tractable in a finite period of time), they truncated the intermolecular potential at 9.5 Å. Since the radius of a C_{60} molecule is 3.5 Å and the lattice constant of the conventional unit cell for the fcc lattice is 14.14 Å; each C_{60} molecule interacts with the nearest face of its neighbour.

Among the results of their simulations, Chen and Klein[14,15] found that the zero pressure lattice constant for the conventional unit cell is 14.14 Å which is in reasonable agreement with the experimental determination. Upon visualisation of the results of their simulations, they noted that the C_{60} molecules rotate rapidly in the disordered phase.

Running the simulation for lower temperatures, Chen and Klein[16] observed an ordering transition of the C_{60} molecules in which the molecules freeze with fixed orientations. However, the resulting structure was not the observed Pa3 structure, but rather a tetragonal one. In the face of these results, Sprik, Cheng, and Klein[8] modified the potential based upon earlier experience with the modelling of interactions in solid nitrogen. In particular, they exploited an observation of David et al.[3,16] who noted that the conjugation of the π-bond network is incomplete as the single bonds (pentagon-hexagon edges) are longer than the double bonds (hexagon-hexagon edges). They initially modelled this assuming that each double bond on one molecule interacts with each double bond on a second molecule together with the carbon atoms on that second molecule via an atom-atom LJ potential. While this proves sufficient to stabilise the Pa3 structure, the resulting transition temperature is far too low and cannot be increased above about 160 K. Returning again to the example of solid nitrogen, they then noted that the electron-rich double bonds carry excess negative charge while the electron-poor single bonds have excess positive charged. They accounted for this by assuming that double bonds possess a charge q_D and the single bonds a positive charge one half of this, i.e. $q_S = -q_D/2$, the interaction being electrostatic.

Keeping the effective diameters of the interaction fixed, $\sigma_{CC} = 3.4$ Å, $\sigma_{CD} = 3.5$ Å and $\sigma_{DD} = 3.6$ Å, they tuned the two parameters ϵ and q_D so as to get agreement with the observed transition temperature, the cohesive energy and the reorientational relaxation time. They found that choosing $q_D = -0.35e$ and $\epsilon = 1.3$

meV yields results in reasonable agreement between simulation and experiment. With these parameters the transition is clearly first order at 215 K (still significantly lower than the observed temperature) exhibiting a hysterisis of 30 K. At the transition they found a jump in the lattice constant of about .02 Å in agreement with the results of Heiney et al.[6] Further, investigating the effects of pressure on the system at 300 K, they observed an ordering of the system at 9 kbar which when isothermally released led to a transition at 5 kbar; they took the transtion at 300 K to be at 7 kbar. With this they calculated that $dT_c/dp = 12 \pm 4$ K/kbar which is in agreement with experimental measurements.

FREE ENERGY APPROACH

Lu, Li, and Martin[9] studied the transition from a different perspective, namely they studied the structural energies of possible structures. For the atom-atom interaction energy, they chose a Lennard-Jones 6-12 potential with $\sigma = 3.407$ Å and $\epsilon = 2.964$ meV. They calculated the Madelung energy for various structures and found that using only the atom-atom potential, an orthorhombic structure is the lowest energy structure and that the calculated cohesive energy is in good agreement with experiment showing that the dominant contribution to the cohesive energy of the buckminsterfullerene system is the van der Waals interaction.

In order to get the correct structure they were forced, in their model, to consider two distinct types of intermolecular interactions: (i) the dominant van der Waals atom-atom interaction between carbon atoms comprising the molecules and (ii) a short-range Coulomb interaction resulting from the charge imbalance between the two types of bonds in the C_{60} molecule. They argued that the strong intra-C_{60} covalent bonds lead to the expectation that the structural properties are dominated by the weak inter-C_{60} interactions. Thus for the C_{60}-C_{60} intermolecular potential they chose:

$$V_{12} = \sum_{\mu,\nu=1}^{60} 4\epsilon \left[\left(\frac{\sigma}{|\mathbf{r}_{1\mu} - \mathbf{r}_{2\nu}|} \right)^{12} - \left(\frac{\sigma}{|\mathbf{r}_{1\mu} - \mathbf{r}_{2\nu}|} \right)^{6} \right] + \sum_{m,n=1}^{40} \frac{q_m q_n}{|\mathbf{b}_{1m} - \mathbf{b}_{2n}|} \quad (2)$$

where $\mathbf{b}_{1m}$ ($\mathbf{b}_{2n}$) are the coordinates of bond-centres on molecule 1 (2) and $q_m = q$ for m = S (a single bond) and $q_m = -2q$ for m = D (a double bond).

Indeed, with this potential they found, upon minimising the total energy for four independent C_{60} molecules per unit cell, that the Pa3 structure had the lowest energy for $q > 0.21e$ and that the angle of rotation is $\phi = 98.7°$ (clockwise) about the proper $\langle 111 \rangle$ axes. They noted, after David et al.,[16] that this structure minimises the short-range coulomb energy.

As the value for the bond charge was the only free parameter in their model, they fixed it in order to give the lattice constant $a = 14.16$ Å finding that $q = 0.27e$. With this value the cohesive energy is -1.990 eV/C_{60} of which 90% is due to the Lennard-Jones interaction and only 10% due to the Coulomb interaction of the bonds.

They estimated the transition temperature in the following way: in the high temperature disordered phase, the molecules rotate almost freely; hence the free energy of the disordered phase consists of the Madelung energy of the system together with the entropy due to the rotation of the molecules. In the ordered phase, the molecules have an optimal orientation about which they vibrate; the free energy of this phase therefore is a sum of the Madelung energy of the Pa3 structure plus the energy due to the vibrations of the molecules (modelled as harmonic oscillators). The free energy of the disordered (fcc) phase is:

$$F_{fcc}(T) = N[E_{fcc} - k_B T \ln 8\pi^2 (Ik_B T/2\pi\hbar^2)^{3/2}] \qquad (3)$$

where $I = 1.0 \times 10^{-43}$ kg m^2 is the moment of inertia of a C$_{60}$ molecule and the Madelung energy E_{fcc} is calculated by averaging over random configurations, while the free energy of the ordered sc phase is

$$F_{sc}(T) = N\left[E_{sc} + 3k_B T \ln\left(1 - e^{-\hbar\omega_0/k_B T}\right)\right] \qquad (4)$$

where $\omega_0 = 1.86 \times 10^{-12}$ s^{-1} is the average frequency of librons and E_{sc} is the Madelung energy of the Pa3 structure.

The transition temperature in this model is the temperature at which the free energies are equal. With the lattice constant $a = 14.16$ Å, they found that

$$E_{fcc} = -1.772 \quad \text{eV}/\text{C}_{60} \qquad E_{sc} = -1.968 \quad \text{eV}/\text{C}_{60}$$

which gives a transition temperature of about 270 K. They state that the transition is first order and that the change in entropy is $\Delta S \approx 6R$. They also calculated the compressibility to be $K_0 = 5.175$ Mbar^{-1} which in turn implies that $da_0/dp = 0.024$Å/kbar. Assuming that the change in pressure affects only the Madelung energy (through the change of lattice constant) and that the rotational and librational motions of the C$_{60}$ molecules are unaffected they calculated the change in transition temperature with pressure to be $dT_c/dp = 11.5 \pm 0.7$K/kbar which is in reasonable agreement with experiment.

Upon the basis of these two investigations it is clear that the dominant contribution to the cohesive energy is the atom-atom interaction between the carbon atoms composing the C$_{60}$ molecules; however, the observed structure is stabilised not by these atom-atom van der Waals interactions but rather by the details of the charge distribution on the molecule and, in particular, the charge imbalance between the single and double bonds of the C$_{60}$ molecules.

LANDAU-TYPE THEORIES

Turning now to the Landau-type theories that have been proposed, it is worthwhile to note that there are two approaches to the Landau theory: the first is microscopic while the other is phenomenological.[17] The microscopic approach beginning from a microscopic discussion of the system in terms of density matrices and interparticle interactions and proceeding through a series of argued approximations (based upon the nature of the system, interaction *etc.*) attempts to arrive at a Landau theory. In contrast, the phenomenological approach begins from the observed structure and proceeds to construct the Landau theory. The strength of the latter is that it begins with the experimental structure and is systematic. It is, however, unable to provide insight into the microscopic physics of the transition. The microscopic approach, on the other hand, can provide this understanding but a very general theory is needed. Simplifying this theory requires that approximations be made and, in general, this is a non-trivial problem and knowledge of the experimental structure is used to guide the approximations.

APPROACH OF HARRIS AND SACHIDANANDAM

Harris and Sachidanandam[10] presented an *ad hoc* molecular field theory of the phase transition following the work that has been done on orientational ordering in solid hydrogen. As a result of the high symmetry of the C_{60} molecule, the orientational probability density distribution can be expanded in spherical harmonics, *i.e.*

$$\sigma(\Omega) = \sum_{\nu=1}^{60} \langle \delta(\Omega - \Omega_\nu)\rangle_T = \sum_{\nu=1}^{60}\sum_{l,m} Y_{lm}^*(\Omega)\langle Y_{lm}(\Omega_\nu)\rangle_T \qquad (5)$$

where $\langle\ldots\rangle_T$ denotes a thermal average at temperature T. Consequently they defined a "molecular" order parameter for a single C_{60} molecule in the following manner:

$$\sigma^{(m)}(I) \propto \langle \sum_{\nu=1}^{60} Y_{6m}(\Omega_\nu)\rangle_T \qquad (6)$$

where the sum is over all atoms ν in the molecule I, and Ω_ν is its orientation. They chose $l = 6$ as this is the lowest non-zero spherical harmonic in the expansion Eq. (5).

Within a molecular field theory, the density matrix which is a function of the Euler angles of the Ith molecule, α_I, β_I, and γ_I, is a product of the single-molecule density matrices, *i.e.*

$$\rho(\{\alpha,\beta,\gamma\}) = \prod_I \rho_I(\alpha_I,\beta_I,\gamma_I). \qquad (7)$$

The free energy is

$$F(\rho) = \sum_I Tr V_I \rho_I + \frac{1}{2}\sum_{I\neq J} Tr V_{IJ}\rho_I\rho_J + k_B T \sum_I Tr\rho_I \ln \rho_I, \qquad (8)$$

where the crystal field V_I is the orientational potential of the molecule I when its neighbours are disordered and V_{IJ} is the C_{60}-C_{60} intermolecular potential between the molecules I and J.

Harris and Sachidanandam noted that the Pa3 structure can be considered to be composed of four sublattices at the wavevectors

$$\begin{aligned}\mathbf{q}_x &= (2\pi/a)(1,0,0), & \mathbf{q}_y &= (2\pi/a)(0,1,0), \\ \mathbf{q}_z &= (2\pi/a)(0,0,1), & \mathbf{q}_0 &= (0,0,0).\end{aligned} \qquad (9)$$

Stating that only the amplitudes at the non-zero wavevectors are "critical", they proceeded to ignore any contribution from $\mathbf{q}_0$. They then introduce a wave-vector dependent order parameter

$$\sigma(\mathbf{q})^{(m)} = \frac{1}{N}\sum_I \sigma^{(m)}(I) e^{i\mathbf{q}\cdot\mathbf{r}_I}, \qquad (10)$$

where N is the total number of C_{60} molecules in the system, $\mathbf{r}_I$ is the position of the centre of mass of the molecule I, and the wave-vector dependent vector order-parameter is

$$\sigma_X(\mathbf{q}_\alpha) = (\sigma(\mathbf{q}_\alpha)^{(-6)}, \sigma(\mathbf{q}_\alpha)^{(-5)}, \ldots, \sigma(\mathbf{q}_\alpha)^{(6)}). \qquad (11)$$

In Eq. (11), $X = \mathcal{A}, \mathcal{B}$ where $\mathcal{A}$ and $\mathcal{B}$ are two of the eight possible structures (any two) which are related to one another by a reflection about $(1,1,0)$.

In keeping with their decision to discard all but the "critical" wavevectors, they take the density matrix to be

$$\rho_I = 1 + C_1 \sum_{X=\mathcal{A},\mathcal{B}} \sum_{\alpha=x,y,z} \xi_X(\mathbf{q}_\alpha) e^{i\mathbf{q}_\alpha \cdot \mathbf{r}_I} \sum_{m=-6}^{6} \sigma_X(\mathbf{q}_\alpha)^{(m)} \sum_{\nu=1}^{60} [Y_{6m}(\Omega_\nu)]^* \quad (12)$$

where $\xi_X(\mathbf{q}_\alpha)$ are the amplitudes of associated with the wave-vector dependent vector order-parameter $\sigma_X(\mathbf{q}_\alpha)$ and $C_1 = 2.666$. As Harris and Sachdanandam are interested in showing the existence of a cubic term in their Landau free energy, they do not expand the free energy Eq. (8) beyond third order (to do so would require a knowledge of the the microscopic intermolecular potential). Ignoring contributions to the free energy which they have identified as non-critical, they find

$$N^{-1} F = \frac{1}{2} s_0 k_B (T - T_c) \sum_{X=\mathcal{A},\mathcal{B}} \sum_{\alpha=x,y,z} |\xi_X(\mathbf{q}_\alpha)|^2 - \frac{1}{6} k_B T C_1^3$$

$$\times \sum_{X,Y,Z=\mathcal{A},\mathcal{B}} \xi_X(\mathbf{q}_x) \xi_Y(\mathbf{q}_y) \xi_Z(\mathbf{q}_z) \sum_{l,m,n=-6}^{6} \sigma_X(\mathbf{q}_x)^{(l)} \sigma_Y(\mathbf{q}_y)^{(m)} \sigma_Z(\mathbf{q}_z)^{(n)}$$

$$\times \sum_{\mu,\nu,\kappa=1}^{60} Tr\left\{[Y_{6l}(\Omega_\nu)]^* [Y_{6m}(\Omega_\mu)]^* [Y_{6n}(\Omega_\kappa)]^*\right\}$$
(13)

where $s_0 = 3.586$. This expression can be further simplified by noting that the trace vanishes unless $l + m + n = 0$. Consequently the third order term is nonzero only when it is composed of an even-parity term in m and two odd-parity terms. As it happens, only two of the wave-vector dependent vector order-parameters are even, so that only four possible combinations of the labels X, Y and Z are possible. Thus Eq. (13) simplifies to

$$N^{-1} F = \frac{1}{2} s_0 k_B (T - T_c) \sum_{X=\mathcal{A},\mathcal{B}} \sum_{\alpha=x,y,z} |\xi_X(\mathbf{q}_\alpha)|^2$$

$$- k_B T w \sum_{X=\mathcal{A},\mathcal{B}} \xi_X(\mathbf{q}_x) \xi_X(\mathbf{q}_y) \xi_X(\mathbf{q}_z) + \mathcal{O}(\xi^4)$$
(14)

with $w = .92854$. They point out that there are eight possible minima of this free energy in agreement with the demands of symmetry. As the free energy Eq. (14) obviously contains a cubic term, Harris and Sachidanandam concluded that the transition is predicted to be first order in agreement with experiment. They also note that the form of their free energy is that of a model with two three-state Potts variables.

APPROACH OF MICHEL

The approach of Michel et al.[11] makes extensive use of symmetry adapted functions. Symmetry adapted functions are linear combinations of spherical harmonics which form a basis for a given irreducible representation of a point group and are defined through

$$S_{l(P)}^\tau(\Omega) = \sum_m \alpha_{l(P)}^{m\tau} Y_{lm}(\Omega) \qquad l = 0, 1, 2, \ldots \quad (15)$$

where $\alpha_{l(P)}^{m\tau}$ are transformation coefficients and $\tau = (\Gamma, \rho, i)$ denotes the irreducible representation Γ of the point group P, ρ labels the representations (should it have a multiplicity greater than one) and i is the row of the representation. Choosing one of the two standard orientations as a reference (see Fig. 3 of Ref. [5]), the orientation of all the molecules in the system can be obtained from this standard orientation by a suitable rotation. Indeed, a molecule with orientation Ω_ν can be obtained from the standard orientation $\bar{\Omega}_\nu$ by a rotation through the Euler angle $\omega = (\alpha, \beta, \gamma)$, i.e.

$$\Omega_\nu = R(\omega)\bar{\Omega}_\nu \qquad (16)$$

where $R(\omega)$ denotes a rotation through ω. The mass distribution of a rotated molecule is

$$\sigma(\Omega; \omega) = R(\omega) \sum_{\nu=1}^{60} \delta(\Omega - \bar{\Omega}_\nu) = \sum_{\nu=1}^{60} \delta(\Omega - R(\omega)\bar{\Omega}_\nu). \qquad (17)$$

Introducing rotator functions $U_l^{1\tau}(\omega)$ the mass distribution of this rotated molecule is

$$\sigma(\Omega; \omega) = \sum_\tau g_l^1 U_l^{1\tau}(\omega) \qquad (18)$$

where

$$g_l^1 = \sum_{\nu=1}^{60} S_{l(I)}^1(\Omega_\nu) \qquad (19)$$

is the molecular form factor. Michel et al.[11] find that for a truncated icosahedron $g_0^1 = 16.92$, $g_6^1 = 2.56$, $g_{10}^1 = 19.35$. Rotator functions $U_{l(P)}^{\lambda\tau}(\omega)$ are to the symmetry adapted functions $S_{l(P)}^{\lambda\tau}(\omega)$ what the Wigner rotator functions $D_{nm}^l(\omega)$ are to the spherical harmonics $Y_{lm}(\omega)$.

The total potential energy for molecules at $\mathbf{r}$ with orientation $\{U_l^{1\tau}(\omega(\mathbf{r}))\}$ situated on a lattice is

$$V = \frac{1}{2} \sum_{\mathbf{r},\mathbf{r}'} \sum_{\nu,\nu'=1}^{60} \int d\Omega_\nu \int d\Omega_{\nu'} V(\mathbf{r},\nu;\mathbf{r}',\nu')\sigma(\Omega_\nu(\mathbf{r}))\sigma(\Omega_{\nu'}(\mathbf{r}'))$$

$$= \frac{1}{2} \sum_{\mathbf{r},\mathbf{r}'} \sum_{l,l'} \sum_{\tau,\tau'} v_{ll'}^{\tau\tau'}(\mathbf{r}-\mathbf{r}')g_l^1 g_{l'}^1 U_l^{1\tau}(\omega(\mathbf{r}))U_{l'}^{1\tau'}(\omega(\mathbf{r}')) \qquad (20)$$

where

$$v_{ll'}^{\tau\tau'}(\mathbf{r}-\mathbf{r}') = \int d\Omega_\nu \int d\Omega_{\nu'} V(\mathbf{r},\nu;\mathbf{r}',\nu')S_l^\tau(\Omega_\nu)S_{l'}^{\tau'}(\Omega_{\nu'}). \qquad (21)$$

Assuming an atom-atom LJ potential between the carbon atoms comprising each C_{60} molecule and restricting themselves to nearest-neighbour interactions they found that the largest matrix elements were obtained when $\tau = \tau'$ belonged to T_{2g} representations. Furthermore, for $l = l' = 6$, $v_{ll'}^{\tau\tau'}(\mathbf{r}-\mathbf{r}')$ is about four times larger than for $l = l' = 10$; however, as the molecular form factor for $l = 10$ is much larger than for $l = 6$, they concluded that it is the $l = 10$ terms that drive the transition. Fourier transforming the potential for $l = 10$ and restricting themselves to consideration of $T_{2g}^{(3)}$ only, Eq. (20) becomes

$$V_{l=10} = \frac{1}{2} \sum_\mathbf{q} \sum_{i,j} v_{ij}(\mathbf{q}) \left(g_{10}^1\right)^2 U_{10}^i(\mathbf{q})U_{10}^j(-\mathbf{q}) \qquad (22)$$

359

where
$$v_{ij}(\mathbf{q}) = \sum_{\kappa \in nn} v_{10,10}^{i,j}(\kappa) \cos(\mathbf{q} \cdot \mathbf{x}(\kappa)) \tag{23}$$

with $\mathbf{x}(\kappa)$ the vector from a molecule to its κth nearest neighbour (nn) and U_{10}^i are the rotator functions for $T_{2g}^{(3)}$, and $v_{10,10}^{i,j}$ the potential corresponding to U_{10}^i and U_{10}^j.

Evaluating the potential for $\mathbf{q}$ at the Γ- and X-points only, they found that at the Γ-point the potential is repulsive whereas at the X-point it is attractive. They, therefore, concluded that this attractive interaction leads to the condensation scheme

$$\text{Fm3m:} \quad (U_{10}^3(2\pi/a,0,0) = U_{10}^1(0,2\pi/a,0) = U_{10}^2(0,0,2\pi/a) = \eta \neq 0$$
$$U_{10}^2(2\pi/a,0,0) = U_{10}^3(0,2\pi/a,0) = U_{10}^1(0,0,2\pi/a) = 0) \to \text{Pa3}.$$

They further concluded that because the product $U_{10}^1 U_{10}^2 U_{10}^3$ has cubic symmetry, there is a cubic term in the free energy expansion and the transition is first order. They then calculated the crystal field due to the nearest neighbours and expanding the free energy to second order in the order parameters they found the transition temperature to be 165 K (which is much lower than the observed transition temperature).

In subsequent papers, Michel[18-20] extended this work studying the effect of the bond-bond interactions on the transition and the implications of the presence of multiple representations at the X-point. The presence of the multiple representations leads him to predict a sequence of five transitions in solid C_{60}.

A difference between the theories of Sachidanandam and Harris, and Michel, Copley and Neuman lies in their claims about which of the modes dominates the transition (the physical significance of the various modes does not seem to be well understood). Irrespective of the difference between the claims about which mode dominates, the symmetry of the free energy will be the same since only one representation of Fm3m at the X-point subduces to Pa3.

APPROACH OF RAPCEWICZ AND PRZYSTAWA

Most recently, Rapcewicz and Przystawa[12] have presented a traditional symmetry analysis and Landau theory of the ordering transition. Denoting the fullerene molecules found at the lattice sites $\mathbf{R}_1 = (0,0,0)$, $\mathbf{R}_2 = (1/2,0,1/2)$, $\mathbf{R}_3 = (1/2,1/2,0)$ and $\mathbf{R}_4 = (0,1/2,1/2)$ by A, B, C and D, respectively, they took one, say A, and applied all of the symmetry operations of the high temperature space group, Fm3m, to it. This generated eight symmetry equivalent molecules A, B, C, D, E, F, G and H. In order to describe the unit cell of the ordered structure, they introduced the mass function

$$\psi_1(\mathbf{r}) = \sum_{\nu=1}^{60} \left[\delta(\mathbf{r} - \mathbf{r}_\nu^A - \mathbf{R}_1) + \delta(\mathbf{r} - \mathbf{r}_\nu^B - \mathbf{R}_2) + \delta(\mathbf{r} - \mathbf{r}_\nu^C - \mathbf{R}_3) + \delta(\mathbf{r} - \mathbf{r}_\nu^D - \mathbf{R}_4)\right]. \tag{24}$$

Applying again the symmetry operations of the disordered phase, Fm3m, to this mass function, they found eight symmetry-equivalent mass functions $\psi_i(\mathbf{r})$ ($i=1,2,\ldots,8$) which describe the eight possible ordered structures. They noted that the disordered phase corresponds to

$$\psi_0(\mathbf{r}) = \frac{1}{8}\sum_{j=1}^{8} \psi_j(\mathbf{r}) \tag{25}$$

and that describing the fcc structure in this way amounts to describing the sphere by 480 points; each occupied with an equal probability.

The order parameter,
$$\delta\psi_1(\mathbf{r}) = \psi_1(\mathbf{r}) - \psi_0(\mathbf{r}), \tag{26}$$
belongs to a seven-dimensional reducible representation of the Fm3m group (denoted by $D(7)$) which they showed is a direct sum of two irreps[21] (ignoring a trivial Γ-point representation which occurs)
$$D(7) = \Gamma_2^+ \oplus X_5^+ \tag{27}$$
where Γ_2^+ is a one-dimensional representation at $\mathbf{k} = (0,0,0)$ (the order parameter belonging to which they denoted ξ) and X_5^+ is a six-dimensional representation at $\mathbf{q}_X = (2\pi/a)(1,0,0)$ (the order parameter belonging to which they denoted $(\eta_1, \eta_2, \ldots, \eta_6)$). Harris and Sachidanandam ignored the presence of this Γ_2^+ representation while Michel et al., after briefly considering it, ignored its implications.

For the free energy expansion they found

$$\begin{aligned}F[\eta,\xi] =& A_1\xi^2 + D_1\xi^4 \\
& + A_2 \sum_{\eta=1}^{6} \eta_i^2 + B(\eta_1\eta_2\eta_3 + \eta_4\eta_5\eta_6) + C\left[\sum_{i=1}^{6}\eta_i^2\right]^2 + D_2\sum_{i=1}^{6}\eta_i^4 \\
& + G_1\left[\eta_1^2\eta_5^2 + \eta_2^2\eta_4^2 + \eta_3^2\eta_6^2\right] + G_2\left[\eta_1^2\eta_4^2 + \eta_2^2\eta_6^2 + \eta_3^2\eta_5^2\right] \\
& + G_3\left[\eta_1^2\eta_6^2 + \eta_2^2\eta_5^2 + \eta_3^2\eta_4^2\right] + g_{41}\xi\left[\eta_1\eta_2\eta_3 - \eta_4\eta_5\eta_6\right] + g_{42}\xi^2\sum_{i=1}^{6}\eta_i^2 \\
& + g_{43}\xi(\eta_1^2 + \eta_2^2 + \eta_3^2 - \eta_4^2 - \eta_5^2 - \eta_6^2).\end{aligned} \tag{28}$$

The solution $\xi \neq 0$, $\eta_1 = \eta_2 = \eta_3 \neq 0$ and $\eta_4 = \eta_5 = \eta_6 = 0$ corresponds to the Pa3 structure, where

$$\psi_1(\mathbf{r}) = \psi_0(\mathbf{r}) + \frac{1}{8}\xi(\mathbf{r})\theta_\xi(T,p) + \frac{1}{4}(\eta_1(\mathbf{r}) + \eta_2(\mathbf{r}) + \eta_3(\mathbf{r}))\theta_\eta(T,p) \tag{29}$$

with
$$\theta_\xi(T,p) = \begin{cases} 0, & \text{for } T > T_\xi; \\ 1, & \text{for full order below } T_\xi \end{cases}$$

and
$$\theta_\eta(T,p) = \begin{cases} 0, & \text{for } T > T_\eta; \\ 1, & \text{for full order below } T_\eta. \end{cases}$$

T_ξ and T_η are the temperatures at which the order parameters ξ and η_i ($i = 1, 2, 3$) become non-zero. Their values will depend upon the particular choice of the parameters A_1, A_2, B, C, D_1, D_2, G_1, G_2, G_3, g_{41}, g_{42}, and g_{43}.

There are two possible ways in which the transition can be accomplished. Firstly, a direct transition to the Pa3 structure which is first order and secondly, a continuous transition to an Fm3 structure which is driven by Γ_2^+ followed by a first order transition to the Pa3 structure driven by X_5^+. This intermediate Fm3 structure, they cautioned, is not the so-called standard structure but rather a structure in which the molecules hop between a restricted set of orientations (A, B, C, and D, or E, F, G, and H). Such a state resulting from C_{60} molecules hopping among a restricted set of possible orientations would be extremely sensitive to impurities, thus they urged a re-examination of the experiments in the light of these theoretical results.

The most important implications of this approach concern the nature of order and the partially ordered state in C_{60} and, more generally, molecular crystals. The order parameter describes the probability that a site is occupied by a given orientation. When θ_ξ and θ_η are equal to unity, the state is fully ordered; when $\theta_\xi, \theta_\eta < 1$, the state is partially ordered. In the partially ordered state, "wrong" orentations are present with certain probabilities. Indeed the nature of order in solid C_{60} is similar to that in alloys where the order parameter describes the probability of occupation of a lattice point by an atom of a given type.

Rasolt[13] presented a Landau theory and renormalisation group study of the transition based upon an analysis of CH_4. However, as the order parameter and Landau free energy is of a different symmetry than the theories already presented, it will not be discussed further.

CONCLUSIONS

Upon the basis of these above theories, it can be concluded that although the dominant contribution to the cohesive energy comes from the atom-atom Lennard-Jones interactions between the carbon atoms of different C_{60} molecules, it is the Coulomb interaction due to the charge imbalance on the single and double bonds that stabilises the low temperature Pa3 structure. The transition to this Pa3 structure is first order and is driven by representations at the X- and Γ-points.

The full symmetry analysis has clarified the nature of order and the form of the order parameter which describes the orientational ordering transition in crystalline C_{60}. The order parameter describes the probability of occupation of a site by a given orientation. The symmetry analysis based upon the fully ordered state ($\theta_\xi = \theta_\eta = 1$) makes a prediction for the partially ordered phase. An analysis of the experimental data upon the basis of this prediction Eq. (29) could prove interesting.

ACKNOWLEDGMENTS

The author is grateful to Dr. Jan Lorenc for a careful reading of the manuscript and many helpful comments upon it.

REFERENCES

1. P. A. Heiney et al., *Phys. Rev. Lett.* **66**, 2911 (1991).
2. R. Sachidanandam and A. B. Harris, *Phys. Rev. Lett.* **67**, 1467 (1991).
3. W. I. F. David et al., *Nature* **353**, 147 (1991)
4. K. Prassides et al., *Carbon* **30**, 1277 (1992).
5. K. Prassides. This volume.
6. P. A. Heiney et al., *Phys. Rev. B* **45**, 4544 (1992).
7. G. A. Samara et al., *Phys. Rev. Lett.* **67**, 3136 (1991).
8. M. Sprik, A. Chen and M. L. Klein, *J. Phys. Chem* **96**, 2027(1992).
9. J. P. Lu, X.-P Li and R. M. Martin, *Phys. Rev. Lett.* **68**, 1551 (1992).
10. A. B. Harris and R. Sachidanandam, *Phys. Rev. B* **46**, 4944 (1992).

11. K. H. Michel, J. R. Copley and D. A. Neumann, *Phys. Rev. Lett.* **68**, 2929 (1992).
12. K. Rapcewicz and J. Przystawa, to be published, 1993.
13. M. Rasolt, *Phys. Rev. B* **46**, 1944 (1992).
14. A. Chen and M. L. Klein, *J. Phys. Chem.* **95**, 6750 (1991).
15. A. Chen and M. L. Klein, *Phys. Rev. B* **45**, 1889 (1992).
16. W. I. F. David *et al.*, *Europhys. Lett.* **18**, 219 (1992)
17. J. Przystawa, *Physica* **114A**, 557 (1982).
18. K. H. Michel, *Z. Phys. B* **88**, 71 (1992).
19. K. H. Michel, *Chem. Phys. Lett.* **193**, 478 (1992).
20. K. H. Michel, *J. Chem. Phys.* **97**, 5155 (1992).
21. S. C. Miller and W. F. Love, *Tables of Irreducible Representations of Space Groups and Co-representations of Magnetic Space Groups* (Boulder, Colorado: Pruett, 1967).

DYNAMICS CLOSE TO THE GLASS TRANSITION

Lukasz A. Turski

Center for Theoretical Physics, Polish Academy of Sciences
Al. Lotników 32/46, 02-668 Warszawa, Poland

Abstract

In spite of enormous efforts, the theoretical understanding of the phenomenon referred to as a *glass transition* is still fragmentary. Laboratory and computer experiments on supercooled liquid and glass properties close to the transition temperature (T_g) show how important a role is played, in these materials, by local geometrical arrangements of constituent particles. In this lecture I shall outline a particular theory of glass transition phenomena focusing on a new dynamical concept — the rheological stress — used to account for several important non-equilibrium properties of a glass and its parent supercooled liquid close to the glass transition T_g.

1. INTRODUCTION

A precious collector item nowadays is the XIX century Gallé glass; before its invention people had to work on glass for almost 70 000 years [1] – the age of oldest man-shaped obsidian pieces known to us. Pliny records that the first piece of man-made glass was "produced" by Phoenician sailors in Syria around 5000 B.C. Sometime around 300 B.C., again a Phoenician, discovered what is now called the blowing iron technique. This technique depends on the fundamental property of a glass - the slowness of atomic rearrangements as compared with the "experimental" time. We shall base much of our further analysis of glass properties on the recognition of this simple fact.

The word amorphous comes from Greek $\alpha\mu\rho\rho\phi\rho\sigma$, which means "shapeless". There are close connections between amorphous solids and glasses. The first report on the formation of amorphous metallic alloys by the vapor deposition technique was due to Kramer [2]. Buckel *et al.* [3] used a cryogenically cooled substrate and were first to examine the structure of amorphous metal by means of electron diffraction. In 1960 Duwez *et al.* demonstrated that amorphous metal films can be created by liquid quenching at cooling rates of 10^6K/s or higher. Nowadays higher cooling rates can be achieved. The pulsed-laser melting technique has achieved cooling rates of the order

10^{12}K/s. Other cooling methods involve electron-beam irradiation and some chemical processes. Even higher cooling rates were reported in heavy ion collision experiments 10^{14}K/s [4]. Glass can be produced by radiation damage [5] and mechanical processes [6].

All these experimental methods are very difficult and therefore clean and completely controllable *computer experiments* are highly desirable. It is due to the computer experiments that we have gained considerable knowledge about the internal structure of glasses and most of what I will tell in the following is based on an analysis of the existing molecular dynamics data.

I owe my understanding of the simulation results to Ray Mountain. The theory of glasses I will present originated in close collaboration with Sushanta Dattagupta. We both have learned a lot during lengthy discussions with Annette Zippelius and Reiner Kree.

2. THE MODEL

Saying that "ice is order and glass is confusion" John Tyndall [1] gave quite a precise definition of a glass. Indeed Tyndall pointed out two essential physical properties of glass, its lack of any long ranged order - orchestral harmony and the presence of frustration, as we now call Tyndall's confusion. For a long time we were more interested in "ice" than "glass". This has changed, nowadays we have spin glasses, orientational and protonic glasses, metallic glasses, and even superconducting glasses.

It is quite impossible to account here for current activities in all areas of glass physics. A general review was written by Jäckle [7]. Binder and Young, [8] have given a comprehensive presentation of spin glass ideas. Here I shall discuss aspects of the glass transition, GT, in classical liquids, using the idea of local orientational order in supercooled liquids, $LORO$, proposed by Tammann, Frenkel and, in particular, Bernal [9].

A sufficiently clean liquid, when nucleation is prevented, can be easily cooled much below the equilibrium freezing point [10]. Before we reach a point at which the liquid entropy is less than that of the solid (the Kautzman paradox) the GT occurs. The liquid "phase diagram" close to the GT point is time dependent; the larger is the cooling rate the higher is T_g. The time scale for quench rates is set by the inverse rate of nucleation. This is why it is so difficult to obtain noble gas glasses in the laboratory. The situation is quite different in computer experiments. Here quench rates can be extremely high and, indeed, one can keep vitreous computer liquid Ar at $T \sim 15$K [11]. For theoreticians this is a source of problems. For we either need to use non-equilibrium statistical mechanics or to modify the equilibrium description of a glass in order to incorporate time scales.

In practice a glass is defined as the state of a system when the viscosity becomes $\eta \sim 10^{13}$poise, a dramatic increase from values of the order 10^{-2} for normal liquids. The increase of the shear viscosity in the vicinity of the glass transition can be fitted

---

[1] To many persons here present a block of ice may seem of no more interest and beauty than a block of glass; but in reality it bears the same relation to glass that orchestral harmony does to the cries of the marketplace. The ice is music, the glass is noise. The ice is order, the glass is confusion. In glass, molecular forces constitute an inextricably entangled skein, in ice they are woven to symmetric texture...(John Tyndall–1863)

quite well by the Vogel–Fulcher law [12]:

$$\eta = \eta_0 \exp\left(\frac{b}{T - T_0}\right), \qquad (1)$$

where b, η_0 and T_0 are empirical fitting parameters. T_0 is found to be close, but somewhat below the actual glass transition temperature. The Vogel–Fulcher law holds very well for temperatures close, but not too close, to the transition temperature. Very close to the transition point, the viscosity dependence seem to change again resulting in the usual Arrhenius ($\sim \exp(-\Delta/T)$) behaviour.

Due to the above facts it is assumed that the *GT* is a dynamical phenomenon caused by a dramatic change in the nature of the one-particle motion in the supercooled liquid - from the Fickean (usual diffusion) into hopping like self-diffusion. This was the key point in the *mode-mode* coupling theory of the supercooled liquids and glasses [13].

The change in the single particle motion and the dependence of T_g on the rate of cooling indicate that certain degrees of freedom of the system cannot relax to equilibrium but are *quenched* in a high temperature configuration. The freezing-out of the motion in certain regions of the phase space makes the glass transition look somewhat similar to the transition in spin glass alloys [14]. This has motivated the application of the theoretical techniques used to study spin glasses to the problem of the glass transformation [15, 16, 17, 18, 19]. However, a crucial fact distinguishes ordinary glasses from spin glasses in that unlike in the latter there are no quenched impurities but certain "dynamically-quenched" fluctuations which are "self-trapped" over macroscopic time scales. This view is supported by recent numerical investigations of apparent non-ergodicity in ordinary glass forming binary liquids by Mountain and Thirumalai [20, 21, 22].

In a fluid in thermodynamic equilibrium, each fluid particle experiences over time nearly the same set of phase space environments as every other particle. In the time–averaged sense each particle is then equivalent to every other and that is what Mountain et al. call statistical symmetry. (This is very similar to the notion of self-averaging). If the concept of LORO is true, then in a supercooled liquid and glass a particle does not sample all of its phase space environments. A particle rarely moves further than one particle diameter, it does not diffuse, and therefore it samples, over long times, different environments from all other particles. When this happens the statistical symmetry is broken and individual contributions of particles to the measurable quantities may be different. To probe the above concepts, Mountain and collaborators invented suitable measures of non-ergodicity called "metrics" based upon the individual particle energies.

In Fig. 1, I have shown recent data for the "energy" metric obtained by Ray Mountain [23]. Indeed, the glass forming system behaves in a non-ergodic fashion.

One of the difficulties with the *GT* is that it is not clear what is the proper language for its description. The liquid state physics work horse, the pair correlation function $g(r) \sim \langle \rho(r)\rho0)\rangle$, cannot really tell us what we need to know about liquid-solid transition. In Figs. 2,3 the behavior of radial correlation function for a binary Lennard–Jones mixture is shown for larger and smaller species and for two different reduced temperatures corresponding to the situation in the liquid and glass phases.

Note that there are no dramatic changes in the $g(r)$ behaviour. Similarly one would fail to describe supercooled liquid and glass properties using concepts, like phonons, from conventional solid state theory.

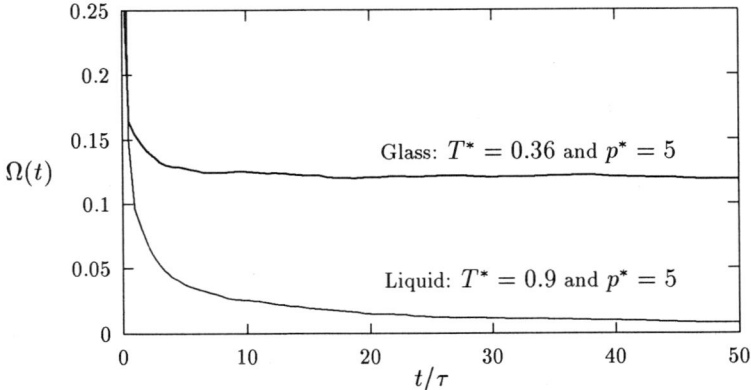

Figure 1. Time behavior of the non-ergodicity metric for the Lennard–Jones binary mixture, after Ref. [23]. T^* and p^* are the Lennard-Jones liquid reduced temperature and pressure, respectively.

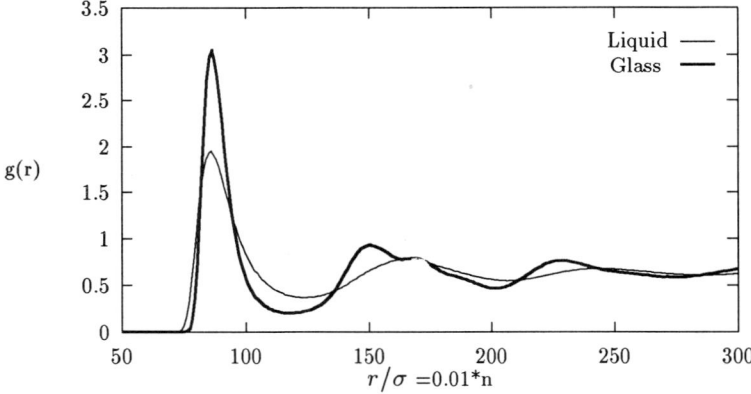

Figure 2. Radial correlation function for smaller component of the Lennard–Jones binary mixture. The radius of the particle $\sigma = 0.8$. Liquid data were taken at reduced temperature and pressure $T^* = 0.9$ and $p^* = 5$. Glass transition temperature for that system is $T^* \simeq 0.5$; after Ref. [23]

These "language" difficulties result in various approaches to the description of the glass transition ranging from the atomic level density functional approach [24, 25] to the memory function formulation [26, 27]. In the *LORO* theory, a dense liquid possesses local orientational order with symmetry akin to the icosahedral symmetries observed in quasicrystals [28, 29, 30]. The *LORO* can be pictured in terms of a "cage" which is a cluster of atoms positioned at the vertices of, say, a Voronoi polyhedron with a central atom at the origin whose diffusivity is thus severely restricted [31, 32]. A similar description of a liquid was used in the early days of neutron scattering experiments [33]. The recent computer experiments of Schober et al. [34] have revealed connections between low temperature glass properties and LORO concepts.

The relevant information about the *LORO* is embodied in higher particle distribution functions; eg., the four particle distribution function $\rho_4(\mathbf{r}_1, \mathbf{r}_2, \mathbf{r}_3, \mathbf{r}_4)$, for three dimensional systems with pair-wise spherically symmetric interactions [28, 29]. How

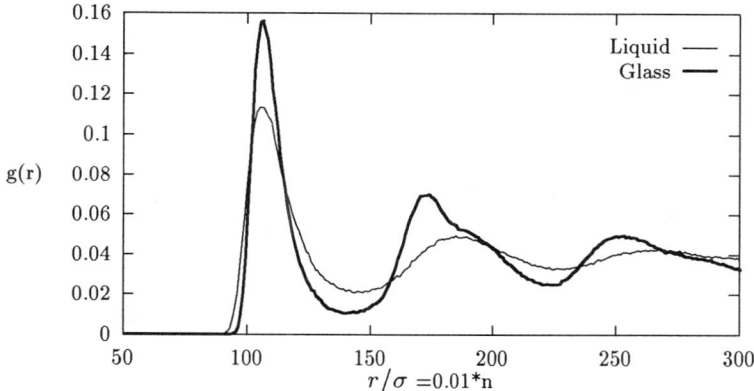

Figure 3. Radial correlation function for larger component of the Lennard–Jones binary mixture. The radius of the particle $\sigma = 1$. Liquid data were taken at reduced temperature and pressure $T^* = 0.9$ and $p^* = 5$. Glass transition temperature for that system is $T^* \simeq 0.5$; after Ref.[23]

ρ_4 is described in terms of an orientational order parameter, a tensor of rank four or higher, has been the subject of the original work of Sushanta Dattagupta and myself, [35] (henceforth referred to as I). In I we have shown that the coupling between the local orientational variables and the deformation tensor leads to an internal stress which physically captures the effect of the local misfit energy due to two misaligned cages. Our objective in this paper was to analyse the influence this internal stress has on the viscoelastic properties of a supercooled liquid [9, 36, 37].

The principal idea presented in I was that $\rho_4(\mathbf{r}_1, \mathbf{r}_2, \mathbf{r}_3, \mathbf{r}_4)$ can be related to the probability density $\rho(\mathbf{r}, \{\mathbf{h}_i\})$ of having, at an arbitrary point $\mathbf{r}$ in the liquid, a local triad of vectors $\mathbf{h}_i$, $(i = 1, 2, 3)$, defining the local coordinates of a cage [38]. For practical purposes this probability density can be expanded into a proper set of irreducible tensors of rank four or higher (Wigner matrices or similar) [31]. We shall denote them as $\Lambda^A(\mathbf{r})$, for the cage located at position $\mathbf{r}$; A may stand for four or six indices in order to describe cubic or icosahedral ordering, respectively [31, 32]. In the standard fashion, Ref. I, we write the liquid Hamiltonian (coarse grained free energy) as a quadratic functional of these tensors [39]. The orientational variables $\Lambda^A(\mathbf{r})$ are expected to couple to the local distortions of the liquid, as in "compressible" lattice models [40].

The considerations above yield the following form of the free energy functional describing orientation-strain interactions [35]:

$$\begin{aligned}
\mathcal{F}\{\Lambda, \epsilon\} &= \frac{1}{2V^2} \int d\mathbf{r} d\mathbf{r}' E^0_{AB}(\mathbf{r}-\mathbf{r}')\Lambda^A(\mathbf{r})\Lambda^B(\mathbf{r}') \\
&+ \frac{1}{2V^2} \int d\mathbf{r} d\mathbf{r}' G^{AB}_{\alpha\beta}(\mathbf{r}-\mathbf{r}')\Lambda^A(\mathbf{r})\Lambda^B(\mathbf{r}')\epsilon_{\alpha\beta}(\mathbf{r}) \\
&+ \frac{1}{2V} \int d\mathbf{r} C_{\alpha\beta\gamma\delta}\epsilon_{\alpha\beta}(\mathbf{r})\epsilon_{\gamma\delta}(\mathbf{r}).
\end{aligned} \quad (2)$$

The first term in Eq.(2) accounts for the pure orientational coupling between two cages centred at $\mathbf{r}$ and $\mathbf{r}'$ with strength E^0_{AB}. This term is known, in mean field theory, to lead to a *first* order phase transformation (when Λ's are treated, for example, as Potts variables) from the $\langle\Lambda\rangle = 0$ to a $\langle\Lambda\rangle \neq 0$ phase. The second term is the most crucial for our further analysis, for it describes the coupling between the orientational degrees of freedom and the local strain in the liquid, with the strength of interaction given by the tensor $G^{AB}_{\alpha\beta}$, which is symmetric in all indices. Finally, the last term accounts for the elastic energy of conventional elasticity theories [41] in which the $C_{\alpha\beta\gamma\delta}$ are the bare elastic coefficients $C_{\alpha\beta\gamma\delta} = \lambda\delta_{\alpha\beta}\delta_{\gamma\delta} + \mu(\delta_{\alpha\gamma}\delta_{\beta\delta} + \delta_{\alpha\delta}\delta_{\beta\gamma})$.

In Eq.(2) we adopt the summation convention for tensor indices and use the symbol V for the volume of the system.

In Ref. I we have demonstrated, for quasistatically cooled liquids, how the jumps in $\langle\Lambda\rangle$ and the Λ correlations $\langle\Lambda(\mathbf{r})\Lambda(\mathbf{r}')\rangle$, cause quantities like the total volume, the heat capacity and the elastic compliance, to change drastically as the liquid undergoes a transition to an orientationally ordered phase. These results agree well with empirical rules [42].

When a liquid is cooled down at a finite rate we have to take into account the presence of quenched fluctuations. The stochastic freezing of the cages manifests itself in the measurable elastic properties of the glassy phase. In particular the glass should exhibit a finite zero–frequency shear modulus.

The presence of quenched fluctuations implies that the direct cage interactions E^0_{AB}, and the coupling between the cage and the deformation field $G^{AB}_{\alpha\beta}$, in Eq. (2) have fluctuating contributions. The variances of these quenched variables, denoted E_f and G_f respectively, measure the amount of disorder in the system and increase with the cooling rate.

Any *LORO* theory which takes quenched fluctuations into account becomes unduly complex unless a certain simplified assumption concerning the orientational degrees of freedom is adopted. The assumption which captures the essential physics of the problem is that in which the local cages are permitted to have discrete orientations [31, 32, 35], and are visualized as a p-state Potts variable.

The random Potts model, without coupling to the elastic degrees of freedom was analyzed in [16]. The transition from a disordered Potts (paramagnetic) phase to a Potts glass phase at the temperature $T_g \sim \sqrt{E_f}$ was found. A very interesting result is that the order of the glass transition depends on the Potts dimensionality p. For $p < 4$ the transition is continuous, whereas for $p > 4$ it is a first order transition. In either case though, there is no latent heat. The specific heat is discontinuous for $p > 4$ at T_g and has a cusp for $p < 4$. Using the replicas technique we were able to show that when the random Potts model is coupled with elastic degrees of freedom the following facts hold [17]:

1. The coupling to the elastic degrees does not destroy the transition. This is similar to the liquid-solid case discussed previously.

2. The glass transition temperature is $T_g = \sqrt{E_f + \frac{1}{4}\mathrm{tr}(G_f C^{-1})^2}$. Thus T_g depends on the variance of the quenched disorder and therefore on the quench rate. This is in agreement with experiments and computer simulations. There is no latent heat for the transition for any value of p. The liquid state does not exist as a metastable phase in the glassy state. This means that one cannot supercool a liquid below the glass transition even though by adjusting the cooling rate one can obtain the glassy phase at various temperatures. There is also no macroscopic volume change at the glass transition.

3. The elastic coefficients are enhanced at the transition. The change of the elastic constants is continuous and independent of the order of the transition. In the weak coupling limit this change is given in terms of the Parisi function measuring the mutual overlap of many ergodic components of the glass [14, 43].

My main task now is to construct a dynamic theory associated with the free energy functional in Eq. (2). Identifying the order parameter $\Psi(\mathbf{r})$ as that particular component of Λ^A which points toward the axis along which the cages would like to order, the free energy in Eq. (2) takes a simpler form:

$$\begin{aligned}
\mathcal{F}\{\Psi, \epsilon\} &= \frac{1}{2V^2} \int d\mathbf{r} d\mathbf{r}' \Delta E(\mathbf{r} - \mathbf{r}') \Psi(\mathbf{r}) \Psi(\mathbf{r}') \\
&+ \frac{1}{2V^2} \int d\mathbf{r} d\mathbf{r}' \Delta G_{\alpha\beta}(\mathbf{r} - \mathbf{r}') \epsilon_{\alpha\beta}(\mathbf{r}) \Psi(\mathbf{r}) \Psi(\mathbf{r}') \\
&+ \frac{1}{V} \int d\mathbf{r} C_{\alpha\beta\gamma\delta} \epsilon_{\alpha\beta}(\mathbf{r}) \epsilon_{\gamma\delta}(\mathbf{r}').
\end{aligned} \qquad (3)$$

where ΔE and ΔG refer to the differences in the energy parameters between mutually parallel and perpendicular orientations of the cages [35].

To write down a kinetic relaxation equation for the orientational variables we imagine "orientational flips" in much the same manner as "spin flips" in kinetic Ising models [44, 45]. These flips represent the reorientational motion of the cages due to interactions with other degrees of freedom. It is well known that in the linearized mean field theory the relaxation equation in the wave vector space looks like [46]:

$$\begin{aligned}
\dot\Psi(\mathbf{q}, t) &= - \nu(\mathbf{q}) \Psi(\mathbf{q}, t) \\
&+ \frac{\beta}{V\tau} K_{\alpha\beta\gamma\delta} \int d\mathbf{q}' \Delta G_{\alpha\beta}(\mathbf{q}') \Psi(\mathbf{q}', t) \sigma_{\gamma\delta}(\mathbf{q} - \mathbf{q}', t).
\end{aligned} \qquad (4)$$

where the quantity $\nu(\mathbf{q})$ is the rate of relaxation associated with the pure orientational part of the free energy (viz. the first term in Eq. (3)), β is the inverse temperature in energy units, and τ is a parameter which sets the basic time scale of the relaxation processes induced by the heat bath. An analysis of computer symulations Ref. [29] suggests a conserved order parameter *LORO* kinetics, akin to the Kawasaki spin-exchange process in the kinetic Ising model [47]. This implies that $\nu(\mathbf{q})$ must vanish as $|\mathbf{q}| \to 0$ with the leading order term in $\mathbf{q}$ proportional to $\mathbf{q}^2$. In fact a mean field analysis yields [48]:

$$\nu(\mathbf{q}, T) = N \frac{\mathbf{q}^2 \xi^2}{\tau} \left[1 - \frac{T_0}{T} + \frac{T_0}{T} R^2 \mathbf{q}^2 \right], \quad (5)$$

where N is the number of nearest neighbor cages, ξ is the mean separation between them, and T_0 is the temperature at which pure orientational order (no coupling to the remaining degrees of freedom) would occur. The quantity R is the typical range of interactions between the cages described by ΔE in Eq. (3).

3. VISCOELASTICITY

To establish a proper dynamical description of a supercooled liquid near T_g we develop a generalisation of the Maxwell model of viscoelasticity theory. Recall that the Maxwell constitutive relation interpolating between solid– and liquid–like behaviour reads [9, 49, 50]:

$$\dot{\sigma}_{\alpha\beta}(t) = -\frac{1}{\tau}\sigma_{\alpha\beta}(t) + C_{\alpha\beta\gamma\delta}\dot{\epsilon}_{\gamma\delta}(t). \quad (6)$$

The quantity τ, the Maxwell relaxation time, determines the medium viscosity through the relation:

$$\eta_{\alpha\beta\gamma\delta} = \tau C_{\alpha\beta\gamma\delta}. \quad (7)$$

Taking the frequency Fourier transform of Eq. (6) we have

$$\tilde{\sigma}_{\alpha\beta}(\omega) = \frac{i\tau\omega C_{\alpha\beta\gamma\delta}}{1 + i\omega\tau}\tilde{\epsilon}_{\gamma\delta}(\omega) \equiv i\eta_{\alpha\beta\gamma\delta}(\omega)\tilde{\epsilon}_{\gamma\delta}(\omega). \quad (8)$$

When the experimental probe frequency is much higher than the inverse of the Maxwell time, $\omega\tau \gg 1$, we obtain the stress-strain relation for a solid. For $\omega\tau \ll 1$, we observe viscous relaxation. In order to make the Maxwell model applicable to a liquid we shall work systematically in the wave vector representation [37], thus $\tau = \tau(\mathbf{q})$.

Our strategy is to search for a mechanism that causes an enhancement of the Maxwell time τ so that the solid-like behaviour persists over longer experimental time scales (i.e. smaller ω) leading to an increase in η when one supercools a liquid. The crucial point is the observation that the presence of local orientational order induces an internal stress tensor the physical origin of which has been discussed earlier:

$$\sigma_{\alpha\beta}^{int}(\mathbf{r}) = \frac{1}{V}\int d\mathbf{r}' \Delta G_{\alpha\beta}(\mathbf{r} - \mathbf{r}')\Psi(\mathbf{r})\Psi(\mathbf{r}'). \quad (9)$$

Since the total stress in the system is additive, we have to replace the Maxwell constitutive relation, Eq. (6), by a generalized one which would lead to a proper superposition of stresses in either of the frequency limits discussed earlier. To do so we introduce the concept of a rheological stress $s_{\alpha\beta}$, a quantity which measures the mismatch between the measured stress in the system $\sigma_{\alpha\beta}$, and the static stress tensor defined customarily as the functional derivative of the free energy Eq. (2) with respect to the strain [36]:

$$s_{\alpha\beta} = \sigma_{\alpha\beta} - \frac{\delta \mathcal{F}}{\delta \epsilon_{\alpha\beta}}. \tag{10}$$

Using the explicit form of $\mathcal{F}\{\Psi, \epsilon\}$ one easily finds that $s_{\alpha\beta}$ is indeed the difference between the measured stress and the internal stress given by Eq. (9) plus the elastic stress given by $\delta \mathcal{F}_{el}/\delta \epsilon_{\alpha\beta}$.

When the local orientational degrees of freedom (the cages) become ordered, the flow of the medium has to stop; thus we shall replace the Maxwell constitutive relation by:

$$\dot{s}_{\alpha\beta} + \frac{1}{\tau(\mathbf{q})} s_{\alpha\beta} = -\frac{1}{\tau(\mathbf{q})} \frac{\delta \mathcal{F}_{el}\{\epsilon\}}{\delta \epsilon_{\alpha\beta}}. \tag{11}$$

The relaxation time $\tau(\mathbf{q})$ in Eq. (11) is the main, bare time scale of all the relevant relaxation processes in our model. Therefore $\tau(\mathbf{q})$ has to be identified with the relaxation time (i.e $1/\nu(\mathbf{q})$) associated with the bare kinetics of the local orientational order parameter $\Psi(\mathbf{q})$, Eq. (4).

The equation (11) is a novel constitutive relation the use of which is not restricted to our current analysis. It can be employed in modelling other physical phenomena wherein the stress tensor is split in a natural way into two parts describing the long wavelength behaviour, and the other describing the short wavelength behaviour, for example in the theory of coupling between acoustic waves and thermal phonons [51].

It is convenient to rewrite Eq. (11) in terms of the measured stress, the internal stress and the rate of strain tensor, thus

$$\frac{d}{dt}\left[\sigma_{\alpha\beta}(\mathbf{q},t) - \sigma_{\alpha\beta}^{int}(\mathbf{q},t)\right] + \frac{1}{\tau(\mathbf{q})}\left[\sigma_{\alpha\beta}(\mathbf{q},t) - \sigma_{\alpha\beta}^{int}(\mathbf{q},t)\right]$$
$$= C_{\alpha\beta\gamma\delta}\dot{\epsilon}_{\gamma\delta}. \tag{12}$$

In the absence of internal stresses, Eq. (12) reduces to the Maxwell constitutive relation.

To check that Eq. (12) leads to the required superposition of stresses for any frequency take the time Fourier transform of it leading to

$$\tilde{\sigma}_{\alpha\beta}(\mathbf{q},\omega) = \tilde{\sigma}_{\alpha\beta}^{int}(\mathbf{q},\omega) + \frac{i\omega\tau(\mathbf{q})}{1+i\omega\tau(\mathbf{q})} C_{\alpha\beta\gamma\delta}\tilde{\epsilon}_{\gamma\delta}(\mathbf{q},\omega). \tag{13}$$

For high frequencies, Eq. (13) gives us a solid-like stress superposition, i.e. $\sigma = \sigma^{int} + \sigma^{el}$. For low frequencies we obtain fluid-like behaviour in which the rheological stress cancels the elastic contribution yielding:

$$\tilde{\sigma}_{\alpha\beta} = \tilde{\sigma}_{\alpha\beta}^{int} + i\omega\eta_{\alpha\beta\gamma\delta}(\mathbf{q})\tilde{\epsilon}_{\alpha\beta}, \tag{14}$$

where we have used the obvious generalization of the definition of viscosity from Eq. (7). As we shall show in the the bare viscosity is renormalized due to a coupling with the orientational degrees of freedom in a fashion characteristic of the viscosity enhancement associated with the glass transformation.

4. ORIENTATION-CUM STRESS RELAXATION

We shall now investigate the coupling between the local orientational order and the remaining degrees of freedom using the generalized viscoelastic model from Section 3. At the outset we will be primarily interested in the slow relaxational motion in the system.

Solving Eq. (4) and substituting it into the definition of the internal stress we obtain:

$$\sigma^{int}(\mathbf{q},t) = \frac{\beta}{\eta_0} \int d\mathbf{q}' \Delta G(\mathbf{q}') \int d\mathbf{q}'' \int dt' \exp[-\lambda(\mathbf{q},\mathbf{q}',t-t')] \Delta G(\mathbf{q}'') \Psi(\mathbf{q}'',t')$$
$$\times [\Psi(\mathbf{q}-\mathbf{q}',t')\sigma(\mathbf{q}'-\mathbf{q}'',t') + \Psi(\mathbf{q}',t')\sigma(\mathbf{q}-\mathbf{q}'-\mathbf{q}'',t')], \tag{15}$$

with the initial condition $\sigma^{int}(\mathbf{q},t=0) = 0$. The choice of this particular initial condition will become clear shortly. In the above I have used the relation between bare viscosity and compliance tensor $1/\eta_0 = K/\tau = 1/\tau C$ and $\lambda(\mathbf{q},\mathbf{q}') = \nu(\mathbf{q}) + \nu(\mathbf{q}-\mathbf{q}')$.

Our aim now is to substitute Eq. (15) into Eq. (12) and consider the resulting equation, with an appropriate noise term added, as the Langevin equation for the stress tensor. For our purposes it is sufficient to calculate the averaged value of the stress tensor over the statistics of the noise. While carrying out this average (denoted by angular braces) we make a random phase-like decoupling:

$$\langle \Psi(\mathbf{q}',t')\Psi(\mathbf{q}'',t')\sigma(\mathbf{q}-\mathbf{q}'-\mathbf{q}'',t') \rangle = $$
$$\langle \Psi(\mathbf{q}',t')\Psi(\mathbf{q}'',t') \rangle \langle \sigma(\mathbf{q}-\mathbf{q}'-\mathbf{q}'',t') \rangle. \tag{16}$$

Since the underlying stochastic process is assumed stationary and the system is translationally invariant, we have:

$$\langle \Psi(\mathbf{q}',t')\Psi(\mathbf{q}'',t') \rangle = \langle \Psi(\mathbf{q}')\Psi(\mathbf{q}'') \rangle \delta(\mathbf{q}+\mathbf{q}'')$$
$$= \chi(\mathbf{q}')\delta(\mathbf{q}'+\mathbf{q}''), \tag{17}$$

where $\chi(\mathbf{q})$ is the "susceptibility" associated with the orientational order. Using now Eqs. (16) and (17), and substituting them into Eq. (15), we obtain:

$$\langle \sigma^{int}(\mathbf{q},t) \rangle = \beta \int_0^t dt' F(\mathbf{q},t-t') \langle \sigma(\mathbf{q},t') \rangle, \tag{18}$$

where the memory kernel $F(\mathbf{q},t)$ is given by:

$$F(\mathbf{q},t) = $$
$$\frac{1}{\eta_0} \int d\mathbf{q}' \chi(\mathbf{q}') \Delta G(-\mathbf{q}') \exp[-t\lambda(\mathbf{q},\mathbf{q}')] (\Delta G(\mathbf{q}') + \Delta G(\mathbf{q}-\mathbf{q}')). \tag{19}$$

Having this we may now solve Eq. (12) for the stress tensor using Laplace transforms. We obtain:

$$\langle \tilde{\sigma}(\mathbf{q}, z) \rangle =$$
$$\frac{1}{D(\mathbf{q}, z)} \frac{\tau_{\mathit{eff}}(\mathbf{q}, z)}{1 + z\tau_{\mathit{eff}}(\mathbf{q}, z)} \times$$
$$[zC\langle \tilde{\epsilon}(\mathbf{q}, z) \rangle - C\langle \epsilon(\mathbf{q}, t=0) \rangle + \langle \sigma(\mathbf{q}, t=0) \rangle], \tag{20}$$

where we define the renormalized Maxwell relaxation time $\tau_{\mathit{eff}}(\mathbf{q}, z)$ as

$$\tau_{\mathit{eff}}(\mathbf{q}, z) = \frac{\tau}{1 - \beta \tilde{F}(\mathbf{q}, z)}. \tag{21}$$

The memory function $\tilde{F}(\mathbf{q}, z)$ is the Laplace transform of expression Eq. (19) and

$$D(\mathbf{q}, z) = 1 - \beta \frac{z\tau_{\mathit{eff}}(\mathbf{q}, z)}{1 + z\tau_{\mathit{eff}}(\mathbf{q}, z)} \tilde{F}(\mathbf{q}, z). \tag{22}$$

Eq. (21) is the main result of this section, which shows that, within our mean field–like analysis, the generalised relaxation time diverges at the temperature T_g such that

$$1 - \beta \tilde{F}(\mathbf{q}, z) = 0. \tag{23}$$

The relaxation processes under consideration are clearly important for low frequencies; in that regime the memory function $\tilde{F}$ is essentially z independent. The Eq. (20) can now be analysed in the limits when $z\tau_{\mathit{eff}}(\mathbf{q}, z=0) \gg 1$ or $z\tau_{\mathit{eff}}(\mathbf{q}, z=0) \ll 1$. The former one applies when the temperature approaches the glass transition temperature defined by Eq. (23). In that case Eq. (20) describes the elastic response of the medium with the elastic coefficients enhanced by the factor $1/(1 - \beta \tilde{F})$. In the low frequency limit, Eq. (20) gives us the typical viscous stress-strain relation with the viscosity given by:

$$\eta(\mathbf{q}, z) = \frac{(C/\tau)}{1 - \beta \tilde{F}(\mathbf{q}, z)} = \frac{\eta_0}{1 - \beta \tilde{F}(\mathbf{q}, z)}. \tag{24}$$

We conclude therefore that the rheological constitutive relation together with the kinetic model for the orientational degrees of freedom provide the necessary input to construct a kinetic theory for the liquid-glass transition.

Eq. (23) defines the glass transition temperature T_g as the one at which the zero-frequency long wavelength component of the viscosity blows up. For systems with inversion symmetry ($\nu(\mathbf{q}) = \nu(-\mathbf{q})$), we obtain the expression for T_g in terms of the microscopic parameters of the model:

$$T_g = \frac{1}{Nk_B\eta_0\xi^2} \int dq \frac{\chi(\mathbf{q}, T_g)(\Delta G(\mathbf{q}))^2}{1 - \frac{T_0}{T_g} + \frac{T_0}{T_g}R^2\mathbf{q}^2}. \tag{25}$$

This result should now be compared with that obtained from the theory of the glass transition based on a spin-glass theory framework [17] as discussed in Sec. II. The structure of both expressions for T_g turns out to be very similar. Indeed Eq. (25) contains the square of the coupling coefficient between the orientational and elastic degrees of freedom which can be identified with the variance of the quenched fluctuations (cf. Ref. [17]). This analogy can be further substantiated by noting, that the assumed vanishing of the initial value of the internal stress tensor, viz. Eq. (21), can be related to the vanishing of $\Delta G(\mathbf{q})$, averaged in the sense of Ref. [17].

5. DYNAMIC MODES

To study the hydrodynamic modes of the system we begin by writing down the usual set of conservation laws for the density and momentum in wave vector space as:

$$\frac{\partial}{\partial t}\rho(\mathbf{q},t) = iq^\alpha j^\alpha(\mathbf{q},t)$$
$$\frac{\partial}{\partial t}j^\alpha(\mathbf{q},t) = iq^\beta \Pi^{\alpha\beta}(\mathbf{q},t), \tag{26}$$

where ρ is the fluid density, j^α particle current and $\Pi^{\alpha\beta}$ is the total stress tensor in the liquid containing also a hydrostatic pressure, $-p(\rho)\delta^{\alpha\beta}$, and kinematic $\sigma^{\alpha\beta}_{kin} = \rho j^\alpha j^\beta$ terms. In what follows we shall be interested in linearized hydrodynamics, therefore we shall neglect σ_{kin} (the so called convective terms). Furthermore, we shall now linearize equations Eq. (26) around $\rho = \rho_0 + \delta\rho$, ρ_0 being the uniform fluid density. Using $p(\rho) = p_0 + (\partial p/\partial \rho)_0 \delta\rho = p_0 + \kappa_T \delta\rho$ where κ_T is the isothermal compressibility, and including the stress tensor of Section 3 in $\Pi_{\alpha\beta}$, we obtain:

$$\frac{\partial}{\partial t}\delta\rho(\mathbf{q},t) = iq_\alpha j_\alpha(\mathbf{q},t),$$
$$\frac{\partial}{\partial t}j^\alpha(\mathbf{q},t) = iq^\beta\left\{\kappa_T \delta^{\alpha\beta}\delta\rho(\mathbf{q},t) + \sigma^{\alpha\beta}(\mathbf{q},t)\right\}. \tag{27}$$

Taking the Laplace transform with respect to the time variable, we have

$$z\tilde{j}^\alpha(\mathbf{q},z) + \frac{\kappa_T}{z}q^\alpha q^\beta \tilde{j}^\beta(\mathbf{q},z) - iq^\beta \tilde{\sigma}^{\alpha\beta}(\mathbf{q},z)$$
$$= j^\alpha(\mathbf{q},0) + i\kappa_T q^\alpha \delta\rho(\mathbf{q},0). \tag{28}$$

In order to analyse Eqs. (26) we shall use for $\sigma^{\alpha\beta}$ the rheological constitutive equation from Section 3, namely Eq. (12). We should emphasize that all the hydrodynamic quantities appearing in this Section are identified with the averaged values in the sense of Section 4. Now, using Eqs. (12),(20) and recalling that in the linearized theory

$$\dot{\epsilon}^{\alpha\beta}(\mathbf{q},t) = (1/2\rho_0)i(q^\alpha j^\beta + q^\beta j^\alpha), \tag{29}$$

we obtain the matrix equation for the current:

$$\mathcal{G}^{\alpha\beta}(\mathbf{q},z)\tilde{j}^\beta(\mathbf{q},z) = f^\alpha(\mathbf{q},z), \tag{30}$$

where

$$\mathcal{G}^{\alpha\beta}(\mathbf{q},z) = \left[z^2\delta^{\alpha\delta} + \kappa_T q^\alpha q^\delta + \frac{z\tau_{\text{eff}}(\mathbf{q},z)}{1+z\tau_{\text{eff}}(\mathbf{q},z)}\frac{C^{\alpha\beta\gamma\delta}q^\gamma q^\delta}{\rho_0 D(\mathbf{q},z)}\right] \tag{31}$$

and the right hand side of Eq. (30) contains the initial values of all the relevant quantities.

Eq. (30) is the central result of this section. The poles of the matrix $\mathcal{G}^{-1}$, $z = z(\mathbf{q})$ determine the hydrodynamic modes of the system. Note that $\tilde{j}_\alpha(\mathbf{q}, z)$ determines the relaxation of the macroscopic current, with an arbitrary initial condition. Hence by the Onsager regression hypothesis, the long time decay of the long wavelength component of the equilibrium current fluctuations, described by the correlation function $\tilde{\mathcal{J}}(\mathbf{q}, z)$, is also governed by the matrix $\mathcal{G}^{-1}$.

Consider first the high frequency regime discussed in Section 4. In that limit the matrix $\mathcal{G}$ simplifies and one obtains the usual expression for the dynamical matrix of an elastic medium with enhanced elastic coefficients. Indeed using the expression for the elastic coefficient tensor of an isotropic medium, one obtains expressions for the frequencies of the longitudinal and transverse waves:

$$\omega_\| = q\sqrt{\kappa_T + \frac{1}{\rho_0}\left(\frac{\lambda + 2\mu}{1 - \beta\tilde{F}}\right)},$$

$$\omega_\perp = q\sqrt{\frac{1}{\rho_0}\frac{\mu}{1 - \beta\tilde{F}}}. \tag{32}$$

Note the occurrence of the denominator $1 - \beta\tilde{F}$ in Eq. (32) indicating an increase of the rigidity of the supercooled liquid when the temperature approaches T_g.

In the opposite limit of low frequencies, the system dynamics becomes fluid-like with enhanced viscosity. For the interesting case of transverse waves we obtain:

$$\omega_\perp = -iq^2 \frac{\eta_0}{1 - \beta\tilde{F}}, \tag{33}$$

the imaginary i indicating diffusive character of these modes.

In a similar fashion we can write down expressions for the longitudinal and transverse current correlation functions, which appear to be essentially the same as in Ref. [37], provided the values of the elastic coefficients and the viscosities are modified by the correlations due to coupling to the local orientational degrees of freedom in the manner described above.

6. CONCLUSIONS

The mean field analysis of the combined viscoelasticity and local orientational order presented in the previous sections shows how the kinetic effects associated with the orientational ordering in a supercooled liquid lead to an enhancement of the fluid viscosity and the elastic moduli of the system as the glass transition is approached from above. The crucial role in the model is played by the renormalized Maxwell relaxation time $\tau_{\it eff}$ given by Eq. (21). The fact that $\tau_{\it eff}$ increases when one approaches the glass transition temperature implies that the viscous behaviour of the system would be observed only after waiting for a time much longer than in the "high" temperature fluid, for example near the triple point. The enhancement of the Maxwell time explains why one can use a crude model of a glass in which the translational degrees of freedom of the system are treated as in an elastic medium. Our results also show that the sluggishness of the system manifests itself in the increase of *both* the viscosity and elastic moduli as one approaches T_g.

The mean field analysis presented here suffers from several shortcomings. For example, it predicts a divergence of the relaxation time at the transition. A more refined analysis would remove this defect of the theory, though at the expense of considerable calculational complexity which, we believe, will not add much to the perceived physical picture. I emphasize that the present, fully dynamical model, correlates with results obtained by methods from the theory of spin glasses.

The important ingredient of our model is the concept of rheological stress, Eq. (10) and the rheological constitutive relation, Eq. (11). We believe that these concepts have a much wider application than envisaged in the present treatment. The rheological stress can also be used in analysing the properties of such systems for which a clear classification of different groups of degrees of freedom can be made based on the considerations of space-time and "internal" symmetries. The possible examples that immediately come to mind are compressible magnetic systems, polymers, deformable dielectrics, etc. in which our ideas can be tested.

ACKNOWLEDGMENTS

I wish to express my sincere appreciation for the invitation to participate in the first *KARPACZ WINTER SCHOOL* ever held in *KUDOWA*. Special thanks are due to Professor Tadeusz Paszkiewicz for providing us with this unique opportunity to exchange ideas, learn some physics and simultaneously relax. The financial help of the School sponsor is kindly acknowledged. This work was supported in part by Polish Science Council (KBN) Statuary Grant awarded to the Center for Theoretical Physics.

REFERENCES

[1] H. Logan, *How Much Do You Know About Glass?*, Dodd, New York 1951

[2] J. Kramer, Ann.Phys. (Paris) **19**, 37 (1934)

[3] W. Buckel, R. Hilsch, Z. Phys.**138**, 109 (1954)

[4] W. I. Johnson, Y. T. Cheng, and M. Van Rossum, and M. A. Nicolet, Nucl. Instrument Methods **7/8**, 657 (1986)

[5] A. Pabst, Am. Mineral, **37**, 137 (1952)

[6] R. B. Schwarz, C. C. Kok, Appl. Phys. Lett. (1988)

[7] J. Jäckle, Prog. Theor. Phys. **49**, 171 (1986)

[8] K. Binder and A. P. Young, Rev. Mod. Phys. **58**, (1986)

[9] G. Tammann, *Der Glasszustand*. L. Voss, Leipzig (1933); J. Frenkel, *Kinetic Theory of Liquids*, Dover, New York (1955); J. D. Bernal, in *Structure and Properties of Liquids* (Edited by T. J. Hughel). Elsevier, London (1965)

[10] L. Bosio and C. G. Windsor, Phys. Rev. Lett. **35**, 1652 (1975)

[11] A. Rahman, M. J. Mandell, and J. P. McTauge, J. Chem. Phys. **64**, 1564 (1976)

[12] H. Vogel, Z. Phys. **22**, 645 (1921)

[13] A. Sjölander and L. A. Turski, J.Phys.**C11**, 1973 (1978)

[14] K.H. Fisher and J.A. Hertz, *Spin Glasses*, Cambridge University Press, Cambridge, 1990

[15] D. Elderfield and D. Sherrington, J. Phys. **C 16**, L497, L971, L1169 (1983)

[16] D.J. Gross, I. Kanter and H. Sompolinsky, Phys. Rev. Lett. **55**, 304 (1985)

[17] R. Kree, L.A. Turski and A. Zippelius, Phys. Rev. Lett. **58**, 1656 (1987)

[18] D. Thirumalai and T.R. Kirkpatrick, Phys. Rev. **B37**, 5342 (1988)

[19] T.R. Kirkpatrick and P.G. Wolynes, Phys. Rev. **B36**, 8552 (1987)

[20] R.D. Mountain and D. Thirumalai, Jour. Chem. Phys. **93**, 6975 (1989)

[21] D. Thirumalai, R.D. Mountain and T.R. Kirkpatrick, Phys. Rev. **A39**, 3536 (1989) J. of. Chem. Phys. **93**, 6975 (1989)

[22] R.D. Mountain and D. Thirumalai, Int. Jour. of Mod. Phys. **C1**, 77 (1990)

[23] R.D. Mountain and D. Thirumalai, Phys. Rev. **E**, (1993) in press

[24] T.V. Ramakrishnan and M. Yussouff, Phys. Rev. **B19**, 2775 (1979); A.D.J. Haymet, Ann. Rep. Phys. Chem. **38**, 89 (1987)

[25] D.W. Oxtoby, in *Liquids, Freezing and Glass Transition* Les Houches 1989, edited by J.P. Hansen, D. Levesque *et.* J. Zinn-Justin, North Holland, Amsterdam (1991)

[26] E. Leutheusser, Phys. Rev. **A29**, 2765 (1984); U. Bengtzelius, W. Götze and A. Sjölander, J. Phys. **C17**, 5915 (1984); for recent updates, see W. Götze in *Proceedings of Discussion Meeting on Relaxation in Complex Systems*, Crete 1990, Jour. Non-Crystalline Solids **131-133**, (1992)

[27] G.F. Mazenko, in *Proceedings of Discussion Meeting on Relaxation in Complex Systems*, Crete 1990, Jour. Non-Crystalline Solids **131-133**, (1992)

[28] S. Hess, Physica (Amsterdam) **127 A**, 509 (1984); C.A. Angell, J.H.R. Clarke and L.V. Woodcock, Adv. Chem. Phys. **48**, 398 (1981); R.D. Mountain and P.K. Basu, Phys. Rev. **A28**, 370 (1983)

[29] M. Kimura and F. Yonezawa, J. of Non-Crystalline Solids **61-62**, 535 (1984); S. Nose and F. Yonezawa, J. Chem. Phys. **84**, 1803 (1986); D.K. Remler and A.D.J. Haymet, Phys. Rev. **B35**, 245 (1987); H. Jonsson and H.C. Andersen, Phys. Rev. Lett. **60**, 2295 (1988)

[30] P.J. Steinhard, D.R. Nelson and M. Ronchetti, Phys. Rev. Lett. **47**, 1297 (1981)

[31] S. Hess, Z. Naturforsch. **35a**, 69 (1979)

[32] A.S. Mitus and A.Z. Patashinski, Zh. Eksp. Teor. Fiz. **80**, 1554 (1981) (Soviet Phys. JETP **53**(4), 798 (1981)); A.D.J. Haymet, Phys. Rev. **B27**, 1725 (1983)

[33] P.A. Egelstaff, *An Introduction to the Liquid State*, Academic Press, London (1967)

[34] H.R. Schober and B.B. Laird, Phys. Rev. Lett. **66**, 636 (1991); Phys. Rev. **B 44**, 6746 (1991)

[35] S. Dattagupta and L.A. Turski, Phys. Rev. Lett. **54**, 2359 (1985); referred to in the text as I

[36] S. Dattagupta and L.A. Turski, Phys. Rev. **E** (1993) in press

[37] J.P. Boon and S. Yip, *Molecular Hydrodynamics*, McGraw Hill, New York, (1980)

[38] L.A. Turski, Acta Phys. Pol. **A75**, 111 (1989); L.A. Turski, Physica Scripta **T13**, 259 (1986)

[39] T.R. Kirkpatrick and D. Thirumalai, J. Phys. **A 22**, L149 (1989)

[40] See for instance, L. Gunther, D.J. Bergmann and Y. Imry, Phys. Rev. Lett. **27**, 558 (1971)

[41] L.D. Landau and E.M. Lifshitz, *Theory of Elasticity*, Pergamon Press, Oxford, (1959)

[42] J.-P. Hansen and I. Mac Donald, *The Liquid State*

[43] *Spin Glass Theory and Beyond*, edited by M. Mezard, G. Parisi and M.A. Virasaro, World Scientific, Singapore (1987)

[44] K.Kawasaki, in *Phase Transitions and Critical Phenomena*, Vol. **2**, edited by C. Domb and M.S. Green, Academic Press, New York, (1972)

[45] M. Załuska-Kotur and L.A. Turski, J. of Phys. **A22**, (Math. Gen), 413 (1989)

[46] S.K. Ma, *Modern Theory of Critical Phenomena*, (W.A.Benjamin, Reading, 1976)

[47] S. Dattagupta, *Relaxation Phenomena in Condensed Matter Physics*, (Academic Press, New York 1987)

[48] S. Dattagupta, V.Heine, S. Marais, and E. Salje, J. Phys. Cond. Matt **3**, 2975 (1991)

[49] L.D. Landau and E.M. Lifshitz, Sov. Phys. JETP **32**, 618 (1957)

[50] A.G. Frederickson, *Principles and Applications of Rheology*, Prentice-Hall, Englewood Cliffs, NJ (1964)

[51] V.L. Gurevich and A. Thellung, Phys. Rev. **B42**, 7345 (1990)

THE SURFACE IMPEDANCE AND THE SLAB CONDUCTIVITY OF METALS BEYOND THE RELAXATION TIME APPROXIMATION

Jerzy Czerwonko,[1] Moisei I. Kaganov,[2] and Grigorii Ya. Lyubarskii[3]

[1] Institute of Physics, Technical University of Wrocław, 50-370 Wrocław, Poland
[2] P. L. Kapitza Institute for Physical Problems, Moscow, Russia
[3] Ukrainian Physico-Technical Institute, Kharkov, Ukraine

1. INTRODUCTION

Kinetic theory of the skin effect for the electrons obeying an isotropic dispersion law has been developed by Reuter and Sondheimer.[1] They applied the so-called τ-approximation (or relaxation time approximation), taking into account only the loss term in the electron scattering integral, $-f_1/\tau$. Here f_1 denotes the deviation of the phase-space electron distribution function from its local equilibrium value.* The electron gas is assumed to be a degenerate one, the derivative of the Fermi function $-\partial f_\varepsilon/\partial\varepsilon = \delta(\varepsilon - \varepsilon_F)$.

The τ-approximation is valid in the limiting cases of the normal and highly anomalous skin effect:

i) The surface impedance and the penetration depth of the electric field into the metal, in the regime of the n o r m a l skin effect, $l \ll \delta$, are both expressed via the static conductivity

$$\sigma_{tr} = ne^2 l_{tr}/p_F, \tag{1.1}$$

where the last quantity is determined by the transport mean free path, l_{tr}, in the following way:

$$1/l_{tr} = \int d\Omega W(\cos\theta)(1-\cos\theta); \quad 1/l = \int d\Omega W(\cos\theta). \tag{1.2}$$

The function $W(\cos\theta)$ is the probability density of the electron scattering on element of the solid angle $d\Omega = \sin\theta d\theta d\varphi$ per unit path. Formulae (1.1), (1.2) are equivalent to the τ-approximation provided that the scattering integral is replaced by $-f_1/\tau_{tr}$, $\tau_{tr} \equiv l_{tr}/v_F$. This replacement, indeed, is valid only within the n o r m a l skin effect regime in the isotropic model, when the angular dependence of the electron distribution function is determined solely by the direction of the electric field $E(f_1 \sim E\mathbf{v} = Ev_F\cos\varphi\sin\theta$, cf. below).

ii) Within the regime of the highly a n o m a l o u s skin effect ($l \gg \delta$) the mean free path does not appear in the expression for the surface impedance, ζ, and the integral (gain) term of

---

* The following symbols will be used: p, $\mathbf{v} \equiv p/m^*$, ε, m^*, e denoting the momentum, velocity, energy, effective mass and the charge of electron, respectively. The quantity $\sigma \equiv ne^2 l/p_F$ is the electron conductivity with n being their density; l is the electron mean free path whereas $\tau \equiv l/v_F$ is the average time between consecutive electron scatterings. The quantity $\delta = c/(2\pi\sigma\omega)^{1/2}$ is the "normal" skin depth with ω and c being the frequency of the electromagnetic wave and the velocity of light, respectively.

Die Kunst of Phonons, Edited by T. Paszkiewicz and K. Rapcewicz, Plenum Press, New York, 1994

the impurity scattering operator becomes irrelevant for the solution of the kinetic equation. This statement is justified by the fact that at $l \gg \delta$ the surface impedance is determined only by such electrons on the Fermi surface that move parallelly to the metal surface $z = 0$; the remaining electrons do not contribute to this quantity. Formally, this is connected with the singularity of the function f_1 at $v_z = 0$. In the gain term of the scattering operator, the function f_1 appears only as an integrand. Hence, the singularity is smeared out as compared to the unintegrated singularity of f_1 in the loss part of the scattering operator.

The theory of the anomalous skin effect is equivalent to the solution of the Maxwell equations. The last problem is possible provided that we know the specific constituent relations expressing the current density j in terms of the electric field E:

$$j_i(z) = \sum_{k=1}^{3} \int_0^\infty K_{ik}(z,z') E_k(z') dz'.$$

The conductivity operator, $K_{ik}(z, z')$, should be obtained from the solution of the kinetic equation for the distribution function of electrons. For an inhomogeneous electric field, $E(z)$, and the electron interaction with the sample surface taken into account, the solution of the integrodifferential kinetic equation becomes possible only with simplifying assumptions for both the scattering integral and the boundary conditions. Following Reuter and Sondheimer,[1] in the final results, we will restrict ourselves to purely specular or diffuse boundary conditions on the metal surface but in contrast to[1] we will consider also the gain part of the scattering integral. Note that the formulae[1] obtained for arbitrary values of the parameter l/δ are exact for $W(\theta) = $ const, only.

The unimportant role of the gain term in the highly anomalous skin effect ($l \gg \delta$) made possible the generalization of the theory in various directions. First of all, was generalized the dispersion law for the electrons, the character of the surface electron reflection was explained and a dc magnetic field was introduced (cf. the monographs [2,3] and the review article[4]).

The formulae for the impedance ζ and other metal characteristics valid, according to the statements of their authors, also for arbitrary values of l/δ were obtained, cf. e.g.[5] As a rule, these formulae were based on the τ-approximation and, hence, they cannot be treated as exact ones.

In a few papers known to us,[6-9] an attempt was made to determine the temperature correction to the surface impedance for the highly anomalous skin effect. In these papers, the electron–phonon scattering for specular[6] and diffuse[7] metal surfaces has been taken into account. Moreover, the role of the interelectron Umklapp-process[8] for the specular or diffuse metal surfaces, as well as the electron scattering on the local oscillations in the specular case[9] have also been discussed. In all these papers, impurity scattering was treated as the predominant one and the scattering integrals corresponding to the quoted mechanisms of electron scattering were treated in a perturbative way. Moreover, impurity scattering was taken in the τ-approximation although the scattering integrals being the perturbations[6-9] were not taken in a such manner.

The problem of the skin effect in metals is one of the fundamental problems of physical kinetics. The exact and consequent solution of such a problem is interesting even if it is obtained under simplifying model assumptions. For an arbitrary value of l/δ, the τ-approximation can be treated as exact only if the function $W(\cos\theta)$ becomes angle-independent.

We also introduce briefly another topic of the paper, namely, the conductivity of metallic slabs. This problem has been solved in τ-approximation by Fuchs,[10] with an arbitrary angle-independent specularity coefficient of the border. One of the authors of the present paper considered already the question out of τ-approximation, for a quite arbitrary function $W(\cos\theta)$ expandable in Legendre polynomials.[11,12] He has shown the equivalence of such a question to the solution of system of nonhomogeneous integral equations. This system can be effectively solved via the Fredholm iteration procedure for thin slabs, $D/l \ll 1$, where D is the slab thickness.[11,12] In reference[12] we find the proof that the asymptotic form of the Fuchs solution for $D/l \ll 1$ is not correct unless we deal with purely diffuse border scattering.

Unfortunately, attempts to solve the problem for a thick slab $l/D \ll 1$ were unsuccessful for a quite general function $W(\cos\theta) > 0$ expandable in Legendre polynomials.[11] This is connected with the necessity of generalizing the Wiener-Hopf procedure[13] to systems of integral equations. We were able to solve the question only for almost isotropic impurity scattering $|W(\cos\theta) - W_0| \ll W_0$, where W_0 denotes the function $W(\cos\theta)$ averaged over values of $\cos\theta$ in the interval $(-1, 1)$, for diffuse surfaces.[11] We remark that the extrapolation of the above procedure[11] to partially diffuse surfaces leads to the solutions being functions of the specularity coefficient not expressable as elementary or well-known special functions. This is the reason for which we will restrict ourselves in considering thick slabs to diffuse surfaces. Moreover, we will assume the simplest function $W(\cos\theta) = W_0(1 + G_1 \cos\theta)$, $|G_1| \leqslant 1$, allowing to distinguish between the mean free path (m. f. p.) and the transport mean free path. In fact, more complicated scattering kernels lead to the necessity to generalize the Wiener-Hopf procedure.[13] The considerations of partially diffuse surfaces lead to system of integral equations intractable for this procedure because its kernels do not depend solely on the difference of coordinates. Let us add that because the average conductivity should be an increasing function of the surface specularity thus our results can serve as an estimation of the maximal influence of the surface on the conductivity of thick slabs for a simple $W(\cos\theta)$. The above discussed results were obtained in our paper.[14]

Let us also describe our results for the surface impedance of a semi-infinite metal,[15] where the problem with diffuse and specular surfaces has been solved, for the function $W(\cos\theta)$ taken as in.[14] Note that such a choice of this function leads to the natural expression of the problem in terms of the electric field $E(z)$ and the current density $j(z)$ for both cases: the slab conductivity and the surface impedance. It has been shown[15] that for diffuse surfaces the generalization of the Wiener-Hopf procedure[13] has to be applied to the last case. Hence, the attempt to enlarge the class of the functions $W(\cos\theta)$, in comparison to the simple function used in,[14,15] rather does not lead to effective values of the surface impedance for diffuse surfaces. On the other hand, as we will show next, a similar procedure is manageable for specular surfaces. It will be also shown that the differential transport equation, fulfilling the boundary condition for the angle-dependent specularity, is equivalent to a system of inhomogeneous integral equations with all $W(\cos\theta)$ expandable in Legendre polynomials. In such a way, for the surface impedance, we will carry out such an analysis as it has been done by one of the authors for the conductivity.[11,12] It is worth noting that it is not a purely academic problem. It can be shown,[16] that for almost glancing electrons, $|\cos\theta| \ll 1$ where θ is now the angle of incidence, the specularity coefficient behaves as $1 - \gamma\cos\theta + \ldots$, for prepared "good" surfaces. This behaviour allows for the appearance of the surface magnetic levels observed in metals; for the evidence,[17] for the theory.[18]

In all the calculations performed in references,[14,15] the $\omega\tau$ term has been neglected. Sometimes a partial prescription of how to include such effects has been presented. Taking into account the time variation of the electric field one should introduce also Landau parameters for the electron liquid.[2,3] In the transport equation, the suitable procedure by means of the forward scattering amplitude of quasi-particles[19] has been performed in.[11,12] Moreover, a general recipe has been found of how, in the conductivity problem, to express the solution for $\omega \neq 0$ via the solution for $\omega = 0$. There, we have also the possibility to switch on a non-quantizing dc magnetic field perpendicular to the metal surface.[20] On the other hand, if we deal with the normal skin regime, i.e., for $l \ll \delta$ then, from the relations $\delta = c/(2\pi\sigma\omega)^{1/2}$, $\sigma = ne^2\tau/m^*$ and the estimation $e^2 \sim \hbar v_F,$[3] one finds that $(c/v_F)(a_0/l) \gg (\omega\tau)^{1/2}$ where a_0 is the interatomic distance. The quantity c/v_F is of the order of 100 for metals. Taking into account the values (a_0/l) for more or less pure metals at low temperature, when impurity scattering prevails, we find that it is extremely hard to attain such $\omega\tau$ that are not negligibly small in the normal skin regime. Let us add that the conductivity appears in this regime, because of the necessity of the current homogeneity in the direction along the slab. This is the reason why the problems of the conductivity and the surface impedance in the normal skin regime will be considered in the quasi-static case ($\omega\tau = 0$).

2. TRANSFORMATION OF THE BASIC EQUATIONS FOR THE SURFACE IMPEDANCE

Let us assume, according to the previous Section, that the electron scattering is elastic and that the density of probability W depends only on the angle between the electron momenta before and after the scattering. Hence, the scattering integral has the form:

$$\hat{I}\{f_1\} = -\int d^3 p' W(\mathbf{nn'})[f_1(\mathbf{p}) - f_1(\mathbf{p'})] \delta(\varepsilon_p - \varepsilon_{p'})(p_F m^*)^{-1}, \tag{2.1}$$

cf. (1.2). Here $\mathbf{n} \equiv \mathbf{p}/p = [\sin\theta\cos\varphi, \sin\theta\sin\varphi, \cos\theta]$, $\mathbf{n}' \equiv \mathbf{p}'/p'$. Let us represent

$$W(\mathbf{n}\mathbf{n}') \equiv W_0[1 + \sum_{l=1} G_l P_l(\mathbf{n}\mathbf{n}')] \geq 0, \tag{2.2}$$

where $P_l(\cos\theta)$ denotes the l-the Legendre polynomial. In this case $1/l = 4\pi W_0$ and $1/l_{tr} = 1 - G_1/3$. It is easy to show that the inequalities $|G_l| \leq 2l+1$ are necessary conditions for the inequality in the formula (2.2). Assume, as usual that the electromagnetic wave of frequency ω falls normally onto the metal surface $z = 0$. Choose the electric field of the wave parallel to the x-axis. Then we find, for symmetry reasons, that the solution for the deviation of the electron distribution function from its local equilibrium value can be chosen as follows:

$$f_1(z, \mathbf{p}) = (-\partial f_\varepsilon/\partial\varepsilon)\tau e v_F \chi(z, \cos\theta)\sin\theta\cos\varphi, \tag{2.3}$$

cf., e.g., refs.[11,12] For the degenerate electron liquids, the only nonvanishing x-th component of the current density has the form,

$$j(z) = [2ev_F/(2\pi\hbar)^3]\int d^3p\cos\varphi\sin\theta f_1(z, \mathbf{p}) =$$
$$= (3\sigma/4)\int_0^\pi d\theta\sin^3\theta\chi(z,\cos\theta) = (3\sigma/2)\langle(1-c^2)\chi(z,c)\rangle_c, \tag{2.4}$$

where for simplicity, $\cos\theta$ and $\cos\theta'$ are denoted by c and c', etc. and the symbol $\langle\ldots\rangle_c$ stands for the average over c on the interval $(-1, 1)$.

Let us modify slightly the applied method,[11,12] to obtain the transport equation in the Landau form, by omitting the inhomogeneous electric field $E(z)$ as a factor in the unknown function χ, cf., e.g., (2.3). Writing the equation for the time Fourier transforms of the functions χ and E, taking into account the relation between deviations from local equilibrium value and the equilibrium value of the electron distribution function[19] one can write

$$c\chi'(z,c) + a\chi(z,c) = E(z) + \langle G_\omega(c,c')\chi(z,c')(1-c'^2)\rangle_{c'}. \tag{2.5}$$

Here, the prime denotes the derivative over the dimensionless z-coordinate taken in the units of the mean free path (m.f.p.) $l = v_F\tau$, $a \equiv 1 + i\omega\tau$ and the function G_ω is defined by:

$$G_\omega(c,c') = \sum_{l=1} G_{\omega l}P'_l(c)P'_l(c'), \quad G_{\omega l} \equiv (G_l + i\omega\tau A_l^s)/l(l+1). \tag{2.6}$$

In eq. (2.6), the Legendre amplitudes G_l are determined by the function describing the impurity scattering probability, eq. (2.2), whereas A_l^s are dimensionless partial wave amplitudes of the spin symmetric forward scattering amplitude of quasiparticles, in the conventional form.[19] Note that the primes over the Legendre polynomials denote differentiation by their arguments and that the expression of A_l^s in terms of the Landau parameters F_l^s has the form:[19] $A_l^s = (2l+1)F_l^s/(2l+1+F_l^s)$.

Let us impose boundary conditions on the solution of eq. (2.5), similar to what was done.[11,12] Introduce first the following representation of the function $\chi(z, c)$:

$$\chi(z, c) = [A(z, c)\exp(-za/c) + E(z)]/(a - 2G_{\omega 1}/3). \tag{2.7}$$

Substituting (2.7) into eq. (2.5), we find:

$$cA'(z,c)\exp(-za/c) + cE'(z) = \sum_{l=1} G_{\omega l}P'_l(c)N_l(z), \tag{2.8}$$

where:

$$N_l(z) = \langle(1-c^2)P'_l(c)A(z,c)\exp(-za/c)\rangle_c. \tag{2.9}$$

In turn, substituting the representation (2.7) into the definition of the current density (2.4), we get:

$$j(z) = \sigma_\omega[(3/2)N_1(z) + E(z)], \quad \sigma_\omega \equiv \sigma/(a - 2G_{\omega 1}/3). \tag{2.10}$$

Hence, we can interpret $N_1(z)$ as two-thirds of the surface contribution to the current density, in units of the dynamical and homogeneous conductivity σ_ω.

Integrating the equation (2.8) we obtain:

$$A(z,c) = -E(z)\exp(za/c) + (a/c)\int_0^z dx\, E(x)\exp(ax/c) + f(c)$$
$$+ \sum_{l=1} G_{\omega l} c^{-1} P'_l(c) \int_0^z dx\, N_l(x)\exp(ax/c). \tag{2.11}$$

Impose the boundary conditions for $\chi(z,c)$ when $z \to +\infty$, assuming that the half-space $z > 0$ is filled by the metal. This condition, because of eqs. (2.7) and (2.11) leads to some restriction on $f(c)$ but only for $c < 0$. From the condition $\lim_{z\to\infty}[A(z,c) + E(z)\exp(az/c)] = 0$ we obtain

$$f(c) = -(a/c)\int_0^\infty dx\, E(x)\exp(ax/c) - \sum_{l=1} G_{\omega l} c^{-1} P'_l(c) \int_0^\infty dx\, N_l(x)\exp(ax/c), \quad c < 0. \tag{2.12}$$

Hence, from the eq. (2.11)

$$A(z,c) = -(a/c)\int_z^\infty dx\, E(x)\exp(ax/c) - E(z)\exp(az/c)$$
$$- \sum_{l=1} G_{\omega l} c^{-1} P'_l(c) \int_z^\infty dx\, N(x)\exp(ax/c), \quad c < 0. \tag{2.13}$$

The boundary conditions on the metal surface, $z = 0$, depend on the structure of the surface. The methods of how to obtain the boundary conditions have been the subject of many papers (cf., e.g., the work[20] and the references therein). As our main purpose is to include the gain term of the scattering integral into the transport equation, we will restrict ourselves to incident angle-dependent Fuchs boundary conditions.[10,16,21] These conditions can be written as follows: $\chi(0,c) = \varepsilon(c)\chi(0,-c)$, $c > 0$, where $\varepsilon(c)$ is the angle-dependent specularity coefficient. For $\varepsilon = 0$ and $\varepsilon = 1$ we have, respectively, diffuse and specular surfaces. From eqs. (2.7), (2.11), (2.13) and from the above boundary conditions we find:

$$f(c) = (\varepsilon(c)a/c)\int_0^\infty dx\, E(x)\exp(-ax/c)$$
$$+ \varepsilon(c)\sum_{l=1} G_{\omega l}(-1)^{l-1} P'_l(c) c^{-1} \int_0^\infty dx\, N_l(x)\exp(-ax/c), \quad c > 0. \tag{2.14}$$

This formula, together with (2.11) and (2.13), establishes unambiguously the function $A(z,c)$ for both signs of variable c. Substituting now $A(z,c)$ into the definition of the function $N_l(z)$, given by eq. (2.9), after rather long but simple transformations we obtain the following system of inhomogeneous integral equations:

$$N_k(z) = -(2/3)E(z)\delta_{k1} + a\int_0^\infty dx\,[W^\varepsilon_{k1}(a(z+x)) + W_{k1}(a|z-x|)\mathrm{sign}(z-x)^{k-1}]E(x)$$
$$+ \sum_{l=1} G_{\omega l} \int_0^\infty dx\,[W^\varepsilon_{kl}(a(z+x))(-1)^{l-1} + W_{kl}(a|z-x|)\mathrm{sign}(z-x)^{k-l}]N_l(x), \tag{2.15}$$

where the superscript ε denotes the functional dependence on the function $\varepsilon(c)$. We have

$$W^\varepsilon_{kl}(q) = (1/2)\int_0^1 dc\,\varepsilon(c)(1-c^2)P'_k(c)P'_l(c)c^{-1}\exp(-q/c), \quad \mathrm{Re}\, q > 0, \tag{2.16}$$

and

$$W_{kl}(q) \equiv W^{\varepsilon=1}_{kl}(q), \quad W^{\varepsilon=0}_{kl}(q) = 0.$$

In order to solve the equations (2.15) of our problem, let us apply a consequence of the Maxwell equations. If the displacement current can be neglected, which is the case of practical interest beyond the metallooptic, we have:

$$(d^2 E(z)/dz^2) + (4\pi i\omega/c^2)j(z) = 0 \tag{2.17}$$

or

$$E''(z) + 2i(l/\delta)^2 j(z)/\sigma = 0, \tag{2.17a}$$

where we introduced the "normal" skin depth δ (cf. the footnote of this paper). The prime denotes the differentiation with respect to the dimensionless z-variable taken in the units of the mean free path. Exploiting eqs. (2.10) and (2.17a) it is easy to rewrite the system of equations (2.15) for the functions $E''(z)$, $E(z)$ and $N_l(z)$ at $l > 1$. We have:

$$E''(z) = \int_0^\infty dx [W^e_{11}(a(z+x)) + W_{11}(a|z-x|)][G_{\omega 1} E''(x) - 3i(l/\delta)^2 E(x)]$$

$$+ \sum_{l=2} G_{\omega l} \int_0^\infty dx [(-1)^{l-1} W^e_{1l}(a(z+x)) + W_{1l}(a|z-x|) \operatorname{sign}(z-x)^{l-1}] M_l(x), \tag{2.18}$$

$$M_k(z) = \int_0^\infty dx [W^e_{1k}(a(z+x)) + W_{1k}(a|z-x|) \operatorname{sign}(z-x)^{k-1}][G_{\omega 1} E''(x) - 3i(l/\delta)^2 E(x)] +$$

$$+ \sum_{l=2} G_{\omega l} \int_0^\infty dx [(-1)^{l-1} W^e_{kl}(a(z+x)) + W_{kl}(a|z-x|) \operatorname{sign}(z-x)^{k-l}] M_l(x), \quad k > 1, \tag{2.19}$$

where $M_l(z) \equiv -3i(l/\delta)^2 (a - 2G_{\omega 1}/3)^{-1} N_l(z)$. In the next two sections, the specular and diffuse boundary conditions will be considered. For specular surfaces, we will generalize our previous results.[14]

3. THE SURFACE IMPEDANCE FOR THE SPECULAR SURFACE

In the relaxation time approximation, i.e., for all amplitudes $G_{\omega l}$ vanishing, one can replace e.g., (2.18) for $\varepsilon = 1$ by the integral equation with the convolution kernel at $-\infty < z < \infty$ via the symmetric continuation of the function $E(z)$. In this case, such an equation can be solved and the surface impedance can be found via the Fourier transformation.[2,3] Now, if $G_{\omega,2k} \neq 0$ then the antisymmetric continuation of the function $M_{2k}(z)$ should be done. It is easy to see that the functions $M_l(z)$ should be symmetric for odd l and antisymmetric for even l. The resulting system of equations, analogous to eqs. (2.18), (2.19) can be written as follows:

$$E''(z) = \int_{-\infty}^{\infty} dx\, W_{11}(a|z-x|)[G_{\omega 1} E''(x) - 3i(l/\delta)^2 E(x)]$$

$$+ \sum_{l=2} G_{\omega l} \int_{-\infty}^{\infty} dx\, W_{1l}(a|z-x|) \operatorname{sign}(z-x)^{l-1} M_l(x), \tag{3.1}$$

$$M_k(z) = \int_{-\infty}^{+\infty} dx\, W_{1k}(a|z-x|) \operatorname{sign}(z-x)^{k-1} [G_{\omega 1} E''(x) - 3i(l/\delta)^2 E(x)] +$$

$$+ \sum_{l=2} G_{\omega l} \int_{-\infty}^{+\infty} dx\, W_{kl}(a|z-x|) \operatorname{sign}(z-x)^{k-l} M_l(x). \tag{3.2}$$

The system of equations (3.1), (3.2) can be solved via the Fourier transformation. Taking into account that the symmetric continuation of the electric field leads to the jump of its derivative at $z = 0$ one finds,[2,3]

$$-2E'(+0)[1 - G_{\omega 1} \tilde{W}_{11}(\varkappa)] = \tilde{E}(\varkappa) \{\varkappa^2 [1 - G_{\omega 1} \tilde{W}_{11}(\varkappa)] - 3i(l/\delta)^2 \tilde{W}_{11}(\varkappa)\} +$$

$$+ \sum_{l=2} G_{\omega l} \tilde{W}_{1l}(\varkappa) \tilde{M}_l(\varkappa), \tag{3.3}$$

$$\tilde{M}_k(\varkappa) - \sum_{l=2} G_{\omega l} \tilde{W}_{lk}(\varkappa) \tilde{M}_l(\varkappa) = - \tilde{W}_{1k}(\varkappa)\{2E'(+0)G_{\omega 1} + [\varkappa^2 G_{\omega 1} + 3i(l/\delta)^2]\tilde{E}(\varkappa)\}, \tag{3.4}$$

where:

$$\begin{bmatrix} \tilde{M}_l(\varkappa) \\ \tilde{E}(\varkappa) \end{bmatrix} = \int_{-\infty}^{\infty} dz\, e^{-i\varkappa z} \begin{bmatrix} M_l(z) \\ E(z) \end{bmatrix}, \quad \tilde{W}_{lk}(\varkappa) \equiv \int_{-\infty}^{+\infty} dz\, e^{-i\varkappa z} W_{lk}(a|z|) \operatorname{sign} z^{l-k}. \tag{3.5}$$

Let us recall that z is dimensionless taken in the units of m.f.p. — l. Hence, the variable $\varkappa$ is dimensionless too being taken in the units l^{-1}. Note that in the proof of eqs (3.3), (3.5) the following formula:

$$\int_{-\infty}^{+\infty} dz\, e^{-i\varkappa z} E''(z) = -2E'(+0) - \varkappa^2 \tilde{E}(\varkappa)$$

has been applied, being obtained by integration by parts, *cf.*, the phrase before eq. (3.3). From eq. (2.16) one finds the integral representation of the functions $\tilde{W}_{kl}(\varkappa)$. We have:

$$\tilde{W}_{kl}(\varkappa) = \int_0^1 dc \begin{bmatrix} a \\ -i\varkappa c \end{bmatrix} (1-c^2) P'_k(c) P'_l(c)/(a^2+c^2), \tag{3.6}$$

with the factor a taken when the parities of l and k agree and the factor $-i\varkappa c$ taken when they disagree. The integrals (3.6) can be expressed via the integrals of the functions $c^n/(a^2+\varkappa^2 c^2)$, $n = 0, 1, 2, \ldots$ obtained in an elementary way.

Let us define the quantities $A_k(\varkappa)$, $k > 1$, as the solutions of the system of equations

$$A_k(\varkappa) - \sum_{l=2} G_{\omega l} \tilde{W}_{lk}(\varkappa) A_l(\varkappa) = \tilde{W}_{1k}(\varkappa). \tag{3.7}$$

Comparison of eq. (3.7) with eq. (3.4) shows that

$$\tilde{M}_k(\varkappa) = -\{2E'(+0)G_{\omega 1} + [\varkappa^2 G_{\omega 1} + 3i(l/\delta)^2]\tilde{E}(\varkappa)\} A_k(\varkappa).$$

Subtituting $\tilde{M}_k$ in the above form into eq. (3.3) one can write

$$\tilde{E}(\varkappa) = -2E'(+0)/[\varkappa^2 - 2i(l/\delta)^2 \mu(\varkappa)], \tag{3.8}$$

where

$$\mu(\varkappa) = (3/2) K(\varkappa)/[1 - G_{\omega 1} K(\varkappa)]$$

with

$$K(\varkappa) = \tilde{W}_{11}(\varkappa) + \sum_{l=2} G_{\omega l} \tilde{W}_{1l}(\varkappa) A_l(\varkappa). \tag{3.9}$$

Using the inverse Fourier transform it is possible to obtain an explicit expression for the field $E(z)$ and, in particular, for $E(0)$. Taking into account that $H(0) = (c/i\omega l) E'(+0)$ we find the following expression for the surface impedance:

$$\zeta_{\varepsilon=1} = (\omega l/\pi i c) \int_{-\infty}^{+\infty} d\varkappa/[\varkappa^2 - 2i(l/\delta)^2 \mu(\varkappa)]. \tag{3.10}$$

Note that the expressions for $\tilde{E}(\varkappa)$ and ζ, eqs. (3.8) and (3.10), coincide with those obtained in reference,[14] whereas the expression (3.9) coincides only if $G_{\omega l} = 0$ for $l \geqslant 2$.

In the case of almost normal skin effect, it is sufficient to consider the static case, $a = 1$ and $G_{\omega l} = G_l/l(l+1)$. For $l \ll \delta$ the investigation of eqs. (3.7), (3.9) and (3.10) for small $\varkappa$ is sufficient because the integral (3.10) is given by $2\pi i$ times the residue of the integrand at the point in the upper half-plane $\varkappa = \varkappa_0 \sim l/\delta$. Expanding the integral (3.6) with respect to $\varkappa^2$ at $a = 1$ and exploiting the recurrence relations of the Legendre polynomials we find:

$$\tilde{W}_{kl}(\varkappa) = k(k+1)\delta_{kl}/(2k+1) - 2i\varkappa[\delta_{k,l-1}(k+2)!(2k-1)!!/(k-1)!(2k+3)!!]^s$$
$$- \varkappa^2\{\delta_{kl}k(k+1)(2k^2+2k-3)/(4k^2-1)(2k+3)$$
$$+ [\delta_{k,l-2}(k+3)!(2k-1)!!/(k-1)!(2k+5)!!]^s\} + 0(\varkappa^3), \tag{3.11}$$

where $[A_{kl}]^s \equiv (A_{kl} + A_{lk})/2$. Taking into account this expression, it is easy to see that only A_2 contributes to the sum in the second of eqs. (3.9) if one neglects $0(\varkappa^4)$ terms. With this accuracy, using eqs. (3.7) and (3.11) for $a = 1$ we have $A_2 = -2i\varkappa/(5-G_2)$. Hence, using (3.9), (3.11) one finds:

$$K(\varkappa) = (2/3)[1 - \varkappa^2/(5-G_2)] + 0(\varkappa^4), \quad a = 1. \tag{3.12}$$

According to the first of eq. (3.9) and (3.12) we get

$$\mu(\varkappa) = [1 - \varkappa^2/(1-G_1/3)(5-G_2)]/(1-G_1/3) + 0(\varkappa^4). \tag{3.13}$$

Substituting the expresion (3.13) into eq. (3.10) and performing the elementary integration, we obtain:

$$\zeta_{\varepsilon=1} = \zeta_{norm}[1 - (l_{tr}/\delta)^2/(5-G_2)] + 0((l/\delta)^4), \tag{3.14}$$

where the value of the impedance for the normal skin effect $\zeta_{norm} = (\omega/4\pi i \sigma_{tr})^{1/2}$ and $\sigma_{tr} = \sigma/(1-G_1/3)$. Moreover, $l_{tr} = l/(1-G_1/3)$. It should be emphasized that the penetration depth δ appearing in the correction term of eq. (3.19) contains σ instead of σ_{tr}. Let us stress that the result (3.14) is exact in a model for $l \ll \delta$ and $a = 1$ without any restrictions imposed on the parameters G_l. The contribution of the series expansion (3.11) leads to the conclusion that all correction terms of the order $(l/\delta)^{2k}$, $k = 1, 2, 3, \ldots$ depend on a finite number of the parameters G_l. The denominators $(2l+1-G_l)$ appearing in eqs. (3.12)–(3.14) for $l = 1$ and 2 are always positive by virtue of the inequality[11] $|G_l| < 2l+1$, and hence $2l+1-G_l < 2(2l+1)$. As a result, the modulus of the coefficient near $(l/\delta)^2$ in the square bracket of eq. (3.14) is always greater than 0.025. The relations (3.13), (3.14) manifestantly coincide with analogous formula[15] for $G_2 = 0$. It is interesting to note that also in the first-order approximation the real and imaginary parts of the surface impedance in the regime of the almost normal skin are equal one to another.

In the opposite limit $l \gg \delta$ the main contribution to the integral (3.10) is provided by the domain of large values of the variable $\varkappa$. For pure metals and low temperatures, the restriction imposed on $(\omega\tau)^{1/2}$ is very weak and one can put even $\omega\tau = 0$ if $|\omega\tau| \ll 1$. On the other hand, one can consider also $\omega\tau \sim 1$ and then a or $l(l+1)G_{\omega l}$ cannot be replaced by 1 or G_l, respectively. Unfortunately, it is impossible now to establish an analogue of eqs. (3.10) for large $\varkappa$. The singular character of the solution for the anomalous skin effect $l \gg \delta$ leads to lack of simple selection rules for $\tilde{W}_{kl}(\varkappa)$, $\varkappa \gg 1$, such as expressed in eqs. (3.10) by the Kronecker's deltas. On the other hand, the main effect does not contain the parameters $G_{\omega l}$, cf. papers.[2,3] Let us extend the results[14] assuming that $G_{\omega l} = 0$ only for $l > 2$. Hence, from eqs. (3.7) and (3.9), we obtain

$$\mu(\varkappa) = [(3/2)\tilde{W}_{11} - G_{\omega 2}\tilde{W}_{22}/a][(1-G_{\omega 1}\tilde{W}_{11})(1-G_{\omega 2}\tilde{W}_{22}) + \varkappa^2 G_{\omega 1}G_{\omega 2}\tilde{W}_{22}^2/9a^2]^{-1}, \tag{3.15}$$

where the following simple identities for the functions $\tilde{W}_{ik}$ have been exploited:

$$\tilde{W}_{12} = -i\varkappa\tilde{W}_{22}/3a, \quad \tilde{W}_{11} + \varkappa^2\tilde{W}_{22}/9a^2 = 2/3a. \tag{3.16}$$

Thus, calculating $\tilde{W}_{11}$ and obtaining its asymptotics for $\varkappa \gg 1$, we are able to calculate $\mu(\varkappa)$ in this limit. We have:

$$\tilde{W}_{11}(\varkappa) = -a/\varkappa^2 + (1/2i\varkappa)(1+a^2/\varkappa^2)\ln[(a+i\varkappa)/(a-i\varkappa)]. \tag{3.17}$$

In this formula, the branch of the logarithmic function is determined uniquely by the condition $|\mathrm{Im}\ln[(a+i\varkappa)/(a-i\varkappa)]| < \pi$. As a result of this condition, the function $\tilde{W}_{11}(\varkappa)$ has a discontinuity on the semi-axes $[i-\omega\tau, i\infty-\omega\tau]$, $[-i+\omega\tau, -i\infty+\omega\tau]$. To make simple the series expansion for large $\varkappa$ one should replace the function $(2i)^{-1}\ln[(a+i\varkappa)/(a-i\varkappa)]$ by the function $\pi/2 - (2i)^{-1}\ln[(1+ia/\varkappa)/(1-ia/\varkappa)]$, according to the above prescription. From the formulae (3.15), (3.16) and (3.17) via the series expansion with respect to inverse $\varkappa$ we get

$$\mu(\varkappa) = (3\pi/4\varkappa)[1+(\pi G_{\omega 1}/2 - 4a/\pi - 8G_{\omega 2}/\pi)/\varkappa] + O(\varkappa^{-3}). \tag{3.18}$$

Substituting the expression (3.18) into eq. (3.10) we obtain:

$$\zeta_{\varepsilon=1} = \zeta_{an}[1+2(\sqrt{3}+i)(2a/\pi - \pi G_{\omega 1}/4 + 4G_{\omega 2}/\pi)/3\lambda + O(\lambda^{-2}\ln\lambda)], \tag{3.19}$$

where

$$\lambda = [3\pi(l/\delta)^2/2]^{1/3} = (3\pi^2 l^2 \sigma\omega/c^2)^{1/3}$$

and

$$\zeta_{an} = (2/9)(\sqrt{3}\omega^2 l/\pi^2 c\sigma)^{1/3}(1-i\sqrt{3}) \tag{3.20}$$

is the value of the impedance for highly anomalous skin effect ($\delta/l \to 0$) and the specular surface reflection. To obtain the result (3.19) we need some techniques. The function $\mu(\varkappa)$ is an even function of $\varkappa$ though it contains also odd powers of $1/\varkappa$ as $\varkappa$ tends to infinity. Hence, the asymptotic form (3.18) cannot be used on the whole interval $(-\infty, \infty)$. Let us express the integral in eq. (3.10) as follows:

$$(1/4\pi i)\oint d\varkappa \ln \varkappa/[\varkappa^2 - 2i(l/\delta)^2 \mu(\varkappa)], \tag{3.21}$$

where the integration contour surrounds in a counterclockwise direction the positive semi-axis. Because the integrand fulfils the assumption of the Jordan lemma, one can fill out the above loop by a large circle beginning at $(+\infty + i\varepsilon)$ and ending at $(+\infty - i\varepsilon)$ where ε is positive and suitably small, without any change of the whole integral. Hence, the value of the expression (3.21) is equal to the minus one half of the sum of residues of the integrand. In the first approximation for large λ, the poles of the integrand appear at $\varkappa = \lambda\exp[\pi i(2k+1)/3]$, $k = 0, 1, 2$. Taking into account that the first correction term to this position of the pole does not depend on k, it can be shown that the terms $O(\lambda^{-1}\ln\lambda)$ in the square bracket of eq. (3.19) vanish. Note that in contrast to the almost normal skin regime, the proportion of the real to the imaginary part of the surface impedance is varied by the correction in the almost highly anomalous skin regime.

4. THE SURFACE IMPEDANCE FOR DIFFUSE SURFACE

For diffuse electron scattering with the surface $\varepsilon(c) = 0$, $W_{kl}^s = 0$, and the restriction $G_{\omega l} = 0$ for $l > 1$, it is necessary to apply the generalized Wiener-Hopf procedure.[13] We will apply it here in a short-hand form and without proof. Moreover, we will assume that the quantity $\omega\tau$ will be negligibly small and, hence, $G_{\omega l} = G_l/l(l+1)$. As we have seen, the limit $\omega = 0$ in a and $G_{\omega l}$ is inevitable for the normal skin regime and quite easy to attain for the anomalous one (cf. the end of the Introduction). The procedure can be found, in detail, in our paper.[15] Let us consider the equation (2.18) for $W_{11}^s = G_{\omega l} = 0$ if $l \geq 2$.

In order to obtain the explicit formula for the Fourier transform $\tilde{E}(\varkappa)$ of the electric field, determined in the manner of (3.5) but with the integration over z only on the interval $(0, \infty)$ let us express it in the form:

$$\tilde{E}(\varkappa) = [iE(0) - (1/3)G_1 H_+(\varkappa)]/(\varkappa + \varkappa_0) \exp[G_+(\varkappa)], \tag{4.1}$$

where the functions H_+ and $G_\pm(\varkappa)$ are given by the algorithm described below. Let us introduce

$$\mathscr{F}(\varkappa) = \varkappa^2 - \tilde{W}_{11}(\varkappa)[(1/2)G_1 \varkappa^2 + 3i(l/\delta)^2] \tag{4.2}$$

and

$$G_\mp(\varkappa) = (1/2\pi i) \int_{\pm i\varepsilon - \infty}^{\pm i\varepsilon + \infty} d\xi \ln[\mathscr{F}(\xi)/(\xi^2 - \varkappa_0^2)]/(\xi - \varkappa), \tag{4.3}$$

$$H_+(\varkappa) = (3/4\pi i) \int_{-i\varepsilon_1 - \infty}^{-i\varepsilon_1 + \infty} d\xi [(iE(0)\xi - E'(0))/(\xi - \varkappa_0)(\xi - \varkappa)] \tilde{W}_{11}(\xi) \exp G_-(\xi). \tag{4.4}$$

In these integrals $\varkappa_0$ denotes the unique zero of the function $\mathscr{F}(\varkappa)$ in the first quadrant of the complex plane and $\varepsilon_1 > \varepsilon > 0$ are sufficiently small that the integrals (4.3), (4.4) are independent of ε and ε_1. This means that within the strip $|\operatorname{Im}\xi| < \varepsilon$, the integrands in eqs. (4.3), (4.4) are not singular.

If we put $\varkappa = 0$ in the definition of the Fourier transform of the electric field (4.1), then we obtain only a trivial identity containing no information. On the other hand, applying the trick proposed by Reuter and Sondheimer,[1] the following relation may be obtained:

$$E'(0) = \lim_{\varkappa \to i\infty} [i\varkappa E(0) - \varkappa^2 \tilde{E}(\varkappa)]. \tag{4.5}$$

Hence, the expression for the impedance may be derived as:

$$\zeta = \omega l i \xi_0 (1 - S^2)/c(1 + S^2), \tag{4.6}$$

where:

$$\xi_0 = (l/\delta)(-6i/G_1)^{1/2}, \quad \operatorname{Im}\xi_0 > 0, \quad S = \xi_0/(\xi_0 + \varkappa_0) \exp G_+(\xi_0). \tag{4.7}$$

In order to determine the impedance $\zeta_{\varepsilon=0}$ it is necessary to calculate one of the integrals (4.3). It can be shown that the limiting value of the expression (4.7) for $G_1 \to 0$ is the one obtained by Reuter and Sondheimer.[1]

Let us write the approximate formulae for the impedance at $l \ll \delta$ and $l \gg \delta$. For $l \ll \delta$, more exactly for $l/G_1 \delta \ll 1$, we have:

$$\zeta_{\varepsilon=0} = \zeta_{norm}[1 + (l/\delta)\Phi(G_1)], \tag{4.8}$$

where:

$$\Phi(G_1) = [2(1+i)/(2\pi i)^2 G_1(1-G_1/3)^{1/2}]$$
$$\times \int_C d\varkappa \varkappa^{-2} \ln(\varkappa + i) \ln\{[1 - (G_1/2)\tilde{W}_{11}^-(\varkappa)]/[1 - (G_1/2)\tilde{W}_{11}^+(\varkappa)]\}, \tag{4.9}$$

and the anticlockwise contour C begins and ends in the point $\varkappa = -i$. Moreover, the contour C contains the interval $(-i, i)$ and all the zeroes of the functions $1 - G_1 \tilde{W}_{11}^\pm(\varkappa)/2$ of the complex plane. The functions $\tilde{W}_{11}^\pm(\varkappa)$ are determined by the formula (3.17) at $a = 1$ but with another choice of the branch of the logarithmic function in the multiplier $\ln[(1+i\varkappa)/(1-i\varkappa)]$. Namely, for the functions $W_{11}^\pm(\varkappa)$ the following conditions should be fulfilled: $0 \leqslant \pm \operatorname{Im} \ln[(1+i\varkappa)/(1-i\varkappa)] < 2\pi$. In order to obtain a unique expression for eq. (4.9), when $G_1 \to 0$, one should take into account that

$$\ln[(1 - G_1 \tilde{W}_{11}^-(\varkappa)/2)/(1 - G_1 \tilde{W}_{11}^+(\varkappa)/2)] = G_1(\tilde{W}_{11}^+(\varkappa) - \tilde{W}_{11}^-(\varkappa))/2 + O(G_1^2),$$

and to use the following formula:

$$\tilde{W}_{11}^+(\varkappa) - \tilde{W}_{11}^-(\varkappa) = \pi(\varkappa^{-1} + \varkappa^{-2}).$$

Hence, $\Phi(0) = 3(1+i)/16$ and the main part of eqs. (4.8), (4.9) for $G_1 \to 0$ is identical with the Dingle formula,[5] obtained under the asumptions that $G_1 = 0$, $l/\delta \ll 1$. This means that the formulae (4.8), (4.9) are valid not only for $l/\delta G_1 \ll 1$ but also for $G_1 = 0$ and $l/\delta \ll 0$. On the other hand, it is impossible to write any simple formula for $l/\delta \ll 1$ but when $l/\delta G_1 \sim 1$. The numerical values of the function $\Phi(G_1)$ are listed below

G_1	−0.9	−0.7	−0.5	−0.1	0	0.1	0.3	0.5	0.7	0.9
$10\text{Re}\Phi(G_1)$	1.69	1.73	1.77	1.86	1.875	1.89	1.93	1.97	2.0	2.02
$-10\text{Im}\Phi(G_1)$	1.28	1.39	1.51	1.74	1.875	1.95	2.14	2.34	2.6	2.82

If $l \gg \delta$, then

$$\zeta_{\varepsilon=0} = (9/8)\zeta_{an}[1 + (3 + \sqrt{3}i)(\ln\lambda + f + gG_1)/\pi^2\lambda + 0(\lambda^{-2}\ln\lambda)], \tag{4.10}$$

where $\lambda \gg 1$ and ζ_{an} are defined by eq. (3.20). Let us define also

$$g = -(\pi^3/48)(1 + 2\sqrt{3}/9 - 2i) \approx -0.895 + 1.292i,$$

and

$$f = (\pi^2/4)\int_1^\infty dx \left\{ \arctan\left[\frac{1}{\pi}\left(\ln\frac{x+1}{x-1} + \frac{2x}{x^2-1}\right)\right] - 4/\pi x \right\} + \pi^2/8 - \pi i/6. \tag{4.11}$$

The integral appearing in the formulae (4.11) calculated numerically gives the value 0.05186 and hence $f \approx 1.403 - 0.524i$. Here, the proportion of the real to the imaginary part of the surface impedance is changed by the correction term, in contrast only to the normal skin regime at $\varepsilon = 1$. Equation (4.10) for $G_1 = 0$ becomes identical with that obtained by Dingle,[5] if we take into account the difference of ω-signs in reference[5] and here.

5. THE CONDUCTIVITY OF SLABS: GENERAL FORMALISM AND THIN SLABS

Thin slabs denote here slabs much thinner than the m.f.p. The following formalism almost coincides with the second section of this paper. In the clasical formulation of the problem,[10] the electric field is homogeneous, i.e., independent of the coordinate perpendicular to the slab surface z, as well as, independent of the coordinates parallel to the surface (x, y). In the original formulation,[10] the electric field is also time-independent. The time dependence was first introduced in references.[11,12,20] On the other hand, the assumed homogeneity of the electric field in the parallel direction means that its frequency fulfils the condition for the normall skin regime. This is the reason why we will put $\omega\tau = 0$ everywhere in this and the next section. Let us add that the expression for the conductivity at $\omega\tau \neq 0$, having an analogous quantity at $\omega\tau = 0$, is a question of a simple similarity transformation.[20]

Let us note that the quantization of electron levels by the finite slab thickness leads to energy subbands distanced by energies of the order ε_F/n, where $\varepsilon_F \sim v_F p_F$ is the Fermi energy and n — the number of atomic layers in the slab. Application of classical mechanics leads to the inequality $T \gg p_F v_F/n$ with T being the temperature in the energy scale ($k_B \equiv 1$). On the other hand, the predominance of impurity scattering over the phonon scattering takes place only for $T \ll T_D$, where T_D is the Debye temperature. As $T_D \sim v_F p_F (m/M)^{1/2}$ with m and M being the electron and ionic masses, respectively,[3] n should be greater than $(M/m)^{1/2}$ by at least two orders of magnitude.

The homogeneity of the electric field E makes it convenient to include the factor E on the right-hand side of the expression for the phase-space electron distribution function (2.3). In such a case, the right-hand side of the formula (2.4), with the function χ defined in the modified manner (2.3), determines the local value of the conductivity $\sigma(z)$. The new function $\chi(z, c)$ fulfils the equation (2.5) at $E(z) = a = 1$, with the unaltered function $G_\omega(c, c')$, (2.6)., taken for $\omega = 0$. In the analogue to the substitution (2.7) into the equation (2.5), instead of $E(z)$ and a we should have, consequently, 1. Hence, in the analogue to eq. (2.8), instead of $E'(z)$ we have zero and $a = 1$. Let us solve this system of equations for two planary borders, parallel to one another. The surfaces are characterized by the same angle-dependent specularity coefficient $\varepsilon(c)$. The case of two different surfaces of the slab can be found in our paper.[11]

Let us asume that the metal lies within the slab restricted by the plane surfaces $z = \pm b$. As it was remarked, the analogue of eq. (10) now has the form

$$A'(z,c)\exp(-z/c) = \sum_{l=1} G_{0l} P'_l(c) N_l(z), \tag{5.1}$$

where $N_l(z)$ is defined by $A(z,c)$ via eq. (2.9) and $G_{0l} = G_l/l(l+1)$ denotes $G_{\omega l}$ at $\omega = 0$. Note that our present functions $N_l(z)$ denote the previous functions divided by the electric field E. Because the question concerning the surface impedance and the slab condiuctivity are considered here separately, this does not lead to any ambiguities.

Now, it will be more convenient to determine the integration constant $f(c)$ for the function $A(z,c)$ in a slightly different way than in eq. (2.11), taking

$$A(z,c) = -f(c) + \sum_{l=1} G_{0l} c^{-1} P'_l(c) \int_{-bc/|c|}^{z} dx N_l(x) \exp(x/c). \tag{5.2}$$

The angle-dependent Fuchs boundary conditions, the same for both surfaces, can be now written on the surfaces $z = \pm b$ as follows:

$$[A(b,c)\exp(-b/c)+1]\varepsilon(c) = A(b,-c)\exp(b/c)+1,$$
$$[A(-b,-c)\exp(-b/c)+1]\varepsilon(c) = A(-b,c)\exp(b/c)+1, \quad c > 0, \tag{5.3}$$

cf. the analogue of eq. (2.7), where $E(z) \to 1$. Substituting the formula (5.2) into the boundary conditions (5.3) we find

$$f(c) = \left\{-\varepsilon(|c|) \sum_{l=1} G_{0l}|c|^{-1} P'_l(|c|) \int_{-b}^{b} dx N_l(x)\exp[(y-b)/|c|] + 1 - \varepsilon(|c|)\right\}$$
$$\times [\exp(b/|c|) - \varepsilon(|c|)\exp(-b/|c|)]^{-1}. \tag{5.4}$$

In the result (5.4), the fact that $P'_l(-c) = (-1)^{l-1} P'_l(c)$ and $N_l(z) = (-1)^{l-1} N_l(z)$ has been exploited. The last relation results from the symmetry of the transport equation under simultaneous c and z reflection. Exploiting the definition of the current (2.4) we get:

$$\sigma(z) = \sigma_{tr}[1 + (3/2)N_1(z)], \tag{5.5}$$

where $\sigma(z)$ is the local, z-dependent conductivity. The conductivity of the slab is given by $\sigma(z)$ integrated over the slab thickness. Substituting eq. (5.4) into (5.2) and this result into eq. (2.9), after simple but boring transformations, we get:

$$N_k(z) = -\sum_{\alpha=\pm 1} \alpha^{k-1} V_k^\varepsilon(b+\alpha z, q) + \sum_{l=1} G_{0l} \int_{-b}^{b} dx N_l(x) \left\{[\text{sign}(z-x)]^{l+k} R_{lk}(|z-x|) \right.$$
$$\left. + \sum_{\alpha=\pm 1} \alpha^{l-1} R_{lk}^\varepsilon(q+\alpha z-x, q)\right\}, \quad q \equiv 2b, \tag{5.6}$$

where:

$$V_k^\varepsilon(u, q) = (1/2) \int_0^1 dc(1-c^2) P'_k(c)[1-\varepsilon(c)]\exp(-u/c)[1-\varepsilon(c)\exp(-q/c)]^{-1},$$

$$R^\varepsilon_{ik}(u,q) = (1/2)\int_0^1 dc(c^{-1}-c)\varepsilon(c)P'_k(c)P'_l(c)\exp(-u/c)[1-\varepsilon(c)\exp(-q/c)]^{-1},$$

$$R_{lk}(u) = (1/2)\int_0^1 dc(c^{-1}-c)P'_k(c)P'_l(c)\exp(-u/c), \quad u \geq 0. \tag{5.7}$$

Evidently, $V^{\varepsilon=1}_k = 0$ and, from eqs. (5.6), all $N_k(z)$ vanish. Hence, and also from eq. (5.5), $\sigma(z) = \sigma_{tr}$ for $\varepsilon(c) = 1$ which has a very simple physical interpretation. In fact, the mirror is not an obstacle to the motion parallel to it and we should deal with the conductivity of the bulk sample. On the other hand, it is easy to find that

$$\lim_{q\to 0} V^\varepsilon_k(u,q) = (1/3)\delta_{k1}, \quad |u| < q \tag{5.8}$$

provided that $\varepsilon(c)-1$ vanishes on the c-interval $(0,1)$ only at points without any limiting point. In such a limit, $N_1(z) \to -2/3$ and $\sigma(z)$ vanishes. This nonanalyticity of the function $\sigma(z)$ in the point $(1,0)$ of the variables (ε, q) has a very simple physical reason. If the slab is very thin, with respect to the m.f.p. being infinite along the surface, then the electron moving not exactly along the surface will collide with the surface and, after the n-th border collision, will continue its motion with the probability ε^n. As a result, the electron will surely be caught by the surface because the distance along it is infinite and the value of the m.f.p. allows for multiple surface scattering. Hence, the conductivity of this slab vanishes as there is no nonzero fraction of electrons moving along the surface.

The equations (5.6) are written, as eqs. (2.15), in terms of dimensionless variables z, x taken in the units of the m.f.p. Hence, for small q, the iteration of such integral equations is an asymptotically good procedure. Let us estimate the kernels R^ε_{ik}, R_{lk} and the functions V^ε_k asymptotically. Apply the Sommerfeld-Watson representation of the following integral:

$$S_n(v) \equiv \int_0^1 dc\, c^n \exp(-v/c) = \oint_L d\zeta \int_0^1 dc\, v^\zeta c^{n-\zeta}/2i\sin(\pi\zeta)\Gamma(\zeta+1)$$
$$= -\oint d\zeta\, v^\zeta/2i\sin(\pi\zeta)\Gamma(\zeta+1)(\zeta-n-1), \tag{5.9}$$

$n = -1, 0, 1,\ldots$, where L is an anticlockwise contour around the points $0, 1, 2, 3,\ldots$ of the complex plane. Taking into account that the integrand in the last integral (5.9) has the first order poles at all the natural numbers k, including zero, if $k \neq n+1$ and a second order pole at $k = n+1$, one finds

$$S_n(v) = \sum_{k=0}^\infty{}' (-v)^k/(n+1-k)k! - (-v)^{n+1}\left(\ln v + C - \sum_{k=1}^{n+1} 1/k\right)\Big/(n+1)!, \tag{5.10}$$

where C is the Euler constant and the prime above the sum denotes that the term with $k = n+1$ should be omitted. The kernel R^ε and the function V^ε can be expressed by the result (5.10) via a series expansion of the subintegral factor $[1-\varepsilon(c)\exp(-q/c)]^{-1}$ with respect to $\varepsilon(c)\exp(-q/c)$, and, next, via a series expansion of $\varepsilon(c)$. This procedure is valid if the c-interval $(0,1)$ lies in the convergence radius of the first series expansion, i.e., for $\varepsilon(c) < 1$ at $c \to 0$. Unfortunately, for $\varepsilon(c) \approx 1 - \gamma c + \ldots$ it is impossible to apply such a method, independently of the practical importance of such an $\varepsilon(c)$, reference.[16] Note that the kernel R_{lk} has an asymptotic expansion determined only by eq. (5.10), independent of $\varepsilon(c)$.

If $\varepsilon(0) < 1$, then eqations (5.8) and (5.10) lead to the following estimate:

$$\sum_{\alpha=\pm} (\alpha)^{l-1} V^\varepsilon_l(b+\alpha z, q) = (2/3)\delta_{l1} + [1-(-1)^l]0(q\ln q) + [1+(-1)^l]0(q),$$
$$R^\varepsilon_{l,2k} \sim R_{l,2k} = 0(1), \quad R^\varepsilon_{2m+1, 2k+1} \sim R_{2m+1, 2k+1} = 0(\ln q). \tag{5.11}$$

As we see, the only term $0(1)$ among the free terms in eqs. (5.6) appears for $k = 1$ and is equal to $-2/3$. Iterating this term and, next, iterating the result of the first iteration completed by the $0(q\ln q)$ remainder of the free terms, appearing only for odd k, and after long but elementary integrations, we get:

$$\sigma(z) = (3/2)\sigma\left\{Q_0(z) + \sum_{n=0} G_{0,2n+1} \int_{-b}^{b} dx Q_n(x)[R_{1,2n+1}(|z-x|) + \sum_{\alpha=\pm 1} R^{\varepsilon}_{1,2n+1}(q+\alpha z-x, q)]\right\} + O(q^2), \quad (5.12)$$

where:

$$Q_n(z) = (2/3)\delta_{n0} - \sum_{\alpha=\pm 1} V^{\varepsilon}_{2n+1}(b+\alpha z, q) = O(q \ln q). \quad (5.13)$$

As a result of the continuous version of Kirchhoff's law concering the parallel junction of resistivities, the conductivity of the slab is equal to the quantity (5.12) integrated over z in the slab thickness. After elementary integrations in the formula (5.12), we can write the average conductivity of the slab in the form

$$\bar{\sigma} = \sigma\{1 - 3/8q + (3/2q)\int_0^1 dc\, c(1-c^2)[1-\varepsilon(c)]^2 \exp(-q/c)/[1-\varepsilon(c)\exp(-q/c)]$$

$$+ (3/2q) \sum_{n=0} G_{0,2n+1} \int_{-b}^{b} dx Q_n^2(x)\} + O(q^2). \quad (5.14)$$

Let us assume that $\varepsilon < 1$ is angle-independent and perform the above procedure to obtain the asymptotic form of the integrals. As a result, applying the properties of Legendre polynomials and eq. (5.10), we have:

$$\bar{\sigma} = (3\sigma/4)\left\{-q(\ln q + C - 1)(1+\varepsilon)/(1-\varepsilon) - q(1-\varepsilon)^2 \xi(\varepsilon) + (1/2)(q \ln q)^2 (1+2\varepsilon)\right.$$

$$\left. \times (1+\varepsilon) \sum_{n=0} G_{0,2n+1}[(2n+1)!!/(1-\varepsilon)(2n)!!]^2\right\} + O(q^2), \quad (5.15)$$

$$\xi(\varepsilon) = \sum_{m=1}^{\infty} \varepsilon^m (m+1)^2 \ln(m+1). \quad (5.16)$$

Note that $(1-\varepsilon)^2 \xi(\varepsilon)$ diverges if $\varepsilon \to 1$. The result for the current profile, $\sigma(z)$, for thin slabs can be found in our paper.[12]

If $\varepsilon(c) = \varepsilon_0 - \gamma c + \ldots$ for $c \ll 1$ and $\varepsilon_0 < 1$ is then the extension of eqs. (5.14) and (5.15) is obvious although it leads to an additional complication of the formulae. If $\varepsilon_0 = 1$ (cf. ref.[16]), then the given estimate,[22] $\bar{\sigma} \sim q^{1/2}$, is reasonable although doubtful.[12] In this case, the ε-expansion of the integral, applied if $\varepsilon(c) < 1$, does not lead to convergent results. Let us also explain that a natural representation of the integral (5.14) in the contour form on the complex plane does not lead to its asymptotic expansion. Namely, let us substitute $c = 1/x$ and replace the suitable integral over the x-interval $(1, \infty)$ by the integral over the clockwise loop surrounding this interval, with the integrand multiplied by $(2\pi i)^{-1}\ln(x-1)$. Unfortunately, such an integrand does not fulfil the asumptions of the Jordan lemma and, hence, the loop cannot be closed by a large circle with vanishing contribution to the integral. We deal with this situation because the poles of the integrand has the limiting point at infinity, as a result of the denominator $[1-(1-\gamma/x)\exp(-qx)]$. These considerations, in contrast, should be compared with our remarks concerning the asymptotic properties of the integral (3.21).

6. CONDUCTIVITY OF THICK SLAB

The slabs will be treated as the thick ones if their thickness is much greater than unity in units of m.f.p. Let us consider the equations (5.6), if $G_l = 0$ for $l \geq 2$ and $\varepsilon = 0$. It will be convenient to change now the reference frame assuming that the metal is placed for $0 \leq z < \infty$. In this case we have

$$N_1(z) = -V_1^0(z) + (G_1/2) \int_0^\infty dx\, R_{11}(|z-x|) N_1(x), \tag{6.1}$$

cf. eq. (5.7), and in the transformations one neglects the exponentially small terms $0[\exp(-D/l)]$, where D is the slab thickness and l — the m.f.p.[11] Note that the expression for the local conductivity (5.5) still remains valid unless $0 < q - z \ll 1$; in this case one reproduces $\sigma(z)$ from symmetry considerations. The conductivity averaged over the slab thickness, according to eq. (5.9), will be formally given by:

$$\overline{\sigma(z)} = \sigma_{tr}[1 + (3/2q) \int_0^\infty dz\, N_1(z)], \quad q = D/l \gg 1. \tag{6.2}$$

In this formula, only the surface $z = 0$ is taken into account, because $N_1(z) \to 0$ if $z \to 0$. If the exponentially small interference of both distinct surfaces is neglected then the two diffuse surfaces lead to the formula

$$\overline{\sigma(z)} = \sigma_{tr}[1 + (3/q) \int_0^\infty dz\, N_1(z)]. \tag{6.3}$$

The equation (6.1) can be solved via the Wiener-Hopf technique.[13] The technical details can be found.[14] One can prove that the integral in eq. (6.3) is given by

$$B(G_1) = -(16/3\pi G_1) \int_0^1 dx\, \arctan \Phi(x, G_1), \tag{6.4}$$

where

$$\Phi(x, G_1) = (\pi/4) G_1 x(1 - x^2) \left\{ 1 - (G_1/3) \left[(1 - x^2) \ln \frac{1+x}{1-x} + 2x \right] \right\}^{-1}. \tag{6.5}$$

On the other hand, the asymptotic form of the function $N_1(z)$ at $z \gg 1$ is given by

$$N_1(z) = -A(G_1)/z^2 \exp z, \tag{6.6}$$

where

$$A(G_1) = [16(1 - G_1/3)/3(1 - G_1/2)^{\frac{1}{2}}] \exp\{\int_0^1 dx [\arctan \Phi(x, G_1)]/\pi x(x+1)\}. \tag{6.7}$$

Both integrals in the formulae (6.5) and (6.7) can be calculated numerically. It should be noted that the quantities $A(G_1)$ and $B(G_1)$ are nonzero at $G_1 = 0$. It can be verified that they coincide with suitable quantities in the relaxation time approximations. For $G_1 = 0$, eq. (6.5) becomes identical with the result obtained by Dingle.[5] Note that the result (6.2) can be treated as the result for one diffuse and one specular surface.

7. CONCLUSIONS

Let us briefly return to our topics, from the point of view of their limitations. Here two topics have been considered: the surface impedance of a semi-infinite sample and the slab conductivity, both beyond the τ-approximation. In the second case, the previous method,[11,12] was reviewed, in the first one — an ad hoc method has been elaborated. For thin slabs, in the units of the m.f.p., the conductivity results are simple consequences of the general formalism. The angle-dependent specularity coefficient of the surface does not lead to any essential difficulty, however, the results become fairly complicated. The essential difficulty in the asymptotic form of suitable integrals appears only for $\varepsilon(c) = 1 - \gamma c - 0(c^2)$ if $0 \leqslant c \ll 1$. It should be emphasized that even in this case the integral representation of the average conductivity (5.14) still remains asymptotically valid if $q \ll 1$. On the other hand, any step away from Fuchs boundary conditions, in the angle-dependent form, has not been performed in the direction of rough surfaces[4] unless the gain part of the scattering integral has been neglected. The conductivity of thick slabs can be found for diffuse surfaces only, with strong restrictions on the impurity scattering kernel beyond the trivial case of fully specular surfaces. These restrictions will be removed only if we will be able to generalize of the Wiener-Hopf method.[13]

In contrast to the conductivity, more detailed results for the surface impedance can be obtained only for fully specular or diffuse boundary conditions. In the first case, this is not a trivial question (cf. the situation with the conductivity), in the second one strong restrictions on the scattering kernel are necessary. The case $\varepsilon(c) = 1 - \gamma c + O(c^2)$, $0 \leqslant c \ll 1$ will be considered in the near future for impurity scattering though only for $\gamma \ll 1$. The analytic expression of the surface impedance via well-known functions can be done only in the almost normal or almost highly anomalous skin regime.

As was noted, for diffuse surface, we are forced to restrict ourselves to the first Legendre harmonic into the impurity scattering kernel. For a specular surface, we are able to include many of them, even though we deal with the important difference between the almost normal ($l \ll \delta$) and almost highly anomalous ($l \gg \delta$) skin regime. In the first case, we are able to obtain the result for a quite general function $W(c)$ expandable in a series of Legendre polynomials, in the second one, we must restrict ourselves to the first few Legendre harmonics. Note that in both cases the results coincide with the well-known ones if we take into account the main term. Hence, the measurements of the surface impedance in the highly normal, as well as, the highly anomalous limit can serve for spectroscopic purposes. Let us add that even with the almost highly anomalous skin effect, the correction term to the surface impedance depends appreciably on the gain part of the scattering integral. All correction terms change, several times at least, when the parameters describing impurity scattering are changed in the region allowed for them. The proportion of the real to the imaginary part of the surface impedance is not varied by the correction term only at almost normal skin regime and $\varepsilon = 1$. In all four possible cases, namely for $l \gg \delta$ or $l \ll \delta$ and $\varepsilon = 1$ or $\varepsilon = 0$, the dependence of the surface impedance on the parameters of the problem is very dissimilar.

It is worth emphasizing that because of the singularity of the solution and its linear dependence on the gain part of the scattering integral for the almost highly anomalous skin effect, (3.19), (4.10), these corrections for the surface impedance should be added directly to the suitable temperature corrections obtained in refs.[6-9] Note that the similar procedure is quite improper in the opposite limit because of the nonlinear dependence of the impedance on the gain part, cf. (3.13), (4.8).

ACKNOWLEDGEMENTS

Two of the authors (J. C. and M. I. K.) are greatly indebted to foreign institutes, where substantial parts of their work have been done. The first author thanks the P. L. Kapitza Institute for Physical Problems, Moscow, Russia, the second one – the International Laboratory of Strong Magnetic Fields, Wrocław, Poland. In the first case, also the support of the Committee of Scientific Research of Poland, the grant No. 2 0943 9101 is gratefully acknowledged.

REFERENCES

1. G.E. Reuter and E.H. Sondheimer. *Proc. Roy. Soc.* 195:336 (1948).
2. I.M. Lifshitz, M.Ya. Azbel, and M.I. Kaganov,"Electron Theory of Metals," Consultants Bureau, New York (1973).
3. A.A. Abrikosov, "Fundamentals of the Theory of Metals," North-Holland, Amsterdam (1988).
4. L.A. Falkovsky, *Adv. Phys.* 32:753 (1983).
5. R.B. Dingle, *Appl. Sci. Res.* B 9:69 (1953).
6. A.Manz, J. Black, Kh. Pashaev, and D.L. Mills, *Phys. Rev.* B 17:1721 (1978).
7. A.Manz, J. Black and D.L. Mills, *Phys. Rev.* B 20:4018 (1979).
8. J. Black and D.L. Mills, *Phys. Rev.* B 21:5860 (1980).
9. A.P. Zhernov and Kh. P. Pashaev, *Fiz. Tverd. Tela* 25:3389 (1983).

10. K. Fuchs, *Proc. Camb. Philos. Soc.* 34:100 (1938).
11. J. Czerwonko, Z. Phys. B 80:225 (1990).
12. J. Czerwonko, Physica A 174:438 (1991).
13. P.M. Morse and H. Feshbach, "Methods of Theoretical Physics," McGraw-Hill, New York(1953).
14. M.I. Kaganov, G.Ya. Lyubarskii, and J. Czerwonko, Zh. Eksp. Teor. Fiz. 102:1563 (1992).
15. M.I. Kaganov, G.Ya. Lyubarskii, and J. Czerwonko, Zh. Eksp. Teor. Fiz. 102:1351 (1992).
16. A.F. Andreev, Uspekhi Fiz. Nauk 105:113 (1971).
17. M.S. Khaikin, Adv. Phys. 18:1 (1969).
18. R.E. Prange and T.W. Nee, Phys. Rev. 168:779 (1968).
19. D. Pines and Ph. Nozières, "Theory of Quantum Liquids," Benjamin, New York (1966).
20. J. Czerwonko and M.I. Kaganov, Phys. Lett. A 152:430 (1991).
21. V.I. Okulov and V.V. Ustinov, Fiz. Nizk. Temp. 5:213 (1979).
22. M.Ya. Azbel, S.D. Pavlov, I.A. Gamalya, and A.I. Vereshchagin, Pis'ma Zh. Eksp. Teor. Fiz. 16:295 (1972).

LINEAR THERMOELASTIC GENERATION OF ULTRASOUND IN METALS

[1]M.I. Kaganov, and [2]A.N. Vasil'ev

[1]P.L. Kapitza Institute for Physical Problems
Russian Academy of Sciences, Moscow, Russia
[2]Low Temperature Physics
and Superconductivity Department
Physics Faculty, Moscow State University
Moscow, Russia

The incidence of an electromagnetic wave on a metal surface is accompanied by the generation of acoustic waves within the metal.[1] These waves are excited both at the frequency of the incident wave, and at its harmonics. Accordingly, there exist physical mechanisms of linear and nonlinear ElectroMagnetic–Acoustic Transformation - EMAT. The inductive interaction,[2] as well as the deformation interaction[3], are considered to be responsible for the linear EMAT, while non-linear EMAT is supposed to originate from the thermoelastic interaction.[4] It was shown recently both experimentally, and theoretically that inductive and deformation potential. forces are capable exciting ultrasound at double the frequency of the electromagnetic wave, i.e. in the nonlinear regime.[5-7] The present work makes this picture complete by deriving formulae which demonstrate the possibility of a linear EMAT due to thermoelastic generation.

To begin consider the origin of quadratic effect of the ultrasonic wave generation under the action of thermoelastic stress. This stress is due to the variation of temperature in a metal's skin layer resulting from Joule heat dissipation:

$$Q = Ej, \qquad (1)$$

where E is the intensity of the alternating electric field in the incident wave, and j is the density of alternating current in the skin layer. It is evident that Q contains not only the permanent additive, but the term proportional to $\cos 2\omega t$. Being the source in the thermoconductivity equation which describes the propagation of heat in a metal, the alternating part of the Joule heat is responsible for the temperature variation at double frequency with amplitude proportional to the squared amplitude of the electromagnetic wave. The temperature oscillations are accompanied by oscillations of thermoelastic stress which, in turn, makes the crystal lattice oscillate, i.e. excites acoustic waves.

As far as the interdependence between the temperature oscillations and elastic oscillations is linear it is evident that to realize the linear thermoelastic mechanism of EMAT it is necessary to find the source of linear oscillations of temperature with respect to the electromagnetic wave amplitude. A source of this type can be the thermoelectric force.[8] It will be shown below that just this force is responsible for the generation of longitudinal ultrasound in a metal at the frequency of the incident electromagnetic wave.

Let's consider at first the propagation of coupled electromagnetic and thermal waves in a monocrystalline metal sample oriented so that the normal to its surface does not coincide with a crystal axis of high symmetry. Being different in principle the electromagnetic and thermal waves nevertheless are characterized by a similar feature, i.e. by the skin effect. If the uniform heat conductivity equation is used to describe the temperature oscillations excited at frequency ω at the sample surface ($z = 0$), i.e.

$$C \partial \theta / \partial t = \kappa \partial^2 \theta / \partial z^2 , \qquad (2)$$

where θ is the oscillating part of the temperature, C is the heat capacity, and κ is the heat conductivity, it can be shown that the thermal skin depth is determined by the expression

$$\delta_T = (2\kappa/\omega C)^{1/2} . \qquad (3)$$

Comparing it with the electromagnetic skin depths:

$$\delta_E = c/(2\pi \sigma \omega)^{1/2} , \qquad (4)$$

where c is the velocity of light, and σ is the electroconductivity, one can see that both δ_T, and δ_E are inversely proportional to the square root of the frequency ω. For estimates it is convenient to use the expressions justified by the simplest assumptions about the conduction electrons :

$$\sigma = ne^2 l / p_F, \qquad \kappa = \pi^2 T \sigma / 3e^2 - C l v_F , \qquad (5)$$

where n, and e are electron concentration and charge, v_F, and p_F are electron velocity and momentum at the Fermi surface and l is the electron mean free path. We assume that the main heat carriers are electrons. However, it is not really important, as the expression for the thermal skin depth enters the combination κ/C, which is not dependent on heat conductivity. Comparing the skin depth δ_T and δ_E, one can see that their ratio is not frequency dependent, in fact,

$$\delta_T / \delta_E = l / \delta_L, \qquad (6)$$

where δ_L is the plasma skin depth, in good metals equal to $10^{-5} \div 10^{-6}$ cm. At room temperature $l < \delta_L$, and $\delta_T < \delta_E$. At low temperature in pure metals the opposite situation takes place, i.e. $l > \delta_L$, and $\delta_T > \delta_E$.

As is evident from Eq. (4), the propagation of coupled electromagnetic and thermal waves in a metal is considered under the conditions of the normal skin effect. It has been demonstrated,[8] that the conditions of anomalous skin effect can not be realized for thermal waves, while in the case of the anomalous skin effect for electromagnetic waves the thermoelastic mechanisms of EMAT are not effective.[7] To take into account the thermoelectric phenomena when considering the propagation of electromagnetic waves in a metal one has to combine the Maxwell equations

$$\text{curl } \mathbf{H} = 4\pi \mathbf{j}/c , \qquad (7)$$

$$\operatorname{curl} \mathbf{E} = i\omega \mathbf{H}/c , \qquad (8)$$

with the equation of heat transfer

$$C\dot{\theta} + \operatorname{div} \mathbf{Q} = 0 \qquad (9)$$

where $\mathbf{H}$ is an alternating magnetic field, and $\mathbf{Q}$ is the density of heat flow. Eqs. (7) - (9) should be augmented by the material equations, which express the current density $\mathbf{j}$ and heat flow $\mathbf{Q}$ via the electric field $\mathbf{E}$ and temperature θ, i.e.

$$E_i = \rho_{ik} j_k + \sigma_{ik} \partial \theta / \partial x_k , \qquad (10)$$

$$Q_i = T \alpha_{ik} j_k - \kappa_{ik} \partial \theta / \partial x_k , \qquad (11)$$

where $\rho_{ik} = \sigma_{ik}^{-1}$ is the resistivity tensor, σ_{ik} is the electroconductivity tensor, κ_{ik} is the thermoconductivity tensor and α_{ik} is the thermoelectricity tensor. Evidently, to couple Eqs. (10) and (11) it is necessary that the nondiagonal components of the thermoelectricity tensor should differ from zero.

Consider the normal incidence of an electromagnetic wave on the halfspace ($z > 0$) occupied by anisotropic metal. The metal is assumed to be uniaxial with its axis of symmetry tilted from the normal to the surface. Let the direction of the alternating electric field coincide with the x axis, while all the variables depend only on the z coordinate. In this case Eqs. (7) - (11) can be rewritten as

$$\partial^2 j_x / \partial_z^2 + 4\pi i \sigma j_x / c^2 \rho_{xx} = -(\alpha_{xz}/\rho_{xx}) \partial^3 \theta / \partial z^3 , \qquad (12)$$

$$\partial^2 \theta / \partial z^2 + i\omega C \theta / \kappa_{zz} = (T\alpha_{xz}/\kappa_{zz}) \partial j_x / \partial z . \qquad (13)$$

The dispersion equation for the coupled electromagnetic and thermal oscillations is

$$\left(1 - k_E^2/k^2\right)\left(1 - k_T^2/k^2\right) = T\alpha_{xz}^2 / \rho_{xx} \kappa_{xx} , \qquad (14)$$

where $k_E^2 = 4\pi i \omega / c^2 \rho_{xx}$, and $k_T^2 = i\omega C / \kappa_{xx}$. It is evident, that the coupling between the thermal and electromagnetic waves is determined by the dimensionless parameter

$$\alpha = T\alpha_{xz}^2 / \rho_{xx} \kappa_{zz} , \qquad (15)$$

which is proportional to $(T/\epsilon_F)^2 \ll 1$. Here ϵ_F is the Fermi energy.

As long as $\alpha \ll 1$, one can use approximate expressions for the roots of the dispersion equation (14), i.e.

$$k_1^2 \approx k_E^2 \left[1 - \alpha k_E^2 / \left(k_T^2 - k_E^2\right)\right] , \qquad (16)$$

$$k_2^2 \approx k_T^2 \left[1 - \alpha \, k_T^2 / \left(k_E^2 - k_T^2\right)\right] . \qquad (17)$$

If $k_T^2 \approx k_E^2$, which is analogous to the resonant interaction and takes place at intermediate temperatures when $l - \delta_L$, then

$$k_{1,2}^2 \approx k^2 \left(1 \pm \sqrt{\alpha}\right) . \qquad (18)$$

Note, that at any value of the parameter α, the squares of the wavenumbers are purely imaginary values and, besides, a strict resonance between these two types of oscillations is impossible, because $k_1^2 = k_2^2$ at $\alpha = 0$ only.

To solve the problem of the propagation of coupled electromagnetic and thermal oscillations, the system of Eqs. (12) - (13), in addition to the natural boundary conditions for the alternating electric and magnetic fields,[1] should be augmented with conditions describing heat transfer at the boundary. Assuming that at metal depth ($z \to \infty$), amplitude of thermal oscillations goes to zero, the boundary conditions at the surface can be written for two limiting cases.

At an isothermal boundary

$$\theta|_{z=0} = 0 , \tag{19}$$

at adiabatic boundary

$$Q_z|_{z=0} = 0, \quad \text{or} \quad \kappa_{zz} d\theta/dz|_{z=0} = T\alpha_{xz} j_x(0) . \tag{20}$$

The solution, that is the dependence of all variables on the z coordinate, can be written by introducing two amplitudes, the ratio of which is determined by thermal boundary conditions.

So,

$$j_x(z) = A_1 \exp(ik_1 z) + A_2 \exp(ik_2 z) \tag{21}$$

$$E_x(z) = (4\pi i\omega/c^2) \left[(A_1/k_1^2) \exp(ik_1 z) + (A_2/k_2^2) \exp(ik_2 z) \right] , \tag{22}$$

$$H_y(z) = (4\pi i/c) \left[(A_1/k_1) \exp(ik_1 z) + (A_2/k_2) \exp(ik_2 z) \right] , \tag{23}$$

$$\theta(z) = (iT\alpha_{xz}/\kappa_{zz}) \left\{ \left[A_1 k_1 / \left(k_T^2 - k_1^2 \right) \right] \exp(ik_1 z) + \left[A_2 k_2 / \left(k_T^2 - k_2^2 \right) \right] \exp(ik_2 z) \right\} , \tag{24}$$

where k_1 and k_2 are the roots of the dispersion equation (14). Using Eqs. (19) and (20), we find at an isothermal boundary

$$(A_2/A_1) = -k_1 \left(k_T^2 - k_2^2 \right) / k_2 \left(k_T^2 - k_1^2 \right) , \tag{25}$$

and at an adiabatic boundary

$$(A_2/A_1) = - \left(k_T^2 - k_2^2 \right) / \left(k_T^2 - k_1^2 \right) . \tag{26}$$

On the basis of the expressions obtained, one can write down the temperature distribution in the sample and calculate the metal's surface impedance $Z = E_x(0)/H_y(0)$. To solve the problem of ultrasonic waves generated under the action of a thermoelectric force it is sufficient to restrict ourselves to the temperature distribution.

It is, at an isothermal boundary,

$$\theta(z) = Re \left[cH_y(0)/4\pi \right] \times$$
$$\times \left\{ T\alpha_{xz} k_1^2 k_2^2 \left[\exp(ik_1 z) - \exp(ik_2 z) \right] / \kappa_{zz} k_T^2 \left(k_1^2 - k_2^2 \right) \right\} , \tag{27}$$

and at an adiabatic boundary

$$\theta(z) = Re \left[cH_y(0)/4\pi \right] \times$$
$$\times \left\{ T\alpha_{xz} k_1 k_2 \left[k_1 \exp(ik_1 z) - k_2 \exp(ik_2 z) \right] / \kappa_{zz} (k_2 - k_1) \left(k_1 k_2 + k_T^2 \right) \right\} . \tag{28}$$

Substituting the roots of the dispersion equation (14) into formulae (27) - (28), one can get the precise expressions for the temperature distribution within the metal $\theta = \theta(z)$.

However, taking into account the small magnitude of the coupling parameter α, it is possible to use approximate expressions for the temperature distribution. It would be sufficient to use the thermoconductivity equation (13), where the term $T\alpha_{xz}\partial j_x/\partial z$ serves as the heat source, and $j_x(z) = j_x(0)\exp(ik_E z)$ is the current density not disturbed by thermoelectric force. Then, at an isothermal boundary ($\alpha \ll 1$):

$$\theta(z) \approx Re\left[cH_y(0)/4\pi\right] \times$$
$$\times \left\{T\alpha_{xz}k_E^2\left[\exp(ik_E z) - \exp(ik_T z)\right]/\kappa_{zz}\left(k_T^2 - k_E^2\right)\right\}, \qquad (29)$$

and at an adiabatic boundary ($\alpha \ll 1$)

$$\theta(z) \approx Re\left[cH_y(0)/4\pi\right] \times$$
$$\times \left\{T\alpha_{xz}k_E\left[k_E\exp(ik_E z) - k_T\exp(ik_T z)\right]/\kappa_{zz}\left(k_T^2 - k_E^2\right)\right\}. \qquad (30)$$

Concluding this part of our consideration, note that at low temperatures when $|k_T^2| \ll |k_E^2|$, the thermal wave penetrates a much greater distance than the electromagnetic skin depth. This means, that in accordance with Eqs. (21) - (23) both the electromagnetic field, and current density penetrate the metal a greater distance.

We turn now to the problem of linear ultrasound generation under the action of a thermoelectric force. Generally, metal crystals are characterized by an anisotropy of both conductive, and elastic properties. However, to get a qualitative description of EMAT it would be sufficient in this case to take into account only the anisotropy of the metal conductivity. We have shown that the amplitude of the temperature oscillations is proportional to α_{xz}, the nondiagonal component of thermoelectric coefficient tensor which is equal to zero in an isotropic conductor. The neglect of elastic anisotropy exclude from consideration the generation of transverse ultrasound.

The excitation of longitudinal ultrasonic waves is given by:

$$k_S^2 U + d^2U/dz^2 = \beta d\theta/dz, \qquad (31)$$

where U is the ultrasound amplitude, $k_S = \omega/S$ is ultrasonic wavevector, S is the sound velocity, and β by order of magnitude coincides with the thermal expansion coefficient of the metal.

In EMAT considerations it is convenient to distinguish between two types of boundary conditions for the elasticity equation. At a free boundary the mechanical stress is absent, i.e.

$$dU/dz\big|_{z=0} = \beta\theta\big|_{z=0}, \qquad (32)$$

while at a fixed boundary

$$U\big|_{z=0} = 0. \qquad (33)$$

Note, that according to Eq. (32) the surface force is equal to the oppositely directed volume force resulting in a net force equal to zero. A similar situation takes place when a deformation force[3] is taken into account and the surface scattering of electrons is diffuse. It is evident, that in the case of a non-isothermal boundary, the surface force differs from zero independently of the surface scattering of electrons due to the thermoelastic mechanism of EMAT. The formulae (31) - (33) augmented with the expressions for temperature distribution (29) and (30) permit the calculation of the amplitude of ultrasound excited. In general, we have to calculate the ultrasonic wave's amplitude in four cases. To put the results in order, let's consider independently the cases of mechanically free and fixed surfaces. There are quite a few parameters responsible for the EMAT, and, in particular, different wavevectors, i.e. the electromagnetic wavevector k_E, the thermal wavevector k_T and the ultrasonic wavevector k_S. EMAT usually is realized at conditions, when the sound wavelength sufficiently exceeds the skin layer depth. That is why we will present results valid for $|k_S| \ll |k_E|, |k_T|$. This means also, that the frequency is not too low and satisfies the following restrictions $\omega \ll 2\pi\sigma S^2/c^2$, $S^2/v_F^2 \tau$, where τ is the electron relaxation time.

So, at a mechanically fixed boundary

$$|U| = \begin{cases} \dfrac{cH_y(0)}{4\pi} \beta \dfrac{T\alpha_{xz}}{\kappa_{zz}} \dfrac{k_E}{k_T(k_T+k_E)}, & \text{at isothermal boundary} \\[2ex] \dfrac{cH_y(0)}{4\pi} \beta \dfrac{T\alpha_{xz}}{\kappa_{zz}} \dfrac{k_S^2}{|k_T^2 k_E|}, & \text{at adiabatic boundary} \end{cases} \quad (34)$$

and at a mechanically free boundary

$$|U| = \begin{cases} \dfrac{cH_y(0)}{4\pi} \beta \dfrac{T\alpha_{xz}}{\kappa_{zz}} \dfrac{k_S}{|k_T^2|}, & \text{at isothermal boundary.} \\[2ex] \dfrac{cH_y(0)}{4\pi} \beta \dfrac{T\alpha_{xz}}{\kappa_{zz}} \dfrac{k_S^2}{|k_T^2 k_E|}, & \text{at adiabatic boundary.} \end{cases} \quad (35)$$

Let's compare now the influence of boundary conditions, both mechanical, and thermal, on EMAT. If the boundary is fixed, Eq. (34) shows, that the case of an isothermal boundary is preferable. If the boundary is mechanically free, the result depends on the value of $|k_T/k_E|$. A quantitative estimate of the thermoelectric mechanism of EMAT shows that it is of the same order of magnitude, which can be obtained due to the deformation potential interaction in pure metals at low temperature. The results obtained should be considered as demonstrating the efficiency of the proposed EMAT mechanism. It is possible to calculate the efficiency for any given value of ultrasound wavelength, as well as thermal and electromagnetic skin depths and then one can compare quantitatively Eqs. (34) and (35) with experiment.

The preliminary estimates of the efficiency (or amplitude of excited ultrasound of linear thermoelastic mechanism of EMAT performed for Zn at intermediate temperature range where $\delta_E \sim \delta_T$, shows that this quantity is of the same order of magnitude as the experimentally measured efficiency (or amplitude of excited ultrasound) of linear deformation potential mechanism of EMAT in good metals.

In conclusion, one of the authors (A.N.V.) acknowledges the warm hospitality of his colleagues in Marburg University while drafting this paper.

REFERENCES

[1] A.N. Vasil'ev and Yu.P. Gaidukov, Electromagnetic generation of ultrasound in metals, Sov. Phys. JETP. **26**, 952 (1983).

[2] E.R. Dobbs, Electromagnetic generation of sound in metals, in: Physical Acoustics. Principles and Methods, V.10, W.P. Mason, ed. Academic Press, New York (1973).

[3] I.E. Aronov, V.L. Falko, Electromagnetic generation of sound in metals in a magnetic field, Phys. Rep. **221**, 81 (1992).

[4] L.D. Favro, Thermal nonlinear excitation of sound in solids, in: IEEE Ultrason. Symp. Proc. **1**, 399 (1986).

[5] A.N. Vasil'ev, M.A. Gulyanskii, and M.I. Kaganov, Nonlinear electromagnetic generation of sound in metals, Sov. Phys. JETP, **64**, 117 (1986).

[6] N.M. Makarov, F.P. Rodriguez, and V.A. Yampol'skii, Nonlinear generation of sound in metals in current state, Sov. Phys. JETP, **69**, 1216 (1989).

[7] A.N. Vasil'ev, Yu.P. Gaidukov, M.I. Kaganov, and E.G. Kruglikov, Nonlinear electromagnetic generation of ultrasound in zinc, Sov. Phys. JETP, **74**, 357 (1992).

[8] M.I. Kaganov, Thermoelectric mechanism of electromagneticacoustic transformation, Sov. Phys. JETP, **71**, 1028 (1990).

ELECTRON-PHONON COUPLING IN STRONGLY CORRELATED ELECTRON SYSTEMS

Z. K. Petru* and N.M. Plakida**

* Institute of Theoretical Physics, University of Wroclaw
Plac Maksa Borna 9, 50-204 Wroclaw, Poland

** Joint Institute for Nuclear Research, 141980 Dubna, Russia

ABSTRACT

The electron-phonon coupling is discussed in the framework of the Bogolubov method. Due to imperfect screening in the high-T_c oxides, fluctuations of the crystal field and fluctuations of the Coulomb field are the main contributions to the electron-phonon interaction in ionic metals with strongly correlated electrons.

1. INTRODUCTION

The high-T_c oxide superconductors have a low density of charged carriers, $n \sim 10^{21}$ cm^{-3}, which results in an inefficient screening of ions [1,2]. It differentiates these new superconductors from the "classical" ones with ordinary metallic screening ($n \sim 10^{22} - 10^{23}$ cm^{-3}). The inefficient screening of the Coulomb interaction leads to a strong coupling of electrons with ion vibrations [1,2]. The aim of this paper is to give a microscopic background for the electron-phonon coupling in oxide superconductors following the above mentioned idea of its ionic origin.

To study this many-body interaction, the microscopic theory of strongly correlated electron systems is presented. It gives, in the framework of the Bogolubov method [3], a systematic description of electrons *in the site representation*. Some microscopic models of the high temperature superconductors, which are generalized versions of the Hubbard model, can be derived as special cases of the theory. The main example is a generalization of the Emery model [4]. The approach presented here makes it possible to generalize the models by including the strong coupling of electrons with vibrations of inefficiently screened ions. We will discuss three types of electron-phonon interactions which might enhance the superconducting temperature:

(i) the ordinary electron-phonon coupling related to the intersite hopping of electrons,

(ii) the crystal field fluctuations generated by ion displacements,

(iii) the fluctuations of the electron-electron Coulomb interaction given by ion vibrations.

2. SITE REPRESENTATION

To describe the many-body interactions mentioned above it is convenient to use the second quantization formalism. Due to the localized character of the electron wave functions in the new oxide superconductors the method of orthogonalized atomic orbitals developed by Bogolubov [3] seems to be the most appropriate for studying strongly correlated electron systems [5]. The Bogolubov method (sometimes called the polar model of a metal) is the systematic way to describe a strongly correlated electron system by the second quantization Hamiltonian written *in the site representation*. To do this rigorously, from a mathematical point of view, the basis of the orthonormal functions should be defined. The orthogonalization condition is necessary if a many-electron function is constructed as a Slater determinant. Otherwise the probabilistic interpretation of it becomes meaningless [3].

Usually, in the site description the basis is formed with nonorthogonal atomic orbitals $\phi_i(\mathbf{r})$. The overlap integrals between atomic orbitals:

$$S_{ij} = (\phi_i, \phi_j), \quad i \neq j, \tag{1}$$

form the off-diagonal matrix S. Here the subscript $i \equiv \{\lambda_i, \mathbf{R}_i, \sigma_i\}$ denotes, in general, a type of the atomic orbital λ_i, an instantaneous position $\mathbf{R}_i$ of the i-th ion and spin σ_i, respectively.

The simplest way to build up a set of orthonormal functions $\theta_i(\mathbf{r})$ is to take a linear combination of the ϕ_i of the following form:

$$\theta_i(\mathbf{r}) = \sum_i \phi_j(\mathbf{r}) u_{ji}. \tag{2}$$

The matrix $\mathcal{U}$ of coefficients u_{ij} is determined by the orthonormalization condition:

$$(\theta_i, \theta_j) = \delta_{ij}. \tag{3}$$

$\mathcal{U}$ can formally be written as a function of S according to the formula [6]

$$\mathcal{U} = (1 + S)^{-\frac{1}{2}} \mathcal{B}, \tag{4}$$

where $\mathcal{B}^+\mathcal{B} = 1$. In what follows we choose $\mathcal{B} = 1$.

The Bogolubov method is based on the assumption that the overlap integrals between atomic orbitals are small. Therefore θ_i can be written [3,6] as a power series in S_{ij}:

$$\theta_i = \phi_i - \frac{1}{2} \sum_j \phi_j S_{ji} + \frac{3}{8} \sum_{jk} \phi_j S_{jk} S_{ki} + \ldots \tag{5}$$

Having the orthonormal set $\{\theta_i\}$, one can build up the Slater function (determinant). Then, using the standard procedure, the Hamiltonian of the electron system can be written down in the site representation [3]:

$$H = \sum_{ii'} L_{ii'} a_i^+ a_{i'} + \frac{1}{2} \sum_{iji'j'} F_{ijj'i'} a_i^+ a_j^+ a_{j'} a_{i'}. \tag{6}$$

Here $L_{ii'}$ is the matrix element of the one-electron part:

$$L_{ii'} \equiv L(\lambda_i \mathbf{R}_i; \lambda_{i'} \mathbf{R}_{i'}) \delta_{\sigma_i \sigma_{i'}}$$
$$\equiv \delta_{\sigma_i \sigma_{i'}} \int \theta^*_{\lambda_i \mathbf{R}_i}(\mathbf{r}) (-\frac{\hbar^2}{2m}\Delta + \sum_j U_{\mathbf{R}_j}) \theta_{\lambda_{i'} \mathbf{R}_{i'}}(\mathbf{r}) d\mathbf{r}. \tag{7}$$

The Coulomb potential $U_{\mathbf{R}_j}(\mathbf{r}) \equiv U_j(\mathbf{r} - \mathbf{R}_j)$ is created by the j-th ion being at the instantaneous position $\mathbf{R}_j$ and acting on valence electrons. It is assumed, as usual, that the nuclear charge is partly screened by the core electrons of the ion.

The last term in (6) is the most important for systems of strongly correlated electrons. It describes the Coulomb interaction between electrons (holes)

$$F_{ijj'i'} \equiv \delta_{\sigma_i\sigma_{i'}}\delta_{\sigma_j\sigma_{j'}}F(\lambda_i\mathbf{R}_i, \lambda_j\mathbf{R}_j; \lambda_{j'}\mathbf{R}_{j'}, \lambda_{i'}\mathbf{R}_{i'})$$
$$\equiv \delta_{\sigma_i\sigma_{i'}}\delta_{\sigma_j\sigma_{j'}}\int \theta^*_{\lambda_i\mathbf{R}_i}(\mathbf{r}_1)\theta^*_{\lambda_j\mathbf{R}_j}(\mathbf{r}_2)U(\mathbf{r}_1-\mathbf{r}_2)\theta_{\lambda_{j'}\mathbf{R}_{j'}}(\mathbf{r}_2)\theta_{\lambda_{i'}\mathbf{R}_{i'}}(\mathbf{r}_1)d\mathbf{r}_1 d\mathbf{r}_2. \tag{8}$$

Eqs. (1)-(8) form a basis for describing the electron system in the site representation. The microscopic Hamiltonian Eq. (6) in the second quantization formalism is convenient for the application of standard many-body methods. In particular, we are interested in describing the electron-phonon interaction within the Bogolubov scheme. To do this we apply now the Bogolubov method Eq. (1)-(8) to a simple model system which, however, is still adequate to describe strongly correlated electrons. Such a model related to the high temperature superconductors was proposed by Emery [4]. In the following section it will be shown that the Emery-type models can be obtained as a special case of the Bogolubov approach.

3. THREE BAND MODEL

It is commonly accepted that the properties of the copper-oxide superconductors are dominated by a strong coupling between $Cu3d$ and $O2p$ electrons in the CuO_2 basal plane. In the simplest model, the three band model of Emery [4], only the $d_{x^2-y^2}$ atomic orbital for Cu and the p_x and p_y orbitals for O are taken into account. Substituting them into Eq. (5) gives us a new basis:

$$\theta_{d,\mathbf{R}} = \phi_{d,\mathbf{R}} - \frac{1}{2}S(\phi_{x,\mathbf{R}+\mathbf{a}_x/2} - \phi_{x,\mathbf{R}-\mathbf{a}_x/2} + \\ -\phi_{y,\mathbf{R}+\mathbf{a}_y/2} + \phi_{y,\mathbf{R}-\mathbf{a}_y/2}) + \cdots \tag{9}$$

$$\theta_{\alpha,\mathbf{R}+\mathbf{a}_\alpha/2} = \phi_{\alpha,\mathbf{R}+\mathbf{a}_\alpha/2} \mp \frac{1}{2}S(\phi_{d,\mathbf{R}} - \phi_{d,\mathbf{R}+\mathbf{a}_\alpha}) + \cdots \tag{10}$$

where $-$ and $+$ signs should be taken for $\alpha = x$ and y, respectively. Here

$$S \equiv (\phi_{d,\mathbf{R}}, \phi_{x,\mathbf{R}+\mathbf{a}_x/2}). \tag{11}$$

and, for simplicity, only the overlap integrals for orbitals of the nearest-neighbour ions were considered.

Having defined the functions θ_i we can write down the Slater many-electron function and calculate the matrix elements $L_{ii'}$ and $F_{ijj'i'}$ of the Hamiltonian Eq. (6). Then by substituting Eqs. (9) and (10) into Eqs. (7) and (8) we get the generalization of the Emery-type model:

$$H = \sum_{i\sigma} E_{di}d^+_{i\sigma}d_{i\sigma} + U_d\sum_i n_{di\uparrow}n_{di\downarrow} \\ + \sum_{\alpha=x,y}\left(\sum_{j\sigma} E_{\alpha j}p^+_{\alpha j\sigma}p_{\alpha j\sigma} + \sum_{<i,j>\sigma}(t_{di,\alpha}d^+_{i\sigma}p_{\alpha j\sigma} + h.c.)\right. \\ \left. + \sum_j U_\alpha n_{\alpha j\uparrow}n_{\alpha j\downarrow} + \sum_{<i,j>} U_{d,\alpha}n_{di}n_{\alpha j}\right) + \cdots \tag{12}$$

where only the leading terms (i.e. the lower order terms in S) of a given type were written down. For example,

$$E_{d,i} \equiv L(d,\mathbf{R}_i; d,\mathbf{R}_i) = E_d^0 + \int |\phi_{d,\mathbf{R}_i}(\mathbf{r})|^2 V'_{\mathbf{R}_i}(\mathbf{r})d\mathbf{r}. \tag{13}$$

Here E_d^0 is the initial energy level of an electron from the $d_{x^2-y^2}$ orbital and

$$V'_{R_i}(\mathbf{r}) = \sum_{R_j(\neq R_i)} U_j(\mathbf{r} - \mathbf{R}_j) \tag{14}$$

is a potential of the crystal field generated by all the neighbours of the i-th copper ion. Similar formulæ hold for the energy levels $E_{\alpha j}$ related to the p_α orbitals ($\alpha = x$ or y) of oxygen.

The expression describing the intersite hopping shows that it is of the same order as the overlap integral

$$t_{di,\alpha j}(\mathbf{R}_i - \mathbf{R}_j) \equiv L(d, \mathbf{R}_i; p_\alpha, \mathbf{R}_j) \sim S_{di,\alpha i}(\mathbf{R}_i - \mathbf{R}_j); \tag{15}$$

therefore, according to the Bogolubov approach, terms like (15) are much less important than crystal field terms (13).

On-site repulsion terms, characteristic for the strongly correlated electron systems, are given here as the diagonal matrix elements of the Coulomb electron-electron interaction (8). For the d electrons (holes) this on-site term is defined as

$$U_d \equiv F(d, \mathbf{R}_i, d, \mathbf{R}_i; d, \mathbf{R}_i, d, \mathbf{R}_i)$$
$$= \int | \phi_{d,R_i}(\mathbf{r}_1) |^2 U(\mathbf{r}_1 - \mathbf{r}_2) | \phi_{d,R_i}(\mathbf{r}_2) |^2 d\mathbf{r}_1 d\mathbf{r}_2. \tag{16}$$

A similar formula can be written down for the on-site repulsion of the p_α electrons, U_α.

The off-diagonal matrix elements describe the intersite Coulomb interactions of the form:

$$U_{di;\alpha j} \equiv F(d, \mathbf{R}_i, p_\alpha, \mathbf{R}_j; p_\alpha, \mathbf{R}_j, d, \mathbf{R}_i)$$
$$= \int | \phi_{d,R_i}(\mathbf{r}_1) |^2 U(\mathbf{r}_1 - \mathbf{r}_2) | \phi_{\alpha,R_j}(\mathbf{r}_2) |^2 d\mathbf{r}_1 d\mathbf{r}_2. \tag{17}$$

4. ELECTRON-PHONON COUPLING

The coefficients (7) and (8) of the Hamiltonian Eq. (6) being matrix elements have the form of integrals and so depend only on a relative distance between the instantaneous positions of ions. This suggests a natural generalization of the Bogolubov method by taking into account the displacements of the ions. We must recall that an ion can be displaced by an amount $\mathbf{u}_i$ from its equilibrium position $\mathbf{R}_i^o$. Then the instantaneous position $\mathbf{R}_i$ of the ion can be written in the usual form:

$$\mathbf{R}_i = \mathbf{R}_i^o + \mathbf{u}_i. \tag{18}$$

Now by expanding the coefficients of the Hamiltonian (6) or (12) into a power series in $\mathbf{u}_i$ we will get terms describing the electron-phonon coupling. There are three main origins of the coupling:

(i) Coupling related to the hopping term.

In a traditional approach, the electron-phonon coupling usually comes from the displacement dependence of the hopping term (15):

$$t_{di,\alpha j}(\mathbf{R}_i - \mathbf{R}_j) \approx t_{di,\alpha j}(\mathbf{R}_i^o - \mathbf{R}_j^o) + \sum_\beta t_{di,\alpha j}^\beta (u_j^\beta - u_i^\beta) + \ldots \tag{19}$$

where $t^{\beta}_{di,\alpha j} \equiv \nabla^{\beta} t_{di,\alpha j}$ is a linear term of the power series for the hopping integral. Hence, this gives us the following electron-phonon term in the Hamiltonian

$$H_1^{el-ph} = \sum_{\alpha,ij} \sum_{\sigma} \left[\sum_{\beta} t^{\beta}_{di,\alpha j}(u_j^{\beta} - u_i^{\beta}) d^+_{i\sigma} p_{\alpha j \sigma} + h.c. \right]. \qquad (20)$$

Here only the leading (linear) term is written down. It should be stressed, however, that such a linear coupling can vanish for some phonon modes. This takes place for the in-plane $d_{x^2-y^2}$ and p_x, p_y orbitals coupled via the tilting mode in La_2CuO_4[7]. For these orbitals the leading coupling term should begin with a quadratic term of the power series (19). The interesting exception is the p_z oxygen orbital, perpendicular to the copper-oxide plane, which couples linearly to the $d_{x^2-y^2}$ orbital but only dynamically via the tilting mode [7]. Both couplings mentioned here should be taken into account when the influence of the tetragonal-to-orthorhombic transition on superconductivity is discussed [2,7].

(ii) Fluctuations of the crystal field

For the localized electrons the hopping term $t_{di,\alpha j}$ is small. It is of the order of the overlap integral S. Rough estimates [2] have shown that the electron-phonon coupling coming from the hopping term, Eq. (20), can be ignored in the first approximation. Because of the ionic character of the copper-oxide superconductors, mentioned in the Introduction, the crystal field fluctuations generated by the vibrations of ions are more important. They come from the power series expansion of the crystal field terms E_{di} and $E_{\alpha j}$. From Eq. (13) one can get:

$$\delta E_{d,i} \approx \sum_j \left[\sum_{\beta}(u_j^{\beta} - u_i^{\beta}) U^{\beta}_{di-j} + \frac{1}{2} \sum_{\beta\gamma}(u_j^{\beta} - u_i^{\beta})(u^{\gamma} - u^{\gamma}) U^{\beta\gamma}_{di-j} + \ldots \right] \qquad (21)$$

where

$$U^{\beta\gamma}_{di-j} \equiv \int |\phi_d(\mathbf{r})|^2 \nabla^{\beta}\nabla^{\gamma} U_j(\mathbf{r} - \mathbf{R}_j + \mathbf{R}_i)|_{u=o} d\mathbf{r} \qquad (22)$$

and U^{β}_{di-j} has a similar form to (22) but with one derivative instead of two. The fluctuations of the crystal field (21) couple to the d-electrons (see the first term of the Hamiltonian (12)) and generate this way an electron-phonon coupling of the form:

$$H_2^{el-ph} = \frac{1}{2} \sum_{i,\sigma} \sum_j \sum_{\beta\gamma} U^{\beta\gamma}_{di-j}(u_j^{\beta} - u_i^{\beta})(u_j^{\gamma} - u_i^{\gamma}) d^+_{i\sigma} d_{i\sigma}. \qquad (23)$$

In normal metals the crystal field fluctuations (21) are screened out by free electrons. However, in the copper-oxides, the screening is much less effective. In particular, it is quite ineffective for the ion displacements perpendicular to the copper-oxide plane. As a consequence, the strong electron-phonon coupling of ionic origin is expected in these compounds.

(iii) Fluctuations of the electron-electron interaction

The third source of the electron-phonon interaction becomes the electron-electron Coulomb interaction (8). It is estimated to be large in the copper-oxides which suggests that its role in the electron-phonon coupling is important. However, in strongly correlated electron systems this kind of a coupling has not been discussed until now. This is due to the fact that our knowledge about the details of Coulomb terms (8) is not sufficient.

Usually, the main attention is directed to the on-site interaction term (16) which is the largest one in the Hubbard-type Hamiltonians. The on-site term, however, does not depend on ion displacements, as is seen from Eq. (16), and, therefore, does not involve phonons.

The opposite takes place for the intersite Coulomb interaction of the type (17) which explicitly depends on ion displacements. Here, by expanding $U_{di;\alpha j}$ one can get fluctuations of the intersite electron-electron Coulomb interaction:

$$\delta U_{di,\alpha j} = \frac{1}{2} \sum_{\beta\gamma} (u_j^\beta - u_i^\beta)(u_j^\gamma - u_i^\gamma) U_{di-\alpha j}^{\beta\gamma} \qquad (24)$$

where:

$$U_{di-\alpha j}^{\beta\gamma} \equiv \int |\phi_d(\mathbf{r}_1)|^2 \, \nabla^\beta \nabla^\gamma U(\mathbf{r}_1 - \mathbf{r}_2 + \mathbf{R}_i - \mathbf{R}_j)|_{u=0} \, |\phi_\alpha(\mathbf{r}_2)|^2 \, d\mathbf{r}_1 d\mathbf{r}_2 \qquad (25)$$

Substitution of (25) into the last term of the Hamiltonian (12) creates an electron-phonon coupling of the form:

$$H_3^{el-ph} = \frac{1}{4} \sum_\alpha \sum_{<i,j>} \sum_{\beta\gamma} (u_j^\beta - u_i^\beta)(u_j^\gamma - u_i^\gamma) U_{di\alpha j}^{\beta\gamma} n_{di} n_{\alpha j}. \qquad (26)$$

This interaction should renormalize both the phonon and electron spectra. However, it is difficult to estimate the matrix element $U_{di-\alpha j}^{\beta\gamma}$ (25). Till now even qualitative knowledge about the distance dependence of $U_{di-\alpha j}(\mathbf{R}_i - \mathbf{R}_j)$ has not been available. There is even some contradictory evidence about the magnitude of $U_{di-\alpha j}(\mathbf{R}_{i-j}^o)$ itself. Usually its value is treated as less then 1 eV [8], although recent XPS and Auger data analysis [9] have yielded much larger values ($\approx$ 2.5 eV). On the other hand it is well known that the intersite interaction can explain the appearance of superconductivity in the strongly correlated electron systems [10].

SUMMARY

The electron-phonon coupling in the systems of strongly correlated electrons was described in the site representation. Three types of interaction were discussed. All of them can lead to superconductivity but the fluctuations of the crystal and Coulomb fields are characteristic for the high temperature superconductors with a perovskite-type structure. These fluctuations are large and the related electron-phonon coupling is strong, due to the imperfect screening in high-T_c oxides, contrary to usual metals where such fluctuations are screened out. We hope that by taking into account this coupling we will be able to explain the appearance of instabilities in the mentioned materials as well as their relationship to high temperature superconductivity.

The authors thank S. Barišić and R. Zeyher for valuable discussions. One of the authors (ZKP) acknowledges a support from the State Committee for Scientific Research under Grant No. 2 0121 91 01.

REFERENCES

1. R. Zeyher, in *Festkörper Probleme* **31**, 19 (1991).
2. S. Barišić, in *'Critical Trends in High-T_c Superconductivity*, B.K. Chakraverty (ed.), World Scientific, Singapore (1991).
3. N.N. Bogolubov, *Lectures on Quantum Mechanics*, Gordon and Breach, New York (1967).
4. V.J. Emery, *Phys. Rev. Lett.* **58**, 2794 (1987).

5. N.N. Bogolubov, V.L. Aksenov, and N.M. Plakida, *Physica C* **153-155**, 99 (1988).
6. P.O. Löwdin, *Adv. Phys.* **5**, 1 (1956).
7. Z.K. Petru, and N.M. Plakida, *Mod. Phys. Lett. B* **4**, 489 (1990).
8. F. Mila, *Phys. Rev. B* B **38**, 11358 (1988).
9. E.B. Stechel, and D.R. Jennison, *Phys. Rev. B* **40**, 6919 (1989).
10. R. Micnas, J. Ranninger and S. Robaszkiewicz, *Rev. Mod. Phys.* **62**, 113 (1990).

MAGNETOPHONON RESONANCES ON THREE PHONON MODES IN $ZN_XCD_YHG_{1-X-Y}TE$ EPITAXIAL LAYERS

J. Polit[1], E. Sheregii[1], A. Andruchiv[2] and P. Sidorchuk[2]

[1]Institute of Physics, Pedagogical University,
Rzeszow, Poland
[2]Dept. of Physics, Pedagogical University,
Drohobych, Ukraine.

Four-component materials $Zn_xCd_yHg_{1-x-y}Te$ have an important advantage. In a crystal lattice only the cationic part is changed: Hg, Cd and Zn while the anionic part remains unchanged. On the one hand this means that it is possible to obtain better quality crystals and epitaxial films; on the other hand it leads to the interesting phenomenon of the mixture of transverse and longitudinal optical oscillations of each mode:HgTe-like, CdTe-like and ZnTe-like.[1,2] The energies of phonons have been presented in Table 1.

Table 1. Energies of LO(T)phonons in $Zn_{0.12}Cd_{0.10}Hg_{0.78}Te$ according [1],[2],[3].

HgTe-like lattice [meV]	CdTe-like lattice [meV]	ZnTe-like latice [meV]
17.0	20.5	25.0

Magnetophonon resonance MPR is a good instrument for studying of the phonon-spectra of such mixed crystals.

The investigation was carried out on perfect epitaxial layers $Zn_xCd_yHg_{1-x-y}Te$ ($x = 0.12, y = 0.10$) with electron mobilities $5\times10^4 cm^2/Vc$. In the temperature range 77-350 K oscillations of the transverse magnetoresistance were observed. These oscillations of the transverse magnetoresistance were obtained in the temperature range from 77 to 300 K using pulsed magnetic fields from 0 to 10 T.

Fig 1-3 shows the experimental records of the second derivative of the transverse magnetoresistance ρ_{xx} as a function of magnetic field for five values of temperature. These curves are presented in different scales for magnetic fields.

Three series of peaks a, b, and c can be seen. Series a: with LO-phonons from a ZnTe-like lattice; series b: with LO-phonons from a CdTe-like lattice; series c: with LO-phonons from a HgTe-like lattice.

The Interpretation of the peaks has been done on the basis of Pidgeon-Brown model modified by Weiler et al.[3] for calculation of Landau levels.

The best agreement of experimental results with theory was obtained for zone parameters Eg $0.22eV$, $Ep = 18eV$ (matrix element of quasi-pulse). The interpretation of the peaks is presented in Table 2.

Table 2. The interpretation of the MPR peaks.

$H_{res}^{exp}[H]$	6.1	4.9	3.6
Transi--tions	$0 \to 1+$ +LO(ZnTe)	$0 \to 1+$ +LO(CdTe)	$0 \to 1+$ +LO(HgTe)
$H_{res}^{theory}[T]$	6.0	4.9	3.55
$H_{res}^{exp}[H]$	3.1	2.3	2.3
Transi--tions	$0 \to 2+$ +LO(ZnTe)	$0 \to 2+$ +LO(CdTe)	$0 \to 3+$ +LO(ZnTe)
$H_{res}^{theory}[T]$	3.0	2.45	2.1
$H_{res}^{exp}[H]$	1.7	1.7	1.7
Transi--tions	$0 \to 2+$ +LO(HgTe)	$0 \to 3+$ +LO(CdTe)	$0 \to 4+$ +LO(ZnTe)
$H_{res}^{theory}[T]$	1.82	1.6	1.55
$H_{res}^{exp}[H]$	1.3	1.15	1.3
Transi--tions	$0 \to 3+$ +LO(HgTe)	$0 \to 4+$ +LO(CdTe)	$0 \to 5+$ +LO(ZnTe)
$H_{res}^{theory}[T]$	1.25	1.12	1.25
New Series			
$H_{res}^{exp}[T]$	0.75	0.35	0.15
Transi--tions	$0 \to 1+$ +LO(CdTe) -LO(HgTe)	$0 \to 2+$ +LO(CdTe) -LO(HgTe)	$0 \to 3+$ +LO(CdTe) -LO(HgTe)
$H_{res}^{theory}[T]$	0.75	0.37	0.20
$H_{res}^{exp}[T]$	1.7	0.75	0.55
Transitions	$0 \to 1+$ +LO(ZnTe) -LO(HgTe)	$0 \to 2+$ LO(ZnTe) -LO(HgTe)	$0 \to 3+$ LO(ZnTe) -LO(HgTe)
$H_{res}^{theory}[T]$	1.6	0.8	0.55

It should be noted thet the above mentioned three series of peaks disappear in the range below 1.0 T. A new series of peaks appears in the range $H \leq 1.6$ T. These series can be caused by magnetophonon resonance by subraction of phonon frequencies.For example LO(CdTe-like)-LO(HgTe-like) or LO(ZnTe-like)-LO(HgTe- like) lattice[4].

In this manner we observed first magneto-phonon resonance with participation of three types of phonons belonging to different sub-lattices of the crystals.

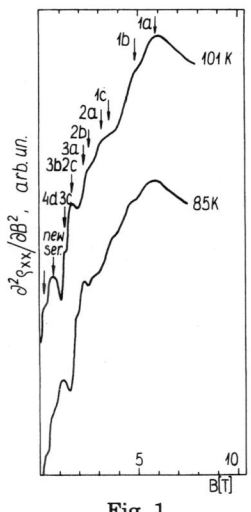

Fig. 1

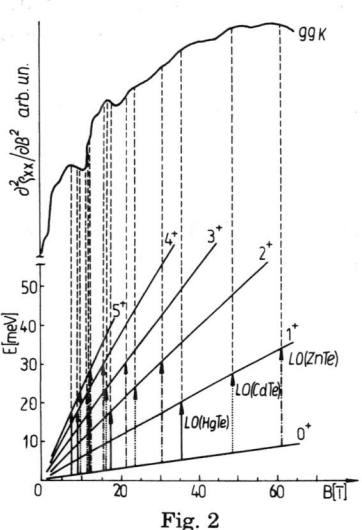

Fig. 2

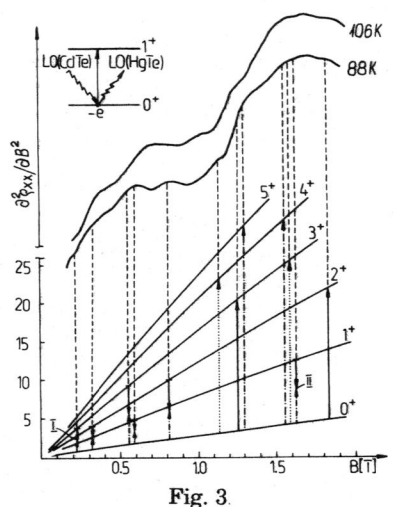

Fig. 3.

REFERENCES

1. J. Baars, F. Sorger, Spectrum of the oscilatione of HgTe and $Cd_xHg_{1-x}Te$, *Solid Stat. Commun.* 10:875(1972).

2. K. Kumazaki, F. Lida, K. Ohno, K. Hatano, K. Imai, Lattice Strain Near Interfance of MBE-Grown ZnTe on GaAs, *Journal of Crystal Growth* 117:285(1992)

3. Weiler M. H., Aggarwal R. L.,Lax B. Interband Magnetoreflectance in semicondacting $Hg_{1-x}Cd_xTe$ alloys, *Phys. Rev.* B 16:3603(1977).

4. E.M. Sheregii, Yu.O. Ugrin, D.D. Shuptar, O.M. Leshko , Magnetophonon Resonance on mixed Phonon Modes in $Cd_xHg_{1-x}Te$, *Pis'ma Ž. Eksp. Teor. Fiz.* 47, 615(1988).

A NOTE ON THE MEASUREMENT OF THE INTERSUBBAND RELAXATION TIME BY AN INFRARED BLEACHING TECHNIQUE

M. Załużny

Institute of Physics, M. Curie–Skłodowska University
PL–20–031, Lublin, pl. M. Curie–Skłodowskiej 1, Poland

INTRODUCTION

The investigation and analysis of electron intersubband transitions within quantum wells (QWs) and multiple quantum well structures (MQWSs) is an important aspect since the successful operation of many proposed electro-optical devices depends on these processes. In particular, the intersubband relaxation time determines to a large degree on the sensitivity of infrared detectors based on intersubband optical absorption.

In "wide" $GaAs/Al_xGa_{1-x}$ As quantum wells where the energy separation (ΔE) between the ground subband (c_1) and excited subband (c_2) is small, only slow intersubband relaxation by acoustic phonon emission is possible. In the "thin" (≤ 100 Å) QWs with ΔE greater than the LO-phonon energy, rapid intersubband relaxation by optical phonon emission is allowed. Several experimental techniques are available to determine the corresponding time constants[1]. One of them is the infrared bleaching technique used by Seilmeier et al[1,2]. This method requires ultra short wavelength-tunable light pulses (with photon energy $\hbar\omega \sim \Delta E$). An intense infrared pulse resonantly excites a considerable number of electrons from the c_1 to the c_2 subband resulting in a bleaching of the intersubband transitions. A second weaker light pulse of the same frequency monitors the absorption change as a function of the time. The absorption recovery reflects the repopulation of the lowest subband.

In this paper we discuss the influence of the electron–electron interaction on the direct determination of the intersubband scattering time by the infrared bleaching technique. We show that there is a situation where this interaction can play a very important role.

THEORY

Seilmeier et al.[1,2] measured the relative change in transmission change of a MQWS ΔT ($= \ln(T/T_0)$) as a function of the time delay t_d between the pump and probe pulses (T_0 is the transmission of the MQWS without excitation by the intense pulse).

To obtain a strong absorption signal they used a special prism geometry. The relative transmission change of the MQWS, normalized to unity for $t_d = 0$, may then be written as:

$$\Delta \bar{T} = \Delta T / \Delta T \Big|_{t_d=0} = \frac{\Lambda_0 - \Lambda}{\Lambda_0 - \Lambda} \Big|_{t_d=0}, \tag{1}$$

where Λ is the absorption (= fraction of the incident energy absorbed) of a single QW and Λ_0 is the absorption without excitation. We assume that in the system considered here only the ground subband is occupied in the absence of the excitation. Thus the instantaneous number of the carriers in the upper subband can be described by

$$N_2 = \Delta N \, \exp(-t_d/\tau), \tag{2}$$

where τ is the intersubband relaxation time, ΔN is the number of carriers generated by an intense first pulse. The absorption Λ is proportional to the real part of the modified two-dimensional conductivity $\tilde{\omega}_{zz}(\omega)$ describing the response of a current to an external (not total) field[3] (the z-axis is perpendicular to the layer). If we neglect, as in Refs. 1 and 2, the Coulomb interaction, then $\tilde{\sigma}_{zz}(\omega)$ is given by[3]

$$\tilde{\sigma}_{zz}(\omega) = (N_1 - N_2)\, e^2 \, z_{21}^2 (-i) \frac{2\omega \, \Delta E}{\Delta E^2 - (\hbar\omega)^2 - 2i\hbar\omega\Gamma}. \tag{3}$$

where N_i is the number of the carriers in the c_i subband, z_{21} is the intersubband dipole matrix element and Γ is a phenomenological parameter describing the line broadening. Near the resonance ($\hbar\omega \sim \Delta E$) the real part of the conductivity can be approximated by

$$Re \, \tilde{\sigma}_{zz}(\omega) \approx (N_1 - N_2)\, e^2 \, z_{21}^2 \, \hbar^{-1} \frac{\Gamma \, \Delta E}{\delta^2 + \Gamma^2}, \tag{4}$$

where $\delta = \hbar\omega - \Delta E$ is the detuning. Note that the complete bleaching of the intersubband absorption takes place when $b \equiv 2\Delta N/N_s = 1$ where $N_s \, (= N_1 + N_2)$ is the total number of carriers. Using Eqs. (1-4) we find that

$$\ln(\Delta \bar{T}) = -t_d/\tau. \tag{5}$$

This formula is equivalent to that used by Seilmeier et al.[1,2]. It shows that when effects induced by the $e - e$ interaction are negligible then the intersubband absorption recovery directly reflects the depopulation of the second subband. Consequently, the intersubband relaxation time can be measured directly by the bleaching technique using Eq. (5). Now we show that inclusion of the $e - e$ interaction complicates the above simple picture. The influence of the Coulomb interaction on the intersubband absorption spectra (of $GaAs \, QWs$) is usually discussed within the Hartree approximation. Let us denote by $\chi_n(z)$ the envelope wave function of the n-th subband and by ΔE the intersubband energy resulting from self-consistent solution of the Schrödinger equation at $N_2 = 0$. The electron redistribution among the subbands by the optical fields leads to a modification of the Hartree potential and consequently to a shift of the intersubband energy (from the $N_2 = 0$ value) by an amount δE_e. In the case of rectangular QWs, dE is much smaller than ΔE and the approximate expression can be used[4]

$$\delta E_e / \Delta E = (2N_2/N_s)\, \lambda_e = (2N_2/N_s)\, e^2 <\delta z>_e N_s/2\Delta \, E \, \epsilon_0 \, \epsilon(0), \tag{6}$$

with

$$<dz>_e = \int_{-\infty}^{\infty} dz \left[\int_{-\infty}^{z} dz' \left(|\chi_1(z')| - |\chi_2(z')| \right) \right]^2 , \qquad (7)$$

where ϵ_0 is the vacuum permitivity and $\epsilon(\omega)$ is the frequency dependent relative dielectric constant. The Hartree approximation predicts also a shift of the intersubband resonant energy (from the intersubband energy) due to the depolarization effect[3]. If we neglect the Coulomb electrostatic effect ($\delta E_e = 0$) then the intersubband resonant energy can be written as[3,5]

$$\Delta \tilde{E} = \Delta E \left(1 + \alpha\right)^{1/2} , \qquad (8)$$

with

$$\alpha = 2e^2 \left(N_1 - N_z\right) L_{12}/\epsilon(\infty) \, \epsilon_0 \, \Delta E , \qquad (9)$$

where

$$L_{12} = \int_{-\infty}^{\infty} dz \left(\int_{\infty}^{z} dz' \chi_1(z') \chi_2(z') \right)^2 , \qquad (10)$$

is the Coulomb matrix element.

Thus, when the depolarization effect is included, redistribution of the electrons among the subbands will reduce the intersubband resonant energy. Since in typical QW's, $\alpha \ll 1$ the reduction δE_d can be approximated by[6]

$$\delta E_d / \Delta E = (2N_2/N_s) \lambda_d = (2N_2/N_s) e^2 <\delta z>_d N_s / 2\Delta \, E \, \epsilon_0 \, \epsilon(\infty) , \qquad (6)$$

where $<\delta z>_d = -2L_{12}$.

RESULTS AND DISCUSSION

In order to estimate how large the corrections δE_e and δE_d can be we have computed E_1, E_2, ΔE, $<dz>_e$ and $<\delta z>_d$ for the simplest model of a rectangular quantum well structure (see inset in Fig. 1). The calculations have been performed for systems with the electron effective mass $m^* = 0.065 \, m_0$ and height of the barrier $V = 260$ and $300 \, meV$. The confining thickness L changes from 50 to 100 Å.

The results (see Fig. 1) lead us to the conclusion that only in the very "thin" layers ($L \simeq 50$ Å) where E_2 is close to V and the electron penetration into the barrier is large, the electrostatic Coulomb correction δE_e is comparable with the depolarization correction δE_d. In "thicker" layers ($L \geq 75$ Å), the electrostatic Coulomb effect is relatively weak and only the depolarization effect can be considered. In the following we concentrate on such a case. If we use GaAs as a "benchmark material" λ_d can be written as

$$\lambda_d \simeq 0.7 \left(\epsilon_{GaAs}/\epsilon(\infty)\right) N_s[10^{12} \, cm^{-2}] <\delta z>_d [\text{Å}]/\Delta E[meV] , \qquad (11)$$

where ϵ_{GaAs} ($\simeq 11$) is the dielectric constant of $GaAs$. Using Eqs. (1,2) and Eq. (4) (with ΔE replaced by $\Delta \tilde{E}$) we find that in the presence of the depolarization effect the expression for the relative transmission change takes the form

$$\Delta \tilde{T} = \frac{1 - [1 - b \, \exp(-t_d/\tau)] \, F(t_d)}{1 - (1-b) \, F(0)} . \qquad (12)$$

with
$$F(t_d) = \frac{\tilde{\delta}^2 + \Gamma^2}{\left[\tilde{\delta} + b\,\exp\left(-t_d/\tau\right)\right]^2 + \Gamma^2}. \qquad (13)$$

where $\tilde{\delta} = \hbar\omega - \Delta\tilde{E}'$ and $\Delta\tilde{E}'$ is the intersubband resonant energy at $N_2 = 0$.

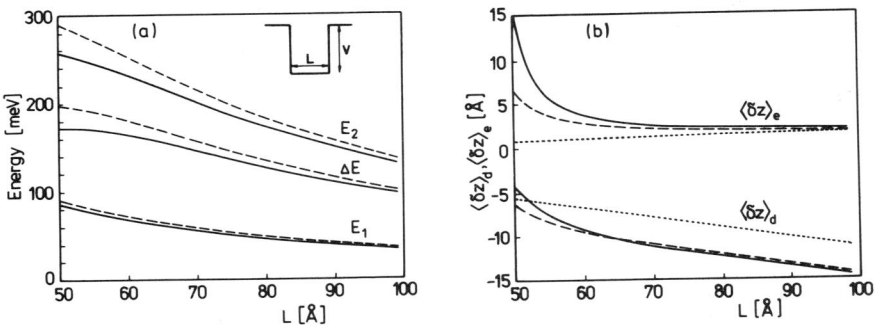

Fig. 1. Dependence of (a) $E_1, E_2, \Delta E$ and (b) $<\delta z>_e$, $<\delta z>_d$ on the well thickness L in $GaAs/Al_xGa_{1-x}As$ QWs shown in the inset. The solid lines are for QWs with $V = 260\ meV$, the dashed are for QWs with $Vt = 300\ meV$ and dotted lines are for infinitely deep QWs.

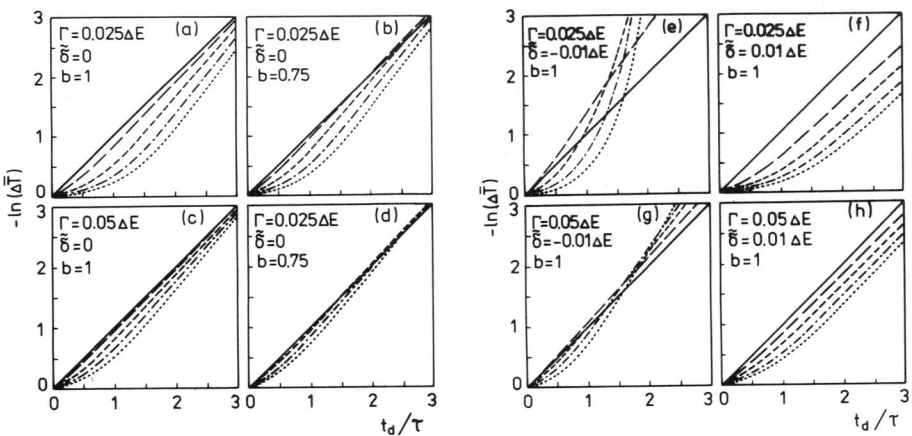

Fig. 2. Dependence of $\ln(\Delta\tilde{T})$ on the delay time t_d in $GaAs/Al_xGa_{1-x}As$ QWs with different values of the parameters Γ, $\tilde{\delta}$ and b. (———)$\lambda_d = 0$, (— —)$\lambda_d = 0.025$, (— — —)$\lambda_b = 0.075$, (— · —)$\lambda_d = 0.05$, $\lambda_d = 0.075$ and (......)$\lambda_d = 0.1$.

To see how strongly the depolarization effect affects the recovery of the intersubband absorption we have calculated the variation of $\ln(\Delta\tilde{T})$ with t_d/τ for different values of the parameters λ_d, b, Γ and $\tilde{\delta}$. From Fig. 2, where the results of the calculations are presented, we find that due to the depolarization effect the relation between $\ln(\Delta\tilde{T})$ and t_d/τ is not linear. A significant deviation from linearity is observed when the depolarization shift ($\lambda_d \Delta E$) is larger than the line width (2Γ). Taking $L = 100$ Å, $b = 1$ and typical (for $GaAs$ QWs) value of the line width ($2\Gamma = 10$ meV) we find from Eq. (11) and Fig. 1 that the above condition is achieved at $N_s \geq 10^{12}$ cm^2.

It is interesting to note that even a small negative detuning enhances substantially the nonlinear behaviour of $\ln(\Delta \tilde{T})$.

The absence of the above type of nonlinearities in the samples studied by Seilmeier et al.[1,2] can be explained by the small thickness of the QWs and the relatively large line width.

In conclusion, we have shown that, in the case of the "thick" QWs with a relatively high electron density, the correct determination of the intersubband relaxation time by the IR bleaching technique must take into account the depolarization effect.

REFERENCES

[1] U. Plodereder, T. Dahinten, A. Seilmeier and G. Wiemann, Influence of doping concentration on intersubband relaxation in modulation-doped quantum well structures Phys. Stat. Sol. b **172**, 373(1992) and references therein.

[2] A. Seilmeier, H. -J.H. Hubner, G. Abstreiter, G. Wieman and W. Schlapp, Intersubband relaxation in GaAs-Al Ga As quantum well structures observed directly by an infrared bleaching technique, Phys. Rev. Lett. **59**, 1345(1987).

[3] T. Ando, A.B. Fowler and F. Stern, Electronic properties of two-dimensional systems, Rev. Mod. Phys. **54**, 437(1982).

[4] J. Khurgin and S. Li, Coulomb enhancement of the third order nonlinearities in the mesoscopic semiconductor structures, Appl. Phys. **A53**, 527(1991).

[5] A. Pinczuk and J.M. Worlock, Light scattering by two–dimensional systems in semiconductors, Surf. Sci. **113**, 69(1982).

[6] M. Zaluzny, Coulomb enhancement of the third-order optical nonlinearities in quantum wells, Appl. Phys. Lett. **61**, 2509(1992).

PARTICIPANTS

Andrey V. Akimov
Kamil L. Aminov
Bartłomiej Andrzejewski
Samvel Badalian
Frank Bagehorn
Romuald Brazis
Paul N. Butcher
Lawrence J. Challis
Stanisław Ciechanowicz
John Cooper
Andrzej Czachor
Jerzy Czerwonko
Borysław Czyżak
Werner Dietsche
Tadeusz Domański
Krzysztof Durczewski
Matvey V. Entin
Arthur G. Every
Ruth H. Eyles
Richard J. Gaitskill
Zygmunt Galasiewicz
Wojciech Gańcza
Rosalind George
Jolanta Góra
Marek Gorzelańczyk
Radii N. Gurzhi
Tomasz Gwizdałła
Grzegorz Harań
George K. Horton
Sergei N. Ivanov
Lucjan Jacak
Czesław Jasiukiewicz
Grzegorz Jastrzębski
Moisei I. Kaganov
Alexander N. Kalinenko
Vytautas Karpus

Dmitrii V. Kazakovtsev
Zbigniew Klusek
Alexander I. Kopeliovich
Arnold M. Kosevich
Tadeusz Kosztołowicz
Mirosław Kozłowski
Hans Kraus
Stanisław Kryszewski
Andrzej Kusy
Dietmar Lehmann
Grzegorz Litak
Jan Lorenc
Tadeusz Łozowski
Jerzy Lukierski
Vadim G. Manzhelii
Miranda Marciniak
Alexei A. Maznev
Michael Meissner
Leonid P. Mezhov-Deglin
Evgenii S. Moskalenko
Jan Mucha
Władysława Nawrocka
Czesław Oleksy
Mariusz Olko
Tadeusz Paszkiewicz
Zygmut Petru
Aleksandra Podolska-Strycharska
Il'ya Ya Polishchuk
Jacek Polit
Kosmas Prassides
Jerzy Przystawa
Krzysztof Rapcewicz
Włodzimierz Salejda
Sergei I. Shevchenko
Alex Shik
Przemysław Siemion

Roman Sikora
Wiesława Sikora
Eugeniusz Soczkiewicz
Alexander Soldatov
Piotr Stachowiak
Jan Stankowski
Małgorzata Sternik
Peter Strehlow
Zbigniew Strycharski
Barbara Sujak-Cyrul
Jerzy Szczepański
Eugeniusz Szeregij
Andrei V. Taranov
Il'ya I. Tartakovskii

Armin Thellung
Nadezhda A. Tulina
Łukasz A. Turski
Alexander N. Vasilev
Michael Warnatz
Ulrich Wenschuh
Marek W. Wilczyński
Michael Willumat
Marek Wolf
Karol Wysokiński
Mirosław Załużny
Elżbieta Zipper
Ryszard Zossel
Franciszka Zossel

INDEX

Acoustic mismatch theory, 119, 202, 264
Amorphous semiconductor films, 121
Anharmonic perturbation theory, 1
   self-consistent phonon, 3, 7
      first-order, 3, 5, 7
      improved, 4-6
      second-order, 4

Bloch diffusion, 205

Conical refraction,
   external, 63
   internal, 63
Compound superlattice, 51
   anisotropic continuous medium approx., 52
   Bragg-type modes in, 53
   interface of, 52
   interface modes of, 51-53
   reststrahlen bands in, 52-53
   slab-guided modes in, 53

Detector
   for β-decay, 298
   bolometric, 55, 141
      superconducting, 56, 89, 140
   calorimetric, 266, 269, 292
   CdS, 57, 166
   cryogenic, 263, 297
   of dark matter, 297
   for heavy ion physics, 264
   magnetic bolometer, 292
   with phase transition thermometer, 267
   phonon, 139
   exciton-cloud, 116, 124-125, 182-183
      superconducting tunnel-junction, 56-57, 69, 89, 266, 277, 286
   x-ray, 264
      micro calorimeter, 267
      superconducting tunnel-junction, 265
   sensor, 280
      superconducting tunnel-junction as, 280, 300-304, 306-316
   sensitivity of, 283, 302
   with superconducting absorber, 315

Elastodynamical Green's function, 56, 73-74

Elastodynamical Green's function *(cont'd)*, 80
Electromagnetic-acoustic transformation, 399, 403-404
   deformation interaction mechanism of, 399, 404
   inductive mechanism of, 399, 403-404
   linear, 399
   nonlinear, 399
   thermoelastic mechanism of, 399, 403-404
Electron capture, 251, 255-256, 261
   acoustic, 259
   bottleneck of, 251
   cascade, 254, 257-259, 261
   coefficient of, 252-260
   inelastic, 252
   rate of, 251-252, 255, 260
      oscillations, 252
   3-D $\to$ 2-D, 251-253, 256, 259-260
   time of, 252, 254-255, 257, 259-260
   2-D $\to$ 2-D, 251-253, 256-257, 260-261
Electron-phonon interaction, 44, 114, 122, 124, 146, 205-206, 209, 213, 235, 243, 247-249, 254-255, 260
   acoustic, 256-257, 259
   deformation potential, 163, 190, 194, 213, 235, 237, 244-245, 248, 258, 399, 403-404
      supression of, 244-245
   in alkali fullerides, 346-350
   linear, 246
   nonlinear, 247 255-256
   piezoelectric coupling, 163, 190, 213, 215, 235-237, 246
      suppression of, 244, 246
   quasielastic, 234
   in strongly correlated electron systems, 407-408
   suppression of, 243, 248-249
   via fluctuations
      of the crystal field, 411
      of the electron-electron interaction, 411
   via hopping term, 410
Electron redistribution
   3-D $\leftrightarrow$ 2-D, 251-255, 257-261

Electron transitions
   isoenergetic, 251, 259
   3-D ↔ 2-D, 251, 257
   3-D → 2-D, 251, 253, 256, 259-261
Enery model of superconductivity, 409
Energy distribution function, 34-35, 38, 44, 49, 78
Frequency distribution function *see* energy distribution function
   of squared frequencies, 34-36
      3-D van Hove singularity of, 40, 78
   partial, 33-44
Epitaxial layer, 415
   magnetophonon resonance in, 415-416
   transverse magnetoresistance, 415
Excitons, 113-117, 121
   bound, 113-114, 117-122, 125
   free, 116-122
   phonon-induced drag, 113-114, 117-119
   surface recombination of, 116, 119

Fountain effect, 190, 199
Fullerenes
   Landau-theory of, 353, 356-362
   molecular dynamics simulations of, 353-354,
   molecular field theory of, 356-357
   μSR studies of, 334-337, 339, 343-345
   neutron scattering studies of, 334-335, 337-338, 340-343, 345-350
   orientational ordering transition in, 333-334, 338, 341, 353, 360
   rotational dynamics of, 333-334, 337-345
   and superconductivity, 333-334, 346-350

GaAs/AlGaAs heterostructure, 159, 164, 179, 198, 201, 203, 205, 207, 209, 211, 215, 219, 223, 225-226, 237,
   interfaces in, 53, 198
   resonant-tunneling diodes, 226
Glass, 105-106, 109, 111, 365-370, 371
   local orientational order in, 367, 372
   local strain in, 370
   orientational degree of freedom in, 370-371, 373, 377
   rheological stress of, 373
   shear viscosity of, 366
   thermalisation of, 109, 111
   transition to, 365-371, 375, 377
   tunnelling model for, 106, 111
   as two-level systems, 111
   visco-elasticity of, 366, 372, 374-375, 377

Heat pulse, 89-90, 105, 107, 109, 111, 116-122, 167
   ballistic, 55, 105, 111, 114, 116, 117, 120-121
Hot electrons, 233, 243
Hot spot, 70, 114-116, 122, 124, 135-137
   degradation of, 136
   formation of, 137
HTS materials, 33, 43-45
Hydrodynamics, 16-17

Hydrodynamics *(cont'd)*,
   phonons in, 15

Kapitza effect, 119, 313

Lagrangian phonons, 15-16, 19-20
Landau level, 160, 162, 174, 190, 198, 237
   line shape, 224
   subbands, 236, 248-249,
Light impurity vibration, 40-41
   local frequency, 41
   local, 40-41, 44-45
   quasilocal, 43-44
Luminescence, 113-115
   band-band, 113
   band-impurity, 113
   exciton, 113-115
   phonon-induced, 113, 117-119

Metal
   conduction electrons, 400
   electroconductivity, 400
   electromagnetic wave in, 399-401
   slab conductivity of, 381-383, 391-396
   specularity coefficient, 383, 385, 392, 395
   surface impedance, 381-383, 387, 389-390, 392, 395-396, 402
   thermal expansion coefficient, 403
   thermal wave in, 400
   thermoelastic stress, 399
   ultrasound in, 400, 404
Metal boundary, 385
   adiabatic, 402, 404
   diffusive, 382-383, 389, 395-396
   free, 403-404
   isothermal, 402-404
   mechanically fixed, 404
   mechanically free, 404
   nonisothermal, 403
   specular, 382-383, 389, 396
Molecular crystals, 321
   librons in, 323
   orientational ordering transitions in, 353
   thermal conductivity of, 325, 327
      inverse temperature law of, 321-323, 325, 329
      isobaric, 322, 325, 328-329
      isochoric, 321-323, 325, 327, 329-331
      orientationally ordered, 330
Momentum
   ordinary, 15, 20, 23, 25, 27
   conservation, 15, 28
   quasi, 15-16, 20-29, 87, 89-91, 93-94,
   conservation, 15-16, 22-24, 26, 28
   density of, 90
   focussing of, 87, 91, 93
   relaxation of, 233
Monte-Carlo method, 1, 4, 5, 7
   applied to ferroelectric, 7
   classical, 1-2
   effective potential, 1-8, 10, 13

Monte-Carlo method *(cont'd)*,
    low-coupling approximation, 4, 10, 12
    of phonon propagation, 56, 58, 61-62, 66-67, 87, 91, 93-94, 129
    path integrals in, 2
    quantum, 2, 4-7, 12-13
Multilayer crystal, 33, 42-45

Neutrino, 263
    coherent elastic scattering of, 263

1-D electron gas, 160, 212-213, 215, 234, 236, 237
    electron temperature measurement, 233, 237
    energy relaxation, 233, 237
    phonon drag in, 211, 214-215
    phonon emission by, 234-235
    screening in, 236
Orthogonalized Atomic Orbital Method, 407

Phonon absorption, 66
Phonon diffusion coefficient, 49, 109
Phonon emission, 66
    optical, 115, 126
Phonon energy density, 90-91, 94
Phonon focusing, 55 59, 61-62, 64, 73, 80, 83, 87-89, 91, 93-94, 170, 194, 198, 205, 215, 304-306
    enhancement factor of, *see* focusing factor, 61, 89, 102
    energy, 87
    focusing factor of, 88-89
    in piezoelectric solids, 64
    quasimomentum, 56
    stationary phase approx., 76, 78
Phonon imaging, 57-58, 61, 63-64, 66, 68-70, 80, 83, 87, 90, 169-170, 192, 199
    anticaustics, 63
    catastrophes in, 62-63
    caustics in, 56, 62-64, 66, 84
    cuspidal edge 85
    of defect structures, 77
    by dispersive phonons, 69
    of energy, 87
    of quasimomentum, 87, 91
    reflection, 68
    of superlattices, 69
    thermal, 55, 73, 80
    3-D tomographic, 70
        diffraction effects in, 56, 76
    transmission, 68
    at ultrasonic frequencies, 74
Phonon localisation, 47
Phonon mode conversion, 67, 121-122
    critical cone for, 67-68
Phonon pulse, 129-130, 213
    ballistic, 89-90, 202, 299
Phonon-phonon interaction,
    decay, 130-131, 135, 143, 277
    scattering, 70, 277

Phonon propagation, 139-140, 143, 149, 189-190, 192-195, 277
    acoustic wave, 69
    ballistic, 90, 99, 105, 107, 111, 136-137, 139-140, 143-144, 168, 189, 282, 304, 306
    diffusive, 105, 109, 111, 143, 146, 282
    head wave, 79, 85
    Monte-Carlo simulation of, 56, 58, 61-62, 66-67, 91, 93-94, 129
    in quartz, 139
    quasiballistic, 145
    quasidiffusive, 135, 148-149, 279
    time-of-flight measurements of, 88, 129-130, 139, 166, 207
Phonon reflection, 66, 114, 119-120, 121
    diffusive, 105, 107
    specular, 105, 107
    at surfaces, 57, 68, 107
Phonon scattering, 70, 105
    by defects, 116, 149-150
    anharmonic, 116, 129, 311-312
    by dislocations, 70
    by impurities, 325
    by isotopes, 70, 114, 121, 129, 279
    by paramagnetic centres, 143
        nonresonant, 143-144, 149
        resonant, 143-144, 146-148, 150
        resonant combinational, 143, 150
    by rotational degrees of freedom, 327-328, 331
    at surfaces, 107
Phonon source, 56, 139
Phonon transmission, 66-67, 119, 121
    in superlattices, 69
    at surfaces, 59, 68
Polarisation vectors
    field of, 59, 67
        sigularities of, 59

Quantum dot, *see* 0-D electron gas
Quantum Hall regime, 162, 166, 170, 177-178, 189, 194, 196-198, 223, 237
    breakdown of, 162, 173
    fractional, 162, 168, 182, 201, 228
    Hall angle, 198
    Hall resistance, 162, 178
    Hall voltage, 191, 198
        phonon-induced, 193, 195, 196
    integer, 161, 168, 171
        magnetoroton excitation, 201
        phonon absorption studies, 201
Quantum point contact
    thermopower of, 220
Quantum well, 120-124, 251-253, 261, 419, 421
    depolarisation effect in, 421, 424
    infrared bleaching technique, 419-420, 424
Quantum wire *see* 1-D electron gas

Self-consistent localisation theory, 49
Si MOSFET, 167, 170, 177, 182, 222-223

Si MOSFET *(cont'd)*,
   phonons absorbed by, 169
   phonons emitted from, 167
Site representation
   electrons, 408-409
Skin effect, 381, 388, 399, 404
   anomalous, 381-382, 389, 396, 400
   electromagnetic depth, 400, 403-404
   normal, 381, 383, 386, 389, 391, 400
   plasma depth in, 400
   thermal depth in, 400
Specific heat,
   logarithmic dependence of, 105, 107
   measurements of, 106, 111
   time-dependent, 105, 107, 109, 111
Spectral density, 36, 38-41
   Peresada's, 35-36
Superconducting films, 301
Superconductivity
   Cooper-Pair breaking rate in, 301
   in High T_c oxides, 407, 409, 411
Superdiffusion, 209-210
Surface
   conical points of, 59, 63-64
   constant energy 88-89
   constant frequency *see* constant energy, 59, 61, 69
   isoenergetic *see* constant energy, 88, 156, 206
   isofrequency, 34, 40
      self-intersection points of, 156
      topological singularities of, 156
      topology of, 40
   ray, 56, 64-66, 79
   roughness of, 107
   slowness, 56, 59, 61, 63-65, 68, 76, 78-79, 88-90, 93
      Gaussian curvature of, 61-63, 76, 88-91, 93
      parabolic lines of, 62, 89, 156
      principal curvature of, 61, 88-89
   tangential degeneracies of, 60
Surface waves, 68
   acoustic, 99-100
      group velocity of, 101
      slowness curve, 101-102
      wavefront, 102
   pseudo-, 68

Thermal conductivity
   due to impurity scattering, 155
   due to Umklapp-processes, 156
   in bulk, 156
   local, 136
   nonlinear, 137
   of thin dielectric slabs, 153-157
      in relaxation time approx., 153-154
Thermal heat sink, 312-314
Thermalisation of phonons, 105, 107, 109, 111
   time of, 107
3-D electron gas, 234-236, 251-254, 258, 260

3-D electron gas *(cont'd)*,
   phonon absorption by, 234
   phonon emission by, 234, 251
2-D electron gas, 114, 122-125, 159, 161-164, 166-168, 170-171, 174, 177, 182, 189-190, 194, 198-199, 205, 207, 219-220, 227, 229, 234-236, 238, 251, 253-254, 260
   Corbino geometry, 198
   diffusion thermopower of, 219
   dissipationless state, 161, 170
   drag current, 206-208
   edge states, 177
   electrical conductivity, 191-192, 221, 237
   electron temperature
      measurement, 233, 237
   energy relaxation, 233, 235, 237, 251, 253, 259
   FIR, 175, 237
   magnetothermopower, 219
   momentum relaxtion, 209, 233
   momentum transfer rate, 191
   Nernst-Etinghaus effect, 193-194
   Peltier coefficient, 220
   phonon absorption by, 66, 122-125, 164, 169, 190, 206
   phonon drag in, 193-196, 205-206, 209, 211
      image of, 192, 196
   phonon drag magnetothermopower of, 223
   phonon emission by, 66, 122-125, 164-165, 174, 176, 189-190, 198, 210, 233, 235
      optical, 233, 237
      cyclotron, 164-166, 174, 177
      primary, 207, 210
      secondary, 207-210
   phonon transmission by, 164
   phonon interaction with
      acoustic, 159, 163
      optical, 166
   screening in, 237
   Seebeck coefficient, 220
   Shubnikov oscillations, 237
   thermal conductivity, 221-222
   thermoelectric tensor, 192
   thermopower, 219, 221-222, 230
   Wigner crystal in, 162-163, 182
2-D hole gas, 159, 163, 166, 168, 180-181, 219, 227, 229
   heat transport tensor, 226
   phonon drag magnetothermopower of, 226, 228
   phonon emission by, 179

Ultrasound
   laser generated, 80

Weakly interacting massive particles, 264, 297

0-D electron gas, 160